橡胶加工工艺

王者辉 孙 红 主编

Elastomer Engineering

·北京·

内容简介

《橡胶加工工艺（Elastomer Engineering）》主要内容包括：第 1 章"高分子化学：橡胶合成（Polymer Chemistry: Elastomer Synthesis）"；第 2 章"橡胶高弹性：基本原理与性质（Polymer Elasticity: Basic Concepts and Behavior）"；第 3 章"橡胶配合（Rubber Compounding）"；第 4 章"橡胶补强剂（Reinforcement of Elastomers by Particulate Fillers）"；第 5 章"橡胶硫化（Vulcanization）"；第 6 章"生胶加工工艺（Processing of Unvulcanized Elastomer）"；第 7 章"橡胶加工流变性（Rheological Behavior in Elastomer Processing）"；第 8 章"热塑性弹性体（Thermoplastic Elastomers）"；第 9 章"橡胶共混（Elastomer Blends）"；第 10 章"橡胶化学改性（Chemical Modification of Elastomers）"；第 11 章"橡胶强度（Strength of Elastomers）"；第 12 章"轮胎工艺（Tire Engineering）"；第 13 章"橡胶再生（Elastomer Recycling）"。本书内容全面，不仅综述性强，而且前沿性强，将橡胶加工新技术与新工艺呈现给读者。

《橡胶加工工艺（Elastomer Engineering）》可供高等学校橡胶工程技术、高分子材料工程技术专业，以及与高分子材料、橡胶工程专业相关的高分子材料与工程、复合材料、材料工程应用技术等本科专业作为教材，也可供橡胶企业工程技术人员和各类橡胶技术培训班学员等参考使用。

图书在版编目（CIP）数据

橡胶加工工艺＝Elastomer Engineering：英文/王者辉，孙红主编.—北京：化学工业出版社，2021.3
ISBN 978-7-122-38365-5

Ⅰ.①橡… Ⅱ.①王…②孙… Ⅲ.①橡胶加工-生产工艺-高等学校-教材-英文 Ⅳ.①TQ330.1

中国版本图书馆CIP数据核字（2021）第017597号

责任编辑：褚红喜　宋林青　　　　　　　　　装帧设计：张　辉
责任校对：宋　夏

出版发行：化学工业出版社（北京市东城区青年湖南街13号　邮政编码100011）
印　　装：涿州市般润文化传播有限公司
787mm×1092mm　1/16　印张31¾　字数796千字　2021年9月北京第1版第1次印刷

购书咨询：010-64518888　　　　　　　　　　售后服务：010-64518899
网　　址：http://www.cip.com.cn
凡购买本书，如有缺损质量问题，本社销售中心负责调换。

定　　价：128.00元　　　　　　　　　　　　　　　　　　　版权所有　违者必究

前　言

橡胶行业是国民经济的重要基础产业之一。它不仅为人们提供日常生活不可或缺的日用、医用等轻工橡胶产品，而且向采掘、交通、建筑、机械、电子等重工业和新兴产业提供各种橡胶制品、生产设备或橡胶部件。橡胶行业的产品种类繁多，服务产业十分广阔。橡胶工业是随着汽车工业发展起来的。二十世纪六十年代汽车工业与石油化学工业高速发展，使橡胶工业生产水平有了很大的提高；进入二十世纪七十年代，为适应汽车的高速、安全和节约能源，消除污染，防止公害等方面的需要，促进了轮胎新品种的不断出现，而原料胶消耗在交通运输领域占有相当大的比重。

橡胶弹性体材料具有长的柔性大分子网状链结构，在外力作用下可以发生大的变形，本体黏度随温度和应变速率的变化而变化，研究弹性体材料的高弹性具有重要的意义。全书重点介绍橡胶弹性体材料的合成、表征、力学性能、流动性能、胶料配合以及硫化成型加工工艺、橡胶弹性理论及测定。

本书第1章由张宏志老师编写，第3章由常海涛老师编写，第4章由巩学勇老师编写，第6章由孙红老师编写，第10章由刘光耀老师编写，其余章节由王者辉老师编写，最后由王者辉老师统一校订全稿。尽管我们常年从事橡胶学科的教学和科研工作，终因学识有限，错误和疏漏之处在所难免，恳请读者指正。

编者
2021年8月

Preface

Rubber industry is one of the important fundamental industries of national economy. It not only provides light industrial rubber products, for both daily and medical use, which are indispensable for people's daily life, but also provides various rubber products, production equipment or rubber parts to heavy industries and emerging industries such as mining, transportation, construction, machinery, electronics, etc. There are many kinds of products in rubber industry, which depends on very broad supporting industries. The rubber industry is developing with the automobile industry. In the 1960s, with the rapid development of automobile industry and petrochemical industry, the production level of rubber industry has been greatly improved. In the 1970s, in order to meet the needs of high rapidity, safety, energy saving, pollution elimination and prevention of public hazards, new tire varieties were promoted, and the consumption of raw rubber accounted for a considerable proportion in the field of transportation.

For elastomers, in particular, it is most instructive to derive the unique features of high elasticity from those of long flexible chain molecules in their matted and netted state and the changes imposed by large deformations, including the key role played by the internal viscosity as a function of temperature and rate. The book takes the reader from an introduction through synthesis, characterization, mechanical behavior, and flow to the major processing steps of filling, compounding, and vulcanization and to the theories and measurement of elastomeric performance, leaning strongly on the "materials" approach.

Mr. Zhang Hongzhi, Chang Haitao, Gong Xueyong, Liu Guangyao and Mrs. Sun Hong edited part of the book (Chapter 1, 3, 4, 6, and 10 respectively) and Mr. Wang Zhehui made the final edition. The authors would like to acknowledge with appreciation the numerous and valuable comments, suggestions, constructive criticisms and praise from evaluators and reviewers.

All chapters, while presenting theory, mechanism, and the author's overview of the internal consistency of the material's pattern of behavior, serve also as substantial sources of data and as guides to the relevant literature and to further self study. As such, this book should be suitable not only as a basis for elastomer course, but also as an instrument of instruction for students, teachers, and workers in all fields of polymer and, indeed, of material science.

Author
December 2020

Content

Chapter 1　Polymer Chemistry: Elastomer Synthesis ………………………………… 1

   1.1　Introduction ……………………………………………………………………… 2
   1.2　Classification of Polymerization Reactions and Kinetic Considerations …… 3
   1.3　Polyaddition/ Polycondensation ………………………………………………… 6
   1.4　Chain Polymerization by Free Radical Mechanism …………………………… 8
   1.5　Emulsion Polymerization ………………………………………………………… 14
   1.6　Copolymerization ………………………………………………………………… 23
   1.7　Chain Polymerization by Cationic Mechanism ………………………………… 28
   1.8　Chain Polymerization by Anionic Mechanism ………………………………… 34
   1.9　Stereospecific Chain Polymerization and Copolymerization by Coordination Catalysts …… 43
   1.10　Graft and Block Copolymerization …………………………………………… 51
   References ……………………………………………………………………………… 57

Chapter 2　Polymer Elasticity: Basic Concepts and Behavior ……………………… 65

   2.1　Introduction ……………………………………………………………………… 66
   2.2　Elasticity of a Single Molecule ………………………………………………… 66
   2.3　Elasticity of a Three-Dimensional Network of Polymer Molecules ………… 69
   2.4　Comparison with Experiment …………………………………………………… 72
   2.5　Continuum Theory of Rubber Elasticity ……………………………………… 74
   2.6　Second-Order Stresses …………………………………………………………… 80
   2.7　Elastic Behavior Under Small Deformations ………………………………… 81
   2.8　Some Unsolved Problems in Rubber Elasticity ……………………………… 84
   References ……………………………………………………………………………… 85

Chapter 3　Rubber Compounding …………………………………………………………… 87

   3.1　Introduction ……………………………………………………………………… 87
   3.2　Polymers …………………………………………………………………………… 87
   3.3　Filler Systems …………………………………………………………………… 98
   3.4　Stabilizer Systems ……………………………………………………………… 108
   3.5　Vulcanization System …………………………………………………………… 113
   3.6　Special Compounding Ingredients …………………………………………… 120
   3.7　Compound Development ………………………………………………………… 123
   3.8　Compound Preparation ………………………………………………………… 126
   3.9　Environmental Requirements in Compounding ……………………………… 126
   3.10　Summary ………………………………………………………………………… 129
   References ……………………………………………………………………………… 129

Chapter 4   Reinforcement of Elastomers by Particulate Fillers ········ 131

4.1   Introduction ········ 131
4.2   Preparation of Fillers ········ 131
4.3   Morphological and Physicochemical Characterization of Fillers ········ 134
4.4   The Mix: a Nanocomposite of Elastomer and Filler ········ 143
4.5   Mechanical Properties of Filled Rubbers ········ 148
References ········ 157

Chapter 5   Vulcanization ········ 158

5.1   Introduction ········ 158
5.2   Definition of Vulcanization ········ 158
5.3   Effects of Vulcanization on Vulcanizate Properties ········ 159
5.4   Characterization of the Vulcanization Process ········ 161
5.5   Vulcanization by Sulfur without Accelerator ········ 164
5.6   Accelerated-Sulfur Vulcanization ········ 165
References ········ 194

Chapter 6   Processing of Unvulcanized Elastomer ········ 195

6.1   Introduction ········ 195
6.2   Two Roll Mill ········ 195
6.3   Internal Mixers ········ 197
6.4   Continuous Mixers ········ 201
6.5   Trouble Shooting the Mixing Process ········ 202
6.6   Extrusion Process ········ 203
6.7   Calendering ········ 206
6.8   Continuous Vulcanization System ········ 211
6.9   Moulding ········ 212
References ········ 217

Chapter 7   Rheological Behavior in Elastomer Processing ········ 219

7.1   Introduction ········ 219
7.2   Basic Concepts of Mechanics ········ 223
7.3   Rheological Properties ········ 225
7.4   Boundary Conditions ········ 241
7.5   Mechanochemical Behavior ········ 244
7.6   Rheological Measurements ········ 246
7.7   Processing Technology ········ 251
7.8   Engineering Analysis of Processing ········ 261
References ········ 270

Chapter 8   Thermoplastic Elastomers ········ 281

8.1   Introduction ········ 281
8.2   Synthesis of Thermoplastic Elastomers ········ 286

8.3	Morphology of Thermoplastic Elastomers	292
8.4	Properties and Effect of Structure	309
8.5	Thermodynamics of Phase Separation	315
8.6	Thermoplastic Elastomers at Surfaces	320
8.7	Rheology and Processing	326
8.8	Applications	329
	References	330

Chapter 9 Elastomer Blends ········ 331

9.1	Introduction	331
9.2	Miscible Elastomer Blends	333
9.3	Immiscible Elastomer Blends	338
9.4	Conclusion	350
	References	350

Chapter 10 Chemical Modification of Elastomers ········ 355

10.1	Introduction	355
10.2	Chemical Modification of Polymers within Backbone and Chain Ends	356
10.3	Esterification, Etherification, and Hydrolysis of Polymers	357
10.4	The Hydrogenation of Polymers	360
10.5	Dehalogenation, Elimination, and Halogenation Reactions in Polymers	361
10.6	Other Addition Reactions to Double Bonds	364
10.7	Oxidation Reactions of Polymers	366
10.8	Functionalization of Polymers	367
10.9	Miscellaneous Chemical Reactions of Polymers	367
10.10	Block and Graft Copolymerization	368
	References	380

Chapter 11 Strength of Elastomers ········ 382

11.1	Introduction	382
11.2	Initiation of Fracture	383
11.3	Threshold Strengths and Extensibilities	388
11.4	Fracture Under Multiaxial Stresses	390
11.5	Crack Propagation	394
11.6	Tensile Rupture	401
11.7	Repeated Stressing: Mechanical Fatigue	406
11.8	Surface Cracking by Ozone	408
11.9	Abrasive Wear	409
	References	412

Chapter 12 Tire Engineering ········ 415

12.1	Introduction	416
12.2	Tire Types and Performance	417
12.3	Basic Tire Design	419

12.4	Tire Engineering	421
12.5	Tire Materials	431
12.6	Tire Testing	444
12.7	Tire Manufacturing	448
12.8	Summary	452
	References	454

Chapter 13　Elastomer Recycling ... 456

13.1	Introduction	456
13.2	Retreading of Tire	458
13.3	Recycling of Rubber Vulcanizates	459
13.4	Use of Recycled Rubber	474
13.5	Pyrolysis and Incineration of Rubber	484
13.6	Concluding Remarks	485
	References	485

Appendix I　Demonstration ... 492

Appendix II　Acronyms for Common Elastomers ... 498

Chapter 1
Polymer Chemistry: Elastomer Synthesis

An elastomer is a polymer with viscoelasticity (i. e., both viscosity and elasticity) and very weak intermolecular forces, and generally low Young's modulus and high failure strain compared with other materials. The term, a portmanteau of elastic polymer, is often used interchangeably with rubber, although the latter is preferred when referring to vulcanisates. Each of the monomers which link to form the polymer is usually a compound of several elements among carbon, hydrogen, oxygen and silicon. Elastomers are amorphous polymers maintained above their glass transition temperature, so that considerable molecular reconformation, without breaking of covalent bonds, is feasible. At ambient temperatures, such rubbers are thus relatively soft ($E \approx 3MPa$) and deformable. Their primary uses are for seals, adhesives and molded flexible parts. Application areas for different types of rubber are manifold and cover segments as diverse as tires, soles for shoes, and damping and insulating elements. The importance of these rubbers can be judged from the fact that global revenues are forecast to rise to CNY¥560 billion in 2020. IUPAC defines the term "elastomer" by "Polymer that displays rubber-like elasticity".

Rubber-like solids with elastic properties are called elastomers. Polymer chains are held together in these materials by relatively weak intermolecular bonds, which permit the polymers to stretch in response to macroscopic stresses. Natural rubber, neoprene rubber, buna-s and buna-n are all examples of such elastomers. Elastomers are usually thermosets (requiring vulcanization) but may also be thermoplastic (see thermoplastic elastomer). The long polymer chains cross-link during curing, i. e., vulcanizing. The molecular structure of elastomers can be imagined as a 'spaghetti and meatball' structure, with the meatballs signifying cross-links. The elasticity is derived from the ability of the long chains to reconfigure themselves to distribute an applied stress. The covalent cross-linkages ensure that the elastomer will return to its original configuration when the stress is removed. As a result of this extreme flexibility, elastomers can reversibly extend from 5%~700%, depending on the specific material. Without the cross-linkages or with short, uneasily reconfigured chains, the applied stress would result in a permanent deformation. Temperature effects are also present in the demonstrated elasticity of a polymer. Elastomers that have cooled to a glassy or crystalline phase will have less mobile chains, and consequentially less elasticity, than those manipulated at temperatures higher than the glass transition temperature of the polymer. It is also possible for a polymer to exhibit elasticity that is not due to covalent cross-links, but instead for thermodynamic reasons.

Polymer chemistry is a sub-discipline of chemistry that focuses on the chemical synthesis, structure, chemical and physical properties of polymers and macromolecules. The principles and methods used for polymer chemistry are common to chemistry sub-disciplines organic chemistry, analytical chemistry, and physical chemistry. Many materials have polymeric structures, from fully inorganic metals and ceramics to DNA and other biological molecules, however, polymer chemistry is typically referred to in the context of synthetic, organic compositions. Synthetic polymers are ubiquitous in commercial materials and products in everyday use, commonly referred to as plastics, rubbers, and composites. Polymer chemistry can also be included in the broader fields of polymer science or even nanotechnology, both of which can be described as encompassing polymer physics and polymer engineering.

1.1 Introduction

The development of synthetic elastomer played a special role in the history of polymerization chemistry. This was due primarily to the fact that attempts to synthesize elastomer were made long before there was even the faintest idea of the nature of polymerization reactions. Such attempts began very soon after the elegant analytical work of Williams in 1860, which clearly demonstrated that Hevea elastomer was "composed" of isoprene. Thus, Bouchardat in 1879 was actually able to prepare a elastomer-like substance from isoprene (which he obtained from elastomer pyrolysis), using heat and hydrogen chloride. Tilden repeated this process in 1884 but used isoprene obtained from pyrolysis of turpentine to demonstrate that it was not necessary to use the "mother substance" of elastomer itself. These explorations were soon followed by the work of Kondakow (1900) with 2,3-dimethylbutadiene, that of Thiele with piperylene in 1901, and finally that of Lebedev on butadiene itself in 1910. Mention should also be made of the almost simultaneous, and apparently independent, discoveries in 1910 by Harries in Germany and Matthews and Strange in England of the efficient polymerization in isoprene by sodium.

Although all of these attempts had a noble purpose indeed, the means used could hardly be considered a contribution to science, as the transformation of the simple molecules of a diene into the "colloidal" substance known as elastomer was then far beyond the comprehension of chemical science. As a matter of fact, the commercial production of synthetic elastomer was already well established, at least in Germany and Russia, before Staudinger laid the basis for his macromolecular hypothesis during the 1920s. Even such relatively modern synthetic elastomers as polychloroprene and the poly (alkylene sulfides) were already in commercial production before 1931. This was, of course, also before Carothers and coworkers' pioneering studies on the polymerization of chloroprene.

Hence, it is apparent that it was not the development of an understanding of polymerization that led to the invention of synthetic elastomer, but perhaps the reverse. In contrast, it was the new science of organic macromolecules, whose foundations were established by Staudinger, which expanded rapidly during the 1930s and 1940s, and pointed the way to the synthesis of a vast array of new polymeric materials, including synthetic fibers and plastics and even new elastomers. This new science included the classical studies of polycondensation by Carothers and Flory and the establishment of the principles governing free radical chain addition reactions by Schulz, Flory, Mayo, and others.

Thus it was that the paths of synthetic elastomer and macromolecular science finally crossed and became one broad avenue. Hence today the design of a new elastomer or the modification of an old one requires the same kind of molecular architecture which applies to any other polymer and is based on an understanding of the principles of polymerization reactions.

1.2 Classification of Polymerization Reactions and Kinetic Considerations

Historically polymers have been divided into two broad classes: condensation polymers and addition polymers. Flory has defined these as follows: condensation polymers, in which the molecular formula of the structural unit (or units) lacks certain atoms present in the monomer from which it is formed, or to which it may be degraded by chemical means, and addition polymers, in which the molecular formula of the structural unit (or units) is identical with that of the monomer from which the polymer is derived.

Thus, an example of a condensation polymer would be a polyester, formed by the condensation reaction between a glycol and a dicarboxylic acid (with the evolution of water), whereas an addition polymer is exemplified by polystyrene, formed by the self-addition of styrene monomers.

Although these earlier definitions were based on the chain structure of the polymers, they were closely related, as just described, to the mode of formation as well. It soon became apparent that such a classification has serious shortcomings, as so-called polycondensates could result from 'addition' polymerization reactions. For example, although Nylon 6 can be prepared by the polycondensation reaction of ε-aminocaproic acid, it is now synthesized by the ring-opening addition polymerization of ε-caprolactam, and this process has a profound effect on the properties of the resulting polymer. This is, of course, due basically to the magnitude of the molecular weight of the final polymer.

Because it is the extraordinarily large size of the macromolecules which leads to their unusual properties, it would be most sensible to classify polymerization reactions in accordance with the way in which they affect the molecular size and size distribution of the final product, i.e., in terms of the mechanism of polymerization. On this basis, there appear to be only two basic processes whereby macromolecules are synthesized:

(1) *step-growth polymerization* (polycondensation and polyaddition);

(2) *chain-growth (chain) polymerization*.

1.2.1 Polyaddition/ Polycondensation

The distinguishing mechanistic feature of step-growth polymerization is that all molecular species in the system can react with each other to form higher-molecular-weight species as shown in Eq. (1-1), where P_i is a species with a number-average number of monomer units per chain equal to i, P_j is a species with a number-average number of monomer units per chain equal to j, and P is a species with a number average of monomer units per $i+j$ chain equal to $i+j$. The kinetic consequence of this mechanism of polymer growth is that chain length increases monotonically with extent of reaction, i.e., with time of reaction, as shown in Fig. 1.1 (A).

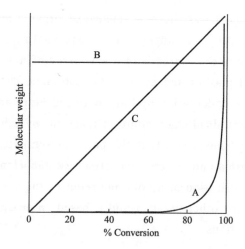

Fig. 1.1 Variation of molecular weight with % conversion for (A) step-growth polymerization; (B) chain-growth polymerization, and (C) living chain-growth polymerization with no chain transfer and no chain termination.

$$P_i + P_j \longrightarrow P_{i+j} \qquad (1\text{-}1)$$

These step-growth polymerization reactions fall into two classes:

① Polycondensation: growth of polymer chains proceeds by condensation reactions between molecules of all degrees of polymerization. A low-molar-mass by-product (AB) is also formed.

$$\text{A-R-A} + \text{B-R-B} \longrightarrow \text{A-R-R-B} + \text{AB} \qquad (1\text{-}2)$$

② Polyaddition: growth of polymer chains proceeds by addition reactions between molecules of all degrees of polymerization.

$$\text{A-R-A} + \text{B-R-B} \longrightarrow \text{A-R-AB-R-B} \qquad (1\text{-}3)$$

Here A and B are the functional end groups which react with each other.

Examples of polycondensation can be seen in the formation of (1) polyesters and (2) polyamides, where the A and B groups would be (1) hydroxyl and carboxyl and (2) amine and carboxyl, respectively, which would combine and split off a molecule of water. On the other hand, a polyaddition reaction [Eq. (1-2)] would be exemplified by the reaction of diisocyanates with glycols to form polyurethanes. In that case, of course, no by-products are formed.

The polymerizations shown in Eq. (1-2) and Eq. (1-3) actually represent well-known reactions of small molecules, the only distinction being the minimum requirements of difunctionality of each molecule for polymer formation, which makes it possible for the product of each reaction to participate in further reactions. As a rule, the functional groups retain their reactivity regardless of the chain length, so that these reactions follow the same kinetic rules as for simple molecules; however, in contrast to polyaddition reactions, polycondensations suffer from the serious problem of reversibility (e.g., hydrolysis, or "depolymerization") as a result of the possible accumulation of the by-product (e. g., water), and this must be taken into account. In general, because of the unfavorable equilibrium constant for polycondensation reactions, the formation of high polymer requires removal of the small molecule by-products.

In both of the foregoing types of reactions, two factors which govern the molecular weight of the polymer are the stoichiometry and the extent of reaction. Thus, it is obvious that an excess of one type of end group will control the maximum chain length attainable, and this can be predicted if the initial molar ratio of functional groups is known. On the other hand, with equivalent amounts of the two types of end groups, the final chain length is theoretically limitless, i. e., infinite in size.

1.2.2 Chain Polymerization

The distinguishing mechanistic feature of chain-growth or chain polymerization is that chain growth (propagation) occurs only by addition of monomer to reactive sites present on the growing polymer molecules as shown in Eq. (1-4), where P^* is a polymer chain with a reactive site ($*$), and n is the degree of polymerization of n, M is a monomer unit and P^*_{n+1} is a polymer chain with a reactive site ($*$) and a degree of polymerization of $n+1$.

$$P^*_n + M \longrightarrow P^*_{n+1} \tag{1-4}$$

This type of polymerization involves the successive addition of monomers to a growing chain, which is initiated by some reactive species (initiation). Such addition reactions may involve either multiple bonds or rings. The reactive species which initiate such chain reactions must be capable of opening one of the bonds in the monomer and may be either a radical, an electrophile, a nucleophile, or an organometallic species. Hence these polymerizations may proceed by a variety of possible mechanisms depending on the electronic nature of the chain-carrying species, viz., free radical, cationic, anionic, and coordination, as illustrated by the following equations for reactions of double bonds with various types of initiating species:

Free radical

$$R\cdot + \hspace{-2pt}\begin{array}{c}\\\end{array}\hspace{-8pt}C=C\hspace{-8pt}\begin{array}{c}\\\end{array} \longrightarrow R-\overset{|}{\underset{|}{C}}-\overset{|}{\underset{|}{C}}\cdot \longrightarrow \cdots \longrightarrow$$

Cationic

$$H^+A^- + \hspace{-2pt}\begin{array}{c}\\\end{array}\hspace{-8pt}C=C\hspace{-8pt}\begin{array}{c}\\\end{array} \longrightarrow H-\overset{|}{\underset{|}{C}}-\overset{|}{\underset{|}{C}}{}^+A^- \longrightarrow \cdots \longrightarrow$$

Anionic

$$A^-M^+ + \hspace{-2pt}\begin{array}{c}\\\end{array}\hspace{-8pt}C=C\hspace{-8pt}\begin{array}{c}\\\end{array} \longrightarrow A-\overset{|}{\underset{|}{C}}-\overset{|}{\underset{|}{C}}{}^-M^+ \longrightarrow \cdots \longrightarrow$$

Coordination

$$R\text{-Met} + \hspace{-2pt}\begin{array}{c}\\\end{array}\hspace{-8pt}C=C\hspace{-8pt}\begin{array}{c}\\\end{array} \longrightarrow R-\overset{|}{\underset{|}{C}}-\overset{|}{\underset{|}{C}}-\text{Met} \longrightarrow \cdots \longrightarrow$$

In these equations, the exact nature of the initiating and chain-carrying species can vary from essentially covalent for transition-metal organometallic species in coordination polymerization to ion pairs or free ions in ionic polymerizations, depending on the structure of the chain-carrying species, the counterion, the solvent, and the temperature.

A significant distinction between step polymerization and chain polymerization is that in the latter, each macromolecule is formed by a "chain reaction" which is initiated by some activation step. Thus, at any given time during the polymerization, the reacting species present consist only of growing chains and monomer molecules, in addition to the "dead" polymer chains formed earlier by chain termination reactions. These growing chains may be very short-lived (e. g., free radicals or free ions) but may attain very long chain lengths during their brief lifetimes as illustrated in Fig. 1.1(B). On the other hand, they may have very long lifetimes (e. g., living polymers), in which case the chain lengths may increase as a direct function of time of reaction as shown in Fig. 1.1(C). Hence, unlike the case of step polymerizations, the molecular weights in chain addition polymerization systems may or may not be directly related to time or extent of reaction [see Fig. 1.1(A)].

1.3　Polyaddition/ Polycondensation

Although, as indicated earlier, polyaddition and polycondensation did not figure prominently in the early explorations of elastomer synthesis, it was one of the earliest general methods used for polymerization, because of its relative simplicity. It is thus not surprising that the earliest truly synthetic resins and plastics were of the polycondensate type, such as phenol formaldehyde and polyester. The concept of linking together reactive end groups to build large molecules is fairly simple to comprehend and also lends itself to a relatively simple mathematical analysis.

As stated previously, the kinetics of polyadditions and polycondensations follow the same rules as the simple monofunctional reactions, as the reactivity of the functional groups is maintained regardless of chain length. The only new feature is, of course, the growth in molecular size, and this has been amenable to a mathematical analysis. Considering the type of reactions defined in Eq. (1-2) and Eq. (1-3), in the normal case, where the number of A and B groups are equal, the chain lengths are easily predictable as a function of the extent of reaction. Thus, if p represents the fraction of end groups consumed at any given time, then the number-average of units per chain (X_n) is given by $1/(1-p)$. Thus,

$$M_n = M_o/(1-p) \qquad (1\text{-}5)$$

where M_n is the number-average molecular weight of the polymer and M_o is the molecular weight of a chain repeating unit. The consequences of this simple relationship are profound. For example, when 50% of the functional groups have reacted, the number-average degree of polymerization is only 2. To prepare polymers with useful properties, molecular weights of at least 10 000 are required; this means that the degree of conversion of the functional groups must be greater than 99% for a repeating unit with a molar mass of 100g ($X_n = 100$). It is obvious that relatively few reactions will qualify in terms of this rigorous requirement because of side reactions.

Because this type of polymerization is a completely random process, with all molecules having equal probability of reacting, the distribution of molecular weights corresponds to the most probable, or binomial, distribution, which is related to the extent of polymerization as follows:

$$W_x = xp^{x-1}(1-p)^2 \tag{1-6}$$

$$N_x = p^{x-1}(1-p) \tag{1-7}$$

where W_x is the weight fraction of x-mers (chains having x units) and N_x is the mole fraction of x-mers. This distribution function can be used to calculate M_w, the weight-average molecular weight, as $M_w = M_o \sum xW_x$. It can be shown that the foregoing summation leads to the relation

$$M_w = \frac{1+p}{1-p}M_n \tag{1-8}$$

which then means that

$$\frac{M_w}{M_n} = 1 + p \tag{1-9}$$

Hence the weight/number ratio of chain lengths in these systems undergoes a steady increase with extent of reaction, approaching an ultimate value of 2. Thus, we see that polyaddition and polycondensation are characterized by the following features:

① All molecules have equal probability of reacting;

② The polymerization rates are essentially described by the concentrations and reactivity of the functional groups;

③ The chain lengths are monotonic functions of the extent of reaction and hence of time of reaction;

④ The attainment of high molecular weights requires a high degree of conversion.

In those cases where at least one of the monomers has more than two functional groups, the added feature of branching chains is introduced, eventually leading to the formation of molecular networks, i.e., gelation. This, of course, complicates the molecular size distribution but does not affect the kinetics of the polymerization.

The foregoing relationships of chain length to extent of reaction would then be expected to apply to such step polymerizations as are involved in the synthesis of poly(alkylene sulfides) from a dihalide and sodium polysulfide (polycondensation) or in the formation of the urethane polymers from glycols and diisocyanates (polyaddition). The polysulfide reaction is actually carried out in a suspension of the dihalide in an aqueous solution of the polysulfide, using a surfactant to stabilize the resulting polymer suspension.

The urethane polymers offer an interesting illustration of the characteristic molecular weights to be expected in this type of polymerization, which can be written as

$$\text{HO-P-OH} + \text{OCN-R-NCO} \longrightarrow \text{HO} \!\!\begin{array}{c}\\[-6pt]\end{array}\!\!\text{P-O-}\underset{\underset{\text{O}}{\|}}{\text{C}}\text{-NH-R-NH-}\underset{\underset{\text{O}}{\|}}{\text{C}}\text{-O}\!\!\begin{array}{c}\\[-6pt]\end{array}\!\!{}_x\text{P-OH} \quad (1\text{-}10)$$

It should be noted that the P in Eq. (1-10) represents a low-molecular-weight polymer of a polyester or polyether type ($M_w = 2\,000$), so that this is really a "chain extension" reaction. It turns out that the reaction between an is cyanate group and a hydroxyl goes to a high conversion, i. e. , to approximately 98% ($p = 0.98$). Hence the value of x in Eq. (1-10) is about 50, and the final molecular weight of the urethane polymer is about 100 000. Such high molecular weights are, of course, due solely to the fact that this reaction goes so far toward completion, i. e. , where the reactive functional groups can be reduced to concentrations of the order of 10^{-2} mol/L.

1.4　Chain Polymerization by Free Radical Mechanism

1.4.1　General Kinetics

The general kinetics for this mechanism involve the usual three primary steps of any chain reaction, i. e. , initiation, propagation, and termination, as shown below. Initiation generally occurs by the formation of free radicals through the homolytic dissociation of weak bonds (e. g. , in peroxides or azo compounds) or by irradiation. Termination reactions for vinyl polymers can occur either by combination (coupling), by disproportionation, or by a combination of both reactions.

Initiation

$$\text{I} \xrightarrow{k_i} 2\text{R}\cdot$$
$$\text{R}\cdot + \text{M} \longrightarrow \text{R-M}\cdot$$

Propagation

$$\text{M}_j\cdot + \text{M} \xrightarrow{k_p} \text{M}_{j+1}\cdot$$

Termination

$$\text{M}_j\cdot + \text{M}_k\cdot \xrightarrow[(k_{tc} + k_{td})]{k_t} \text{dead chains}$$

where I = initiator, M = monomer, R = initial free radical, and $\text{M}_j\cdot$ = propagating free radical.

Combination

$$\text{RO}\!\!\begin{array}{c}\\[-6pt]\end{array}\!\!\text{CH}_2\text{-}\underset{\text{X}}{\text{CH}}\!\!\begin{array}{c}\\[-6pt]\end{array}\!\!{}_i \text{CH}_2\underset{\text{X}}{\text{CH}}\cdot + \text{RO}\!\!\begin{array}{c}\\[-6pt]\end{array}\!\!\text{CH}_2\text{-}\underset{\text{X}}{\text{CH}}\!\!\begin{array}{c}\\[-6pt]\end{array}\!\!{}_j \text{CH}_2\underset{\text{X}}{\text{CH}}\cdot \xrightarrow{k_{tc}}$$

$$\text{RO}\!\!\begin{array}{c}\\[-6pt]\end{array}\!\!\text{CH}_2\text{-}\underset{\text{X}}{\text{CH}}\!\!\begin{array}{c}\\[-6pt]\end{array}\!\!{}_i \text{CH}_2\underset{\text{X}}{\text{CH}}\text{-CHCH}_2\!\!\begin{array}{c}\\[-6pt]\end{array}\!\!\underset{\text{X}}{\text{CH}}\text{-CH}_2\!\!\begin{array}{c}\\[-6pt]\end{array}\!\!{}_j\text{OR}$$

Disproportionation

$$RO\!-\!(CH_2\!-\!CH)_i\!-\!CH_2CH\cdot \ + \ RO\!-\!(CH_2\!-\!CH)_j\!-\!CH_2CH\cdot \ \xrightarrow{k_{td}}$$
$$\hspace{4cm} | \hspace{0.5cm} | \hspace{3cm} | \hspace{0.5cm} |$$
$$\hspace{4cm} X \hspace{0.5cm} X \hspace{3cm} X \hspace{0.5cm} X$$

$$RO\!-\!(CH_2\!-\!CH)_i\!-\!CH_2CH_2 \ + \ RO\!-\!(CH_2\!-\!CH)_j\!-\!CH\!=\!CH$$
$$| \hspace{1cm} | \hspace{3cm} | \hspace{1cm} |$$
$$X \hspace{1cm} X \hspace{3cm} X \hspace{1cm} X$$

This sequence of steps then leads to the following simple kinetic treatment:

Rate of initiation $\hspace{2cm} R_i = 2k_i[I]$ \hfill (1-11)

Rate of propagation $\hspace{1.6cm} R_p = k_p[M_j\cdot][M]$ \hfill (1-12)

Rate of termination $\hspace{1.7cm} R_t = 2k_t[M_j\cdot]^2$ \hfill (1-13)

Assuming a steady-state condition where the rate of formation of radicals is equal to their rate of disappearance, i.e., $R_i = R_t$,

$$[M_j\cdot] = k_i^{1/2} k_t^{-1/2} [I]^{1/2} \tag{1-14}$$

and $\hspace{2cm} R_p = k_p k_i^{1/2} k_t^{-1/2} [M][I]^{1/2}$ \hfill (1-15)

Eq. (1-15) thus illustrates the dependency of the overall rate of polymerization on the concentrations of initiator and monomer. The half-power dependence of the rate on the initiator concentration appears to be a universal feature of the free radical mechanism and has been used as a diagnostic test for the operation of this mechanism.

Another important aspect of free radical polymerization is the dependency of the number-average degree of polymerization on initiator and monomer concentrations as shown in Eq. (1-16). Comparison with Eq. (1-15) shows that increasing the rate of initiation, by increasing the initiator concentration, increases the rate of polymerization but decreases the degree of polymerization, X_n, which corresponds to the number-average number of units per chain.

$$X_n = k_p k_i^{-1/2} k_t^{-1/2} [M][I]^{-1/2} \tag{1-16}$$

The general nature of free radical chain polymerization deserves some special attention. Because of the high reactivity of the propagating chain radical, it can only attain a very short lifetime, several seconds at best. This results in a very low stationary concentration of propagating chain radicals (about 10^{-8} mol/L in a homogeneous medium). During this short lifetime, however, each growing radical may still have the opportunity to add thousands of monomer units.

Hence the chain length of the macromolecules formed in these systems has no direct relation to the extent of reaction, i.e., to the degree of conversion of monomer to polymer [see Eq. (1-16) and Fig. 1.1]. At all times during the polymerization, the reaction mixture contains only monomer, a very small concentration of propagating chains, and dead (nonpropagating) polymer, the latter usually of high molecular weight.

To illustrate more clearly the nature of free radical polymerization, it is instructive to examine the values of the individual rate constants for the propagation and termination steps. A number of these rate constants have been deduced, generally using nonstationary state measurements such as rotating sector techniques and emulsion polymerization. Recently, the IUPAC Working Party on "Modeling of kinetics and processes of polymerization" has recommended the analysis of molecular weight distributions of polymers produced in pulsed-laser-initiated polymerization (PLP) to determine values of propagation rate constants. Illustrative values of propagation and termination rate constants are listed in Table 1.1. Thus, although the chain growth step can be a very fast reaction (several orders of magnitude faster than the rates of the step reactions of the end groups), it is still several orders of magnitude slower than the termination step, i.e., the reaction of two radicals. It is this high ratio of k_t/k_p which leads to the very low stationary concentration of growing radicals ($\sim 10^{-8}$ mol/L) in these systems.

Table 1.1 Propagation and termination rate constants in radical polymerization

Monomer	k_p at 60℃ /[L/(mol·s)]	$k_t (\times 10^{-7})$ at 60℃ /[L/(mol·s)]
Styrene	176	3.6
Methyl methacrylate	367	1.0
Methyl acrylate	2100	0.5
Vinyl acetate	3700	7.4
Butadiene	100	~100
Isoprene	50	—
Chloroprene	1270	—

Although the three individual steps which combine to make up the chain reaction act as the primary control of the chain lengths [see Eq. (1-16)], "chain transfer" reactions can occur whereby one chain is terminated and a new one is initiated, without affecting the polymerization rate; such reactions will also, of course, affect the chain length. Chain transfer usually involves the homolytic cleavage of the most susceptible bond in molecules of solvent, monomer, impurity, etc., by the propagating radical, e.g.,

$$\sim\sim\sim—CH_2—CH(C_6H_5)\cdot + CCl_4 \xrightarrow{k_{tr}} \sim\sim\sim—CH_2—CHCl(C_6H_5) + \cdot CCl_3$$

and can be designated as follows:

Monomer transfer $\qquad M_j\cdot + M \xrightarrow{k_{trM}} M_j + M\cdot$ (1-17)

Solvent transfer $\qquad M_j\cdot + S \xrightarrow{k_{trS}} M_j + S\cdot$ (1-18)

$\qquad\qquad\qquad\qquad S\cdot + M \xrightarrow{k_p'} SM\cdot$ (1-19)

Hence the chain length of the polymer being formed at any given instant can be expressed as the ratio of the propagation rate to the sum of all the reactions leading to termination of the chain as

$$X_n = \frac{k_p[M_j\cdot][M]}{(k_{tc}+2k_{td})[M_j\cdot]^2 + k_{trM}[M_j\cdot][M] + k_{trS}[M_j\cdot][S]}$$

or

$$\frac{1}{X_n} = \frac{(k_{tc}+2k_{td})R_p}{k_p^2[M]^2} + \frac{k_{trM}}{k_p} + \frac{k_{trS}[S]}{k_p[M]} \qquad (1\text{-}20)$$

where X_n is the number-average number of units per chain, k_{tc} is the rate constant for termination by combination, and k_{td} is the rate constant for termination by disproportionation.

1.4.2 Molecular Weight Distribution

The chain length distribution of free radical addition polymerization can also be derived from simple statistics. Thus, for polymer formed at any given instant, the distribution will be the "most probable" and will be governed by the ratio of the rates of chain growth to chain termination,

$$W_x = xp^{x-1}(1-p)^2 \tag{1-21}$$

where p is the probability of propagation, and $(1-p)$ is the probability of termination (by disproportionation or transfer). This expression is of course identical to Eq. (1-6), except for the different significance of the term p. Unlike Eq. (1-6), however, it expresses only the instantaneous chain length for an increment of polymer, not the cumulative value for the total polymer obtained.

From Eq. (1-21) it follows that the number- and weight-average chain lengths X_n and X_w are expressed by

$$X_n = \frac{1}{1-p} \quad \text{or} \quad X_w = \frac{1+p}{1-p} \approx \frac{2}{1-p} \tag{1-22}$$

as p must always be close to unity for high polymers. Hence it follows again that

$$X_w/X_n = 2 \tag{1-23}$$

The value of X_w/X_n for the cumulative polymer may, of course, be much higher, depending on the changes in the value of p with increasing conversion. It should be noted, however, that this is valid only where the growing chains terminate by disproportionation or transfer, not by combination. It can be shown in the latter case that the increment distribution is much narrower, i.e.,

$$X_w/X_n = 1.5 \tag{1-24}$$

Thus, in summary, the kinetics of free radical polymerization are characterized by the following features:

① Rate is directly proportional to the half-power of the initiator concentration;

② Molecular weight is inversely proportional to the half-power of initiator concentration;

③ The lifetime of the growing chain is short (several seconds) but a high molecular weight is obtained, leading to formation of high polymer at the outset of reaction;

④ No direct relation exists between extent of conversion and chain length;

⑤ Instantaneous chain length is statistical, but the cumulative value can be considerably broader because of changes in relative rates of propagation and termination.

1.4.3 Special Case of Diene Polymerization

As polydienes still constitute the backbone of the synthetic elastomer industry, it is important to consider the special features which dienes exhibit in free radical polymerization. Despite the fact that this type of polymerization has played and is still playing the major role in industrial production of various polymers, it has never been successful in bulk or solution polymerization of dienes. This is an outcome of the kinetic features of the free radical polymerization of dienes, as indicated in Table 1.2. Thus, the relatively high k_t/k_p ratio (as compared with the other monomers shown) leads to very low molecular weights and very slow rates for polydienes prepared in homogeneous systems, as illustrated in Table 1.2. It can be seen from these data that even in the case of these thermal uncatalyzed polymerizations, where the molecular weight would be at a maximum compared with catalyzed systems, it is still too low by at least an order of magnitude. These systems are also complicated by a competitive Diels-Alder reaction, leading to low-molecular-weight compounds, i.e., "oils".

Table 1.2 Thermal polymerization of Dienes

Temperature/℃	Time/h	Isoprene			2,3-Dimethylbutadiene		
		Yield /%		M_w, Elastomer	Yield /%		M_w, Elastomer
		Oil	Elastomer		Oil	Elastomer	
85	100	7.9	16.3	4600	—	—	—
85	250	—	—	—	2.7	19.6	3500
85	900	—	35.3	5700	—	49.7	3500
145	12.5	54.7	15.6	4000	11.1	15.6	2100

It is therefore not surprising that the early investigators saw no promise in this mechanism of polymerization of butadiene, isoprene, etc., either by pure thermal initiation or by the use of free radical initiators, such as the peroxides. Instead they turned to sodium polymerization, which, although also rather slow and difficult to reproduce, at least yielded high-molecular-weight elastomery polymers from the dienes. Later, in the 1930s, when emulsion polymerization was introduced, it was found that this system, even though it involves the free radical mechanism, leads to both fast rates and high molecular weights, conducive to the production of synthetic elastomer. The special features of emulsion polymerization which lead to such surprising results are discussed later.

1.4.4 Controlled Radical Polymerization

There has been a revolution in free radical polymerization chemistry that began in the 1980s with the seminal patent of Solomon, Rizzardo, and Cacioli. These scientists found that it was possible to obtain controlled radical polymerization of monomers such a styrene and alkyl(meth)acrylates by affecting free radical polymerization in the presence of stable nitroxyl radicals as shown below.

$$\text{(structure: N-OR)} \rightleftarrows \text{(structure: N-O•)} + R•$$

$$\downarrow n \overset{}{=}\!\!\!\!\diagup_{CO_2R}$$

$$\text{(N-O-CHCH}_2\text{[CH-CH}_2\text{]}_n\text{R, with CO}_2\text{R groups)} \longleftarrow \text{(N-O•)} + R\text{[CH}_2\text{CH]}_n\text{CH}_2\text{CH•, with CO}_2\text{R groups}$$

(TEMPO)

It has been found that these controlled polymerizations carried out in the presence of stable nitroxyl radicals, such as the tetramethylpyridinyloxy radical (TEMPO) shown above, lead to the synthesis of polymers with controlled molecular weight, narrow molecular weight distributions, end-group functionality, architecture, and block copolymer composition. The key requirements for this type of controlled polymerization are:

(a) a thermally labile bond that undergoes homolysis reversibly to form reactive radicals capable of initiating or propagating polymerization of vinyl monomers;

(b) simultaneous formation of a stable radical that rapidly and reversibly combines with propagating radicals but which does not add to vinyl monomers;

(c) an equilibrium constant between radicals and covalent, dormant species that favors the dormant species.

In order for a system of this type to be useful, the ratio of the concentration of active radical species to dormant species must be less than 10^{-5}. This implies that the majority of the lifetime of the chain is spent in the dormant stage. Successful systems must maintain an optimum amount of nitroxide such that polymerization can occur at an appreciable rate. It should be noted that radical-radical coupling can still occur, but it is minimized by the low concentration of propagating radicals (e. g., 10^{-8} mol/L). Because termination still occurs, it is obviously inappropriate to call these polymerizations living, although these types of controlled radical polymerizations are often referred to as living in the literature. The kinetics of the stable free radical polymerization are controlled by the persistent radical effect which has been clearly elucidated by Fischer.

Careful and extensive investigations of these nitroxide mediated polymerizations (also referred to as stable free radical polymerization) have established optimum conditions for controlled radical polymerization of a variety of vinyl monomers. Variables examined include the structure of the nitroxide and the presence of other additives to control spontaneous polymerization of monomers such as Styrene. It is noteworthy that in place of alkoxyamine initiators, a mixture of a normal free radical initiator such as an azo compound or a peroxide can also be used.

The application of these procedures to 1,3-dienes has presented problems. The rates of polymerization were observed to decrease and then stop due to a buildup of excess nitroxide. An effective procedure for the controlled polymerization of isoprene at 145℃ involved the addition of a reducing sugar such as glucose in the presence of sodium bicarbonate to react with the excess nitroxide. After four hours, polyisoprene with $M_n = 21000$ and $M_w/M_n = 1.33$ was obtained in 25% yield. The reaction of TEMPO-terminated polystyrene with either butadiene or isoprene resulted in the formation of the corresponding diblock copolymers that were characterized by ^{1}H-NMR and SEC. No evidence for either polystyrene or polydiene homopolymers was reported.

An alternative procedure to reduce the concentration of excess nitroxide radicals has been reported by Hawker and coworkers. They used the initiator shown below to successfully effect the controlled polymerization of isoprene.

It was reported that the corresponding nitroxide has α-hydrogens that can decompose via disproportionation, thereby preventing buildup of excess nitroxide. Using this initiating/nitroxide system, it was possible to prepare a variety of polyisoprenes with controlled molecular weight, high 1,4-microstructure, and polydispersities that ranged from 1.07 for low molecular weights (e.g., $M_n = 5000$) to 1.20 for number average molecular weights of 100000. However, the required reaction conditions were 130°C and reaction times up to 48 hours. Well-defined copolymers of Isoprene and Styrene or (meth)acrylates were also prepared at 120°C ($M_n \approx 17000$; $M_w/M_n = 1.1 \sim 1.2$).

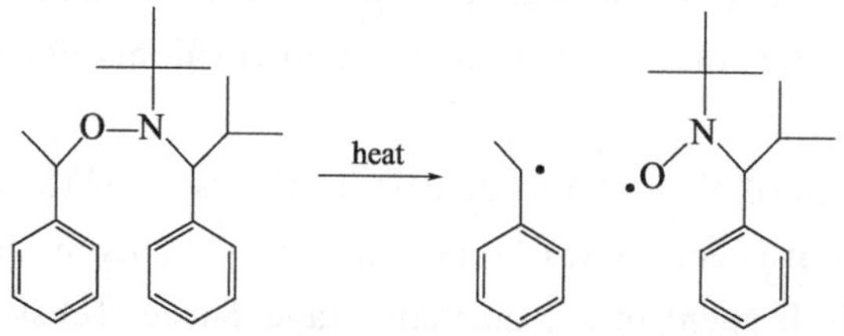

Several other methods for controlled radical polymerization have been developed and should be applicable to elastomer synthesis. One of the other most important systems for controlled radical polymerization is atom transfer radical polymerization (ATRP). A transition metal (Mt) catalyst participates in an oxidation-reduction equilibrium by reversibly transferring an atom, often a halogen, from a dormant species (initiator or polymer chain) as shown below. Although a variety of transition metal salts are effective, copper salts have been most extensively investigated.

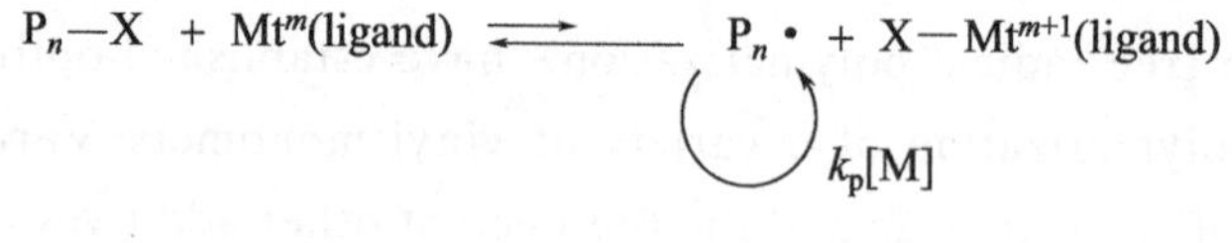

$$P_n—X + Mt^m(\text{ligand}) \rightleftharpoons P_n\cdot + X—Mt^{m+1}(\text{ligand})$$
$$k_p[M]$$

1.5 Emulsion Polymerization

1.5.1 Mechanism and Kinetics

Polymerization in aqueous emulsions, which has been widely developed technologically, represents a special case of free radical chain polymerization in a heterogeneous system. Most emulsion polymerization systems comprise a water-insoluble monomer in water with a surfactant and a free radical initiator. Although it might be thought that polymerization of water insoluble monomers in an emulsified state simply involves the direct transformation of a dispersion of monomer into a dispersion of polymer, this is not really the case, as evidenced by the following features of a true emulsion polymerization:

① The polymer emulsion (or latex) has a much smaller particle size than the emulsified monomer, by several orders of magnitude;

② The polymerization rate is much faster than that of the undiluted monomer, by one or two orders of magnitude;

③ The molecular weight of the emulsion polymer is much greater than that obtained from bulk polymerization, by one or two orders of magnitude.

It is obvious from the foregoing facts that the mechanism of emulsion polymerization involves far more than the mere bulk polymerization of monomer in a finely divided state. In fact, the very small particle size of the latex, relative to that of the original monomer emulsion, indicates the presence of a special mechanism for the formation of such polymer particles.

The mechanism of emulsion polymerization, as originally proposed by Harkins, can best be understood by examining the components of this system, as depicted in Fig. 1.2, for a typical "water insoluble" monomer such as styrene (solubility = 0.07g/L). The figure shows the various loci in which monomer is found, and which compete with each other for the available free radicals. Thus, in the initial stages, the monomer is found in three loci: dissolved in aqueous solution, as emulsified droplets, and within the soap micelles.

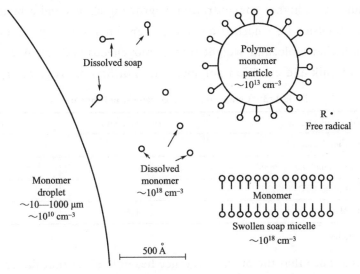

Fig. 1.2 Loci in mechanism of emulsion polymerization.

Both the dissolved monomer and the relatively large monomer droplets represent minor loci for reaction with the initiator radicals (except, of course, in the case of highly water-soluble monomers). The large number of soap micelles containing imbibed monomer, however, represents a statistically important locus for initiation of polymerization. It is thus not surprising that most of the polymer chains are generated within the monomer-swollen soap micelles. The very large number ($\sim 10^{15}$/mL) of very small polymer particles thus formed which are stabilized by adsorbing monolayers of soap deplete the available molecularly dissolved soap, thus destroying the soap micelles at an early stage of the polymerization ($\sim 10\%$ conversion in the usual recipe). As all the available soap is distributed, and redistributed, over the surface of the growing particles, the amount of soap is the main factor controlling latex particle size.

During the second stage of the emulsion polymerization, therefore, the loci for available monomer consist of the dissolved monomer, the free monomer droplets, and the monomer imbibed by the numerous polymer particles. As before, the first two of these loci make a minor contribution, whereas the polymer-monomer particles provide a major locus for reaction with the initiator radicals diffusing from the aqueous phase. The major portion of the polymerization reaction apparently occurs within this large number of latex particles which are isolated from each other by electrostatic repulsion and kept saturated with monomer diffusing from the monomer droplets. It is this aspect which leads to the unique characteristics of this system. Thus, once an initiator radical enters a polymer monomer particle and initiates a chain, the latter must continue to propagate with the available monomer until another radical enters the same particle. In this way, the rate of chain termination is actually controlled by the rate of entry of radicals into the particles, and this generally increases the lifetime of the growing chains, and hence the chain length. Furthermore, because the growing chains are all located in different particles, they are unable to terminate each other, leading to a higher concentration of growing chains and a hence faster rate.

In this way, emulsion polymerization systems can simultaneously achieve a much faster rate and a much higher molecular weight than homogeneous systems. A comparison of the kinetic features of bulk and emulsion polymerization of Styrene is given in Table 1.3.

Table 1.3 Comparison of free radical polymerization

Methods of Styrene	Homogeneous	Emulsion
Monomer concentration/(mol/L)	5	5[a]
Radical concentration/(mol/L)	10^{-8}	10^{-6}
Rate of polymerization at 60℃/(%/hr)	~ 2	100
Molecular weight (M_n)	105	107

[a] Within latex particles.

It is obvious at once that the main difference lies in the fact that the emulsion system is capable of raising the steady-state concentration of growing chain radicals by two to three orders of magnitude but not at the expense of increasing the termination rate which occurs in homogeneous solution [see Eq. (1-16)]!

The situation described earlier, i.e., where radicals entering individual latex particles successively initiate and terminate growing chains, is referred to as ideal emulsion polymerization, as defined by the Smith-Ewart theory. Under these conditions, the concentration of growing chains per unit volume of latex is easily predictable, because at any given time, half of the particles will contain a growing chain. In other words, the number of growing chains will be one-half the number of particles. As the latter is of the order of 10^{18} per liter, the concentration of growing chains is of the order of 10^{-6} mol/L compared with 10^{-8} mol/L for homogeneous polymerization systems. Because such growing chains are in an environment rich in monomer (within the monomer polymer particles), it is not surprising that emulsion polymerization rates are one or two orders of magnitude higher than those of bulk polymerization, as shown in Table 1.3. Furthermore, this high radical concentration does not affect the radical lifetime, i.e., the chain size, which is governed solely by the availability of another free radical for termination and, thus, by the period between successive entries of radicals into particles. For a given rate of initiation, the time between radical entry depends on the number of latex particles; i.e., the radical lifetime (and molecular weight) increases with increasing numbers of particles. According to the theory of Smith and Ewart, the number of polymer particles depends on both the initiator concentration and the surfactant concentration,

$$N \propto [I]^{2/5}[S]^{3/5} \tag{1-25}$$

where $[I]$ is the concentration of initiator; and $[S]$ is the concentration of surfactant.

The foregoing situation, of course, holds only for the ideal case, as defined earlier. If the growing chain within the latex particle undergoes some side reaction which transfers the radical activity out of the particle before the next radical enters, or if termination is not rapid when two radicals occupy the same particle, then the number of growing chains at any given time will be, respectively, either smaller or larger than one-half the number of particles. The latter case (more than one radical per particle) can occur, for example, if the particle size is sufficiently large and the termination rate too slow. The rate and molecular weight will then also be governed by other considerations than the interval between entry of successive radicals. Diagnostically, these situations can be distinguished from the ideal case by the effect of added initiator on the rate of polymerization after formation of particles is complete. Thus, in either case, an increase in initiator concentration will lead to a faster rate of entry of radicals into particles and hence an increase in the number of radicals per particle, leading to an increase in polymerization rate. In contrast, in the ideal case, an increase in frequency of radical entries into particles should not affect the rate, as the particles will still contain a radical only half the time, even though the periods of chain growth will be shorter, leading to a lower molecular weight.

The ideal case of the Smith-Ewart treatment actually proposes a rather elegant method for obtaining the absolute value of the propagation rate constant k_p from emulsion polymerization systems, as shown in Eq. (1-26), where N is the number of particles per unit volume.

$$R_p = k_p[M]N/2 \tag{1-26}$$

Eq. (1-26) leads to a solution for k_p from available knowledge of the rate R_p, the concentration of monomer in the monomer-polymer particles [M], and the number of particles N. This method has been applied to several monomers and has been especially useful in the case of the dienes, where the classical method of photoinitiation poses difficulties. Some of these results are shown in Table 1.4 in the form of the usual kinetic parameters. The results obtained for Styrene by photoinitiation techniques are included for comparison.

Table 1.4 Propagation rate constants from emulsion polymerization

Monomer	k_p at 60℃/[L/(mol·sec)]	E_p/(kcal/mol)	$A_p(\times 10^{-7})$/[L/(mol·s)]
Butadiene	100	9.3	12
Isoprene	50	9.8	12
2,3-Dimethylbutadiene	120	9.0	9
Styrene	280	7.9	4
Photoinitiation	176	7.8	2.2

It can be seen that the agreement is remarkably good, considering the widely different experimental methods used. Recent studies of the emulsion polymerization of butadiene have shown that the rate constant for propagation is even higher than previously estimated (see Table 1.1). The data in Table 1.4 provide evidence that the slow rates and low molecular weights obtained in homogeneous free radical polymerization of these dienes are not due to a low rate constant for propagation but rather must be caused by a high rate constant for termination (as indicated in Table 1.1). Hence, under the special conditions of emulsion polymerizations, where the termination rate is controlled by the rate of entry of radicals into particles, it becomes possible to attain both faster rates and higher molecular weights. It is this phenomenon which led to the rise of the emulsion polymerization system for the production of diene-based synthetic elastomers.

1.5.2 Styrene-Butadiene elastomer

The most successful method developed for the production of a general purpose synthetic elastomer was the emulsion copolymerization of butadiene and styrene (SBR), which still represents the main process in use today.

1.5.2.1 Kinetics and Molecular Weights

The types of recipes used are seen in Table 1.5. The recipes shown are to be considered only as typical, as they are subject to many variations. It should be noted that the initiator in the 50℃ recipe (hot elastomer) is the persulfate, whereas in the 5℃ recipe (cold elastomer) the initiator consists of a redox system comprising the hydroperoxide-iron(Ⅱ)-sulfoxylate-EDTA. In the latter case, the initiating radicals are formed by the reaction of the hydroperoxide with the ferrous iron, whose concentration is controlled by the EDTA complexing agent; the sulfoxylate is needed to convert the oxidized ferric(Ⅲ) back to ferrous iron. The phosphate salt serves as a stabilizing electrolyte for the latex.

Table 1.5 Typical SBR emulsion polymerization recipes

	Property	SBR-1000*	SBR-1500
	Polymerization temperature /℃	50	5
	Time /h	12	12
	Conversion/%	72	60~65
Ingredients	Butadiene	71	71
	Styrene	29	29
	Water	190	190
	Soap (fatty or rosin acid)	5	4.5~5
	Potassium persulfate	0.3	—
	n-Dodecanethiol	0.5	—
	t-Dodecanethiol	—	0.2
	p-Menthane hydroperoxide or pinane hydroperoxide	—	0.08
	Trisodium phosphate ($Na_3PO_4 \cdot 10H_2O$)	—	0.5
	Ferrous sulfate ($FeSO_4 \cdot 7H_2O$)	—	0.4
	Sodium formaldehyde sulfoxylate	—	0.10
	Tetrasodium salt of ethylenediamine tetraacetic acid (EDTA)	—	0.06

* Commercial grade numbers assigned by the International Institute of Synthetic elastomer Producers to "hot" and "cold" SBR, respectively.

In both recipes, the thiol acts as a chain transfer agent to prevent the molecular weight from attaining the excessively high values possible in emulsion polymerization systems (see Table 1.6).

Table 1.6 Typical properties of emulsion-polymerized SBR

	Property	Hot SBR	Cold SBR
	Styrene content	24	24
Molecular weight	Viscosity average	$1.5 \times 10^5 \sim 4.0 \times 10^5$	2.8×10^5
	Weight average	—	5×10^5
	Number average	$0.3 \times 10^5 \sim 1.0 \times 10^5$	$1.1 \times 10^5 \sim 2.6 \times 10^5$

It acts in an analogous fashion to the solvent in Eq. (1-27) and Eq. (1-28), except that the sulfur-hydrogen bond is extremely susceptible to attack by the growing chain radical, which is thus terminated by a hydrogen atom, forming the RS· radical which initiates growth of a new chain:

$$P\cdot + RS\text{—}H \xrightarrow{k_{tr}} P\text{—}H + RS\cdot \qquad (1\text{-}27)$$

$$RS\cdot + M \xrightarrow{k'_j} RS\text{—}M\cdot \qquad (1\text{-}28)$$

These thiols, which are known as "regulators", have transfer constants greater than 1, e.g., k_{tr}/k_p may be 3~4, so that only a small proportion is needed to reduce the molecular weight from several million to several hundred thousand. Diene-based polymers can undergo crosslinking reactions during the polymerization, which leads to the formation of insoluble "gel" elastomer when the molecular weight becomes too high. Hence, thiol is used as "modifier" to prevent gel formation and keep the elastomer processible. It is also necessary to stop the reaction at intermediate levels of conversion to minimize undesirable gel formation (see Table 1.7).

Table 1.7 Crosslinking parameters for polybutadiene

Temperature /℃	Relative crosslinking rate, $r_x (\times 10^4)$	$X_w (\times 10^4)$ of primary chains at gel point
60	1.98	2.15
50	1.36	3.13
40	1.02	4.18
0 (cal.)	0.16	26.3

Note:

$r_x = k_x/k_p$, where k_x is the crosslinking rate constant and k_p is the propagation rate constant.

Shortly after Word War Ⅱ, the American synthetic elastomer industry began production of "cold" SBR, from which, it was found, superior tire elastomer, especially as regards tread wear, could be prepared. Subsequent studies showed that the reduction in temperature from 50℃ to 5℃ had little or no effect on the microstructure of the polydiene units (cis-1,4- versus trans-1,4- versus 1,2-), or on the comonomer composition, but did exert a marked influence on the molecular weight distribution (Table 1.6). It was also shown that the crosslinking reaction i.e., addition of growing chains to polymer double bonds (mainly with 1,2-side chain units), was substantially reduced at these lower temperatures, thus reducing the tendency for gel formation at any given molecular weight.

Table 1.7 shows the maximum molecular weights of polybutadiene attainable at different polymerization temperatures, prior to gelation, expressed as the critical weight-average chain length, X_w, of the primary chains at the gel point. Thus it can be seen that it is possible to increase the chain length by a factor of 9, without forming gel, by decreasing the polymerization temperature from 50℃ to 0℃. Hence, the amount of thiol chain transfer agent can also be reduced, and this improves the overall chain length distribution by avoiding the formation of the very low molecular weight fraction which results from the rapid reaction of the thiol in the early stages of the polymerization. Furthermore, this possibility of producing gel-free higher molecular weight SBR at reduced polymerization temperature enabled the preparation of a high Mooney viscosity (~100) polymer which could be plasticized by low-cost petroleum oils ("oil-extended" elastomer), and still retain its advantageous mechanical properties. As a result, the cold SBR process accounts for more than 85% of the emulsion SBR produced.

1.5.2.2 Chain Microstructure

As might be expected, the emulsion polymerization system does not alter the basic mechanism of free radical polymerization as regards the chain unit structure. The latter is, of course, independent of the type of free radical initiator used, in view of the "free" nature of the growing chain end radical. The temperature of polymerization does exert some influence, as shown by the data in Table 1.8, but not to a very great extent. It can be seen that the 1,2-side-chain vinyl content is rather insensitive to the temperature, whereas the *trans*-1,4-content increases with decreasing temperature, at the expense of the *cis*-1,4-content. The latter almost vanishes, in fact, at low temperatures and the polymer then attains its highest *trans*-1,4-content of about 80%. Hence this type of polybutadiene is sufficiently stereoregular to undergo a substantial amount of crystallization on cooling. However, the introduction of the styrene comonomer is sufficient to destroy the chain regularity necessary for crystallization. Furthermore, it is the high *cis*-1,4-polybutadiene which is desirable and not the *trans*-1,4-form, since the latter has a crystalline melting point of about 150°C and is not an elastomer at ambient temperature. As can be seen from Table 1.8, the possibility of attaining a high *cis*-1,4-content at a reasonably high polymerization temperature is quite remote.

Table 1.8 Chain structure of emulsion polybutadiene and SBR

Polymers	Polymerization temperatures /°C	Isomer (w, %)		
		cis-1,4-	*trans*-1,4-	1,2-
Polybutadiene	-33	5.4	78.9	15.6
	5	13.0	69.9	16.5
	50	19.0	62.7	18.8
	70	20.8	59.4	19.8
SBR	-33	5.4	80.4	12.7
	5	12.3	71.8	15.8
	50	18.3	65.3	16.3
	70	20.0	63.0	17.3
	100	22.5	60.1	17.3

Hence, it appears that these minor effects of temperature on the microstructure of the butadiene units cannot be expected to have any real influence on the properties of SBR.

1.5.2.3 Emulsion Polymerization of Chloroprene

1.5.2.3.1 Kinetics

The only other diene that has been used extensively for commercial emulsion polymerization is chloroprene (2-chloro-1,3-butadiene). The chlorine substituent apparently imparts a marked reactivity to this monomer, since it polymerizes much more rapidly than butadiene, isoprene, or any other dienes (see Tables 1.4); $k_p(35°C) = 595L/(mol \cdot s)$. In fact, chloroprene is even more susceptible to spontaneous free radical polymerization than styrene, and requires a powerful inhibitor for stabilization. It polymerizes extremely rapidly in emulsion systems, so that its rate must be carefully controlled.

Various recipes can be used for emulsion polymerization of chloroprene, with potassium persulfate as a popular initiator. A basic recipe which illustrates several interesting features about this monomer is shown in Table 1.9. Two aspects of this recipe are especially noteworthy: the use of a rosin soap, and the presence of elemental sulfur. Rosin soaps are notorious as retarders in emulsion polymerization, as are most polyunsaturated fatty acids.

Table 1.9 Basic recipe for neoprene GN[a]

Ingredient Parts	by weight	Ingredient Parts	by weight
Chloroprene	100	Sodium hydroxide	0.8
Water	150	Potassium persulfate	0.1～0.2
N wood rosin	4	Latex stabilizer[b]	0.7
Sulfur	0.6		

[a] Temperature, 40℃; time, several hours; conversion, 90%.
[b] A sodium salt of naphthalenesulfonic acid-formaldehyde condensation product.

Yet complete conversion can be attained within a few hours. With saturated fatty acid soaps, the reaction is almost completed within one hour at 40℃!

Sulfur copolymerizes with the chloroprene, forming di- and polysulfide linkages in the chain [as illustrated in Eq. (1-29)]. The latex is then treated with the well-known vulcanization accelerator, tetraethyl thiuram disulfide, which, by sulfur-sulfur bond interchange, degrades the crosslinked polychloroprene "gel" and renders it soluble and processible as schematically shown in Eq. (1-30). In this way, it serves a purpose analogous to the thiol chain transfer agents in SBR polymerization. As a matter of fact, the newer grades of polychloroprene are prepared with the use of thiols and other chain transfer agents. The thiols have been found to yield narrower molecular weight distributions for chloroprene than for butadiene or isoprene, due to their much slower rate of disappearance in the presence of chloroprene. These mercaptan grades represent the most common, standard grades.

$$CH_2=C-CH=CH_2+S_8 \longrightarrow \begin{matrix} +CH_2 & CH_2\!\!\!\!+_m S_x +CH_2 & CH_2\!\!\!\!+_n \\ C=C & & C=C \\ Cl & H & Cl & H \end{matrix} \quad (1\text{-}29)$$
$$|$$
$$Cl$$

$$\begin{matrix} +CH_2 & CH_2\!\!\!\!+_m S_x +CH_2 & CH_2\!\!\!\!+_n \\ C=C & & C=C \\ Cl & H & Cl & H \end{matrix} \xrightarrow{(Et)_2NCS_4CN(Et)_2} \begin{matrix} +CH_2 & CH_2\!\!\!\!+_m S_y-SCN(Et)_2 \\ C=C \\ Cl & H \end{matrix} \quad (1\text{-}30)$$

1.5.2.3.2 Chain Structure

Another feature of the emulsion polymerization of chloroprene that distinguishes it from that of the other dienes is the fact that it leads to a predominantly *trans*-1,4-chain microstructure. Thus, even at ambient polymerization temperature, the polychloroprene contains over 90% *trans*-1,4-units, as shown in Table 1.10, which illustrates the effect of polymerization temperature on stereoregularity of the chain. As expected, lower polymerization temperatures lead to a more stereoregular *trans*-1,4-polychloroprene.

Table 1.10 Effect of polymerization temperature on polychloroprene chain microstructure

Temperature/°C	Isomeric chain microstructure /%					
	trans-1,4-		Isomerized			
	Total	Inverted[①]	1,2-	1,2-[②]	3,4-	cis-1,4-
−150	~100	2.0	—	—	—	—
−40	97.4	4.2	0.8	0.6	0.5	0.8
−20	97.1	4.3	0.9	0.6	0.5	0.8
0	95.9	5.5	1.2	1.0	1.1	1.8
20	92.7	8.0	1.5	0.9	1.4	3.3
40	90.8	9.2	1.7	0.8	1.4	5.2
90	85.4	10.3	2.3	0.6	4.1	7.8

[①] 1,4-enchainment.

[②] $-CH_2-\underset{|}{C}=CHCH_2Cl$.

Because of the higher crystal melting point of the *trans*-1,4-polychloroprene (T_m = 105°C), as compared with that of the *cis*-1,4-polyisoprene in Hevea elastomer (~20°C), even the polymer containing as little as 80% *trans* units crystallizes readily on cooling, or on stretching. The melting point of emulsion polychloroprene is generally in the range of 40°C ~ 50°C. Hence emulsion polychloroprene is the only latex polymer which resembles natural elastomer, in that it is sufficiently stereoregular to exhibit strain induced crystallization. This, then, results in high tensile strength in gum vulcanizates, without the need of reinforcing fillers, just as in the case of Hevea elastomer. This makes possible the use of polychloroprene in a variety of gum elastomer products, endowing them with superior oil and solvent resistance (because of its polarity), as well as high strength.

1.6 Copolymerization

1.6.1 Kinetics

Copolymerization involves the simultaneous chain polymerization of a mixture of two or more monomers. Aside from the general kinetic considerations which govern these chain reactions, as described earlier, there is imposed an additional feature, i.e., the relative participation of the different monomers during the growth of the chain. This new parameter is most important, since it controls the composition of the copolymer. Systems involving more than two monomers are difficult to resolve in this respect, but it has been found possible to treat the case of a pair of monomers with relative ease.

In the chain addition polymerization of two monomers, regardless of the mechanism involved, the growing chain always must make a choice of reacting with one of the two monomers. Furthermore, there are two kinds of growing chains, depending on which type of monomer unit occupies the growing end. Thus, four types of propagation steps can be written as follows for any chain copolymerization of two monomers assuming that the reactivity of the chain end depends only on the chain end monomer unit (terminal model):

$$[M_1*] + [M_1] \xrightarrow{k_{11}} [M_1*] \tag{1-31}$$

$$[M_1*] + [M_2] \xrightarrow{k_{12}} [M_2*] \tag{1-32}$$

$$[M_2*] + [M_2] \xrightarrow{k_{22}} [M_2*] \tag{1-33}$$

$$[M_2*] + [M_1] \xrightarrow{k_{21}} [M_1*] \tag{1-34}$$

where M_1* and M_1 refer to the growing chain and the monomer, respectively, as before, while the subscripts refer to the two kinds of monomer in the mixture. It can be seen that these four propagation reactions lead to four propagation rate constants, as shown above. Hence the rate of consumption of each monomer may be expressed by the following equations:

$$\frac{d[M_1]}{dt} = k_{11}[M_1*][M_1] + k_{21}[M_2*][M_1] \tag{1-35}$$

$$\frac{d[M_2]}{dt} = k_{12}[M_1*][M_2] + k_{22}[M_2*][M_2] \tag{1-36}$$

Since it is the relative rate of consumption of the two monomers which will decide the composition of the chain, it can be expressed by dividing Eq. (1-35) by Eq. (1-36) leading to Eq. (1-37).

$$\frac{d[M_1]}{d[M_2]} = \frac{k_{11}[M_1*][M_1] + k_{21}[M_2*][M_1]}{k_{12}[M_1*][M_2] + k_{22}[M_2*][M_2]} \tag{1-37}$$

It is obvious at once that Eq. (1-37) is quite intractable for direct use. However, it is possible to simplify it considerably by utilizing the "steady state" treatment, analogous to the one previously described. This is done by assuming the rate of Eq. (1-32) to be equal to that of Eq. (1-34), and this leads to the equivalence

$$[M_2*] = \frac{k_{12}}{k_{21}}[M_1*]\frac{[M_2]}{[M_1]}$$

which, when inserted into Eq. (1-37), yields Eq. (1-38), after appropriate rearrangements are made,

$$\frac{d[M_1]}{d[M_2]} = \frac{[M_1]}{[M_2]} \frac{r_1[M_1] + [M_1]}{r_2[M_2] + [M_2]} \tag{1-38}$$

where $r_1 = k_{11}/k_{12}$ and $r_2 = k_{22}/k_{21}$. The parameters r_1 and r_2 are known as the monomer reactivity ratios, since they express the relative reactivity of each of the two kinds of growing chain ends with their "own" monomer as compared with the "other" monomer. They may in fact be considered as expressing the "homopolymerization" tendency of each type of monomer relative to cross over with the comonomer.

Eq. (1-38), which relates the instantaneous composition of the copolymer ($d[M_1]/d[M_2]$) to the prevailing monomer concentrations, can be used to determine the values of r_1 and r_2. Many such values have been recorded. Typical values of these parameters for styrene copolymerizations are shown in Table 1.11, which illustrates the wide variations that prevail. The relative reactivity actually expresses the relative reactivity of each of the monomers shown toward the styrene radical compared to the reaction with styrene monomer.

Table 1.11 Monomer reactivity ratios for free radical copolymerizations with Styrene (M_1)

Monomer (M_2)	r_1	r_2	Relative reactivity ($1/r_1$)
Maleic anhydride	0.097	0.001	10.3
2,5-Dichlorostyrene	0.268	0.810	3.73
Methyl methacrylate	0.585	0.478	1.71
Methyl acrylate	0.871	0.148	1.15
Vinylidene chloride	1.839	0.087	0.54
Diethyl maleate	6.07	0.01	0.16
Vinyl acetate	18.8	0.02	0.05

Thus, the r_1 and r_2 values permit some conclusions about the expected composition of the copolymer obtained at any given monomer ratio. For example, it can be deduced from Table 1.11 that a copolymer of styrene and maleic anhydride ($r_1 r_2 \geqslant 0$) would be strongly "alternating", since it would be improbable to have a sequence of two styrene unit, and highly improbable to have a sequence of two maleic anhydride units. Also, it would obviously be extremely difficult to prepare a copolymer of styrene and vinyl acetate, since the latter monomer would be virtually excluded from the styrene polymerization.

It is also obvious, from Eq. (1-38), that the copolymer composition would not necessarily correspond to the comonomer charge, depending on the values of r_1 and r_2. A desirable system would, of course, be one in which this were the case, i.e., where the comonomers enter into the copolymer in the ratio of their concentrations; i.e., where

$$\frac{d[M_1]}{d[M_2]} = \frac{[M_1]}{[M_2]} \tag{1-39}$$

This is defined as an "azeotropic" copolymerization, by analogy to the distillation of two miscible liquids. Eq. (1-37) would apply under the conditions where

$$\frac{r_1[M_1]+[M_2]}{r_2[M_2]+[M_1]} = 1 \tag{1-40}$$

and this would be valid, for example, where $r_1 = r_2 = 1$. In that case, Eq. (1-39) could apply for all charge ratios, i.e., the two types of growing chains show no particular preference for either of the two monomers; this is described as a random copolymerization. Also Eq. (1-40) would be valid when $r_1 = r_2$ and $[M_1] = [M_2]$, i.e., an azeotropic copolymerization would result only at equimolar charge ratios. In general, Eq. (1-40) is valid when

$$\frac{[M_1]}{[M_2]} = \frac{r_2 - 1}{r_1 - 1} \tag{1-41}$$

This means that any copolymerization will be of an azeotropic type at the particular comonomer charge ratio indicated by Eq. (1-41). However, it also means that an azeotropic copolymerization is only possible when both r_1 and r_2 either greater than 1 or less than 1.

It is important to emphasize that this kinetic treatment is valid for any chain polymerization mechanisms, i.e., free radical, cationic, anionic, and coordination. However, in the case of the ionic mechanisms, the type of initiator used and the nature of the solvent medium may influence the r_1 and r_2 values. This is due to the fact that the growing chain end in ionic systems is generally associated with a counterion, so that the structure and reactivity of such chain ends can be expected to be affected by initiator and the solvent.

1.6.2 Emulsion Copolymerization of Dienes

The three cases which involve copolymerizations leading to commercial synthetic elastomers are styrene-butadiene (SBR), butadiene-acrylonitrile (NBR), and chloroprene with various comonomers.

1.6.2.1 Styrene-Butadiene (SBR)

A large number of studies have been made of the reactivity ratios in this copolymerization, both in homogeneous and emulsion systems, and the average r values have been computed for butadiene and styrene, respectively, as $r_B = 1.6$, $r_S = 0.5$.

These values apply to solution and emulsion polymerization, presumably because neither monomer is particularly soluble in water, and both are quite insensitive to temperature. It appears, therefore, that the butadiene must enter the chain substantially faster than its charging ratio, and that each increment of polymer formed contains progressively more styrene. This is confirmed by the change in composition of the copolymer with conversion, as shown in Table 1.12. It can be seen that, at high conversion, the increment, or differential, composition becomes quite high in styrene content with concomitant loss of elastomery properties, even though the cumulative, or integral, composition still shows a low styrene content. This indicates the advisability of stopping the reaction at conversions not much higher than 60%.

Table 1.12 Comonomer composition of SBR[a]

Conversion/%	Styrene in copolymer (w, %)	
	Differential	Integral
0	17.2	—
20	18.8	17.9
40	20.6	18.7
60	23.3	19.7
80	29.5	21.2
90	36.4	22.5
95	45.0	—
100	(100)	(25.0)

[a] Charge weight ratio, 75 : 25, butadiene : styrene; 50℃.

It should be noted, too, that the r values for this system do not permit an azeotropic polymerization, as predicted by Eq. (1-39). With respect to the distribution of styrene monomer units in the copolymer, the monomer reactivity ratio product, $r_B r_S = 0.8$, is close to a value of 1.0, which would correspond to an "ideal" copolymerization which would correspond to a random distrubution of styrene units along the chain. For an "ideal" copolymerization, the relative rates of incorporation of the two monomers are independent of the chain end unit as predicted by Eq. (1-42).

$$r_B = \frac{1}{r_S}, \text{therefore } \frac{k_{BB}}{k_{BS}} = \frac{k_{SB}}{k_{SS}} \tag{1-42}$$

It is reported that the number average number of styrene units in a sequence is 1.2 as determined by high resolution gel permeation chromatography of ozonolysis products. The observed sequence distribution of monomer units was in accord with calculated values based on the monomer reactivity ratios.

1.6.2.2 Butadiene-Acrylonitrile (Nitrile elastomer)

According to Hofmann, the reactivity ratios of this pair of monomers at 50℃ in emulsion polymerization are $r_B = 0.4$, $r_{AN} = 0.04$ and they decrease somewhat at lower temperatures, but not to a great extent. These ratios are no doubt influenced by the marked water solubility of the acrylonitrile compared to that of butadiene.

The foregoing r values lead to the following situation. In accordance with Eq. (1-39), an azeotropic copolymer is formed when the acrylonitrile charge is 35% to 40% by weight (or by mole), so that a constant composition is maintained throughout the polymerization. If the acrylonitrile charge is below this value, the initial copolymer is relatively rich in acrylonitrile, which progressively decreases with increasing conversion. However, if the acrylonitrile charge is higher than the "azeotrope", the initial copolymer contains less acrylonitrile than charged, but the acrylonitrile content increases with conversion. Since the commercial nitrile elastomers have nitrile contents from 10% to 40%, these considerations have a very practical significance. With respect to the distribution of comonomer units in the copolymer, the monomer reactivity ratio product, $r_B r_{AN} = 0.016$, is close to zero, which would correspond to an alternating distribution of comonomer units.

1.6.3 Chloroprene

Polychloroprene is generally prepared commercially as a homopolymer, although small amounts of a comonomer are included in several grades of these elastomers. There are two good reasons for the paucity of chloroprene based copolymers. In the first place, the homopolymer, as stated previously, has a high *trans*-1,4-chain structure and is therefore susceptible to straininduced crystallization, much like natural elastomer, leading to excellent tensile strength. It also has other favorable mechanical properties. Furthermore, chloroprene is not very susceptible to copolymerization by the free radical mechanism, as indicated by the r values in Table 1.13. Thus, except for 2,3-dichloro-1,3-butadiene, chloroprene does not efficiently undergo copolymerization with other monomers. Hence, it is not surprising that the few copolymers of chloroprene available commercially contain only minor amounts of comonomers, which are included for their moderate effects in modifying the properties of the elastomers.

Table 1.13 Monomer reactivity ratios in copolymerization of chloroprene

Monomer M_2	r_1	r_2
Styrene	5.98	0.025
Isoprene	2.82	0.06
Acrylonitrile	5.38	0.056
Methyl methacrylate	6.33	0.080
2,3-Dichloro-1,3-butadiene	0.31	1.98

1.7 Chain Polymerization by Cationic Mechanism

1.7.1 Mechanism and Kinetics

In these chain addition reactions, the active species is cationic in nature, initiated by strong acids, either of the protic or Lewis variety. Since most of these ionic polymerizations are carried out in nonaqueous solvents with low dielectric constants, it is unlikely that the active species is a "free" ion, analogous to a free radical. A multiplicity of active species may be involved as propagating species as shown below by the spectrum of cationic species, one or more of which may be involved as active propagating species, especially in more polar solvents. Unfortunately, very little information is available about the exact nature of the propagating species in cationic systems. This is mainly due to the inherent experimental difficulties, caused by high reactivity and sensitivity to impurities, especially to traces of water.

$$R-X \rightleftharpoons R^{\oplus}, X^{\ominus} \rightleftharpoons R^{\oplus} \| X^{\ominus} \rightleftharpoons R^{\oplus} + X^{\ominus}$$

$$\text{covalent} \quad\quad \text{tight ion pair} \quad\quad \text{loose ion pair} \quad\quad \text{free ions}$$

$$M \Big| k_p^1 \quad\quad M \Big| k_p^2 \quad\quad M \Big| k_p^3 \quad\quad M \Big| k_p^4$$

The most common initiators of cationic polymerization are Lewis acids, such as $AlCl_3$, BF_3, $SnCl_4$, and $TiCl_4$, although strong protic acids such as H_2SO_4 may also be used. Cationic polymerization is restricted to vinyl monomers with electron-donating or electron-delocalizing substituents, e.g., isobutylene, vinyl alkyl ethers, vinyl amines, styrene, and other conjugated hydrocarbons.

These polymerizations are characterized by rapid rates at very low temperatures, e.g., isobutylene is polymerized almost instantaneously at $-100°C$ by $AlCl_3$. The presence of a hydrogen donor, such as water or a protic acid, as a cocatalyst, is usually a prerequisite, as has been shown in the case of isobutylene. On this basis, the Evans-Polanyi mechanism proposed the following reaction sequence for the polymerization of isobutylene by BF_3 monohydrate:

Initiation

$$BF_3 \cdot H_2O + CH_2=C(CH_3)_2 \longrightarrow CH_3-\underset{CH_3}{\overset{CH_3}{C^{\oplus}}}-BF_3OH^{\ominus} \tag{1-43}$$

Propagation

$$CH_3-\underset{CH_3}{\overset{CH_3}{C^{\oplus}}}BF_3OH^{\ominus} + CH_2=\underset{CH_3}{\overset{CH_3}{C}} \longrightarrow \cdots \longrightarrow CH_3-\underset{CH_3}{\overset{CH_3}{C}}\Big[CH_2-\underset{CH_3}{\overset{CH_3}{C}}\Big]_x CH_2-\underset{CH_3}{\overset{CH_3}{C^{\oplus}}}BF_3OH^{\ominus} \tag{1-44}$$

Termination

$$CH_3-\underset{CH_3}{\underset{|}{C}}\underset{CH_3}{\overset{CH_3}{\left[-CH_2-\underset{CH_3}{\underset{|}{C}}-\right]_x}}CH_2-\underset{CH_3}{\overset{CH_3}{\underset{|}{C}}}{}^{\oplus}BF_3OH^{\ominus} \longrightarrow$$

$$CH_3-\underset{CH_3}{\underset{|}{C}}\overset{CH_3}{\underset{CH_3}{\left[-CH_2-\underset{|}{C}-\right]_x}}CH=C\underset{CH_3}{\overset{CH_3}{\diagdown}} + CH_3-\underset{CH_3}{\underset{|}{C}}\overset{CH_3}{\underset{CH_3}{\left[-CH_2-\underset{|}{C}-\right]_x}}CH_2-\underset{CH_3}{\overset{CH_2}{\diagup}}C + BF_3 \cdot OH_2$$

(1-45)

Chain-transfer to monomer is also an integral step in the cationic polymerization of isobutylene, and it is this reaction which controls the molecular weight.

Chain Transfer

$$CH_3-\underset{CH_3}{\underset{|}{C}}\overset{CH_3}{\underset{CH_3}{\left[-CH_2-\underset{|}{C}-\right]_x}}CH_2-\overset{CH_3}{\underset{CH_3}{\underset{|}{C}}}{}^{\oplus}BF_3OH^{\ominus} + CH_2=C\underset{CH_3}{\overset{CH_3}{\diagdown}} \xrightarrow{k_{tr}}$$

(1-46)

$$CH_3-\underset{CH_3}{\underset{|}{C}}\overset{CH_3}{\underset{CH_3}{\left[-CH_2-\underset{|}{C}-\right]_x}}CH_2-\underset{CH_3}{\overset{CH_2}{\diagup}}C + (CH_3)_3C^{\oplus}BF_3OH^{\ominus}$$

This mechanism has been supported by infrared and ^{1}H-NMR spectroscopic data which provided evidence for the presence of the polymer end groups proposed by this mechanism, i. e. ,

$$(CH_3)_3C- \qquad -CH_2-\underset{CH_3}{\underset{|}{C}}=CH_2 \qquad -CH_2=C\underset{CH_3}{\overset{CH_3}{\diagdown}}$$

Furthermore, the use of the tracer complex $BF_3 \cdot D_2O$ showed that the polymer contained deuterium while the initiator became converted to $BF_3 \cdot H_2O$. This mechanism actually involves initiation by addition of a proton [Eq. (1-43)], from the Brønsted acid formed by the Lewis acid and a coinitiator, to the monomer and subsequent termination of chain growth by loss of a proton to the initiator anion [Eq. (1-45)] or by chain transfer to monomer [Eq. (1-46)]. The chain growth, therefore, occurs during the brief lifetime of the carbenium ion, and the initiator or a new carbenium ion is constantly regenerated. The chain transfer step would be expected to have no effect on rate, since the trimethylcarbenium ion should rapidly reinitiate chain growth, but the chain length will decrease. It should be pointed out again, at this point, that it is improbable, in view of the low dielectric constants of the solvents employed, that the active propagating species in these systems are free ions; therefore, the counterion has been depicted as being associated with the carbenium ion in each mechanistic step. This simplified mechanism does not consider the actual nature of the propagating cationic species.

The foregoing simplified mechanism is amenable to kinetic analysis, using the steady state method, as in the case of the free radical mechanism. Using the notation HA to designate the acid initiator, we can write

Initiation $\qquad$ $HA + M \xrightarrow{k_i} HM^+ A^-$ $\qquad$ (1-47)

Propagation $\qquad$ $HM^+ A^- + M \xrightarrow{k_p} \cdots \longrightarrow HM_x M^+ A^-$ $\qquad$ (1-48)

Termination $\qquad$ $HM_x M^+ A^- \xrightarrow{k_t} M_{x+1} + HA$ $\qquad$ (1-49)

Transfer $\qquad$ $HM_x M^+ A^- + M \xrightarrow{k_{tr}} M_{x+1} + HM^+ A^-$ $\qquad$ (1-50)

Rate of initiation $\qquad$ $R_i = k_i [HA][M]$ $\qquad$ (1-51)

Rate of termination $\qquad$ $R_t = k_t [HM_x M^+ A^-]$ $\qquad$ (1-52)

Rate of propagation $\qquad$ $R_p = k_p [HM_x M^+ A^-][M]$ $\qquad$ (1-53)

Rate of transfer to monomer $\qquad$ $R_{tr} = k_{tr} [HM_x M^+ A^-][M]$ $\qquad$ (1-54)

Here R_t has been assumed to be a first-order reaction, since the counterion A is considered to be specifically associated with the carbenium ion as shown in Eq. (1-45) and not as a separate species. Using the steady-state assumption, we equate R_i and R_t and thus obtain Eq. (1-55) for the steady state concentration of growing chains.

$$[HM_x M^+ A^-] = (k_i/k_t)[HA][M] \qquad (1-55)$$

Hence, $\qquad R_p = -d[M]dt = (k_p - k_i/k_t)[HA][M]^2 \qquad (1-56)$

Here again, the propagation rate is virtually the polymerization rate, since the consumption of monomer by the initiation step is negligible. Unlike the free radical case, the rate here is first-order in initiator concentration, obviously due to the first-order termination step.

Experimental verification of the foregoing kinetic scheme has been obtained in the case of the cationic polymerization of styrene and vinyl alkyl ethers, where the polymerization rate was indeed found to be dependent on the first power of the initiator and on the square of the monomer concentration. However, it should be noted that this simple kinetic scheme is not general for cationic polymerizations; even the steady-state assumption is not valid in many cationic polymerizations.

The molecular weight of the polymer can again be expressed in terms of X_n, the number average of units per chain, which can be defined here as the ratio of the propagation rate to the sum of the rates of all processes leading to chain termination (including transfer). Hence from Eq. (1-52) to Eq. (1-54),

$$X_n = \frac{R_p}{R_t + R_{tr}} = \frac{k_p [HM_x M^+ A^-][M]}{k_t [HM_x M^+ A^-] + k_{tr} [HM_x M^+ A^-][M]}$$

or $\qquad \dfrac{1}{X_n} = \dfrac{k_{tr}}{k_p} + \dfrac{k_t}{k_p}[M] \qquad (1-57)$

Eq. (1-57) provides a means of determining the relative contribution of the termination and transfer steps. Thus, if k_{tr} is largely relative to k_t, the molecular weight will be virtually independent of monomer concentration, but if the reverse is true, X_n will be directly proportional to [M]. Hence this relation lends itself to a simple experimental test,

i. e. , a plot of $1/X_n^\circ$ (the reciprocal of the initial X_n value) against $1/[M]^\circ$ (the reciprocal of the initial monomer concentration). It has actually been found that the polymerization of styrene by $SnCl_4$ in ethylene dichloride, and of vinyl alkyl ethers by $SnCl_4$ in m-cresol, showed a dominance of termination over transfer, i. e. , $X_n \propto [M]$; however, for isobutylene polymerization catalyzed by $TiCl_4$ in n-hexane, the observed polymer molecular weights were independent of monomer concentration, i. e. , transfer appeared to predominate.

It is interesting to compare the nature of the individual steps in cationic polymerization with those of the free radical mechanism. Thus, unlike the situation in the latter case, the termination step in cationic polymerization may be expected to require a greater energy than that of propagation, since it involves σ-bond rupture compared to the low-energy attack of the growing carbenium ion on the π bond of the monomer. If indeed the termination step has a higher activation energy than that of propagation, then a rise in temperature should lead to an increase in termination relative to propagation and thus to a lower steady-state concentration of growing chains. The net result would thus be a decrease in polymerization rate and molecular weight, i. e. , an apparent "negative" overall activation energy for polymerization. This might, of course, be partially or wholly offset if the activation energy of the initiation step were sufficiently high. However, in the majority of cases it appears that this is not the case, so that faster rates (and higher molecular weights) are indeed obtained at reduced temperatures (about $-100°C$). Kinetic studies have shown that the ion pair propagation rate constant is $5 \times 10^8 \sim 6 \times 10^8 L/(mol \cdot s)$ for the polymerization of isobutylene with $EtAlCl_2$ in hexanes/methyl chloride $(60:40, V/V)$ at $-80°C$.

1.7.2 Butyl elastomer

The only important commercial elastomer prepared by a cationic polymerization is butyl elastomer, i. e. , a copolymer of isobutene and isoprene. The latter monomer is incorporated in relatively small proportions (~ 1.5 mole %) in order to introduce sufficient unsaturation for sulfur vulcanization. The slurry process with aluminum chloride at $-98°C$ to $-90°C$ in methyl chloride diluent can be described by the accompanying "flow sheet". In this process the polymerization is almost instantaneous and extensive cooling by liquid ethylene is required to control the reaction.

```
95.5%~98.5% Isobutene ⎫                          0.2% AlCl₃      Copolymer ──→  Flash
                      ⎬ 30 vol. % sol in CH₃Cl  ──────────→      crumb          tank
1.5%~4.5% Isoprene   ⎭                           in CH₃Cl        suspension     (hot water)
                                                  -100°C                         │
                                                                                 │
                                                                 Polymer         ↓
                                                                 filtration      Monomer
                                                                 and drying      and solvent
                                                                                 recovery
```

The molecular weights of butyl elastomer grades are in the range of 300000~500000, and they are very sensitive to polymerization temperature above $-100°C$. For example, a rise in polymerization temperature of 25℃ can result in a fivefold or 10-fold decrease in molecular weight, presumably due to the kinetic factors discussed previously. The molecular weight distribution of butyl elastomers can be as high as $M_w/M_n = 3 \sim 5$, presumably because of the heterogeneous nature of the polymerization process.

Isoprene is used as the comonomer in butyl elastomer (0.5~2.5mol %) because the isobutene-isoprene reactivity ratios are more favorable for inclusion of the diene than those of the isobutene-butadiene pair. Thus, for the former pair, the r (isobutene) = 2.5, and r (isoprene) = 0.4. It should be noted, however, that as discussed previously, such r values can be markedly influenced by the nature of the initiator and solvent used in the polymerization. The values just quoted are applicable to the commercial butyl elastomer process, as described earlier. It has been shown that the isoprene unit enters the chain predominantly in a 1,4-configuration.

1.7.3 Living Cationic Polymerizations

The controlled/living polymerization of alkyl vinyl ethers was reported in 1984 using the HI/I_2 initiating system in nonpolar solvents. These polymerizations produced polymers with controlled molecular weights, narrow molecular weight distributions, and number average molecular weights that increased linearly with conversion. Shortly thereafter, Faust and Kennedy reported the discovery of the living cationic polymerization of isobutylene by initiation with cumyl acetate/boron trichloride in mixtures of chlorinated solvents plus n-hexane to obtain a homogeneous polymerization.

Subsequent investigations have discovered a variety of living cationic polymerization systems. For controlled/living polymerization of isobutylene, tertiary halides are generally used in conjunction with strong Lewis acid co-initiators (BF_3, $SnCl_4$, $TiCl_4$, $EtAlCl_2$, and Met_2AlCl). A key ingredient in many of these systems is a proton trap, such as 2,6-di-*tert*-butylpyridine, to suppress initiation by protons. The general features of living polymerization systems are analogous to those of controlled radical polymerization, i.e., a predominant, unreactive dormant species (covalent species, R—X or P—X) in equilibrium with a small concentration of reactive, propagating species (cationic ion pairs, P^+X^-) as shown below.

$$R-X + A \text{ (Lewis acid)} \rightleftharpoons R^{\oplus}XA^{\ominus}$$

$$R^{\oplus}XA^{\ominus} + \text{monomer} \xrightarrow{k_p} P^{\oplus}XA^{\ominus}$$

$$\underset{\text{active form}}{P^{\oplus}XA^{\ominus}} \longrightarrow \underset{\text{dormant form}}{P-X + A}$$

1.7.4 Other Cationic Polymerizations: Heterocyclic Monomers

Although butyl elastomer is by far the most important commercial elastomer to be synthesized by cationic polymerization, several heterocyclic monomers provide useful elastomeric materials via this mechanism also. Epichlorohydrin can be polymerized to high molecular weight using a complex catalyst formed from a trialkylaluminum compound and water as shown in Eq. (1-58). For copolymerizations with ethylene oxide, a catalyst formed from a trialkylaluminum compound, water, and acetylacetone is useful. The mechanism proposed for these polymerizations is illustrated in Eq. (1-59), where coordination to two aluminum sites has been invoked to explain the stereochemical course of these polymerization. The average molecular weight of the homopolymer is 500000, while the equimolar copolymer with ethylene oxide has molecular weights approaching 1000000.

$$CH_2\text{---}CHCH_2Cl \xrightarrow{R_3Al/H_2O} \text{---}[OCH_2\text{---}CH(CH_2Cl)]_n\text{---} \quad (1\text{-}58)$$
$$\underset{O}{\diagdown\diagup}$$

$$(1\text{-}59)$$

The cationic polymerization of tetrahydrofuran is used commercially to produce α,ω-dihydroxypoly(tetramethylene oxide) (PTMO glycol). Although this polymer is not used by itself as an elastomer, it is used as one of the elastomeric block components for preparation of segmented thermoplastic polyurethane and thermoplastic polyester elastomers. The cationic polymerization of tetrahydrofuran (THF) is a living polymerization under proper experimental conditions, i.e., it does not exhibit any termination step, very much like the analogous anionic polymerizations. However, these polymerizations are complicated by the fact that the ceiling temperature, where the free energy of polymerization is equal to zero, is estimated to be approximately $(83 \pm 2)\text{°C}$ in bulk monomer solution; therefore, the polymerization is reversible and incomplete conversion is often observed, especially in the presence of added solvent. For example, at equilibrium in the bulk at 30°C, conversion is 72%; in a mixture of 37.5 vol % CH_2Cl_2, conversion is only 27%. These factors limit the ability to prepare polytetrahydrofurans with controlled molecular weight and narrow molecular weight distributions which are often associated with living polymerizations. To the extent that equilibrium is approached, the polymer molecular weight distribution would broaden towards a statistical value of 2. The most commonly used catalyst for the commercial polymerization of tetrahydrofuran is fluorosulfuric acid as shown in Eq. (1-60).

$$(n+1)\, \square_O \xrightarrow{FSO_3H} \xrightarrow{H_2O} HO\text{---}[CH_2CH_2CH_2CH_2O]_n\text{---}CH_2CH_2CH_2CH_2OH \quad (1\text{-}60)$$

The mechanism of this cationic polymerization is quite different from the polymerization of isobutene [Eq. (1-43) ~ Eq. (1-46)] in that the growing chain end is an oxonium ion intermediate in which the positive charge is located on oxygen atom rather than on carbon as shown in the following:

Initiation

$$\text{(THF)} + FSO_3H \rightleftharpoons HO^\oplus\text{(THF)} \; SO_3F^\ominus \xrightarrow{THF} HO(CH_2)_4-O^\oplus\text{(THF)} \; SO_3F^\ominus \quad (1\text{-}61)$$

Propagation

$$HO(CH_2)_4-O^\oplus\text{(THF)} \; SO_3F^\ominus + n \text{(THF)} \rightleftharpoons HO(CH_2)_4-[OCH_2CH_2CH_2CH_2]_n-O^\oplus\text{(THF)} \; SO_3F^\ominus$$

$$(1\text{-}62)$$

Another interesting aspect of this polymerization is the observation that the covalent ester is in equilibrium with the oxonium ion [Eq. (1-63)] and that both of these species can participate in propagation by reaction with monomer.

$$HO(CH_2)_4-[OCH_2CH_2CH_2CH_2]_n-O^\oplus\text{(THF)} \; SO_3F^\ominus \rightleftharpoons HO(CH_2)_4-[OCH_2CH_2CH_2CH_2]_n-(CH_2)_4SO_3F$$

$$(1\text{-}63)$$

The commercial α,ω-dihydroxypoly(tetramethylene oxide) (PTMO glycol) polymers have molecular weights in the range of 600~3 000 with molecular weight distributions in the range of 1.2~1.6 which is consistent with the equilibrium nature of these polymerizations.

1.8 Chain Polymerization by Anionic Mechanism

1.8.1 Mechanism and Kinetics

An anionic mechanism is proposed for those polymerizations initiated by alkali metal organometallic species, where there is good reason to assume that the metal is strongly electropositive relative to the carbon (or other) atom at the tip of the growing chain. However, analogous to the discussion of the active species in cationic polymerization, a multiplicity of active species may be involved as propagating species in anionic polymerization as shown below.

$$(R-M)_n \rightleftharpoons R-M \rightleftharpoons R^\ominus M^\oplus \rightleftharpoons R^\ominus \| M^\oplus \rightleftharpoons R^\ominus + M^\oplus$$

| aggregates | covalent | tight ion pair | loose ion pair | free ions |

$$M \downarrow k_p^1 \quad M \downarrow k_p^2 \quad M \downarrow k_p^3 \quad M \downarrow k_p^4 \quad M \downarrow k_p^5$$

In contrast to cationic polymerization, however, there is experimental evidence for the involvement of many of these species under certain experimental conditions.

Although the ability of alkali metals, such as sodium, to initiate polymerization of unsaturated organic molecules has long been known (the earliest record dating back to the work of Matthews and Strange and of Harries, around 1910, on polymerization of dienes), the mechanism had remained largely obscure due to the heterogeneous character of this type of catalysis. The pioneering work of Higginson and Wooding on the homogeneous polymerization of styrene by potassium amide in liquid ammonia, and that of Robertson and Marion on butadiene polymerization by sodium in toluene, merely showed the important role of the solvent in participating in chain transfer reactions.

The true nature of homogeneous anionic polymerization only became apparent through studies of the soluble aromatic complexes of alkali metals, such as sodium naphthalene. These species are known to be radical anions, with one unpaired electron stabilized by resonance and a high solvation energy, and are therefore chemically equivalent to a "soluble sodium". They initiate polymerization by an "electron transfer" process, just as in the case of the metal itself, except that the reaction is homogeneous and therefore involves a much higher concentration of initiator. The mechanism of polymerization initiated by alkali metals (or their soluble complexes) can therefore be written as follows, using styrene as an example:

$$\text{Initiation} \quad Na + \underset{C_6H_5}{CH_2=CH} \rightleftharpoons [\underset{C_6H_5}{CH_2=CH}]^{\bullet-} Na^{\oplus} \quad (1\text{-}64)$$

$$2[\underset{C_6H_5}{CH_2=CH}]^{\bullet-} Na^{\oplus} \longrightarrow Na^{\oplus} {}^{\ominus}\underset{C_6H_5}{CH}-CH_2-CH_2-\underset{C_6H_5}{CH}{}^{\ominus} Na^{\oplus} \quad (1\text{-}65)$$

Propagation

$$Na^{\oplus} {}^{\ominus}\underset{C_6H_5}{CH}-CH_2-CH_2-\underset{C_6H_5}{CH}{}^{\ominus}Na^{\oplus} + (2n)\ \underset{C_6H_5}{CH_2=CH} \longrightarrow \cdots \longrightarrow$$

$$Na^{\oplus} {}^{\ominus}\underset{C_6H_5}{CH}-CH_2-[\underset{C_6H_5}{CH}-CH_2]_n-[CH_2-\underset{C_6H_5}{CH}]_n-CH_2-\underset{C_6H_5}{CH}{}^{\ominus}Na^{\oplus}$$

(1-66)

Thus the first step in the initiation reaction [Eq. (1-64)] involves a reversible electron transfer reaction from the alkali metal to the styrene monomer to form the styryl radical anion; in a rapid subsequent reaction, two radical anions couple to form a di-anion which can grow a polymer chain at both ends. In the case of the soluble alkali metal aromatic complexes, the overall initiation reaction is extremely fast, due to the high concentrations of radical anion ($\sim 10^{-3}$ mol/L) and monomer (~ 1 mol/L), and so is the subsequent propagation reaction. However, in the case of the alkali metal initiators, the electron transfer step [Eq. (1-64)] is very much slower, due to the heterogeneous nature of the reaction, so that the buildup of radical anions is much slower. In fact, there is evidence that, in such cases, a second electron transfer step can occur between the metal and the radical anion to form a di-anion, rather than coupling of the radical anions. In either case, the final result is a di-anion, i. e. , a difunctional growing chain.

However, it was investigations of the homogeneous systems initiated by sodium naphthalene in polar solvents which demonstrated the special nature of anionic polymerization, i. e. , the fact that a termination step may be avoided under certain circumstances, leading to the concept of "living" polymers. Since these are homogeneous systems, the stoichiometry of the reaction becomes apparent, i. e. , two molecules of sodium naphthalene generate one chain. Furthermore, since all the chains are initiated rapidly and presumably have an equal opportunity to grow, their molecular weight distribution becomes very narrow, approximating the Poisson distribution. These aspects are obscured in the metal-initiated polymerizations owing to the continued slow initiation over a long period of time, leading to a great difference in the "age" of the growing chains and hence in their size distribution.

Polymerization initiated by electron transfer from a metal, or by an aromatic radical anion, represents only one of the anionic mechanisms. It is, of course, possible to consider separately those polymerizations initiated directly by organometallic compounds. Of the latter, the organolithium compounds are probably the best examples, since they are soluble in a wider variety of solvents and are relatively stable. Furthermore, it is these organometallic compounds which are used commercially for the preparation of synthetic elastomers. The mechanism of these polymerizations is somewhat simpler than in the case of sodium naphthalene, since there is no electron transfer step; thus

Initiation

$$R^{\ominus}Li^{\oplus} + CH_2{=}CH(C_6H_5) \longrightarrow R-CH_2-CH^{\ominus}(C_6H_5) Li^{\oplus} \tag{1-67}$$

Propagation

$$R-CH_2-CH^{\ominus}(C_6H_5)Li^{\oplus} + nCH_2{=}CH(C_6H_5) \longrightarrow R{\hbox to 0pt{$-$\hss}}[CH_2-CH(C_6H_5)]_n{-}CH_2-CH^{\ominus}(C_6H_5)Li^{\oplus} \tag{1-68}$$

Termination by impurity or deliberate termination

$$R{-}[CH_2-CH(C_6H_5)]_n{-}CH_2-CH^{\ominus}(C_6H_5)Li^{\oplus} + H_2O \longrightarrow R{-}[CH_2-CH(C_6H_5)]_n{-}CH_2-CH_2(C_6H_5) + LiOH \tag{1-69}$$

Hence each organolithium molecule generates one chain, and there is no termination of the growing chains or chain transfer reactions in the absence of adventitious impurities, such as water and acids, and if higher temperatures are avoided to prevent side reactions.

Unlike sodium naphthalene, which requires the presence of highly solvating solvents, such as tetrahydrofuran (THF), the organolithium systems can operate in various polar and nonpolar solvents such as ethers or hydrocarbons. However, the rates are much slower in the latter than in the former solvents.

Hence, if the initiation reaction [Eq. (1-67)] is very much slower than the propagation reaction, the molecular weight distribution may be considerably broadened. This does not, of course, vitiate the "living" polymer aspect of the polymerization, which has been shown to operate in these systems, regardless of type of solvent, if side reactions do not intervene.

The absence of chain termination and chain transfer reactions in homogeneous anionic polymerization can lead to many novel synthetic routes. Thus, since each chain continues to grow when additional monomer is added, it is possible to synthesize block polymers by sequential addition of several monomers. Another possibility is the synthesis of linear chains with various functional end groups, by allowing the anionic polymer chain end to react with various electrophilic agents, e. g., with CO_2 to form —COOH groups. In addition, linking reactions of polymer chains with multifunctional electrophilic reagents leads to the formation of "star-branched" polymers. These possibilities are, of course, of considerable industrial interest.

In view of the unusual mechanism of anionic polymerization, especially the absence of termination and chain transfer reactions, the kinetics of these systems can be treated quite differently than for the other mechanisms. Thus it is possible, by suitable experimental techniques, to examine separately the rates of the initiation and propagation reactions, since the stable organometallic chain ends are present in concentrations $[10^{-3} \sim 10^{-5} \text{mol/L}]$ which are easily measured by ultraviolet-visible spectroscopy. The propagation reaction is, of course, of considerable main interest and can be studied by making sure that initiation is complete. In this way, the kinetics of homogeneous anionic polymerization have been extensively elucidated with special reference to the nature of counterion and role of the solvent.

It has been found universally that, in accordance with Eq. (1-66) and Eq. (1-68), the propagation rate is always first order with respect to monomer concentration, regardless of solvent system or counterion. However, in contradiction to the foregoing equations, the propagation rate dependency has generally been found to be lower than first order with respect to the concentration of growing chains, and the order was found to be strongly dependent on the nature of the solvent and counterion. Strongly solvating solvents, such as ethers and amines, lead to much faster rates than nonpolar solvents and affect the kinetics of these polymerizations quite differently than the hydrocarbon media, because more dissociated ionic species such as loose ion pairs and free ions are involved as propagating species. However, since the anionic synthesis of elastomers requires the use of lithium as counterion in hydrocarbon media, the following discussion will focus on the kinetics of these processes.

It would be expected that the kinetics of organolithium-initiated polymerization in hydrocarbon solvents would be simplified because of the expected correspondence between the initiator concentration and the concentration of propagating anionic species, resulting from the lack of termination and chain transfer reactions. However, in spite of intensive study, there is no general agreement on many kinetic aspects of these polymerizations. The complicating feature is that organolithium compounds are associated into aggregates in hydrocarbon solution, and the degree of aggregation depends on the structure of the organolithium compound, the concentration of organolithium compound, the solvent, and the temperature. In general, simple alkyllithium compounds are associated into hexamers or tetramers in hydrocarbon solution.

The kinetics of initiation for styrene and diene polymerization by alkyllithium compounds generally exhibit a fractional kinetic order dependence (e. g. , 1/4 or 1/6) on the concentration of alkyllithium initiator. This can be rationalized in terms of the following steps:

Initiation

$$(RLi)_n \xrightleftharpoons{K_n} n\ RLi \tag{1-70}$$

$$RLi + CH_2=CH(C_6H_5) \xrightarrow{k_i} R-CH_2-CHLi(C_6H_5) \tag{1-71}$$

Thus, it is assumed that only the unassociated alkyllithium compound [formed by dissociation of the aggregate, Eq. (1-70)] reacts with monomer in the initiation step [Eq. (1-71)] so the the rate of initiation can be expressed by Eq. (1-72).

$$R_i = k_i [RLi][M] \tag{1-72}$$

The equilibrium concentration of unassociated alkyllithium can be expressed in terms of Eq. (1-73).

$$[RLi] = K^{1/n} [(RLi)_n]^{1/n} \tag{1-73}$$

When this expression for [RLi] is substituted into Eq. (1-72), Eq. (1-74) is obtained.

$$R_i = k_i K^{1/n} [(RLi)_n]^{1/n} [M] \tag{1-74}$$

A good example of this kinetic behavior was found in the study of the n-butyllithium-styrene system in benzene, in which a kinetic order dependency on n-butyllithium concentration was observed, consistent with the predominantly hexameric degree of association of n-butyllithium. However, this expected correspondence between the degree of association of the alkyllithium compound and the fractional kinetic order dependence of the initiation reaction on alkyllithium concentration was not always observed. One source of this discrepancy is the assumption that only the unassociated alkyllithium molecule can initiate polymerization. With certain reactive initiators, such as sec-butyllithium in hexane solution, the initial rate of initiation exhibits approximately a first order dependence on alkyllithium concentration, suggesting that the aggregate can react directly with monomer to initiate polymerization. A further source of complexity is the cross-association of the initiator with the initiated polymer chain; in general, the cross-associated species exhibits a different degree of association and reactivity from the alkyllithium initiator. As a result of cross-association, only the initial rates of initiation can be used to to determine the kinetic order dependence on initiator concentration. Unfortunately, these considerations have not always been recognized. It is interesting to note that the general reactivity of alkyllithiums as initiators is inversely related to the degree of aggregation, i. e. , sec-butyllithium (tetramer) $>$ n-butyllithium (hexamer).

The kinetics of the propagation reaction in organolithium polymerization of styrenes and dienes in nonpolar solvents (i. e. , hydrocarbons) have also been subjected to intensive study. For styrene polymerizations, a kinetic order 1/2 dependence on chain-end concentration is observed [Eq. (1-75)].

$$R_p = k_p K_n^{1/2} [(PsLi)_2]^{1/2} [M] \tag{1-75}$$

Since it has been determined that poly(styryl)lithium is associated predominantly into dimers in hydrocarbon solution, the observed kinetic order can be explained in terms of Eq. (1-76) and Eq. (1-77), using the same reasoning as delineated for the initiation kinetics [Eq. (1-70) ~ Eq. (1-74)]. This explanation is based on the assumption that only the dissociated chain ends are active.

$$(PsLi)_2 \xrightleftharpoons{K_n} 2PsLi \tag{1-76}$$

$$PsLi + CH_2=CH\underset{C_6H_5}{|} \xrightarrow{k_p} Ps-CH_2-\underset{C_6H_5}{\overset{}{C}}HLi \tag{1-77}$$

The propagation kinetic order dependence on poly(dienyl)lithium chain end concentration for alkyllithium-initiated polymerization of dienes varies from 1/4 to 1/6 for butadiene and from 1/2 to 1/4 for isoprene. However, attempts to relate these kinetic orders to proposed higher states of association of poly(dienyl)lithium chain ends have proven to be complicated, because conflicting physical measurements using the same techniques by different groups have shown that such chain ends are associated into dimers and tetramers in hydrocarbon media. These physical measurements include solution viscosity, cryoscopy, and light scattering measurements. Hence, these findings bring into question the whole theory of the nonreactivity of the associated complex, and suggest that a direct interaction between the monomer and the associated polymer chain end may be contributing. Further complicating this situation are the results of Fetters and coworkers indicating that higher-order aggregates are in equilibrium with dimers.

In conclusion, it should be noted that the molecular weights and their distribution follow the rules originally discussed under living polymers. This means that, regardless of the solvents and counterions used, if no termination, chain transfer, or side reactions occur, and if the initiation reaction is fast relative to the propagation reaction, then the molecular weight distribution will approach the Poisson distribution, i.e.,

$$X_w/X_n = 1 + 1/X_n \tag{1-78}$$

where X_n is the number average of monomer units and X_w is the weight average number of monomer units. This means that, in principle, a polymer chain of 100 units should have an X_w/X_n ratio of 1.01. This is, of course, impossible to prove experimentally, and it can be assumed that the real distribution would be somewhat broader, due for one thing to imperfect mixing in the reaction mixture. However, values of 1.05 for X_w/X_n are commonly found in these systems.

1.8.2 Chain Microstructure of Polydienes

Although the alkali metals, unlike the Ziegler-Natta systems, do not generally polymerize unconjugated olefins and are not known to lead to any tacticity, they do affect the chain microstructure of polydienes. Thus, the proportion of *cis*-1,4- and *trans*-1,4- addition versus the 1,2- (and 3,4-for polyisoprene) mode can be markedly affected by the nature of the counterion as well as the solvent. Ever since the discovery that lithium polymerization of isoprene can lead to a high *cis*-1,4-structure, close to that of natural elastomer, there have been many studies of these effects. Table 1.14 shows some of these results for anionic polymerization of isoprene and butadiene. It is obvious from these data that the stereospecific high *cis*-1,4-polyisoprene is obtained only in the case of lithium in hydrocarbon solvents; the highest *cis* microstructure is also favored by high ratios of monomer to chain end. Other solvents and/or counterions exert a dramatic effect in altering the chain microstructure to form 1,2- and 3,4-enchainments. Similar effects are observed with butadiene and other dienes. However, in the case of butadiene, the maximum *cis*-1,4-content attainable is much less than for isoprene; typical commercial polybutadienes prepared in hydrocarbon solution with butyllithium initiators have microstructures in the range of 36%～44% *cis*-1,4-, 48%～50% *trans*-1,4-, and 8%～10% 1,2-microstructure. The effect of polar solvents, or of the more electropositive alkali metals, is to produce a high-1,2-polybutadiene.

Table 1.14 Microstructure of polydienes prepared by anionic polymerization

	Solvent	Cation	Chain microstructure (mole %)			
			cis-1,4-	*trans*-1,4-	1,2-	3,4-
Butadiene	Hexane[a]	Li$^+$	30	62	8	—
	Cyclohexane[b]	Li$^+$	68	28	4	—
	None	Li$^+$	86	9	5	—
	Tetrahydrofuran (THF)	Li$^+$	6	6	88	—
	Pentane	Na$^+$	10	25	65	—
	THF	Na$^+$	0	9.2	90.8	—
	Pentane	K$^+$	15	40	45	—
	Pentane	Rb$^+$	7	31	62	—
	Pentane	Cs$^+$	6	35	59	—
Isoprene	Cyclohexane[b]	Li$^+$	94	1	—	5
	Cyclohexane[a]	Li$^+$	76	19	—	5
	None	Li$^+$	96	0	—	4
	THF	Li$^+$	12[c]		29	59
	Cyclohexane	Na$^+$	44[c]		6	50
	THF	Na$^+$	11[c]		19	70
	Cyclohexane	K$^+$	59[c]		5	36
	Cyclohexane	Cs$^+$	69[c]		4	27

[a] At monomer/initiator ratio of ～17.
[b] At monomer/initiator ratio of 5 × 10^4.
[c] Total of *cis*- and *trans*-forms.

This marked sensitivity of the stereochemistry of anionic polymerization to the nature of the counterion and solvent can be traced to the structure of the propagating chain end. The latter involves a carbon-metal bond which can have variable characteristics, ranging all the way from highly associated species with covalent character to a variety of ionic species. The presence of a more electropositive metal and/or a cation-solvating solvent, such as ethers, can effect a variety of changes in the nature of the carbanionic chain end: (a) the degree of association of the chain ends can decrease or be eliminated; (b) the interaction of the cation with the anion can be decreased by cation solvation; (c) a more ionic carbon-metal bond will increase delocalization of the π electrons; and (d) polar solvents will promote ionization to form ion pairs and free ions. Direct evidence for these effects has been obtained from concentrated solution measurements, ^{1}H- and ^{13}C-NMR spectroscopy, ultraviolet-visible spectroscopy and electrolytic conductance measurements.

The control of chain structure and molecular weight afforded by the organolithium polymerization of dienes, has, of course, been of great technological interest. Such product developments have been mainly in the form of ①polybutadiene elastomers of various chain structures and functional end groups, ②liquid polybutadienes, ③butadiene-styrene copolymers (solution SBR), and ④styrene-diene triblock copolymers (thermoplastic elastomers).

1.8.3 Copolymers of Butadiene

The possibilities inherent in the anionic copolymerization of butadiene and styrene by means of organolithium initiators, as might have been expected, have led to many new developments. The first of these would naturally be the synthesis of a butadiene-styrene copolymer to match (or improve upon) emulsion-prepared SBR, in view of the superior molecular weight control possible in anionic polymerization. The copolymerization behavior of butadiene (or isoprene) and styrene is shown in Table 1.15.

Table 1.15 Monomer reactivity ratios for organolithium copolymerization of styrene and dienes

Monomer 1	Monomer 2	Solvent	r_1	r_2
Styrene	Butadiene	Toluene	0.004	12.9
Styrene	Butadiene	Benzene	0.04	10.8
Styrene	Butadiene	Triethylamine	0.5	3.5
Styrene	Butadiene	Tetrahydrofuran	4.0	0.3
Styrene	Isoprene	Benzene	0.26	10.6
Styrene	Isoprene	Tetrahydrofuran	9.0	0.1
Butadiene	Isoprene	Hexane	1.72	0.36

As indicated earlier, unlike the free radical type of polymerization, these anionic systems show a marked sensitivity of the reactivity ratios to solvent type (a similar effect is noted for different alkali metal counterions). Thus, in nonpolar solvents, butadiene (or isoprene) is preferentially polymerized initially, to the virtual exclusion of the styrene, while the reverse is true in polar solvents. This has been ascribed to the profound effect of solvation on the structure of the carbon-lithium bond, which becomes much more ionic in such media, as discussed previously. The resulting polymer formed by copolymerization in hydrocarbon media is described as a tapered block copolymer; it consists of a block of polybutadiene with little incorporated styrene comonomer followed by a segment with both butadiene and styrene and then a block of polystyrene. The structure is schematically represented below:

[butadiene]-[B/S]-[styrene]

The datas in Table 1.15 illustrate the problems encountered in such copolymerizations, since the use of polar solvents to assure a random styrene-diene copolymer of desired composition will, at the same time, lead to an increase in side vinyl groups (1,2- or 3,4-) in the diene units (see Table 1.14). This is of course quite undesirable, since such chain structures result in an increase in the glass transition temperature (T_g) and therefore to a loss of good elastomery properties. Hence, two methods are actually used to circumvent this problem: ①the use of limited amounts of polar additives such as tetrahydrofuran to accomplish a reasonable compromise between diene structure and monomer sequence distribution; and ②the addition of small amounts of potassium t-alkoxides.

As mentioned earlier, the "living" nature of the growing chain in anionic polymerization makes this mechanism especially suitable for the synthesis of block copolymers, by sequential addition of different monomers.

1.8.4 Terminally Functional Polydienes

Another characteristic of these homogeneous anionic polymerizations, as mentioned earlier, is their potential for the synthesis of polymer chains having reactive end groups. It was recently reported that chain-end functionalization of high molecular weight polybutadiene and solution styrene-co-butadiene elastomers (SBR) with a derivative of Michler's ketone, 4,4'-bis(diethyl amino)benzophenone, leads to tire tread formulations which have lower rolling resistance and good wet-skid resistance. These effects were observed in spite of the low concentration of chain ends in these polymers (molecular weights>100000).

The production of liquid short-chain difunctional polymers by anionic polymerization is of considerable technological interest and importance, and has attracted much attention in recent years, since it offers an analogous technology to that of the polyethers and polyesters used in urethane polymers. Such liquid "telechelic" polydienes could thus lead, by means of chain extension and crosslinking reactions, directly to "castable" polydiene networks.

The most direct method of preparing telechelic polydienes utilizes a dilithium initiator which is soluble in hydrocarbon solution. The most expedient method of preparing such a dilithium initiator is to react two moles of an alkyllithium compound with a divinyl compound which will not homopolymerize. Unfortunately, because of the association behavior of organolithium compounds in hydrocarbon media, many potential systems fail because they associate to form an insoluble network-like structure. Expediencies such as addition of Lewis bases can overcome solubility problems of dilithium initiators, however, such additives tend to produce large amounts of 1,2- and 3,4-microstructures (see Table 1.4). One exception is the adduct formed from the addition of two equivalents of *sec*-butyl lithium to

1,3-bis(1-phenylethenyl)benzene as shown in Eq. (1-79). Although this is a hydrocarbon-soluble, dilithium initiator, it was found that biomodal molecular weight distributions are obtained; monomodal distributions can be obtained in the presence of lithium alkoxides or by addition of Lewis base additives. This initiator has also been used to prepare telechelic polymers in high yields.

$$2\,sec\text{-}C_4H_9Li + \underset{\text{(structure)}}{} \longrightarrow \underset{\text{(structure)}}{} \tag{1-79}$$

1.9 Stereospecific Chain Polymerization and Copolymerization by Coordination Catalysts

1.9.1 Mechanism and Kinetics

The term "Ziegler-Natta catalysts" refers to a wide variety of polymerization initiators generally formed from mixtures of transition metal salts of Group IV to VIII metals and base metal alkyls of Group II or III metals. It arose from the spectacular discovery of Ziegler et al. that mixtures of titanium tetrachloride and aluminum alkyls polymerize ethylene at low pressures and temperatures; and from the equally spectacular discovery by Natta that the Ziegler catalysts can stereospecifically polymerize monoolefins to produce tactic, crystalline polymers. As can be imagined, these systems can involve many combinations of catalyst components, not all of which are catalytically active or stereospecific. However, we shall be concerned here only with polymerizations involving the commercial elastomers, principally polyisoprene, polybutadiene, and the ethylene-propylene copolymers.

The mechanism of polymerization of alkenes using Ziegler-Natta-type catalysts is described as a coordination or insertion polymerization process. The coordination terminology assumes that the growing polymer chain is bonded to a transition metal atom and that insertion of the monomer into the carbon-metal bond is preceded by, and presumably activated by, the coordination of the monomer with the transition metal center. Since coordination of the monomer may or may not be a specific feature of these polymerizations, the insertion terminology focuses on the proposal that these reactions involve a stepwise insertion of the monomer into the bond between the transition metal atom and the last carbon atom of the growing chain. It is important to note that the bonding of carbon atoms and transition metals is described as substantially covalent, in contrast to anionic organometallic species, such as organoalkali metal species, which are highly ionic.

Typical soluble catalysts for copolymerization of ethylene and propylene are formed from mixtures of vanadium salts with alkylaluminum chlorides, e.g., VCl_4 with either AlR_2Cl or $AlRCl_2$ where R = alkyl group. A possible hexacoordinated metal structure for the resulting active catalyst is shown below.

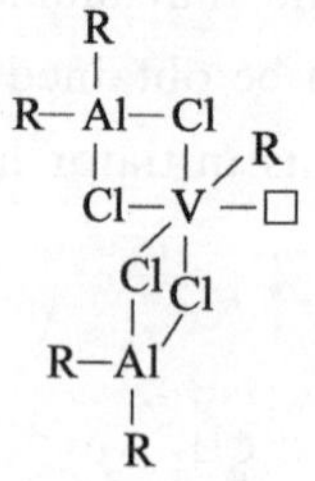

$$\begin{array}{c} R \\ | \\ R-Al-Cl \\ | \quad \diagup | \diagdown \\ Cl-V-\square \\ | \diagdown | \diagup \\ Cl \; Cl \\ | \diagup \\ R-Al \\ | \\ R \end{array}$$

The important features of the active center in accord with the general model of Arlman and Cossee are: ①an alkylated vanadium center, i. e., an RV bond; and ②an empty orbital on vanadium, represented by —□ in the structure, which can be used to bond to the incoming monomer; and an oxidation state of +3 for vanadium.

The formation of the active catalytic center from the reaction of the transition metal compound and an organoaluminum derivative is shown schematically in Eq. (1-80). Reduction to a lower valence state may accompany this alkylation reaction since it is generally considered that the active catalytic center has an oxidation state of +3.

1.9.1.1 Active center formation

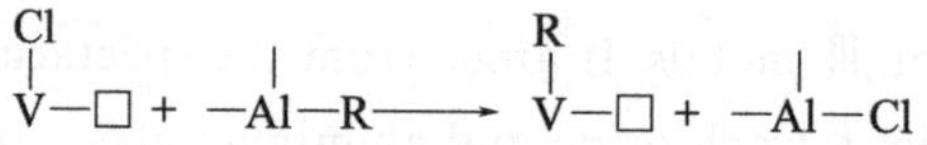

$$\begin{array}{c} Cl \\ | \\ V-\square \end{array} + \begin{array}{c} \\ -Al-R \end{array} \longrightarrow \begin{array}{c} R \\ | \\ V-\square \end{array} + \begin{array}{c} \\ -Al-Cl \end{array} \tag{1-80}$$

The steps involved in the chain polymerization of alkenes using this type of catalyst are shown in Eq. (1-81) ~Eq. (1-85).

1.9.1.2 Initiation

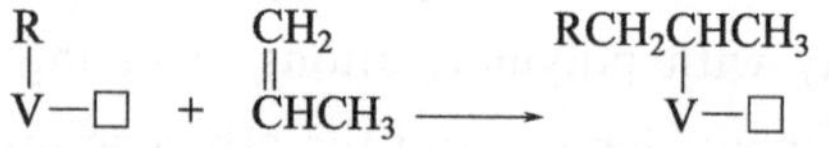

$$\begin{array}{c} R \\ | \\ V-\square \end{array} + \begin{array}{c} CH_2 \\ \| \\ CHCH_3 \end{array} \longrightarrow \begin{array}{c} RCH_2CHCH_3 \\ | \\ V-\square \end{array} \tag{1-81}$$

1.9.1.3 Propagation

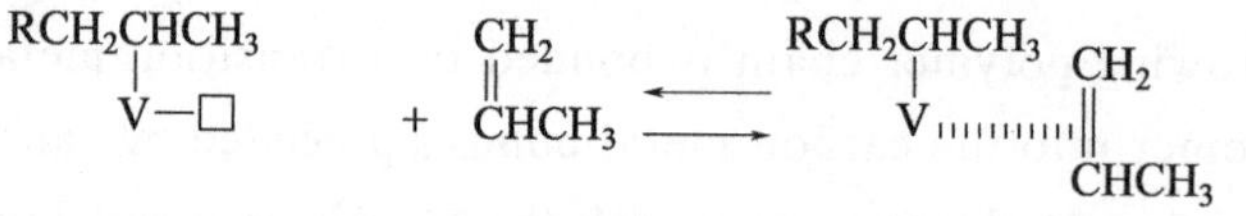

$$\begin{array}{c} RCH_2CHCH_3 \\ | \\ V-\square \end{array} + \begin{array}{c} CH_2 \\ \| \\ CHCH_3 \end{array} \rightleftarrows \begin{array}{c} RCH_2CHCH_3 \quad CH_2 \\ | \quad \quad \| \\ V\text{........}\| \\ \quad \quad CHCH_3 \end{array} \tag{1-82}$$

$$\begin{array}{c} CH_3 \\ | \\ RCH_2CH \\ | \quad CH_2 \\ V\text{........}\| \\ \quad \quad CHCH_3 \end{array} \longrightarrow \left[\begin{array}{c} CH_3 \\ | \\ RCH_2CH\text{......}CH_2 \\ \vdots \quad \quad \quad \vdots \\ V\text{......}CHCH_3 \end{array} \right]^{\ddagger}$$

$$\downarrow$$

$$\begin{array}{c} CH_3 \\ | \\ RCH_2CHCH_2CHCH_3 \\ | \\ V-\square \end{array} \longleftarrow \begin{array}{c} RCH_2CHCH_3 \\ | \quad CH_2 \\ \square \quad | \\ | \quad CHCH_3 \\ V- \end{array} \tag{1-83}$$

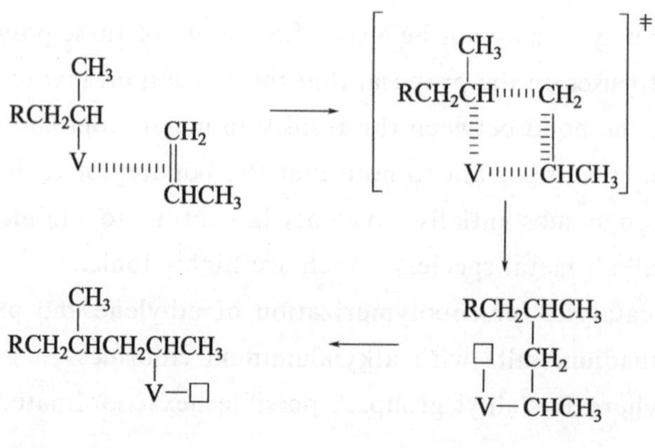

1.9.1.4 Termination

$$P-V \longrightarrow \underset{\text{inactive polymer chain}}{P} + \underset{\text{inactive catalytic center}}{V_i} \qquad (1\text{-}84)$$

1.9.1.5 Spontaneous transfer

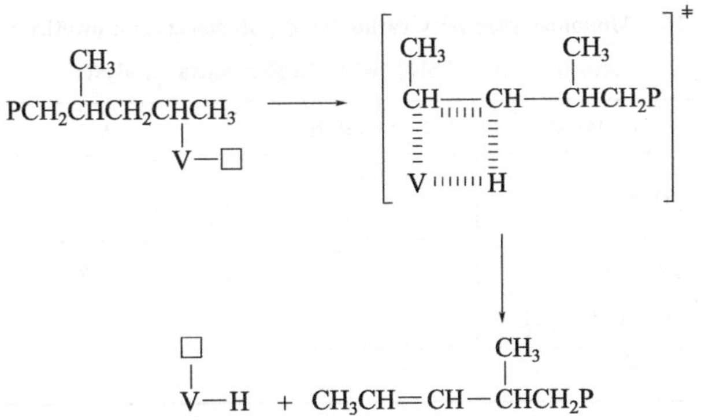

$$\underset{V-\square}{PCH_2CHCH_2CHCH_3} \longrightarrow \left[\begin{array}{c} CH_3 \quad CH_3 \\ CH\!\!-\!\!CH\!\!-\!\!CHCH_2P \\ V\!\!\cdots\!\!H \end{array} \right]^{\ddagger} \qquad (1\text{-}85)$$

$$\downarrow$$

$$\underset{V-H}{\square} + CH_3CH=CH-\underset{CH_3}{CHCH_2P}$$

It should be noted that the monomer coordination step shown in Eq. (1-82) may not be a distinct step as discussed previously. An important feature of this mechanism which affects the stereospecificity of olefin polymerizations using these types of soluble catalysts is the fact the the insertion of the monomer into the transition metal-carbon bond involves a secondary insertion reaction, i. e. , the more substituted carbon of the double bond in the monomer becomes bonded to the transition metal. In contrast, a primary insertion mechanism to form a transition metal bond to the less substituted carbon on the double bond of the monomer [Ti-CH$_2$CHR-P] is involved in polymerizations using typical heterogeneous catalysts, e. g. , from titanium halides and alkylaluminum compounds.

One of the models proposed to explain the stereospecificity for soluble vanadium-based catalysts postulates that it is the minimization of steric effects in the four-center transition state for monomer insertion [see Eq. (1-83)] which is responsible for the stereospecificity of the polymerization. Thus, it is considered that the trans-configuration minimizes steric effects in the transition state and this leads to a syndiotactic configuration of the polymer chain as shown below. In general, the kinetics of alkene polymerizations using Ziegler-Natta-type catalysts are complicated by the multiplicity of active species, catalyst aging and deactivation effects, multiplicity of chain transfer processes, and often by the relatively rapid rates of polymerization.

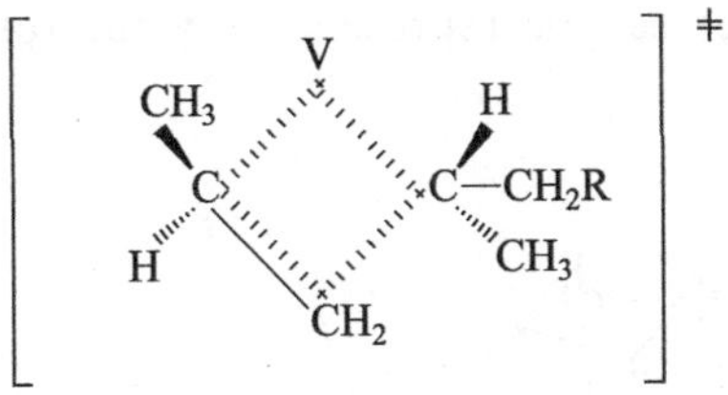

favored *trans* configuration 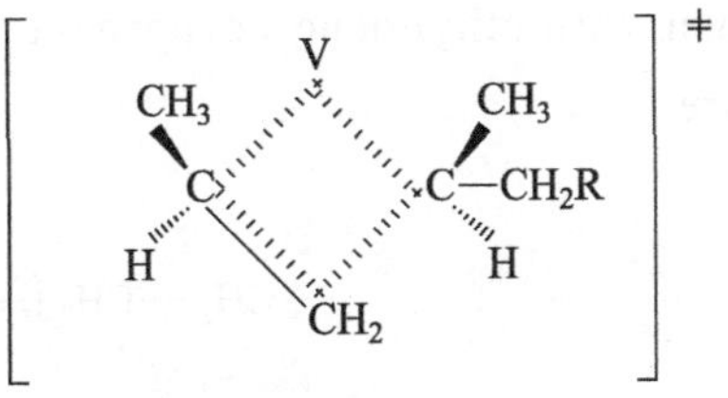   more hindered *cis* configuration

1.9.2 Ethylene-Propylene Rubbers

The copolymerization of propylene with ethylene is complicated by the very unfavorable monomer reactivity ratios for propylene and other monomers with ethylene as shown in Table 1.16.

Table 1.16 Monomer reactivity ratios for copolymerization of ethylene (M_1) and propylene (M_2) with Ziegler-Natta catalysts

Catalyst	Cocatalyst	Temperature/℃	r_1	r_2
VCl_4	$Al(C_2H_5)_2Cl$	−10	13.7	0.021
		21	3.0	0.073
$VOCl_3$	$Al(i\text{-}C_4H_9)_2Cl$	30	16.8	0.052
$V(acac)_3$ [a]	$Al(i\text{-}C_4H_9)_2Cl$	30	16	0.04
$\gamma\text{-}TiCl_3$	$Al(C_2H_5)_2Cl$	60	~8	0.05

[a] Vanadium acetylacetonate.

In general, the less hindered ethylene monomer is favored in Ziegler-Natta copolymerizations by as much as two orders of magnitude for certain catalyst combinations. To obtain homogeneous copolymers, continuous processes are required utilizing incomplete conversions of the propylene comonomer. A further aspect of the commercial preparation of ethylene-propylene rubbers is the inclusion of a third diene comonomer which introduces unsaturation into the final polymer to facilitate peroxide crosslinking reactions and to permit sulfur vulcanization; these terpolymers are called EPDM in contrast to the binary copolymers, which are designated as EPM. The following nonconjugated diene monomers are used commercially because they generate side-chain unsaturation rather than in-chain unsaturation which could lead to oxidative chain scission:

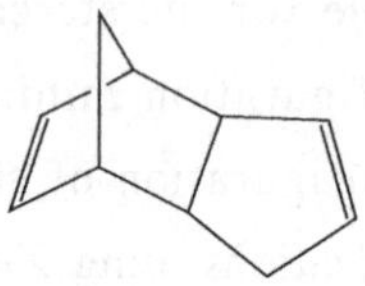

$$CH_2=CH-CH_2-CH=CH-CH_3$$

Dicyclopentadiene   Ethylidene norbornene   1,4-Hexadiene

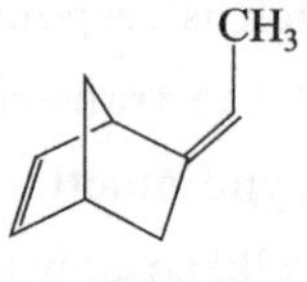

The compositions for the more than 150 grades of EPDM elastomers are in the ranges of 40 to 90mole % ethylene and 0 to 4mole % diene. Thus, the structure of a typical EPDM elastomer with ethylidene norbornene as termonomer can be represented by the following structure:

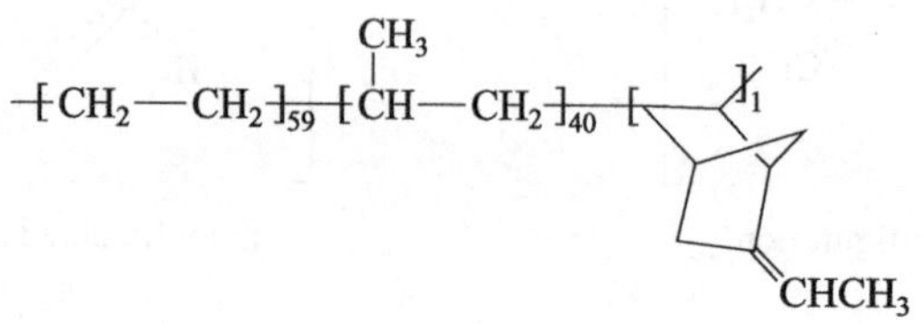

$$-[CH_2-CH_2]_{59}-[CH(CH_3)-CH_2]_{40}-[\ldots]_1-$$

CHCH$_3$

The compositions of EPDM elastomers are controlled by using the appropriate monomer feed ratio [see Eq. (1-38)] to obtain the desired composition in a continuous polymerization process. In general the excess propylene required is recycled. The molecular weights of EPDM polymers are controlled primarily by chain transfer reactions with added molecular hydrogen [Eq. (1-86) and Eq. (1-87)], as is common with other Ziegler-Natta polymerizations.

$$P-V + H_2 \xrightarrow{k_{tr}} P-H + V-H \tag{1-86}$$

$$V-H + CH_2=CHCH_3 \xrightarrow{k_i} V-\underset{\underset{CH_3}{|}}{CH}-CH_3 \tag{1-87}$$

In the past 30 years, there has been a revolution in the field of Ziegler-Nata and related catalysts for olefin polymerization. This revolution has resulted from the discovery of single-site, homogeneous metallocene catalysts that exhibit higher activity and the ability to readily incorporate more hindered comonomers with ethylene more uniformly along the polymer chain. Metallocene catalysts contain one or two cyclopentadienyl rings coordinated to a transition metal such as titanium, zirconium, or hafnium. The higher activity of metallocene catalysts means that processes can be designed without the need for a catalyst removal step. The structure of a high activity, single-site, metallocene catalyst is shown below. It is noteworthy that a metallocene cation is the proposed catalytically active species. For this type of catalyst generated with a different counterion, the monomer reactivity ratios for copolymerization of ethylene and propylene are r(ethylene) = 1.35 and r(propylene) = 0.82, indicating an almost random copolymerization behavior. These results can be compared with the copolymerization parameters for standard Ziegler-Natta catalysts in Table 1.16.

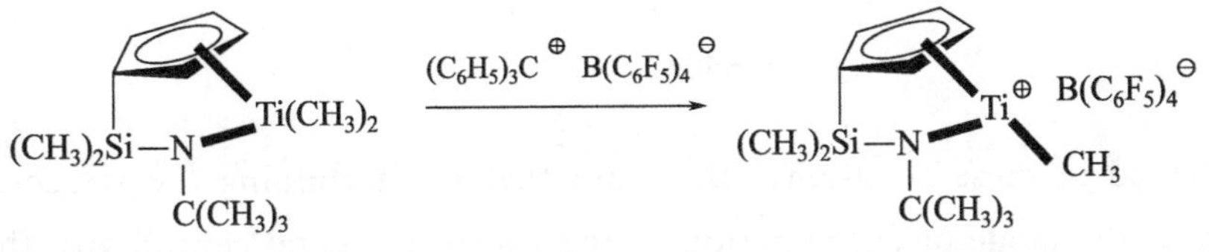

1.9.3 Polydienes

Shortly after the discovery of the Ziegler-Natta catalysts, it was found that analogous transition metal catalysts could also effect the stereospecific polymerization of dienes. The wide range of stereoregular polybutadienes which can be prepared with these catalysts is indicated in Table 1.17. The stereochemistry of polymerization is dependent upon the transition metal salt, the metal alkyl, temperature, additives, and the stoichiometry of the components. Commercial polybutadienes with high *cis*-1,4-microstructure are prepared using a wide range of transition metal catalysts, of which the most important are those derived from cobalt, nickel, neodymium, and titanium, analogous to those listed in Table 1.17.

Table 1.17 Microstructure of polydienes from transition metal-initiated polymerization

Transition metal salt/metal alkyl		Chain microstructure (mole %)			
		cis-1,4-	trans-1,4-	1,2-	3,4-
Isoprene	$TiCl_4/AlR_3$ (1 : 1)	97			3
	$\alpha\text{-}TiCl_3/Al(C_2H_5)$		98~100		
	$Ti(OR)_4/AlR_3$ (1 : 7 or 1 : 8)	36			
Butadiene	$TiI_4/Al(i\text{-}C_4H_9)_3$ (1 : 4 or 1 : 5)	92~93	2~3	4~6	
	$Ni(octanoate)/Al(C_2H_5)_3/BF_3$ (1 : 17 : 15)	96~97	2~3		
	$CoCl_2/Al(C_2H_5)_2Cl/pyridine \cdot H_2O$ (1/1000/100)	98	1	1	
	$NdCl_3/Al(i\text{-}C_4H_9)_3 \cdot nL^a$	97	2.7	0.3	
	$VCl_3/Al(C_2H_5)_3$			99~100	
	$Co(acac)_3/AlR_3/CS_2$				99~100[b]

[a] L = Electron donor such as tetrahydrofuran or pyridine.
[b] Syndiotactic.

The mechanism of stereospecific polymerization of 1,3-dienes is also categorized as an insertion polymerization and simplified representations of the stereoselectivity for cis- [Eq. (1-88)] and trans- [Eq. (1-89)] enchainments are shown below.

$$\text{cis-stereospecificity} \tag{1-88}$$

As indicated in these equations, the main factor determining the stereochemistry of enchainment is the mode of coordination of the transition metal center with the monomer to form either a *syn* or *anti* p-allyl type of intermediate. In general, the coordination with two double bonds of the 1,3-diene in an *s-trans* conguration [see (b) Eq. (1-89)] is less common than the coordination in an *s-cis* conguration shown in Eq. (1-88). This interpretation is complicated by the fact that the *syn* and *anti* p-allyl complexes are in equilibrium. These simple mechanistic representations are reinforced by the observations that the stereochemistry of diene polymerizations can be altered by the addition of electron donors such as $N(C_2H_5)_3$, $P(OC_6H_5)_3$, or C_2H_5OH. Thus, addition of these electron donors changes the stereochemistry from highly cis-stereospecific to highly trans-stereospecific for butadiene with catalysts such as $Co(acac)_2/Al(C_2H_5)_2Cl$. This is explained by assuming that the electron donor occupies one of the two coordination sites required for *cis*-enchainment [see Eq. (1-88)] which forces the monomer to only coordinate with one site [(b) in Eq. (1-89)].

The most recent developments in catalysts for stereospecific polymerization of dienes have been in the area of the rare earth or Lanthanide catalysts, specifically the neodymium complexes. The advantages of these systems are high stereospecificity, high activity, control of molecular weight, and no gel formation.

trans-stereospecificity

$$\text{(1-89)}$$

1.9.4 Polyalkenamers

Cyclic olefins undergo a very unusual type of ring-opening polymerization in the presence of certain transition metal catalysts. This is illustrated in Eq. (1-90) for the ring-opening metathesis polymerization (ROMP) of cyclooctene to form polyoctenamer. Quite surprisingly, the double bond is maintained in the polymer, i.e., it is not a normal addition polymer.

$$n \bigcirc \xrightarrow[\substack{C_2H_5AlCl_2 \\ C_2H_5OH}]{WCl_6} -\!\!\!-\!\!\!\left[\, CH_2CH\!=\!CHCH_2CH_2CH_2CH_2CH_2 \,\right]_{\!n}\!\!-\!\!\!- \qquad (1\text{-}90)$$

The generally accepted mechanism for these polymerizations proposes that the active propagating species is a metal carbene intermediate which undergoes a cycloaddition reaction with the cycloalkene to form a four-membered ring intermediate, i.e., a metallocyclobutane. The metallocyclobutane then undergoes ring-opening to form a new metallocarbene propagating species as shown in the scheme below for polymerization of cyclopentene, where Mt represents the transition metal center, —☐ represents an empty orbital which is available for coordination with the double bond of the monomer, and P_n is the growing polymer chain with number-average number of monomer units equal to n. As indicated, these are reversible polymerizations, and an equilibrium distribution of monomer, cyclic oligomers, and high molecular weight polymer is produced.

1.9.5 Scheme for Metathesis Ring Opening Polymerization:

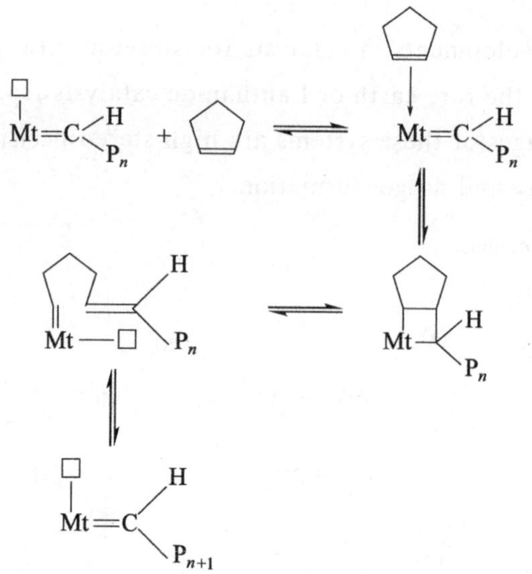

A possible reaction sequence for formation of the metal carbene is shown in Eq. (1-91), where [W] represents a tungsten catalyst center with its attendant ligands not specifically shown.

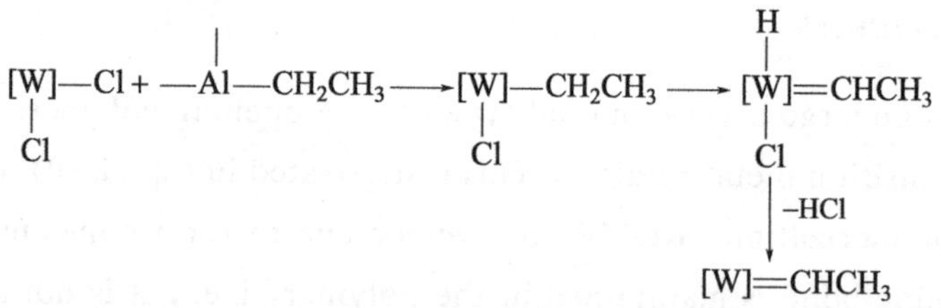

$$[W]-Cl + -Al-CH_2CH_3 \longrightarrow [W]-CH_2CH_3 \longrightarrow [W]=CHCH_3 \atop Cl \qquad\qquad\qquad\qquad Cl \qquad\qquad Cl$$
$$\downarrow -HCl$$
$$[W]=CHCH_3 \qquad (1-91)$$

Although a variety of transition metal compounds can catalyze these ring-opening polymerizations, the most active catalysts are based on molybdenum, tungsten, and rhenium derivatives. These compounds are often used with organometallic cocatalysts, analogous to other transition metal catalysts for olefin and diene polymerization described in previous sections. The $WCl_6/EtAlCl_2/EtOH$ catalyst system has been described as a commercially useful type of catalyst. In general, the stereochemistry of the polymerization varies with the catalyst and reaction time. Polymerizable monomers of importance for elastomer synthesis include cyclopentene, cyclooctene, and 1,5-cyclooctadiene; it is noteworthy that cyclohexene is not polymerizable by this method, presumably because there is no ring strain to drive the polymerization. Another monomer of commercial significance is norbornene, which is very reactive; however, the resulting polymer has a relatively high glass transition temperature ($T_g = 35°C$ for 80% *trans* polymer), but the glass transition temperature can be lowered to $-60°C$ with plasticizers.

Since the ring-opening metathesis polymerization is a reversible polymerization, an equilibrium molecular weight distribution of cyclic oligomers and high molecular weight polymer is ultimately obtained; for example, polyoctenamer generally consists of 10% to 15% cyclic oligomer and 85% to 90% polymer. At short reaction times and high monomer concentrations, relatively high molecular weight polymer is formed as a result of kinetic control; the molecular weight decreases with increasing reaction time. The equilibration process also equilibrates the configuration of the double bonds in the polymer such that eventually an equilibrium distribution of configuration results also. Molecular weight control in ring-opening methathesis polymerization is achieved by addition of acyclic alkenes, which react with the growing chain to terminate chain growth and generate a new metal carbene initiator as shown in Eq. (1-92). Commercially available polyoctenamers have weight-average molecular weights of approximately 105g/mol, variable *trans*-double bond contents (62% to 80%) and glass transition temperatures of $-75°C$ to $-80°C$. An interesting aspect of the physical properties of polyoctenamers is that they undergo stress-induced crystallization. The commercial polymers described above have approximately 8% and 30% crystallinity for samples with 62% and 80% *trans*-double bond contents, respectively.

$$\text{Mt}=C{<}^H_{P_n} + CH_2=CHCH_2CH_3 \longrightarrow \text{Mt}=C{<}^H_H + CH_3CH_2CH=C{<}^H_{P_{n+1}} \quad (1\text{-}92)$$

1.10 Graft and Block Copolymerization

The idea of graft or block copolymerization probably first arose as a means of modifying naturally occurring polymers, such as cellulose (cotton), rubber, or wool. Graft copolymerization, by analogy to the botanical term, refers to the growth of a "branch" of different chemical composition on the "backbone" of a linear macromolecule. In contrast, the related term, block copolymerization, refers to the specific case of growth of a polymer chain from the end of a linear macromolecule, thus leading to a composite linear macromolecule consisting of two or more "blocks" of different chemical structure.

The importance of these types of polymer structures is basically due to the fact that polymer chains of different chemical structure, which are normally incompatible and form separate phases (because the small entropy of mixing is insufficient to overcome the mostly positive enthalpy of mixing), are chemically bonded to each other. This leads to the formation of microheterogeneities, which can have a profound effect on the mechanical properties of these heterophase systems when compared with the two homopolymers or with a physical mixture of the two polymers.

As one might expect, graft and block copolymerization can be accomplished by means of each of the three known mechanisms, i.e, radical, cationic, and anionic, each of which shows its own special characteristics. Hence these mechanisms have been used wherever appropriate for the polymer and monomer involved. The examples quoted in the following discussions will deal primarily with elastomers.

1.10.1 Graft Copolymerization by Free Radical Reactions

This has been the most widely applied system for the formation of graft copolymers, since it provides the simplest method and can be used with a wide variety of polymers and monomers. It has not been very useful in the synthesis of block copolymers, as will become obvious from an examination of the methods used. These can be listed as follows.

1.10.1.1 Chemical Initiation

This is still the most popular method for graft copolymerization of elastomers via free radicals. Free radicals (I·) are generated from the same types of initiators which are used for free radical polymerization and copolymerization. In general, these radicals are formed in the presence of a polydiene elastomer and a monomer; therefore, there are several possible reactions of these initiator-derived radicals which can occur as shown in Eq. (1-93)~ Eq. (1-96). The competition between initiation of monomer polymerization [Eq. (1-93)] and reactions to form polymer-derived radicals [Eq. (1-94)~Eq. (1-96)] is dependent on the reactivity of the initiating radical. Thus, no graft copolymer formation with styrene monomer is observed for either polybutadiene or polyisoprene when azobisisobutryonitrile is used to generate radicals, although good grafting efficiency was observed for benzoyl peroxide-generated radicals.

$$I \longrightarrow I\cdot$$

$$I\cdot + M \xrightarrow{k_1} I-M\cdot \qquad (1-93)$$

$$I\cdot + -\!\!\!-\!\!\![CH_2CH=CHCH_2]\!-\!\!\!-\!\!\! \xrightarrow{k_2} I-H + -\!\!\!-\!\!\![\overset{\cdot}{C}HCH=CHCH_2]\!-\!\!\!-\!\!\! \qquad (1-94)$$

$$I\cdot + -\!\!\!-\!\!\![CH_2CH=CHCH_2]\!-\!\!\!-\!\!\! \xrightarrow{k_3} -\!\!\!-\!\!\![CH_2CH-\overset{\cdot}{C}HCH_2]\!-\!\!\!-\!\!\!\underset{I}{|} \qquad (1-95)$$

$$I\cdot + -\!\!\!-\!\!\!\underset{\underset{CH=CH_2}{|}}{[CH_2CH]}\!-\!\!\!-\!\!\! \xrightarrow{k_3} -\!\!\!-\!\!\!\underset{\underset{\cdot CH-CH_2-I}{|}}{[CH_2CH]}\!-\!\!\!-\!\!\! \qquad (1-96)$$

This result also indicates that growing polystyryl radicals do not abstract hydrogen from these polydienes to generate polymer-derived radicals. The competition between addition of initiator radicals to the double bonds in the polydiene [Eq. (1-95) and Eq. (1-96)] and hydrogen abstraction [Eq. (1-94)] is also dependent on the initiator. Thus, t-butoxy radicals [$(CH_3)_3CO\cdot$] exhibit an unusual preference for hydrogen abstraction compared to alkyl radicals as shown in Table 1.18.

Table 1.18 Radical reactivity toward hydrogen abstraction versus addition to double bonds

Radical	$k_{abstraction}/k_{addition}$ [a]
$t\text{-}(CH_3)_3CO\cdot$	30
$ROO\cdot$	1.0
$H_3C\cdot$	0.25
$RS\cdot$	Exclusive addition

[a] Ratio of rate constant for hydrogen abstraction [see Eq. (1-94)] versus addition to a double bond [see Eq. (1-95) and Eq. (1-96)].

For radicals derived from benzoyl peroxide, the competition between the rate of hydrogen abstraction from polydiene [Eq. (1-94)] compared to addition of the initiator radical to styrene monomer [Eq. (1-93)], e.g., k_{abstr}/k_{ad}^S, was found to be 1.2 for polyisoprene and 0.63 for polybutadiene. With respect to the addition of initiating radicals to the double bond of the polydiene, it is reported that grafting is favored by higher 1,2-microstructure in the polydiene; the rate of addition to a vinyl side chain [Eq. (1-96)] is faster than addition to an in-chain double bond [Eq. (1-95)].

The formation of graft copolymer from the polymer-derived radicals generated in Eq. (1-93)~Eq. (1-96) are shown in Eq. (1-97)~Eq. (1-98), where $P\cdot$ represents the polymeric backbone radicals. Finally, in order to control the molecular weight of the graft chains, chain transfer agents such as long chain alkyl thiols can be added [see Eq. (1-27)].

$$P\cdot + nM \longrightarrow P-[M]_n-M\cdot$$
$$\text{graft copolymer} \quad (1\text{-}97)$$

$$P\cdot + I-[M]_n-M\cdot \longrightarrow \text{graft copolymer} \quad (1\text{-}98)$$

Since all of these reactions are occurring simultaneously during the graft copolymerization, there is always the possibility of formation of homopolymer during the grafting reaction, via reaction of the initiator radical with monomer [Eq. (1-93)] and also by chain transfer of the growing chain with species other than the polymer backbone (e.g., monomer, solvent, initiator). Therefore, in general, the graft copolymer will be contaminated with both the original back-bone homopolymer as well as the monomer-derived homopolymer.

This type of graft copolymerization has been applied to the grafting of monomers like styrene and methyl methacrylate to natural rubber, directly in the latex. Similar methods have been developed for grafting the foregoing monomers, and many other vinyl monomers, to synthetic rubbers like SBR, leading to a variety of plastic-reinforced elastomers and rubber-reinforced high-impact plastics. In this case, grafting can also occur by the "copolymerization" of the monomer with the unsaturated bonds (mainly vinyl) in the polymer as described previously [see Eq. (1-96)]; thus

$$RM_x\cdot + \sim CH_2-CH\sim \atop CH_2=CH \quad \longrightarrow \quad \sim CH_2-CH\sim \atop RM_x-CH_2-CH\cdot \quad \longrightarrow \cdots \longrightarrow$$

This reaction can, of course, also lead to crosslinking of the polymer chains, and this must be controlled.

1.10.1.2 Other Methods

Other methods of generating free radicals can also be used to initiate graft polymerization with elastomers, both natural and synthetic. These include irradiation of polymer-monomer mixtures by ultraviolet light, high-energy radiation, and mechanical shear. The latter is of particular interest because of its unique mechanism, and has been extensively investigated.

Thus it has been convincingly demonstrated that elastomers, when subjected to several mechanical shearing forces, undergo homolytic bond scission to form free chain radicals. The latter, when in the presence of oxygen, may then undergo various reactions, either becoming stabilized ruptured chains or reacting with other chains to form branched or crosslinked species.

When blends of different elastomers are masticated, "interpolymers" are formed by the interaction of the radicals formed from the copolymers. A further extension of such mechanochemical processes occurs when elastomers are masticated in the presence of polymerizable monomers, the chain radicals initiating polymerization and leading to formation of block and graft structures. This was clearly demonstrated in the case of natural rubber and of other elastomers. It should be noted that living anionic polymerizations and controlled/living radical polymerizations such as atom transfer radical polymerization (ATRP) have been used to make well-defined graft copolymers. A very useful method is to copolymerize a well-defined macromonomer bearing a polymerizable chain end group with another monomer to generate a graft copolymer with a random distribution of well-defined graft branches.

1.10.2 Block Copolymers by Anionic Mechanism

It is, of course, the anionic mechanism which is most suitable for the synthesis of block copolymers, since many of these systems are of the living polymer type, as described previously. Thus it is possible to use organoalkali initiators to prepare block copolymers in homogeneous solution by sequential addition of monomers, where each block has a prescribed molecular weight, based on monomer-initiator stoichiometry, as well as a very narrow molecular weight distribution (Poisson). As would be expected, such block copolymers are very "pure", due to the absence of any side reactions during the polymerization (e.g., termination, monomer transfer, branching).

Organolithium initiators have been particularly useful in this regard, since they are soluble in a variety of solvents and since they can initiate the polymerization of a variety of monomers, such as styrene and its homologs, the 1,3-dienes, alkyl methacrylates, vinylpyridines, cyclic oxides and sulfides, and cyclic siloxanes. Various block copolymers of these monomers have been synthesized, some commercially, but the outstanding development in this area has been in the case of the ABA type of triblock copolymers, and these deserve special mention.

The ABA triblocks which have been most exploited commercially are of the styrene-diene-styrene type, prepared by sequential polymerization initiated by alkyllithium compounds as shown in Eq. (1-99)~Eq. (1-101). The behavior of these block copolymers illustrates the special characteristics of block (and graft) copolymers, which are based on the general incompatibility of the different blocks. Thus for a typical "thermoplastic elastomer", the polystyrene end blocks (about $15000 \sim 20000 M_w$) aggregate into a separate phase, which forms a microdispersion within the matrix composed of the polydiene chains ($50000 \sim 70000 M_w$). A schematic representation of this morphology is shown in Fig. 1.3. This phase separation, which occurs in the melt (or swollen) state, results, at ambient temperatures, in a network of elastic polydiene chains held together by glassy polystyrene microdomains. Hence these materials behave as virtually crosslinked elastomers at ambient temperatures, but are completely thermoplastic and fluid at elevated temperatures.

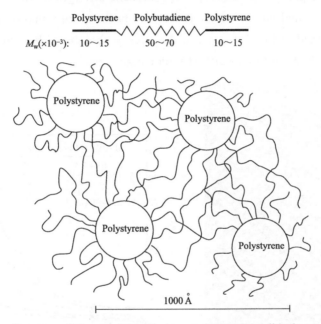

Fig. 1.3 Structure of thermoplastic elastomers from ABA triblock copolymers.

$$C_4H_9Li + nCH_2=CH(Ph) \xrightarrow{RH} C_4H_9-[CH_2-CH(Ph)]_{n-1}-CH_2-CHLi(Ph) \equiv PsLi \quad (1\text{-}99)$$

$$PsLi + mCH_2=CH-CH=CH_2 \longrightarrow Ps-[CH_2CH=CHCH_2]_{m-1}CH_2CH=CHCH_2Li$$
$$\equiv Ps\text{-}PBDLi \quad (1\text{-}100)$$

$$Ps\text{-}PBDLi + nCH_2=CH(Ph) \longrightarrow Ps\text{-}PBD\text{-}PsLi \xrightarrow{ROH} Ps\text{-}PBD\text{-}Ps \quad (1\text{-}101)$$

It is important to note that the morphology of ABA block copolymers is dependent primarily on the relative composition of the block components. For example, as the styrene content increases, the morphology changes from spherical polystyrene domains to cylindrical; further increases in styrene content result in lamellar arrays of both phases and eventually phase inversion to form a continuous polystyrene phase. The properties of the ABA triblock copolymers are dependent and vary with the composition.

Such "thermoplastic elastomers" are very attractive technologically, since they can be heat-molded like thermoplastics, yet exhibit the behavior of rubber vulcanizates. As would be expected, their structure, morphology, and mechanical properties have been studied extensively. An electron photomicrograph of a typical styrene-isoprene-styrene (SIS) triblock film is shown in Fig. 1.4, while the tensile properties of a series of such triblock copolymers are shown in Fig. 1.5. The unusually high tensile strength of these elastomers, better than that of conventional vulcanizates, is ascribed both to the remarkable regularity of the network, as illustrated in Fig. 1.5, and to the energy-absorbing characteristics of the polystyrene domains, which yield and distort under high stress.

Fig. 1.4 Transmission electron photomicrograph (×100000) of an ultrathin film of a styrene-isoprene-styrene triblock copolymer (M_w 16200~75600~162000).

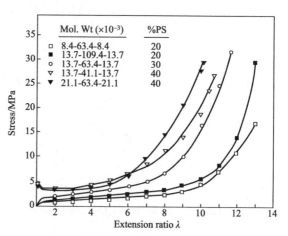

Fig. 1.5 Tensile properties of styrene-isoprene-styrene triblock copolymers.

This interesting behavior of the ABA triblock copolymers is not a unique feature of the styrene-diene structure, but can be found in the case of other analogous chemical structures. Thus thermoplastic elastomers have been obtained from other triblock copolymers, where the dienes have been replaced by cyclic sulfides, cyclic siloxanes, or alkyl acrylates; poly(alkyl methacrylate) end blocks have also been investigated. Furthermore, the advent of a number of different types of living polymerization with transition metal, cationic, and radical propagating centers provides new mechanisms for synthesis of ABA-type block copolymers utilizing a wide variety of monomer types.

It is important to note that any molecular architecture that provides a thermoplastic block chemically bonded to an elastomeric block, which is in turn bonded to another thermoplastic segment, should exhibit the properties of a thermoplastic elastomer. For example, grafting thermoplastic branches onto an elastomeric backbone produces thermoplastic elastomer behavior. Other examples are the segmented-type polymers—$[AB]_n$—with alternating hard and soft segments; thus, a variety of segmented polyesters and polyurethanes with polyether or polyester soft segments exhibit properties of thermoplastic elastomers.

References

[1] G. Williams, Proc. R. Soc. London 10, 516 (1860).
[2] G. Bouchardat, C. R. H. Acad. Sci. 89, 1117 (1879).
[3] W. A. Tilden, J. Chem. Soc. 45, 411 (1884).
[4] I. Kondakow, J. Prakt. Chem. 62, 66 (1900).
[5] J. Thiele, Justus Liebig's Ann. Chem. 319, 226 (1901).
[6] S. V. Lebedev, Zh. Russ. Fiz. -Kim. Ova. Chast. 42, 949 (1910).
[7] C. D. Harries, Justus Liebig's Ann. Chem. 383, 190 (1911).
[8] F. E. Matthews and E. H. Strange, British Patent 24, 790 (1910).
[9] H. Staudinger, Chem. Ber. 53, 1073 (1920).
[10] W. H. Carothers, J. Williams, A. M. Collins, and J. E. Kirby, J. Am. Chem. Soc. 53, 4203 (1931).
[11] P. J. Flory, "Principles of Polymer Chemistry," Cornell Univ. Press, Ithaca, NY, 1953, Chap. 1.
[12] H. Morawetz, "Polymers: The Origins and Growth of a Science," Wiley, New York, 1985.
[13] M. Morton, Rubber Plast. Age 42, 397 (1961).
[14] W. H. Carothers, J. Am. Chem. Soc. 51, 2548 (1929).
[15] W. H. Carothers, Chem. Rev. 8, 353 (1931).
[16] D. Braun, H. Cherdron, and W. Kern, "Practical Macromolecular Organic Chemistry," Harwood Academic Publishers, New York, 1984, p. 255.
[17] S. R. Sandler and W. Karo, "Polymer Syntheses," 2nd ed. , Academic Press, San Diego, 1992, p. 127.
[18] G. C. Eastmond, A. Ledwith, S. Russo, and P. Sigwalt (Eds.), "Comprehensive Polymer Science," Vol. 5: "Step Polymerization," Pergamon Press, Oxford, 1989.
[19] J. A. Moore (Ed.), "Macromolecular Synthesis, Collective Volume 1," Wiley, New York, 1978.
[20] J. H. Saunders and F. Dobinson, in "Comprehensive Chemical Kinetics," C. H. Bamford and C. F. H. Tipper (Eds.), Elsevier, New York, Vol. 15, 1976, Chap. 7.
[21] G. C. Eastmond, A. Ledwith, S. Russo, and P. Sigwalt (Eds.), "Comprehensive Polymer Science," Vol. 3: "Chain Polymerization I;" Vol. 4: "Chain Polymerization II," Pergamon Press, Oxford, 1989.
[22] G. Odian, "Principles of Polymerization," 4th ed. Wiley-Interscience, New York, 2004.
[23] S. A. Penczek, J. Polym. Sci. ; Part A: Polym. Chem. 40, 1665 (2002).
[24] A. D. Jenkins, P. Kratochvil, R. F. T. Stepto, and U. W. Suter, Pure Appl. Chem. 68, 2287 (1996).
[25] D. H. Solomon (Ed.), "Step-Growth Polymerizations," Dekker, New York, 1972.
[26] G. C. Eastmond, in "Comprehensive Chemical Kinetics," Vol. 14A: "Free Radical Polymerization," C. H. Bamford and C. F. H. Tipper (Eds.), Elsevier, New York, 1976, p. 1.
[27] R. G. Gilbert, Pure Appl. Chem. 64, 1563 (1992).
[28] R. G. Gilbert, Pure Appl. Chem. 68, 1491 (1996).

[29] M. Buback, R. G. Gilbert, R. A. Hutchinson, B. Klumperman, F. D. Kuchta, B. G. Manders, K. F. O'Driscoll, G. T. Russell, and J. Schweer, J. Macromol. Chem. Phys. 196, 3267 (1995).

[30] M. S. Matheson, E. E. Auer, E. B. Bevilacqua, and E. J. Hart, J. Am. Chem. Soc. 71, 497, 2610 (1949); ibid., 73, 1700, 5395 (1951).

[31] M. Morton, P. Salatiello, and H. Landfield, J. Polym. Sci. 8, 215 (1952).

[32] M. Morton and S. D. Gadkary, 130th Meet., Am. Chem. Soc., Atlantic City, N. J., 1956; S. D. Gadkary, M. S. Thesis, Univ. of Akron, Akron, Ohio, 1956.

[33] S. Beuermann, M. Buback, T. P. Davis, R. G. Gilbert, R. A. Hutchinson, O. F. Olaj, G. T. Russell, J. Schweer, and A. M. van Herk, Macromol. Chem. Phys. 198, 1545 (1997).

[34] P. A. Weerts, A. L. German, and R. G. Gilbert, Macromolecules 24, 1622 (1991).

[35] R. A. Hutchinson, M. T. Aronson, and J. R. Richards, Macromolecules 26, 6410 (1993).

[36] G. V. Schulz, Z. Phys. Chem., Abt., B 43, 25 (1939).

[37] G. S. Whitby and R. N. Crozier, Can. J. Res. 6, 203 (1932).

[38] D. H. Solomon, E. Rizzardo, and P. Cacioli, U. S. Patent 4, 581, 429 (1986).

[39] M. K. Georges, R. P. N. Veregin, P. M. Kazmeier, and G. K. Hamer, Macromolecules 29, 5245 (1993).

[40] C. J. Hawker, J. Am. Chem. Soc. 116, 11185 (1994).

[41] T. Fukuda, T. Terauchi, A. Goto, Y. Tsuhii, T. Miyamoto, and Y. Shimizu, Macromolecules 29, 3050 (1996).

[42] C. J. Hawker, A. W. Bosman, and E. Harth, Chem. Rev. 101, 3661 (2001).

[43] T. Fukuda and A. Goto, in "Controlled/Living Radical Polymerization," K. Matyjaszewski (Ed.), ACS Symposium Series 768, American Chemical Society, Washington, D. C., 2000, p. 27.

[44] B. Keoshkerian, M. Georges, M. Quinlan, R. Veregin, and B. Goodbrand, Macromolecules 31, 7559 (1998).

[45] H. Fischer, Macromolecules 30, 5666 (1997).

[46] H. Fischer, J. Polym. Sci., Part A: Polym. Chem. 37, 1885 (1999).

[47] K. Matyjaszewski (Ed.), "Controlled/Living Radical Polymerization," ACS Symposium Series 768, American Chemical Society, Washington, D. C., 2000.

[48] K. Matyjaszewski (Ed.), "Controlled/Living Radical Polymerization," ACS Symposium Series 685, American Chemical Society, Washington, D. C., 1998.

[49] M. K. Georges, G. K. Hamer, and N. A. Listigovers, Macromolecules 31, 9087 (1998).

[50] D. Benoit, E. Harth, P. Fox, R. M. Waymouth, and C. J. Hawker, Macromolecules 33, 363 (2000).

[51] K. Matyjaszewski and J. Xia, Chem. Rev. 101, 2921 (2001).

[52] G. W. Poehlein, in "Encyclopedia of Polymer Science and Engineering," J. I. Kroschwitz (Ed.), Wiley-Interscience, New York, Vol. 6, 1986, p. 1.

[53] I. Piirma (Ed.), "Emulsion Polymerization," Academic Press, New York, 1982.

[54] D. C. Blackley, "Emulsion Polymerization," Applied Science Publishers, London, 1975.

[55] D. H. Napper and R. G. Gilbert, in "Comprehensive Polymer Science," G. C. Eastmond, A. Ledwith, S. Russo, and P. Sigwalt (Eds.), Pergamon, Oxford, Vol. 4, 1989, p. 171.

[56] R. G. Gilbert, "Emulsion Polymerization: A Mechanistic Approach," Academic Press, San Diego, 1995.

[57] K. Tauer, in "Encyclopedia of Polymer Science and Technology," J. I. Kroschwitz (Ed.), Wiley-Interscience, New York, Vol. 6, 2003, p. 410.

[58] P. A. Lovell and M. S. El-Aasser (Eds.), "Emulsion Polymerization and Emulsion Polymers," Wiley, New York, 1997.

[59] W. D. Harkins, J. Am. Chem. Soc. 69, 1428 (1947).

[60] J. L. Gardon, in "Polymerization Processes," C. E. Schildknecht (Ed.), Wiley-Interscience, New York, 1977, Chap. 6.

[61] M. Morton, S. Kaizerman, and M. W. Altier, J. Colloid Sci. 9, 300 (1954).

[62] W. V. Smith and R. H. Ewart, J. Chem. Phys. 16, 592 (1948).

[63] M. Morton and W. E. Gibbs, J. Polym. Sci., Part A 1, 2679 (1963).

[64] W. Hofmann, "Rubber Technology Handbook," Hanser Publishers, Munich, Germany, 1989.

[65] C. M. Blow (Ed.), "Rubber Technology and Manufacture," Hewnes-Butterworths, London, 1971.

[66] J. A. Brydson, in "Developments in Rubber Technology-2," A. Whelan and K. S. Lee (Eds.), Applied Science Publishers, London, 1981, p. 21.

[67] R. G. Bauer, in "Kirk-Othmer Encyclopedia of Chemical Technology," 3rd ed., Wiley, New York, Vol. 8, 1979, p. 611.

[68] H. N. Sun and J. P. Wusters, in "Kirk-Othmer Encyclopedia of Chemical Technology," 4th ed., J. I. Kroschwitz (Ed.), Wiley-Interscience, New York, Vol. 4, 2004, p. 365.

[69] M. Demirors, in "Encyclopedia of Polymer Science and Technology," J. I. Kroschwitz (Ed.), Wiley-Interscience, New York, Vol. 4, 2003, p. 229.

[70] M. Morton and P. P. Salatiello, J. Polym. Sci. 6, 225 (1951).

[71] J. L. Binder, Ind. Eng. Chem. 46, 1727 (1954).

[72] K. E. Beu, W. B. Reynolds, C. F. Fryling, and H. L. McMurry, J. Polym. Sci. 3, 465 (1948).

[73] A. W. Meyer, Ind. Eng. Chem. 41, 1570 (1949).

[74] P. R. Johnson, Rubber Chem. Tech. 49, 650 (1976).

[75] C. A. Stewart, Jr., T. Takeshita, and M. L. Coleman, in "Encyclopedia of Polymer Science and Engineering," J. I. Kroschwitz (Ed.), Wiley, Vol. 3, 1985, p. 441.

[76] D. C. Blackley, "Synthetic Rubbers: Their Chemistry and Technology," Applied Science Publishers, Essex, 1983, p. 175.

[77] R. Musch and H. Magg, in "Polymeric Materials Encyclopedia," J. C. Salamone (Ed.), CRC Press, Boca Raton, FL, Vol. 2, 1996, p. 1238.

[78] D. I. Christie, R. G. Gilbert, J. P. Congalidis, J. R. Richards, and J. H. McMinn, Macromolecules 34, 5158 (2001).

[79] M. Morton, J. A. Cala, and M. W. Altier, J. Polym. Sci. 19, 547 (1956).

[80] M. Morton and I. Piirma, J. Polym. Sci. 19, 563 (1956).

[81] A. M. Neal and L. R. Mayo, in "Synthetic Rubber," G. S. Whitby (Ed.), Wiley, New York, 1954, p. 770.

[82] W. E. Mochel and J. H. Peterson, J. Am. Chem. Soc. 71, 1426 (1949).

[83] C. A. Hargreaves, in "Polymer Chemistry of Synthetic Elastomers," J. P. Kennedy, E. Tornqvist (Eds.), Wiley-Interscience, New York, 1968, p. 233.

[84] M. M. Coleman and E. G. Brame, Jr., Rubber Chem. Tech. 51, 668 (1978).

[85] M. M. Coleman, D. L. Tabb, and E. G. Brame, Jr., Rubber Chem. Tech. 50, 49 (1977).

[86] J. R. Ebdon, Polymer 19, 1232 (1978).

[87] R. Petiaud and Q. T. Pham, J. Polym. Sci., Polym. Chem. Ed. 23, 1333 (1985).

[88] R. R. Garrett, C. A. Hargreaves, and D. N. Robinson, J. Macromol. Sci. Chem. A 4, 1679 (1970).

[89] A. E. Hamielec, J. F. MacGregor, and A. Penlidis, in "Comprehensive Polymer Science," G. C. Eastmond, A. Ledwith, S. Russo, P. Sigwalt (Eds.), Pergamon, Oxford, Vol. 3, 1989, p. 17.

[90] G. Odian, "Principles of Polymerization," 4th ed., Wiley-Interscience, New York, 2004, Chap. 6.

[91] G. E. Ham (Ed.), "Copolymerization," Wiley-Intersceince, New York, 1964.

[92]　D. A. Tirrell, in "Encyclopedia of Polymer Science and Engineering," 2nd ed., J. I. Kroschwitz (Ed.), Wiley, New York, Vol. 4, 1986, p. 192.

[93]　D. A. Tirrell, in "Comprehensive Polymer Science," G. C. Eastmond, A. Ledwith, S. Russo, and P. Sigwalt (Eds.), Pergamon, Oxford, Vol. 3, 1989, p. 195.

[94]　F. R. Mayo and C. Walling, Chem. Rev. 46, 191 (1950).

[95]　F. R. Mayo and F. M. Lewis, J. Am. Chem. Soc. 66, 1594 (1944).

[96]　R. Z. Greenley, in "Polymer Handbook," 4th ed., J. Brandrup and E. H. Immergut (Eds.), Wiley-Interscience, New York, 1999, II-181.

[97]　G. C. Eastman and E. G. Smith, in "Comprehensive Chemical Kinetics," C. H. Bamford and C. F. H. Tipper (Eds.), Elsevier, New York, Vol. 14A, 1976, p. 333.

[98]　C. A. Uraneck, in "Polymer Chemistry of Synthetic Elastomers," Part I, J. P. Kennedy and E. Tornqvist (Eds.), Wiley-Interscience, New York, 1968, p. 158.

[99]　C. F. Fryling, in "Synthetic Rubber," G. S. Whitby (Ed.), Wiley, New York, 1954, p. 257.

[100]　Y. Tanaka, H. Sato, and J. Adachi, Rubber Chem. Tech. 59, 16 (1986).

[101]　Y. Tanaka, H. Sato, Y. Nakafutami, and Y. Kashiwazaki, Macromolecules 16, 1925 (1983).

[102]　W. Hofmann, Rubber Chem. Tech. 37, 1 (1964).

[103]　R. Greenley, J. Macromol. Sci. Chem. 14, 445 (1980).

[104]　J. P. Kennedy and B. Ivan, "Designed Polymers by Carbocationic Macromolecular Engineering: Theory and Practice," Hanser Publishers, Munich, 1992.

[105]　J. P. Kennedy and E. Marechal, "Carbocationic Polymerization," Wiley-Interscience, New York, 1982.

[106]　J. P. Kennedy, "Cationic Polymerization of Olefins: A Critical Inventory," Wiley-Interscience, New York, 1975.

[107]　M. Sawamoto and T. Higashimura, in "Encyclopedia of Polymer Science and Engineering," J. I. Kroschwitz (Ed.), Wiley-Interscience, New York, 1989, Supplemental Volume, p. 399.

[108]　G. Odian, "Principles of Polymerization," 4th ed., Wiley-Interscience, New York, 2004, Chap. 5.

[109]　A. Gandini and H. Cheradame, in "Encyclopedia of Polymer Science and Engineering," J. I. Kroschwitz (Ed.), Wiley-Interscience, New York, Vol. 2, 1985, p. 729.

[110]　K. Matyjaszewski (Ed.), "Cationic Polymerizations: Mechanisms, Synthesis, and Applications," Marcel Dekker, New York, 1996.

[111]　J. E. Puskas and G. Kaszas, in "Encyclopedia of Polymer Science and Technology," J. I. Kroschwitz (Ed.), Wiley-Interscience, New York, Vol. 5, 2003, p. 382.

[112]　S. Winstein, E. Clippinger, A. H. Fainberg, and G. C. Robinson, J. Am. Chem. Soc. 76, 2597 (1954).

[113]　R. G. W. Norrish and K. E. Russell, Trans. Faraday Soc. 48, 91 (1952).

[114]　A. G. Evans and M. Polanyi, J. Chem. Soc. 252 (1947).

[115]　E. N. Kresge, R. H. Schatz, and H. C. Wang, in "Encyclopedia of Polymer Science and Engineering," J. I. Kroschwitz (Ed.), Wiley-Interscience, New York, Vol. 8, 1987, p. 423.

[116]　R. N. Webb, T. D. Shaffer, and A. H. Tsou, in "Encyclopedia of Polymer Science and Technology," J. I. Kroschwitz (Ed.), Wiley-Interscience, New York, Vol. 5, 2003, p. 356.

[117]　D. C. Pepper, Sci. Proc. R. Dublin Soc. 25, 131 (1950).

[118]　F. S. Dainton and G. B. B. M. Sutherland, J. Polym. Soc. 4, 347 (1949).

[119]　I. Puskas, E. M. Banas, and A. G. Nerheim, J. Polym. Sci., Symp. 56, 191 (1976).

[120]　D. C. Pepper, Trans. Faraday Soc. 45, 404 (1949).

[121]　D. D. Eley and A. W. Richards, Trans. Faraday Soc. 45, 425, 436 (1949).

[122]　P. H. Plesch, J. Chem. Soc. 543 (1950).

[123] L. Sipos, P. De, and R. Faust, Macromolecules 36, 8282 (2003).
[124] J. P. Kennedy, in "Polymer Chemistry of Synthetic Elastomers," J. P. Kennedy and E. Tornqvist (Eds.), Wiley-Interscience, New York, 1968, Part I, p. 291.
[125] H. Y. Che and J. E. Field, J. Polym. Sci., Part B 5, 501 (1957).
[126] M. Miyamoto, M. Sawamoto, and T. Higashimura, Macromolecules 17, 265 (1984).
[127] R. Faust and J. P. Kennedy, Polym. Bull. (Berlin) 15, 317 (1986).
[128] S. Hadjikyriacou, M. Acar, and R. Faust, Macromolecules 37, (2004).
[129] K. Matyjaszewski (Ed.), "Cationic Polymerizations: Mechanisms, Synthesis, and Applications," Dekker, New York, 1996.
[130] R. W. Body and V. L. Kyllingstad, in "Encyclopedia of Polymer Science and Engineering," J. I. Kroschwitz (Ed.), Wiley-Interscience, New York, Vol. 6, 1986, p. 307.
[131] E. J. Vandenberg, J. Polym. Sci. 7, 525 (1969).
[132] E. J. Vandenberg, in "Coordination Polymerization," C. C. Price and E. J. Vandenberg (Eds.), Plenum Press, New York, 1983, p. 11.
[133] W. Meckel, W. Goyert, W. Wieder, and H. G. Wussow, in "Thermoplastic Elastomers," 3rd ed., G. Holden, H. R. Kricheldorf, and R. P. Quirk (Eds.), Hanser Publishers, Munich, 2004, p. 15.
[134] R. K. Adams, G. K. Hoeschele, and W. K. Witsiepe, in "Thermoplastic Elastomers," 3rd ed., G. Holden, H. R. Kricheldorf, and R. P. Quirk (Eds.), Hanser Publishers, Munich, 2004, p. 183.
[135] S. Inoue and T. Aida, in "Ring-Opening Polymerization," K. J. Ivin and T. Saegusa (Eds.), Elsevier, New York, Vol. 1, 1984, p. 185.
[136] P. Dreyfuss, "Poly (tetrahydrofuran)," Gordon and Breach, New York, 1982.
[137] S. Penczek, P. Kubisa, and K. Matyjaszewski, Adv. Polym. Sci. 68/69, 1 (1985).
[138] S. Penczek and P. Kubisa, in "Comprehensive Polymer Science," G. C. Eastmond, A. Ledwith, S. Russo, and P. Sigwalt (Eds.), Pergamon Press, Oxford, Vol. 3, 1989, p. 751.
[139] P. Dreyfuss, M. P. Dreyfuss, and G. Pruckmayr, in "Encyclopedia of Polymer Science and Engineering," J. I. Kroschwitz (Ed.), Wiley-Interscience, New York, Vol. 16, 1989, p. 649.
[140] M. P. Dreyfuss and P. Dreyfuss, J. Polym. Sci. A-1 4, 2179 (1966).
[141] G. Pruckmayr and T. K. Wu, Macromolecules 11, 662 (1978).
[142] K. Matyjaszewski, P. Kubisa, and S. Penczek, J. Polym. Sci., Polym. Chem. Ed. 13, 763 (1975).
[143] R. P. Quirk, in "Encyclopedia of Polymer Science and Technology," 3rd ed., J. I. Kroschwitz (Ed.), Wiley-Interscience, New York, Vol. 5, 2003, p. 111.
[144] M. Morton, "Anionic Polymerization: Principles and Practice," Academic Press, New York, 1982.
[145] M. Szwarc, "Carbanions, Living Polymers and Electron Transfer Processes," Interscience, New York, 1968.
[146] P. Rempp, E. Franta, and J. E. Herz, Adv. Polym. Sci. 86, 145 (1988).
[147] R. N. Young, R. P. Quirk, and L. J. Fetters, Adv. Polym. Sci. 56, 1 (1984).
[148] M. Szwarc, Adv. Polym. Sci. 49, 1 (1983).
[149] M. van Beylen, S. Bywater, G. Smets, M. Szwarc, and D. J. Worsfold, Adv. Polym. Sci. 86, 87 (1988).
[150] H. L. Hsieh and R. P. Quirk, "Anionic Polymerization," Dekker, New York, 1996.
[151] M. Szwarc, "Ionic Polymerization Fundamentals," Hanser, Cincinnati, 1996.
[152] W. C. E. Higginson and N. W. Wooding, J. Chem. Soc. 760 (1952).
[153] R. E. Robertson and L. Marion, Can. J. Res., Sect. B. 26, 657 (1948).
[154] D. E. Paul, D. Lipkin, and S. I. Weissman, J. Am. Chem. Soc. 78, 116 (1956).

[155] B. J. McClelland, Chem. Rev. 64, 301 (1964).

[156] M. T. Jones, in "Radical Ions," E. T. Kaiser and L. Kevan (Eds.), Wiley, New York, 1968.

[157] M. Szwarc and J. Jagur-Grodzinski, in "Ions and Ion Pairs in Organic Reactions," M. Szwarc (Ed.), Wiley, 1974, p. 1.

[158] N. L. Holy, Chem. Rev. 74, 243 (1974).

[159] M. Szwarc, M. Levy, and R. Milkovich, J. Am. Chem. Soc. 78, 2656 (1956).

[160] P. J. Flory, J. Am. Chem. Soc. 62, 1561 (1940).

[161] H. L. Hsieh, R. C. Farrar, and K. Udipi, CHEMTECH 626 (1981).

[162] I. G. Hargis, R. A. Livigni, and S. L. Aggarwal, in "Developments in Rubber Technology-4," A. Wheland and K. S. Lee (Eds.), Elsevier, Essex, UK, 1987, p. 1.

[163] H. L. Hsieh and O. F. McKinney, J. Polym. Sci., Polym. Lett. 4, 843 (1966).

[164] M. Morton, A. Rembaum, and J. L. Hall, J. Polym. Sci. A 1, 461 (1963).

[165] R. P. Quirk, D. J. Kinning, and L. J. Fetters, in "Comprehensive Polymer Science," G. Allen and J. C. Bevington (Eds.), Pergamon Press, Oxford, Vol. 7, 1989, p. 1.

[166] N. Hadjichristidis, S. Pispas, and G. A. Floudas, "Block Copolymers: Synthetic Strategies, Physical Properties, and Applications," Wiley-Interscience, New York, 2003.

[167] R. P. Quirk, in "Comprehensive Polymer Science," First Supplement, S. L. Aggarwal and S. Russo (Eds.), Pergamon Press, Oxford, 1992, p. 83.

[168] H. L. Hsieh, Rubber Chem. Tech. 49, 1305 (1976).

[169] B. J. Bauer and L. J. Fetters, Rubber Chem. Tech. 51, 406 (1978).

[170] S. Bywater, Adv. Polym. Sci. 30, 89 (1979).

[171] N. Hadjichristidis, M. Pitsikalis, S. Pispas, and H. Iatrou, Chem. Rev. 101, 3747 (2001).

[172] S. Bywater, in "Comprehensive Chemical Kinetics," C. H. Bamford and C. F. H. Tipper (Eds.), Elsevier, New York, Vol. 15, 1976, p. 1.

[173] A. H. Müller, in "Comprehensive Polymer Science," Vol. 3: "Chain Polymerization I," G. C. Eastmond, A. Ledwith, S. Russo, and P. Sigwalt (Eds.), Pergamon Press, Oxford, 1989, p. 387.

[174] S. Bywater, A. F. Johnson, and D. J. Worsfold, Can. J. Chem. 42, 1255 (1964).

[175] M. Morton, in "Vinyl Polymerization," Part II, G. Ham (Ed.), Marcel Dekker, New York, 1969, p. 211.

[176] J. L. Wardell, in "Comprehensive Organometallic Chemistry: The Synthesis, Reactions and Structures of Organometallic Compounds," G. Wilkinson, F. G. A. Stone, and E. W. Abel (Eds.), Pergamon Press, Oxford, Vol. 1, p. 43.

[177] T. L. Brown, Acc. Chem. Res. 1, 23 (1968).

[178] B. L. Wakefield, The Chemistry of Organolithium Compounds, Pergamon Press, Oxford, 1974.

[179] D. J. Worsfold and S. Bywater, Can. J. Chem. 38, 1891 (1960).

[180] S. Bywater and D. J. Worsfold, J. Organomet. Chem. 10, 1 (1967).

[181] F. Schue and S. Bywater, Macromolecules 2, 458 (1969).

[182] M. Morton, R. A. Pett, and L. J. Fetters, Macromolecules 3, 333 (1970).

[183] C. M. Selman and H. L. Hsieh, J. Polym. Sci., Polym. Lett. Ed. 9, 219 (1971).

[184] S. Bywater and D. J. Worsfold, in "Recent Advances in Anionic Polymerization," T. E. Hogen-Esch and J. Smid (Eds.), Elsevier, New York, 1987, p. 109.

[185] L. J. Fetters and M. Morton, Macromolecules 7, 552 (1974).

[186] J. B. Smart, R. Hogan, P. A. Scherr, M. T. Emerson, and J. P. Oliver, J. Organomet. Chem. 64, 1 (1974).

[187] P. D. Bartlett, S. J. Tauber, and W. P. Weber, J. Am. Chem. Soc. 91, 6362 (1969).

[188] P. D. Bartlett, C. V. Goebel, and W. P. Weber, J. Am. Chem. Soc. 91, 7425 (1969).

[189] R. N. Young, L. J. Fetters, J. S. Huang, and R. Krishnamoorti, Polym. Int. 33, 217 (1994).

[190] S. Bywater, Polym. Int. 38, 325 (1995).

[191] L. J. Fetters, J. S. Huang, and R. N. Young, J. Polym. Sci.: Part A: Polym. Chem. Ed. 34, 1517 (1996).

[192] S. Bywater, Macromol. Chem. Phys. 199, 1217 (1998).

[193] J. Stellbrink, J. Allgaier, L. Willner, D. Richter, T. Slawecki, and L. J. Fetters, Polymer 43, 7101 (2002).

[194] R. P. Quirk and B. Lee, Polym. Int. 27, 359 (1992).

[195] M. Morton and L. J. Fetters, J. Polym. Sci., Macromol. Rev. 2, 71 (1967).

[196] F. W. Stavely and coworkers, Ind. Eng. Chem. 48, 778 (1956).

[197] L. E. Foreman, in "Polymer Chemistry of Synthetic Elastomers," Part II, J. P. Kennedy and E. Tornqvist (Eds.), Wiley, New York, 1968, p. 491.

[198] E. W. Duck and J. M. Locke, in "The Stereo Rubbers," W. M. Saltman (Ed.), Wiley-Interscience, New York, 1977, p. 139.

[199] S. Bywater, in "Comprehensive Polymer Science," Vol. 3: "Chain Polymerization I," G. C. Eastmond, A. Ledwith, S. Russo, and P. Sigwalt (Eds.), Pergamon Press, Oxford, 1989, p. 433.

[200] M. Morton and J. R. Rupert, in "Initiation of Polymerization," F. E. Bailey, Jr. (Ed.), ACS Symposium Series 212, American Chemical Society, Washington, D. C., 1983, p. 283.

[201] D. J. Worsfold and S. Bywater, Macromolecules 11, 582 (1978).

[202] A. V. Tobolsky and C. E. Rogers, J. Polym. Sci. 40, 73 (1959).

[203] A. Rembaum, et al., J. Polym. Sci. 61, 155 (1962).

[204] S. Bywater and D. J. Worsfold, Can J. Chem. 45, 1821 (1967).

[205] M. Morton and L. J. Fetters, J. Polym. Sci., Part A 2, 3311 (1964).

[206] M. Morton, L. J. Fetters, R. A. Pett, and J. F. Meier, Macromolecules 3, 327 (1970).

[207] E. R. Santee, Jr., L. O. Malotky, and M. Morton, Rubber Chem. Tech. 46, 1156 (1973).

[208] T. Hogen-Esch, Adv. Phys. Org. Chem. 15, 153 (1977).

[209] A. Halasa, Rubber Chem. Tech. 54, 627 (1981).

[210] N. Nagata, T. Kobatake, H. Watanabe, A. Ueda, and A. Yoshioka, Rubber Chem. Tech. 60, 837 (1987).

[211] R. Luxton, Rubber Chem. Tech. 54, 596 (1981).

[212] G. Holden and D. R. Hansen, in "Thermoplastic Elastomers" 3rd ed., G. Holden, H. R. Kricheldorf, and R. P. Quirk (Eds.), Hanser Publishers, Munich, 2004, p. 45.

[213] R. Ohlinger and F. Bandermann, Makromol. Chem. 181, 1935 (1980).

[214] M. Morton and L. K. Huang, unpublished data; L. K. Huang, Ph. D. Dissertation, The University of Akron, Akron, Ohio, 1979.

[215] F. R. Ells, Ph. D. Dissertation, The University of Akron, Akron, Ohio, 1963.

[216] Y. L. Spirin, A. A. Arest-Yakubovich, D. K. Polyakov, A. R. Gantmakher, and S. S. Medvedev, J. Polym. Sci. 58, 1161 (1962).

[217] D. J. T. Hill, J. H. O'Donnell, P. W. O'Sullivan, J. E. McGrath, I. C. Wang, and T. C. Ward, Polym. Bull. 9, 292 (1983).

[218] T. A. Antkowiak, A. E. Oberster, A. F. Halasa, and D. P. Tate, J. Polym. Sci., Part A-1 10, 1319 (1972).

[219] C. F. Wofford and H. L. Hsieh, J. Polym. Sci., Part A-1 7, 461 (1969).

[220] M. Fontanille, in "Comprehensive Polymer Science," Vol. 3: "Chain Polymerization I," G. C. Eastmond, A. Ledwith, S. Russo, and P. Sigwalt (Eds.), Pergamon Press, Oxford, 1989, p. 376.

[221] F. Bandermann, H. D. Speikamp, and L. Weigel, Makromol. Chem. 186, 2017 (1985).

[222] L. H. Tung and G. Y. Lo, in "Advances in Elastomers and Rubber Elasticity," J. Lal and J. E. Mark (Eds.), Plenum Press, New York, 1986, p. 129.

[223] T. E. Long, A. D. Broske, D. J. Bradley, and J. E. McGrath, J. Polym. Sci., Polym. Chem. Ed. 27, 4001 (1989).

[224] R. P. Quirk and J. J. Ma, Polym. Inter. 24, 197 (1991).

[225] R. P. Quirk, T. Yoo, Y. Lee, J. Kim, and B. Lee, Adv. Polym. Sci. 153, 67 (2000).

[226] De, Sadhan K. Rubber Technologist's Handbook, Volume 1 (1st ed.). Smithers Rapra Press. p. 287 (1996).

[227] F. Ciardelli, in "Comprehensive Polymer Science," First Supplement, S. L. Aggarwal and S. Russo (Eds.), Pergamon Press, Oxford, 1992, p. 67.

[228] J. Boor, Jr., "Ziegler-Natta Catalysts and Polymerizations," Academic Press, New York, 1979.

[229] K. Ziegler, E. Holzkamp, H. Breil, and H. Martin, Angew. Chem. 67, 541 (1955).

[230] G. Natta, J. Polym. Sci. 16, 143 (1955).

[231] L. Porri and A. Giarrusso, in "Comprehensive Polymer Science," Vol. 4: "Chain Polymerization II," G. C. Eastmond, A. Ledwith, S. Russo, and P. Sigwalt (Eds.), Pergamon Press, Oxford, 1989, p. 53.

[232] W. Cooper, in "The Stereo Rubbers," W. M. Saltman (Ed.), Wiley-Interscience, New York, 1977, p. 21.

[233] Ph. Teyssie, P. Hadjiandreou, M. Julemont, and R. Warin, in "Transition Metal Catalyzed Polymerizations," R. P. Quirk (Ed.), Cambridge University Press, Cambridge, UK, 1988, p. 639.

[234] P. J. T. Tait and I. G. Berry, in "Comprehensive Polymer Science," Vol. 4: "Chain Polymerization II," G. C. Eastmond, A. Ledwith, S. Russo, and P. Sigwalt (Eds.), Pergamon Press, Oxford, 1989, p. 575.

[235] F. P. Baldwin and G. Ver Strate, Rubber Chem. Tech. 42, 709 (1972).

[236] G. Ver Strate, in "Encyclopedia of Polymer Science and Engineering," J. I. Kroschwitz (Ed.), Wiley-Interscience, New York, Vol. 6, 1986, p. 522.

[237] S. C. Davis, W. Von Hellens, H. A. Zahalka, and K. P. Richter, in "Polymeric Materials Encyclopedia," J. C. Salamone (Ed.), CRC Press, Boca Raton, 1996, p. 2264.

[238] J. W. M. Noordermeer, in "Encyclopedia of Polymer Science and Technology," J. I. Kroschwitz (Ed.), Wiley-Interscience, New York, Vol. 6, 2003, p. 178.

[239] E. J. Vandenberg, in "Encyclopedia of Polymer Science and Engineering," J. I. Kroschwitz (Ed.), Wiley-Interscience, New York, Vol. 4, 1986, p. 174.

Chapter 2
Polymer Elasticity: Basic Concepts and Behavior

Following its introduction to Europe from the New World in the late 15th century, natural rubber (polyisoprene) was regarded mostly as a fascinating curiosity. Its most useful application was its ability to erase pencil marks on paper by rubbing, hence its name. One of its most peculiar properties is a slight (but detectable) increase in temperature that occurs when a sample of rubber is stretched. When the sample is allowed to quickly retract, an equal amount of cooling is observed. This phenomenon caught the attention of the English physicist John Gough. In 1805 he published some qualitative observations on this characteristic and also described the dependence of the tensile stress on temperature. By the mid nineteenth century, the theory of thermodynamics was being developed and within this framework, the English mathematician and physicist Lord Kelvin showed that the change in mechanical energy required to stretch a rubber sample should be proportional to the increase in temperature. Later, this would be associated with a change in entropy. The connection to thermodynamics was firmly established in 1859 when the English physicist James Joule published the first careful measurements of the temperature increase accompanying extension, confirming the theoretical predictions of Lord Kelvin. It was not until 1838 that the American inventor Charles Goodyear found that its properties could be immensely improved by adding a few percent sulphur. The short sulfur chains produced chemical cross-links. Before it is cross-linked, the liquid natural rubber consists of very long linear chains, containing thousands of isoprene backbone units, connected head-to-tail. Every chain follows a random path through the liquid and is in contact with thousands of other nearby chains. When heated, a cross-linker molecule (such as sulfur or dicumyl peroxide) can create a chemical bond (a network node) between two adjacent chains in the liquid, resulting in a three dimensional network. All of the original separate linear chains are connected together at multiple points to form a single giant molecule. The sections between two cross-links on the same chain are called network chains and can contain up to several hundred isoprene units. In natural rubber, each cross-link produces a network node with four chains emanating from it. The network is the sine qua non of elastomers. Because of the enormous economic and technological importance of rubber, determining how a molecular network responds to mechanical strains has been of enduring interest to scientists and engineers. To understand the elastic properties of rubber, theoretically, it is necessary to know both the physical mechanisms that occur at the molecular level and how the random-walk nature of the chain produces the network. The physical mechanisms that occur within short sections of the polymer chains produce the elastic forces and the network morphology determines how these forces combine to produce the macroscopic stress that we observe when a rubber sample is deformed, e. g. subjected to tensile strain.

2.1 Introduction

The single most important property of elastomers—that from which their name derives—is their ability to undergo large elastic deformations, that is, to stretch and return to their original shape in a reversible way. Theories to account for this characteristic high elasticity have passed through three distinct phases: the early development of a molecular model relating experimental observations to the known molecular features of rubbery polymers; then generalization of this approach by means of symmetry considerations taken from continuum mechanics which are independent of the molecular structure; and now a critical reassessment of the basic premises on which these two quantitative theories are founded. In this chapter, the theoretical treatment is briefly outlined and shown to account quite successfully for the observed elastic behavior of rubbery materials. The special case of small elastic deformations is then discussed in some detail because of its technical importance. Finally, attention is drawn to some aspects of rubber elasticity which are still little understood.

2.2 Elasticity of a Single Molecule

The essential requirement for a substance to be rubbery is that it consists of long flexible chainlike molecules. The molecules themselves must therefore have a backbone of many noncolinear single valence bonds, about which rapid rotation is possible as a result of thermal agitation. Some representative molecular subunits of rubbery polymers are shown in Fig. 2.1; thousands of these units linked together into a chain constitute a typical molecule of the elastomers listed in Fig. 2.2. Such molecules change their shape readily and continuously at normal temperatures by Brownian motion. They take up random conformations in a stress-free state but assume somewhat oriented conformations if tensile forces are applied at their ends (Fig. 2.2). One of the first questions to consider, then, is the relationship between the applied tension f and the mean chain end separation r, averaged over time or over a large number of chains at one instant in time.

Although the terms configuration and conformation are sometimes used interchangeably, the former has acquired a special meaning in organic stereochemistry and designates specific steric structures. Conformation is used here to denote a configuration of the molecule which is arrived at by rotation of single-valence bonds in the polymer backbone.

Chains in isolation take up a wide variety of conformations, governed by three factors: ①the statistics of random processes; ②a preference for certain sequences of bond

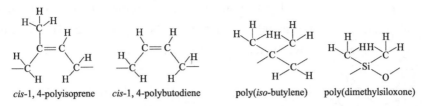

Fig. 2.1　Repeat units for some common elastomer molecules.

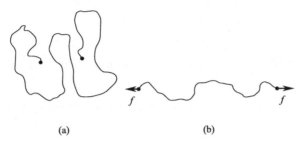

Fig. 2.2　(a) Random chain and (b) oriented chain.

arrangements because of steric and energetic restraints within the molecule; ③ the exclusion of some hypothetical conformations which would require parts of the chain to occupy the same volume in space.

In addition, cooperative conformations are preferred for space-filling reasons in concentrated solutions or in the bulk state.

Flory has argued that the occupied-volume exclusion (repulsion) for an isolated chain is exactly balanced in the bulk state by the external (repulsive) environment of similar chains, and that the exclusion factor can therefore be ignored in the solid state. Direct observation of single-chain dimensions in the bulk state by inelastic neutron scattering gives values fully consistent with unperturbed chain dimensions obtained for dilute solutions, although intramolecular effects may distort the local randomness of chain conformation in theta solvents. These are (poor) solvents in which repulsion between different segments of the polymer molecule is balanced by repulsion between polymer segments and solvent molecules. These are (poor) solvents in which repulsion between different segments of the polymer molecule is balanced by repulsion between polymer segments and solvent molecules.

Flory has again given compelling reasons for concluding that the chain end-to-end distance r in the bulk state will be distributed in accordance with Gaussian statistics for sufficiently long chains, even if the chains are relatively stiff and inflexible over short lengths. With this restriction to long chains it follows that the tension — displacement relation becomes a simple linear one,

$$f = Ar \tag{2-1}$$

where f is the tensile force, r is the average distance between the ends of the chain, and A is inversely related to the mean square end-to-end distance r_0^2 for unstressed chains,

$$A = 3kT/r_0^2 \tag{2-2}$$

where k is Boltzmann's constant and T is the absolute temperature.

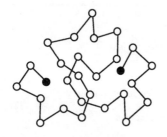

Fig. 2.3 Model chain of freely jointed links.

If the real molecule is replaced by a hypothetical chain consisting of a large number n of rigid, freely jointed links, each of length l (Fig. 2.3), then

$$r_0^2 = nl^2 \tag{2-3}$$

is independent of temperature because completely random link arrangements are assumed. The tension f in Eq. (2-1) then arises solely from an entropic mechanism, i.e., from the tendency of the chain to adopt conformations of maximum randomness, and not from any energetic preference for one conformation over another. The tension f is then directly proportional to the absolute temperature T.

For real chains, consisting of a large number n of primary valence bonds along the chain backbone, each of length l,

$$r_0^2 = C_\infty nl^2 \tag{2-4}$$

where the coefficient C_∞ represents the degree to which this real molecule departs from the freely jointed model. C_∞ is found to vary from 4 to 10, depending on the chemical structure of the molecule and also on temperature, because the energetic barriers to random bond arrangements are more easily overcome may thus be regarded as the effective bond at higher temperatures. $C_\infty^{1/2} l$ length of the real chain, a measure of the stiffness of the molecule.

Eq. (2-1) is reasonably accurate only for relatively short distances r, less than about one-third of the fully stretched chain length. Unfortunately, no good treatment exists for the tension in real chains at larger end separations. We must therefore revert to the model chain of freely jointed links, for which

$$f = (kT/t) L^{-1} (r/nl) \tag{2-5}$$

where L^{-1} denotes the inverse Langevin function. An expansion of this relation in terms of r/nl,

$$f = (3kTr/nl^2)[1 + (3/5)(r/nl)^2 + (99/175)(r/nl)^4 + (513/817)(r/nl)^6 + \cdots] \tag{2-6}$$

gives a useful indication of where significant departures from Eq. (2-1) may be expected.

Eq. (2-5) gives a steeply rising relation between tension and chain end separation when the chain becomes nearly taut (Fig. 2.4), in contrast to the Gaussian solution, Eq. (2-1), which becomes inappropriate for $r > (1/3) nl$. Rubber shows a similar steeply rising relation between tensile stress and elongation at high elongations. Indeed, experimental stress-strain relations closely resemble those calculated using Eq. (2-5) in place of Eq. (2-1) in the network theory of rubber elasticity (outlined in the following section). The deformation at which a small but significant departure is first found between the observed stress and that predicted by small-strain theory, using Eq. (2-1), yields a value for the effective length l of a freely jointed link for the real molecular chain. This provides a direct experimental measure of molecular stiffness. The values obtained are relatively large, of the order of 5~15 main-chain bonds for the only polymer which has been examined by this method so far, *cis*-1,4-polyisoprene.

Eq. (2-5) has also been used to estimate the force at which a rubber molecule will become detached from a particle of a reinforcing filler, for example, carbon black, when a filled rubber is deformed. In this way, a general semiquantitative treatment has been achieved for stress-induced softening (Mullins effect) of filled rubbers (shown in Fig. 2.5).

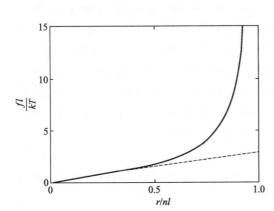

Fig. 2.4 Tension-displacement relation for a freely jointed chain [Eq. (2-5)], ---, Gaussian solution [Eq. (2-1)].

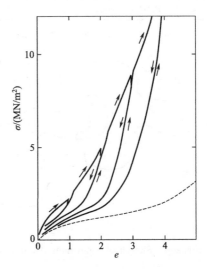

Fig. 2.5 Stress-induced softening of a carbon black-filled vulcanizate of a copolymer of styrene and butadiene (25 : 75); ---, stress-strain curve of a corresponding unfilled vulcanizate.

2.3 Elasticity of a Three-Dimensional Network of Polymer Molecules

Some type of permanent structure is necessary to form a coherent solid and prevent liquidlike flow of elastomer molecules. This requirement is met by incorporating a small number of intermolecular chemical bonds (crosslinks) to make a loose three-dimensional molecular network. Such crosslinks are generally assumed to form in the most probable positions, so that the long sections of molecules between them have the same spectrum of end-to-end lengths as a similar set of uncrosslinked molecules would have. Under Brownian motion each molecular section takes up a wide variety of conformations, as before, but now subject to the condition that its ends lie at the crosslink sites. The elastic properties of such a molecular network are treated later. We consider first another type of interaction between molecules.

High-molecular-weight polymers form entanglements by molecular intertwining, with a spacing (in the bulk state) characteristic of the particular molecular structure. Some representative values of the molecular weight M between entanglement sites are given in Table 2.1. Thus, a high-molecular weight polymeric melt will show transient rubberlike behavior even in the absence of any permanent intermolecular bonds.

Table 2.1 Representative values of the average molecular weight M_e between entanglements for polymeric Melts[a]

Polymer	M_e	Polymer	M_e
Polyethylene	4000	Poly(dimethylsiloxane)	29000
Poly(isobutylene)	17000	cis-1,4-Polyisoprene	14000
cis-1,4-Polybutadiene	7000	Polystyrene	35000

[a] Obtained from flow viscosity measurements.

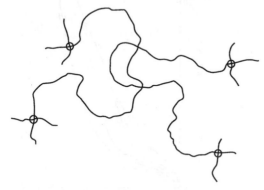

Fig. 2.6 Sketch of a permanent entanglement.

In a crosslinked rubber, many of these entanglements are permanently locked in (Fig. 2.6), the more so the higher the degree of crosslinking. If they are regarded as fully equivalent to crosslinks, the effective number N of network chains per unit volume may be taken to be the sum of two terms N_e and N_c, arising from entanglements and chemical crosslinks, respectively, Where

$$N_e = \rho N_A / M_e, \quad N_c = \rho N_A / M_c$$

and ρ is the density of the polymer, N_A is Avogadro's number, and M_e and M_c denote the average molecular weights between entanglements and between crosslinks, respectively. The efficiency of entanglements in constraining the participating chains is, however, somewhat uncertain, particularly when the number of chemical crosslinks is relatively small. Moreover, the force-extension relation for an entangled chain will differ from that for crosslinked chain, being stiffer initially and nonlinear in form. The effective number N of molecular chains which lie between fixed points (i.e., crosslinks or equivalent sites of molecular entanglement) is therefore a somewhat ill-defined quantity, even when the chemical structure of the network is completely specified.

It is convenient to express the elastic behavior of the network in terms of the strain energy density W per unit unstrained volume. The strain energy W for a single chain is obtained from Eq. (2-1) as

$$W = Ar^2/2 \tag{2-7}$$

For a random network of N such chains under a general deformation characterized by extension ratios λ_1, λ_2, λ_3 (deformed dimension/undeformed dimension) in the three principal directions (Fig. 2.7), W is given by

$$W = NAr_f^2(\lambda_1^2 + \lambda_2^2 + \lambda_3^2 - 3)/6 \tag{2-8}$$

where r_f^2 denotes the mean square end-to-end distance between chain ends (crosslink points

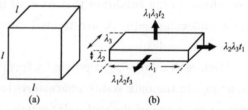

Fig. 2.7 (a) Undeformed and (b) deformed states.

or equivalent junctions) in the undeformed state. The close similarity of Eq. (2-7) and Eq. (2-8) is evident, especially since $r^2 = (r_f^2/3)(\lambda_1^2 + \lambda_2^2 + \lambda_3^2)$.

For random crosslinking r_f^2 may be assumed to be equal to r_0^2, the corresponding mean square end-to-end distance for unconnected chains of the same molecular length. Because A is inversely proportional to r_0^2 [Eq. (2-2)], the only molecular parameter which then remains in Eq. (2-8) is the number N of elastically effective chains per unit volume. Thus, the elastic behavior of a molecular network under moderate deformations is predicted to depend only on the number of molecular chains and not on their flexibility, provided that they are long enough to obey Gaussian statistics.

Although r_f^2 and r_0^2 are generally assumed to be equal at the temperature of network formation, they may well differ at other temperatures because of the temperature dependence of r_0^2 for real chains [Eq. (2-4)]. Indeed, the temperature dependence of elastic stresses in rubbery networks has been widely employed to study the temperature dependence of r_0^2.

Another way in which r_f^2 and r_0^2 may differ is when the network is altered after formation. For example, when the network imbibes a swelling liquid, r_2 for the swollen network will be increased by a factor λ_s^2 in comparison to its original value, where λ_s is the linear swelling ratio. At the same time the number of chains per unit volume will be decreased by a factor λ_s^{-3}. Thus, the strain energy density under a given deformation will be smaller for a swollen network by a factor λ_s^{-1}.

From the general relation for strain energy, Eq. (2-8), the elastic stresses required to maintain any given deformation can be obtained by means of virtual work considerations (Fig. 2.7),

$$\lambda_2 \lambda_3 t_1 = \partial W / \partial \lambda_1$$

with similar relations for t_2 and t_3. Because of the practical incompressibility of rubbery materials in comparison to their easy deformation in other ways, the original volume is approximately conserved under deformation. The extension ratios then obey the simple relationship

$$\lambda_1 \lambda_2 \lambda_3 = 1 \tag{2-9}$$

As a result, the stress-strain relations becomes

$$t_1 = \lambda_1 (\partial W / \partial \lambda_1) - p$$

where p denotes a possible hydrostatic pressure (which has no effect on an incompressible solid). Thus, only stress differences can be written explicitly

$$t_1 - t_2 = (NAr^2/3)(\lambda_1^2 - \lambda_2^2) \tag{2-10}$$

For a simple extension, say in the 1-direction, we set $\lambda_1 = \lambda$, and $\lambda_2 = \lambda_3 = \lambda^{-1/2}$ [from Eq. (2-9)], and $t_2 = t_3 = 0$. Hence,

$$t = t_1 = (NAr^2/3)(\lambda^2 - \lambda^{-1}) \tag{2-11}$$

It is customary to express this result in terms of the tensile force f acting on a test piece of cross-sectional area A_0 in the unstrained state, where

$$f/A_0 = t/\lambda$$

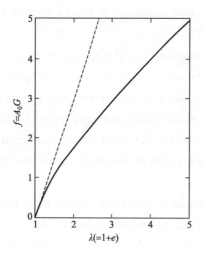

Fig. 2.8 Force-extension relation for simple extension; ---, Linear relation obtaining at infinitesimal strains.

The corresponding relation is shown in Fig. 2.8. It illustrates a general feature of the elastic behavior of rubbery solids: although the constituent chains obey a linear force-extension relationship [Eq. (2-1)], the network does not. This feature arises from the geometry of deformation of randomly oriented chains. Indeed, the degree of nonlinearity depends on the type of deformation imposed. In simple shear, the relationship is predicted to be a linear one with a slope (shear modulus G) given by

$$t_{12} = G\gamma, \quad G = NAr^2/3 \qquad (2\text{-}12)$$

where γ is the amount of shear, e.g., dx/dy.

Because rubbery materials are virtually incompressible in bulk, the value of Poisson's ratio is close to 0.5. Young's modulus E is therefore given by $3G$ to good approximation; however, the predicted relation between stress and tensile strain (extension), e ($e = \lambda^{-1}$), is linear only for quite small extensions (Fig. 2.8), so that Young's modulus is applicable only for extensions or compressions of a few percent.

All of the stress relations given above are derived from Eq. (2-8). They are therefore valid only for moderate deformations of the network, i.e., for deformations sufficiently small for the chain tensions to be linearly related to their end-to-end distances r [Eq. (2-1)]. Unfortunately, no correspondingly simple expression can be formulated for W using Eq. (2-5), the relationship for large strains of the constituent chains, in which the molecular stiffness parameter reappears. Instead, a variety of series approximations must be used, as in Eq. (2-6), to give close approximations to the behavior of rubber networks under large strains.

2.4 Comparison with Experiment

Although the treatment of rubber elasticity given in the preceding section is generally rather successful, certain discrepancies are found to occur. The first consists of observed stresses higher than predicted, e.g., by Eq. (2-11), and is often expressed by an additional contribution referred to as the C_2 term. This contribution is relatively large at small strains (although it is always the smaller part of the observed stress) and decreases in importance as the strain increases.

It also decreases as the network is dilated by swelling with an inert liquid becoming zero at a swelling ratio of about 5. Thus, the "stress" C_2 appears to reflect a non-Gaussian characteristic of network chains, which is important only at small values of the chain end-to-end distance r. Indeed, Thomas has shown that the magnitude of the C_2 stress and its complex dependence on type and degree of strain, and on degree of swelling, can all be accurately described by a simple additional term in the relation for the strain energy W for a single network chain, Eq. (2-7), which becomes

$$2W = Ar^2 + Br^{-2} \tag{2-13}$$

The second term clearly becomes insignificant at large values of r.

Further evidence bearing on the physical nature of the discrepancy is provided by two other observations: C_2 does not appear to be strongly dependent on temperature and therefore does not appear to be associated with the energetics of chain conformations; and it is closely correlated with the tendency of the polymer chains to form molecular entanglements. For example, those polymers that have a high density of entanglements in the bulk state (Table 2.1) yield rubbery networks with a relatively high C_2 stress component.

Finally, there is no evidence that isolated chains in theta solvents fail to conform to Gaussian statistics, so that the C_2 discrepancy appears to arise only when the molecular chains are tied into a network.

These varied aspects of the C_2 stress suggest that it is associated with entangled chains in networks (Fig. 2.6) and specifically that it arises from restrictions on the conformations available to entangled chains, different from those operating at crosslink sites. Prager and Frisch have pointed out that chains involved in model entanglements are governed by different statistics; their conclusions are quite consistent with what is known of the C_2 stress.

A second discrepancy between theory and experiment is found when the Gaussian part of the measured stresses is compared with the theoretical result for an ideal network. Numerical differences of up to 50% are obtained between the density of effective chains calculated from the observed stresses and that calculated from the chemistry of crosslinking. This discrepancy may be due to an error in the theoretical treatment as given here. James and Guth arrived at stresses only half as large as those given in Eq. (2-10), from a somewhat different theoretical standpoint.

A third and major discrepancy, already referred to, is found at large deformations when the network chains fail to obey Gaussian statistics, even approximately. Considerable success is achieved in this case by using Eq. (2-5) in place of Eq. (2-1) for chain tensions in the network.

Not with standing these discrepancies, the simple treatment of rubber elasticity outlined in this chapter has proved to be remarkably successful in accounting for the elastic properties of rubbers under moderate strains, up to about 300% of the unstrained length (depending on the length and flexibility, and hence the extensibility, of the constituent chains). It predicts the general form of the stress-strain relationships correctly under a variety of strains, the approximate numerical magnitudes of the stresses for various chemical structures, and the effects of temperature and of swelling the rubber with an inert mobile liquid on the elastic behavior. It also predicts novel second-order stresses, discussed later, which have no counterpart in classical elasticity theory. In summary, it constitutes a major advance in our understanding of the properties of polymeric materials.

2.5 Continuum Theory of Rubber Elasticity

A general treatment of the stress-strain relations of rubberlike solids was developed by Rivlin, assuming only that the material is isotropic in elastic behavior in the unstrained state and incompressible in bulk. It is quite surprising to note what far-reaching conclusions follow from these elementary propositions, which make no reference to molecular structure. Symmetry considerations suggest that appropriate measures of strain are given by three strain invariants, defined as

$$J_1 = \lambda_1^2 + \lambda_2^2 + \lambda_3^2 - 3$$
$$J_2 = \lambda_1^2 \lambda_2^2 + \lambda_2^2 \lambda_3^2 + \lambda_3^2 \lambda_1^2 - 3$$
$$J_3 = \lambda_1^2 \lambda_2^2 \lambda_3^2 - 1$$

where λ_1, λ_2, λ_3 are the principal stretch ratios (the ratios of stretched to unstretched lengths, Fig. 2-7). Moreover, for an incompressible material, J is identically zero, and hence only two independent measures of strain, J_1 and J_2, remain. It follows that the strain energy density W is a function of these two variables only:

$$W = f(J_1, J_2) \tag{2-14}$$

Furthermore, to yield linear stress-strain relations at small strains, W must be initially of second order in the strains e_1, e_2, e_3. Therefore, the simplest possible form for the strain energy function is:

$$W = C_1 J_1 + C_2 J_2 \tag{2-15}$$

where C_1 and C_2 are elastic coefficients with a sum $2(C_1 + C_2)$ equal to the small-strain shear modulus G. Eq. (2-15) was originally proposed by Mooney and is often called the Mooney-Rivlin equation. It is noteworthy that the first term corresponds to the relation obtained from the molecular theory of rubber elasticity, Eq. (2-8), if the coefficient C_1 is identified with $Nar^2/6 = \frac{1}{2}NkT\,(r/r_0)^2$.

On expanding Eq. (2-15) as a power series in strains e, where $e = \lambda^{-1}$, it is found to include all terms in e^2 and e^3. Thus it necessarily gives good agreement with experiment at small strains, say for values of e up to 10% to 20%, where higher powers of e are negligibly small. However considerable confusion has arisen from its application at larger strains, for values of e of 100% or more, when it no longer holds. It is rather unfortunate that experimental stress-strain relations in simple extension appear to be in accord with Eq. (2-15) up to moderately large strains. This fortuitous agreement arises because the particular strain energy function obeyed by rubber, discussed later, depends on strain in such a way that the two stress-strain relations in tension are similar in form. Relations for other types of strain are quite different, even at modest strains.

2.5.1 Stress-Strain Relation

Stresses can be obtained from the derivatives of the strain energy function W:

$$t_1 = \lambda_1 \frac{\partial W}{\partial \lambda_1} - p \qquad (2\text{-}16)$$

Rewriting Eq. (2-16) in terms of the generic derivatives $\partial W/\partial J_1$ and $\partial W/\partial J_2$ yields

$$t_1 = 2\left[\lambda_1^2 \frac{\partial W}{\partial J_1} - \frac{1}{\lambda_1^2} \frac{\partial W}{\partial J_2}\right] - p \qquad (2\text{-}17)$$

The functions $\partial W/\partial J_1$ and $\partial W/\partial J_2$ are denoted W_1 and W_2 hereafter. Experimental measurements indicate that W is approximately constant. However, the second term is far from constant even at moderate strains. Good agreement is obtained when it is expressed as a logarithmic function of J_2:

$$W = C_1 J_1 + C_2' \ln(1 + J_2/3) \qquad (2\text{-}18)$$

where C_2' is a constant. This form of the second term is in reasonably good numerical agreement with the predictions of Thomas's additional term in the strain energy function for a single chain, Eq. (2-13), and simpler in form.

Values of C_1 and C_2' are similar in magnitude, 0.25MPa to 0.5MPa, for typical soft rubber vulcanizates. However, whereas C_1 is approximately proportional to the number of network strands per unit volume, C_2' appears to be rather constant, independent of the degree of crosslinking, and thus it is relatively more important for lightly crosslinked materials. As mentioned earlier, it appears to reflect physical restraints on molecular strands like those represented in the "tube" model of restricted configurations in the condensed state—restraints that diminish in importance as the deformation increases or the strands become more widely separated.

2.5.2 Strain-Hardening at Large Strains

Rubber becomes harder to deform at large strains, probably because the long flexible molecular strands that comprise the material cannot be stretched indefinitely. The strain energy functions considered up to now do not possess this feature and therefore fail to describe behavior at large strains. Strain-hardening can be introduced by a simple modification to the first term in Eq. (2-18), incorporating a maximum possible value for the strain measure J_1, denoted J_m:

$$W = -C_1 J_m \ln\left(1 - \frac{J_1}{J_m}\right) + C_2' \ln(1 + J_2/3) \qquad (2\text{-}19)$$

Eq. (2-19) reduces to Eq. (2-18) when the strains are relatively small, i.e., the ratio J_1/J_m is small. Thus Eq. (2-19) is probably the simplest possible strain energy function that accounts for the elastic behavior to good approximation over the entire range of strains. It requires three fitting parameters, two of which are related to the small strain shear modulus G:

$$G = 2(C_1 + C_2'/3) \qquad (2\text{-}20)$$

The molecular theory of rubberlike elasticity predicts that the first coefficient, C_1, is proportional to the number N of molecular strands that make up the three-dimensional network. The second coefficient, C_2', appears to reflect physical restraints on molecular strands like those represented in the "tube" model and is in principle amenable to calculation. The third parameter, J_m, is not really independent. When the strands are long and flexible, it will be given approximately by $3\lambda_m^2$, where λ_m is the maximum stretch ratio of an average strand. But λ_m is inversely proportional to N for strands that are randomly arranged in the unstretched state. J_m is therefore expected to be inversely proportional to C_1. Thus the entire range of elastic behavior arises from only two fundamental molecular parameters.

Considerable success has also been achieved in fitting the observed elastic behavior of rubbers by strain energy functions that are formulated directly in terms of the extension ratios λ_1, λ_2, λ_3 instead of in terms of the strain invariants I_1, I_2, I_3. Although experimental results can be described economically and accurately in this way, the functions employed are empirical and the numerical parameters used as fitting constants do not appear to have any direct physical significance in terms of the molecular structure of the material.

On the other hand, the molecular elasticity theory, supplemented by a simple non-Gaussian term whose molecular origin is in principle within reach, seems able to account for the observed behavior at small and moderate strains with comparable success.

At moderate strains, the value of J_2 is often large enough for terms to be neglected. Some stress-strain relations are now derived involving W_2 using this approximation to illustrate how such calculations are carried out and to deduce under what conditions the deformations become unstable. Instabilities are interesting from a theoretical point of view because they occur suddenly, at a well-defined deformation, and they are often unexpected on the basis of classical elasticity theory. Moreover, a comparison of the observed onset of instability with the predictions of various strain energy functions W provides, at least in principle, a critical test for the validity of a proposed form for W. From a practical standpoint, unstable states are quite undesirable because the deformation becomes highly nonuniform, leading to premature failure.

2.5.3 Inflation of a Thin-Walled Tube

Inflation of a tube is described by extension ratios of λ_1 in the circumferential direction and λ_2 in the axial direction, with the wall thickness h becoming $h/\lambda_1\lambda_2$ because the rubber volume remains constant. The inflation pressure p gives rise to stresses in the circumferential and axial directions:

$$t_1 = \lambda_1^2 \lambda_2 rp/h$$
$$t_2 = \lambda_1^2 \lambda_2 rp/2h \quad (2\text{-}21)$$

where r is the tube radius in the unstrained state.

From Eq. (2-21), on putting the stress $t_3 = 0$, the undefined pressure p is obtained as:

$$p = -2W_1/\lambda_1^2\lambda_2^2 \quad (2\text{-}22)$$

In a thin-walled tube of large radius the inflating pressure p is much smaller than the stresses t_1 and t_2 that it generates, and thus p can be neglected in comparison with the stress t_3 in determining p. Inserting this result for p in Eq. (2-21) and Eq. (2-22) yields a relation between the extension ration λ_2 and the expansion ratio $\nu(\nu = \lambda_1^2 \lambda_2)$ of the internal volume of the tube:

$$2\lambda_2^3 = (\nu^2 + 1)/2\nu^2 \tag{2-23}$$

The relation between inflating pressure p and internal volume of the tube is then obtained as:

$$pr/hC_1 = 2(\nu^2 - 1)[2\nu/(\nu^2 + 1)]^{1/3}/[\nu^2(1 - J_1/J_m)] \tag{2-24}$$

This relation is plotted in Fig. 2.9 for various values of the limiting strain measure. The inflating pressure is seen to pass through a maximum at a volume J_m expansion ratio between about 58% and 66%, depending on the value assumed for J_m. This feature suggests that larger expansions will be unstable. Indeed, m thin-walled tubes undergo a strikingly non-uniform deformation at a critical inflation pressure, shown schematically in Fig. 2.10. One portion of the tube becomes highly-distended as a bubble or aneurysm while the rest is lightly inflated. The two stable deformations that can coexist at the same inflation pressure after the critical state is reached are shown schematically by the horizontal broken line in Fig. 2.9. However, when J_m is infinitely large, the aneurysm m is unbounded and failure would then occur immediately on reaching the critical pressure.

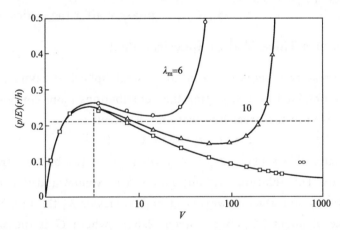

Fig. 2.9 Pressure-volume relations for a thin-walled tube from Eq. (2-24) using various values for the maximum possible extension ratio λ_m. The vertical broken line denotes the onset of instability. ($E = 6C_1$.)

Fig. 2.10 Sketch of an aneurysm in an inflated tube.

2.5.4 Inflation of a Thin-Walled Spherical Balloon

In this case, if the balloon radius expands by a factor λ, equibiaxial extensions of ratio λ will be set up in the balloon, with a shrinkage ratio λ^{-2} of the wall thickness to maintain the rubber volume constant. The circumferential stresses t_1 and t_2 are equal and given by:

$$t_1 = t_2 = 2W_1(\lambda^2 - \lambda^{-4})/(1 - J_1/J_m) \tag{2-25}$$

from Eq. (2-17), when the stress $t_3 = 0$ and $W_2 = 0$. The inflation pressure p is then given by:

$$pr/hC_1 = 4(\lambda^{-1} - \lambda^{-7})/(1 - J_1/J_m) \tag{2-26}$$

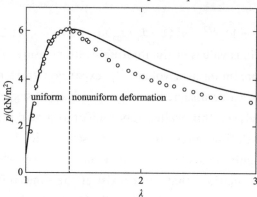

Fig. 2.11 Inflation of a thin-walled spherical rubber balloon [Solid curve: Eq. (2-26) with $J_m = \infty$].

where r and h are the unstrained radius and wall thickness of the balloon. In this case the potential instability occurs even earlier, at a radial expansion ratio between 38% and 50% depending on the value chosen for J_m. In practice, the deformation becomes quite complex (Fig. 2.11). The balloon remains roughly spherical in shape but one part is lightly stretched while the remainder is highly stretched. The two states of strain resemble the two deformations that are predicted at a given pressure after the critical point is reached.

2.5.5 Inflation of a Thick-Walled Spherical Shell

The internal pressure p required to inflate a small spherical cavity in the center of a thick block can be obtained by integrating the contributions from concentric shells (thin-walled balloons) given in the preceding section. The result is

$$p/C_1 = 4\lambda_0^{-1} + \lambda_0^{-4} - 5 \tag{2-27}$$

for an infinitely extensible rubber ($J_m = \infty$), where λ_0 is the biaxial stretch ratio at the surface of the cavity. This relation does not exhibit a maximum and thus does not indicate that the deformation is unstable. However, at high values of λ_0 the pressure p asymptotes to a constant value of about $5C_1$, i.e., about $2.5G$, where G is the small strain shear modulus. For typical rubbery materials where G is about 0.5MPa, the maximum pressure is thus about 1.2MPa, or about 12 bar. Any small cavity will expand greatly at this rather modest inflation pressure. Internal fracture is therefore likely to occur in soft rubbery solids at inflation pressures or equivalently, triaxial tensions, of this amount. In practice, all rubbery solids are found to develop internal fractures when supersaturated with gases or liquids at pressures or triaxial tensions about equal to $2.5G$.

Note that the initial radius of the spherical cavity does not appear in Eq. (2-27). Thus, cavities of all sizes are predicted to inflate equally. However, we have neglected surface energy contributions that will tend to stabilize small cavities. When they are taken into account it appears that only cavities having radii greater than about 100nm will expand dramatically at the low pressures predicted by Eq. (2-27). Internal fractures suggest that vulcanized rubber must contain many precursor cavities of this effective size or larger.

2.5.6 Surface Instability of Compressed or Bent Blocks

Biot showed that the surface of an elastic half space will become unstable at critical values of strain ratios λ_1, λ_3 set up in two perpendicular directions in the surface. The critical condition is

$$\lambda_1^2 \lambda_3 = 0.2956 \tag{2-28}$$

When the block is subjected to unidirectional compression parallel to the surface, with free expansion permitted in the other two directions, then $\lambda_3 = \lambda_1^{-1/2}$ and Eq. (2-28) yields a critical value for λ_1 of 0.444. A large block of rubber in simple compression is therefore predicted to show a surface instability at a compression of 55.5%. It is noted that various buckling and bulging modes of deformation are generally encountered before this, depending on the slenderness of the block. If the block is subjected to equibiaxial compression parallel to the surface, then $\lambda_1 = \lambda_3$ and the critical compression becomes 33.3%.

When a thick elastic block (cuboid) is bent, the inner surface becomes compressed while the extension ratio λ_3 along the width is largely unchanged (at unity). Thus, from Eq. (2-28) an instability would be expected on the inner surface when λ_1 is 0.544, i.e., when the surface is compressed by about 46%. Experimentally, sharp folds or creases appear suddenly in the inner surface of a bent block at a critical degree of bending, see Fig. 2.12. However, the critical compression of the inner surface was considerably smaller than predicted by Biot's theory, 35% instead of 46%. It is not known why the instability occurred so much sooner than expected. Although rubber follows a more complex strain energy function than the simple form assumed here, it is unlikely that the difference would have such a large effect.

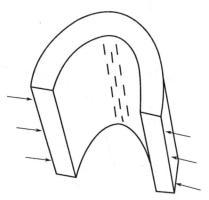

Fig. 2.12 Sketch of a bent block showing creases that appear on the inner surface where the compressive strain is greatest.

Rubber articles are often subjected to rather severe bending deformations, for example, in tires. Folds and creases in the interior may pass undetected. Nevertheless, they represent lines of high stress concentration and sites of possible failure. Folds (called Schallamach waves) also appear when soft rubber slides over a rigid counter surface. They appear to be Biot creases caused by frictional compression of the surface.

2.5.7 Resistance of a Compressed Block to Indentation

When a block is subjected to a sufficiently large equibiaxial compression in the surface plane it becomes unstable to small indentations. Green and Zerna expressed the relation between indentation force N and amount of indentation d as:

$$N/G = 8/3 R^{1/2} d^{3/2} f(\lambda) \tag{2-29}$$

where G is the shear modulus of the half space material, R is the radius of the indentor, and $f(\lambda)$ is a function of the equibiaxial compression ratio λ, given by

$$f(\lambda) = (\lambda^9 + \lambda^6 + 3\lambda^3 - 1)/\lambda^4(\lambda^3 + 1) \tag{2-30}$$

Values of indentation force N for a given small indentation, from Eq. (2-27), are plotted in Fig. 2.13 against the equibiaxial strain e parallel to the surface, where $e = \lambda - 1$. N_0 denotes the value for an initially unstrained block, when $f(\lambda) = 2$. The resistance to indentation is seen to decrease sharply as the compressive strain is increased, becoming zero at a compressive strain of 0.333, in agreement with Biot's result.

2.5.8 Torsional Instability of Stretched Rubber Rods

Another unstable state is encountered when a stretched rubber rod is subjected to large torsions. A kink suddenly appears at one point along the rod, Fig. 2.14, and more kinks form on twisting the rod further.

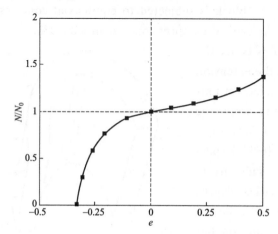

Fig. 2.13 Force N for a small indentation vs. equibiaxial strain e parallel to the surface of a half-space. N_0 denotes the force when $e = 0$.

Fig. 2.14 Sketch of a "kink" that appears on twisting a stretched rubber rod.

Minimization of the total elastic strain energy suggests that the rod will become unstable at a critical amount of torsion: part of the rod will unwind and form a tight ring while the remainder of the rod will become slightly more stretched. A simple criterion can be derived on this basis for the onset of "kinks". For a neo-Hookean material, Eq. (2-8), the condition for forming a kink becomes:

$$4(1 - 1/\lambda^3) = -(a^2\varphi^2/\lambda) + 2\pi(a^2\varphi^2/\lambda)/(\pi - a\varphi/\lambda^{1/2}) \tag{2-31}$$

where φ is the critical amount of torsion at which uniform torsion becomes unstable, in terms of the imposed extension ratio λ and the rod radius a. Measured values for rods of different radius, stretched to extensions of up to 250%, were found to be in reasonably good agreement with Eq. (2-31), indicating that the sudden formation of kinks in twisted rubber rods is, indeed, a consequence of an elastic instability.

2.6 Second-Order Stresses

Because the strain energy function for rubber is valid at large strains, and yields stress-strain relations which are nonlinear in character, the stresses depend on the square and higher powers of strain, rather than the simple proportionality expected at small strains. A striking example of this feature of large elastic deformations is afforded by the normal stresses t_{11}, t_{22}, t_{33} that are necessary to maintain a simple shear deformation of amount γ (in addition, of course, to simple shear stresses). These stresses are predicted to increase in proportion to γ^2.

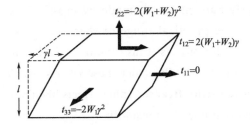

Fig. 2.15 Stresses required to maintain a simple shear deformation of amount γ (the normal stress t_{11} is set equal to zero).

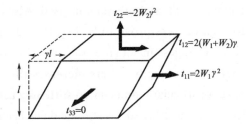

Fig. 2.16 Stresses required to maintain a simple shear deformation of amount γ (the normal stress t_{33} is set equal to zero).

They are represented schematically in Fig. 2.15 and Fig. 2.16 for two different choices of the arbitrary hydrostatic pressure p, chosen so as to give the appropriate reference (zero) stress. In Fig. 2.15, for example, the normal stress t_{11} in the shear direction is set equal to zero; this condition would arise near the front and rear surfaces of a sheared block. In Fig. 2.16, the normal stress t_{33} is set equal to zero; this condition would arise near the side surfaces of a sheared block. In each case a compressive stress t_{22} is found to be necessary to maintain the simple shear deformation. In its absence the block would tend to increase in thickness on shearing.

When the imposed deformation consists of an inhomogeneous shear, as in torsion, the normal forces generated (corresponding to the stresses t_{22} in Fig. 2.15 and Fig. 2.16) vary from point to point over the cross-section (Fig. 2.17). The

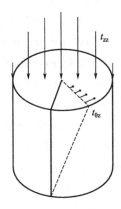

Fig. 2.17 Sketch of a cylindrical rod under torsion showing the distribution of normal stress t_{zz} (corresponding to $-t_{22}$ in Fig. 2.15 and Fig. 2.16) over the cross-section of the rod.

exact way in which they are distributed depends on the particular form of strain energy function obeyed by the rubber, i.e., on the values of W_1 and W_2 which obtain under the imposed deformation state.

2.7 Elastic Behavior Under Small Deformations

Under small deformations rubbers are linearly elastic solids. Because of high modulus of bulk compression, about 2000MPa, compared with the shear modulus G, about 0.2MPa to 5MPa, they may be regarded as relatively incompressible. The elastic behavior under small strains can thus be described by a single elastic constant G. Poisson's ratio is effectively $1/2$, and Young's modulus E is given by $3G$, to good approximation.

A wide range of values for G can be obtained by varying the composition of the elastomer, i. e., by changing the chemistry of crosslinking, oil dilution, and filler content; however, soft materials with shear modulus of less than about 0.2MPa prove to be extremely weak and are seldom used. Also, particularly hard materials made by crosslinking to high degrees prove to be brittle and inextensible. The practical range of shear modulus, from changes in degree of crosslinking and oil dilution, is thus about 0.2MPa to 1MPa. Stiffening by fillers increases the upper limit to about 5MPa, but those fillers, which have a particularly pronounced stiffening action, also give rise to stress-softening effects like those shown in Fig. 2.5, so that the modulus becomes a somewhat uncertain quantity.

It is customary to characterize the modulus, stiffness, or hardness of rubbers by measuring their elastic indentation by a rigid die of prescribed size and shape under specified loading conditions. Various nonlinear scales are employed to derive a value of hardness from such measurements. Corresponding values of shear modulus G for two common hardness scales are given in Fig. 2.18.

Many rubber products are normally subjected to fairly small deformations, rarely exceeding 25% in extension or compression or 75% in simple shear. A good approximation for the corresponding stresses can then be obtained by conventional elastic analysis assuming linear relationships.

One particularly important deformation is treated here: the compression or extension of a thin rubber block, bonded on its major surfaces to rigid plates (Fig. 2.19). A general treatment of such deformations has been reviewed.

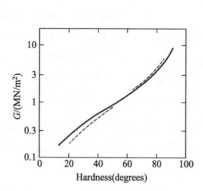

Fig. 2.18 Relations between shear modulus G and indentation hardness: —, Shore A Scale; ---, International Rubber Hardness Scale.

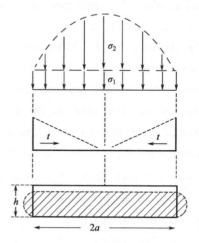

Fig. 2.19 Sketch of a bonded rubber block under a small compression. The distributions of normal stress σ and shear stress t acting at the bonded surfaces are represented by the upper portions of the diagram.

It is convenient to assume that the deformation takes place in two stages: a pure homogeneous compression or extension of amount e, requiring a uniform compressive or tensile stress $\sigma_1 = Ee$, and a shear deformation restoring points in the planes of the bonded surfaces to their original positions in these planes. For a cylindrical block of radius a and thickness h, the corresponding shear stress t acting at the bonded surfaces at a radial distance r from the cylinder axis is given by

$$t = Eer/h$$

This shear stress is associated with a corresponding normal stress or pressure, given by

$$\sigma_2 = Ee(a^2/h^2)(1 - r^2/a^2) \tag{2-32}$$

These stress distributions are shown schematically in Fig. 2.19. Although they must be incorrect right at the edges of the block, because the assumption of a simple shear deformation cannot be valid at these points of singularity, they appear to provide satisfactory approximations over the major part of the bonded surfaces.

By integrating the sum of the normal stresses $\sigma_1 + \sigma_2$ over the bonded surface, the total compressive force F is obtained in the form

$$F/\pi a^2 e = E(1 + a^2/2h^2) = E' \tag{2-33}$$

Clearly, for thin blocks of large radius the effective value of Young's modulus E given by the right-hand side of Eq. (2-33) is much larger than the real value E because of the restraints imposed by the bonded surfaces. Indeed, for values of the ratio a/h greater than about 10, a significant contribution to the observed displacement comes from volume compression or dilation because E' is now so large that it becomes comparable to the modulus of bulk compression (Fig. 2.20).

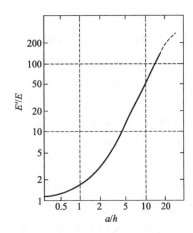

Fig. 2.20 Effective value of Young's modulus E' for bonded blocks versus ratio of radius to thickness a/h.

A more accurate treatment of the compression of bonded blocks has been given by Horton et al. without invoking the assumption that a simple shear deformation holds right up to the bonded edges. They obtained a result of the same form as Eq. (2-33) but with the bracketed term on the right-hand side replaced by $[1.2 + (a^2/2h^2)]$. However, this term does not yield the correct value of unity for tall blocks, i.e., when a/h is small, and it is equivalent to Eq. (2-33) for thin blocks of large radius, when a/h is large. It should therefore be regarded as a better approximation for blocks of intermediate size.

When a thin bonded block is subjected to tensile loading, a state of approximately equal triaxial tension is set up in the central region of the block. The magnitude of the stress in each direction is given by the tensile stress, or negative pressure, σ_2 at $r = 0$, i.e., Eea^2/h^2, from Eq. (2-32). Under this outwardly directed tension a small cavity in the central region of the block will expand indefinitely at a critical value of the tension, of about $5E/6$. Thus, if cavities are present in the interior of a bonded block, they are predicted to expand indefinitely, i. e. , rupture, at a critical tensile strain e_c, given approximately by

$$e_c = 5h^2/6a^2 \tag{2-34}$$

and at a corresponding critical value of the applied tensile load, obtained by substituting this value of e in Eq. (2-33). To avoid internal fractures of this kind it is thus necessary to restrict the mean tensile stress applied to thin bonded blocks to less than about $E/3$.

In compression, on the other hand, quite large stresses can be supported. A stress limit can be calculated by assuming that the maximum shear stress, developed at the bonded edges, should not exceed G; i. e. , the maximum shear deformation should not exceed about 100%. This yields a value for the allowable overall compressive strain of $h/3a$, corresponding to a mean compressive stress of the order of E for disks with a/h between about 3 and 10. This calculation assumes that the approximate stress analysis outlined earlier is valid right at the edges of the block, and this is certainly incorrect. Indeed, the local stresses in these regions depend strongly on the detailed shape of the free surface in the neighborhood of the edge.

2.8 Some Unsolved Problems in Rubber Elasticity

Temperature affects the elasticity of elastomers in an unusual way. When the elastomer is assumed to be in a stretched state, heating causes them to contract. Vice versa, cooling can cause expansion. This can be observed with an ordinary rubber band. Stretching a rubber band will cause it to release heat (press it against your lips), while releasing it after it has been stretched will lead it to absorb heat, causing its surroundings to become cooler. This phenomenon can be explained with the Gibbs free energy. Rearranging $\Delta G = \Delta H - T\Delta S$, where G is the free energy, H is the enthalpy, and S is the entropy, we get $T\Delta S = \Delta H - \Delta G$. Since stretching is nonspontaneous, as it requires external work, $T\Delta S$ must be negative. Since T is always positive (it can never reach absolute zero), the ΔS must be negative, implying that the rubber in its natural state is more entangled (with more microstates) than when it is under tension. Thus, when the tension is removed, the reaction is spontaneous, leading ΔG to be negative. Consequently, the cooling effect must result in a positive ΔH, so ΔS will be positive there.

The result is that an elastomer behaves somewhat like an ideal monatomic gas, in as much as (to good approximation) elastic polymers do not store any potential energy in stretched chemical bonds or elastic work done in stretching molecules, when work is done upon them. Instead, all work done on the rubber is "released" (not stored) and appears immediately in the polymer as thermal energy. In the same way, all work that the elastic does on the surroundings results in the disappearance of thermal energy in order to do the work (the elastic band grows cooler, like an expanding gas). This last phenomenon is the critical clue that the ability of an elastomer to do work depends (as with an ideal gas) only on entropy-change considerations, and not on any stored (i.e., potential) energy within the polymer bonds. Instead, the energy to do work comes entirely from thermal energy, and (as in the case of an expanding ideal gas) only the positive entropy change of the polymer allows its internal thermal energy to be converted efficiently (100% in theory) into work.

Invoking the theory of rubber elasticity, one considers a polymer chain in a cross-linked network as an entropic spring. When the chain is stretched, the entropy is reduced by a large margin because there are fewer conformations available. Therefore, there is a restoring force, which causes the polymer chain to return to its equilibrium or unstretched state, such as a high entropy random coil configuration, once the external force is removed. This is the reason why rubber bands return to their original state. Two common models for rubber elasticity are the freely-jointed chain model and the worm-like chain model.

We turn now to some features of the elastic response of rubbery materials which are still not fully understood. As normally prepared, molecular networks comprise chains of a wide distribution of molecular lengths. Numerically, small chain lengths tend to predominate. The effect of this diversity on the elastic behavior of networks, particularly under large deformations, is not known. A related problem concerns the elasticity of short chains. They are inevitably non-Gaussian in character and the analysis of their conformational statistics is likely to be difficult. Nevertheless, it seems necessary to carry out this analysis to be able to treat real networks in an appropriate way.

It is also desirable to treat network topology in greater detail, i.e., to incorporate the functionality of crosslinks, their distribution in space, and loop formation. The effect of mutual interaction between chains in the condensed state appears to be accounted for satisfactorily by the "tube" model for uncrosslinked polymers, but its application to networks seems incomplete. But the problem in greatest need of attention is the response of highly filled elastomers to stress. Filled elastomers are not really elastic; their stress-strain relations are irreversible (see Fig. 2.5), and it is therefore inappropriate to describe their response to stress by a strain energy function. Moreover, they appear to become anisotropic on stretching and to some degree after release. At present, the molecular processes that occur on deformation and the mathematical framework suitable for describing them are both unclear.

References

[1] P. J. Flory, Statistical Mechanics of Chain Molecules, Wiley-Interscience, New York, 1969.
[2] D. E. Hanson and J. L. Barber, Modelling and Simulation in Materials Science and Engineering 21 (2013).
[3] M. C. Morris, J. Appl. Polm. Sci. 8, 545 (1964).
[4] F. Bueche, J. Appl. Polym. Sci. 4, 107 (1960); 5, 271 (1961).

[5] A. V. Tobolsky and H. F. Mark (Eds.), Polymer Science and Materials, Wiley, New York 1971, Chap. 13.
[6] L. J. Fetters, D. J. Lohse, and W. W. Graessley, J. Polym. Sci.: Part B: Polym. Phys. 37, 1023 (1999).
[7] D. E. Hanson and J. L. Barber, Phys. Chem. Chem. Phys. 20, 8460 (2018)
[8] T. A. Vilgis, in Elastomeric Polymer Networks, J. E. Mark and B. Erman (Eds. Prentice-Hall, Englewood Cliffs, NJ, 1992, Chap. 5.
[9] W. W. Graessley, Polymeric Liquids and Networks: Structure and Properties, Taylor a Francis Books, New York, 2004.
[10] S. Prager and H. L. Frisch, J. Chem. Phys. 46, 1475 (1967).
[11] L. R. G. Treloar, The Physics of Rubberlike Elasticity, 3rd ed., Clarendon Press, Oxford 1975.
[12] E. M. Arruda and M. C. Boyce, J. Mech. Phys. Solids 41, 389 (1993).
[13] A. G. Thomas, Trans. Faraday Soc. 51, 569 (1955).
[14] H. M. James and E. Guth, J. Chem. Phys. 11, 455 (1943); J. Polym. Sci. 4, 153 (1949).
[15] R. S. Rivlin, Philos. Trans. Roy. Soc. (London) Ser. A 241, 379 (1948).
[16] R. S. Rivlin, in Rheology, Theory and Application, F. R. Eirich (Ed.), Academic Press, New York, Vol. 1, 1956, Chap. 10.
[17] M. Mooney, J. Appl. Phys. 11, 582 (1940).
[18] R. S. Rivlin and D. W. Saunders, Philos. Trans. Roy. Soc. (London) Ser. A 243, 251 (1951).
[19] A. N. Gent and A. G. Thomas, J. Polym. Sci. 28, 625 (1958).
[20] A. N. Gent, Rubber Chem. Technol. 69, 59 (1996).
[21] E. Pucci and G. Saccomandi, Rubber Chem. Technol. 75, 839 (2002).
[22] R. W. Ogden, Non-Linear Elastic Deformations, Ellis Harwood, Chichester, UK, 1984; Dover Publications, Mineola, NY, 1997, Chap. 7.
[23] A. N. Gent and P. B. Lindley, Proc. Roy. Soc. (London) A 249, 195 (1958).
[24] A. N. Gent and D. A. Tompkins, J. Appl. Phys. 40, 2520 (1969).
[25] M. Biot, Mechanics of Incremental Deformations, Wiley, New York, 1965
[26] M. F. Beatty, in Finite Elasticity, R. S. Rivlin (Ed.), AMD Vol. 27, American Society Mechanical Engineers, New York, 1977, p. 125.
[27] A. N. Gent and I. S. Cho, Rubber Chem. Technol. 72, 253 (1999).
[28] A. Schallamach, Wear 17, 301 (1971).
[29] A. E. Green and W. Zerna, Theoretical Elasticity, 2nd ed., Clarendon Press, Oxford, 1975, Section 4.6, p. 135.
[30] A. N. Gent and K. C. Hua, Int. J. Non-Linear Mech. 39, 483 (2004).
[31] R. S. Rivlin, J. Appl. Phys. 18, 444 (1947).
[32] A. L. Soden, A Practical Manual of Rubber Hardness Testing, Maclaren, London, 1952.
[33] A. N. Gent, Rubber Chem. Technol. 67, 549 (1994).
[34] J. M. Horton, G. E. Tupholme, and M. J. C. Gover, ASME, J. Appl. Mech. 69, 836 (2002).
[35] J. E. Mark and B. Erman, Rubberlike Elasticity: A Molecular Primer, John Wiley & Sons, New York, 1988.
[36] J. E. Mark and B. Erman (Eds.), Elastomeric Polymer Networks, Prentice-Hall, Englewood Cliffs, NJ, 1992.
[37] A. N. Gent, J. Polym. Sci. Poly. Symp. 28, 625 (1958).

Chapter 3
Rubber Compounding

3.1 Introduction

Compounding, a term that has evolved within the tire and rubber industry, is the materials science of modifying a rubber or elastomer or a blend of polymers and other materials to optimize properties to meet a given service application or set of performance parameters. Compounding is therefore a complex multidisciplinary science necessitating knowledge of materials physics, organic and polymer chemistry, inorganic chemistry, and chemical reaction kinetics. The materials scientist, when designing a rubber formulation, has a range of objectives and restrictions within which to operate. Product performance requirements will dictate the initial selection of formula ingredients. These materials must be environmentally safe, meet occupational health and safety requirements, be processable in the product manufacturing facilities, and be cost effective.

Compounded rubber has many unique characteristics not found in other materials, such as dampening properties, high elasticity, and abrasion resistance. Hence rubber has found use in applications such as tires, conveyor belts, large dock fenders, building foundations, automotive engine components, and a wide range of domestic appliances. The ingredients available to the materials scientist for formulating a rubber compound can be divided into five categories:

① Polymers: Natural rubber, synthetic polymers
② Filler systems: Carbon blacks, clays, silicas, calcium carbonate
③ Stabilizer systems: Antioxidants, antiozonants, waxes
④ Vulcanization system components: Sulfur, accelerators, activators
⑤ Special materials: Secondary components such as pigments, oils, resins, processing aids, and short fibers.

Each class of materials is reviewed in this chapter.

3.2 Polymers

World rubber usage of around 18 million metric tons is split between natural rubber, which constitutes about 46% of global consumption, and synthetic rubber, of which styrene-butadiene rubber (SBR) accounts for about 18%. The balance of synthetic rubbers (47%) consists of polybutadiene rubber and a range of speciality polymers such as urethanes, halogenated polymers, silicones, and acrylates. Traditionally, the growth of synthetic and natural rubber consumption is virtually in line with the gross national product of, collectively, North America, the European Community, and the northwest Pacific rim.

3.2.1 Natural Rubber

Global natural rubber consumption is split among tires (75%), automotive mechanical products (5%), nonautomotive mechanical products (10%), and miscellaneous applications such as medical and health-related products (10%). Since the 1960s, the quality and consistency of natural rubber has improved, primarily because of the implementation of standard specifications defining a range of grades of rubber. Natural rubber is available in three basic types: technically specified rubbers, visually inspected rubbers, and specialty rubbers.

The American Society for Testing and Materials (ASTM) describes six basic grades of coagulated technically specified natural rubber which is processed and compacted into 34-kg blocks (Table 3.1). These six general grades of technically specified natural rubber are defined in more detail by the respective producing countries. Standard Malaysian Rubber (SMR), Standard Indonesian Rubber (SIR), and Thai Technical Rubber (TTR) expand the range of rubbers available. For example, two constant-viscosity Standard Malaysian Rubber CV grades are available, SMR CV50 and CV60. SMR 10 and SMR 20 grades are also available as viscosity stabilized (SMR 10CV and SMR 20CV).

Table 3.1 Specifications for technically graded natural rubber

Property	Rubber grade					
	L	CV	5	10	20	50
Dirt/% maximum	0.050	0.050	0.050	0.100	0.200	0.500
Ash/% maximum	0.60	0.60	0.60	0.750	1.00	1.50
Volatile matter/%	0.80	0.80	0.80	0.80	0.80	0.80
Nitrogen/%	0.60	0.60	0.60	0.60	0.60	0.60
Plasticity	30	—	30	30	30	30
Plasticity retention index	60	60	60	50	40	30
Color index	6.0	—	—	—	—	—
Mooney Viscosity	—	60	—	—	—	—

The Rubber Manufacturers Association has a further set of standards for quality and packing of latex natural rubber grades. Table 3.2 defines the eight types of rubber covered in their specifications. Here, coagulated latex is sheeted, dried, and packed into bales of up to 113.5kg. Grading is by visual inspection. Quality assurance laboratories have sets of visual standards for inspections.

The third category of natural rubbers are the specialty materials, which include liquid low molecular weight rubber, methyl methacrylate grafted polymers, oil-extended natural rubber, deproteinized natural rubber, epoxidized natural rubber, and superior-processing natural rubber.

Table 3.2 International natural rubber type and grade specification

Type	Natural rubber	Description
1	Ribbed smoked sheet	Coagulated sheets, dried, and smoked latex. Five grades available (RSS1-5)
2	White and pale crepe	Coagulated natural liquid latex milled to produce a crepe
3	Estate brown crepe	Fresh lump and other high-quality scrap generated on the plantation
4	Compo crepe	Lump, tree scrapes, and smoked sheet cuttings are milled into a crepe
5	Thin brown crepe	Unsmoked sheets, wet slab, lump, and other scrap from estates and small holdings
6	Thick blanket crepe	Wet slab, lump, and unsmoked sheets milled to give a crepe
7	Flat bark crepe	All types of scrap natural rubber including earth scrap
8	Pure smoked blanket crepe	Milled smoked rubber derived exclusively from ribbed smoked sheets

Natural rubber usage has increased substantially in modern radial tires. Bernard and coworkers compared the natural rubber levels of heavy-duty radial truck tires to those of the equivalent bias tire and noted the following increase:

Natural rubber/%	Bias	Radial
Tread	47	82
Skim coat	70	100
Side wall	43	58

The reasons for the increase have been attributed to improved green strength, increase in component-to-component adhesion, improved tear strength, lower tire temperatures generated under loaded dynamic service conditions, and lower tire rolling resistance to improve vehicle fuel efficiency.

The increase in natural rubber usage translates into approximately 21kg per tire for a radial construction compared with approximately 9kg found in a bias truck tire. Natural rubber compounds also tend to find use in covers of high-performance conveyor belts where a similar set of performance parameters such as those of a truck tire tread compound are found. Low hysteretic properties, high tensile strength, and good abrasion resistance are required for both products.

3.2.2 Synthetic Elastomers

Classification of synthetic rubber is governed by the International Institute of Synthetic Rubber Producers (IISRP). In the case of styrene-butadien rubber, polyisoprene rubber, and polybutadiene, a series of numbers have been assigned which classify the general properties of the polymer. For example, the IISRP 1500 series defines cold emulsion-polymerized (i.e., below 10℃), nonpigmented SBR. The 1700 series of polymers describes oil extended cold emulsion SBR. Table 3.3 illustrates the general numbering used by IISRP. The numbering system for solution-polymerized stereo elastomers is given in Table 3.4.

Table 3.3　Classification of synthetic rubbers by IISRP

Class number	Description
1000 series	Hot nonpigmented emulsion SBR (polymerized above 38℃)
1500 series	Cold nonpigmented emulsion SBR (polymerized below 10℃)
1600 series	Cold polymerized/carbon black master batch/14phr oil (max)SBR
1700 series	Oil extended cold emulsion SBR
1800 series	Cold emulsion-polymerized/carbon black master batch/more than 14phr oil SBR
1900 series	Emulsion resin rubber master batches

Table 3.4　IISRP solution-polymerized stereo elastomers

Name	Butadiene and copolymers	Isoprene and copolymers
Dry polymer	1200~1249	2200~2249
Oil extended	1250~1299	2250~2299
Black master batch	1300~1349	2300~2349
Oil-black master batch	1350~1399	2350~2399
Latex	1400~1449	2400~2449
Miscellaneous	1450~1499	2450~2499

Tire production consumes approximately 60% of the global synthetic rubber production (Table 3.5). Of this, SBR is the largest volume polymer, representing over 65% of the synthetic rubber used in tires. Polybutadiene (BR) ranks second in production output. Table 3.6 and Table 3.7 illustrate the consumption of synthetic rubber by product group. Styrene butadiene rubber finds extensive use in tire treads because it offers wet skid and traction properties while retaining good abrasion resistance. Polybutadiene (BR) is frequently found in treads, sidewalls, and some casing components of the tire because it offers good abrasion resistance, and tread wear performance and enhances resistance to cut propagation. BR can also be blended with natural rubber, and many authors have reported that such compositions give improved fatigue and cut growth resistance.

Table 3.5　Synthetic rubber consumption

Name	Consumption
Tires	60%
Automotive parts	10%
Nonautomotive mechanical goods	9%
Thermoplastic elastomer composites	6%
Footwear	4%
Construction	3%
Wire and cable	2%
Adhesives	1%
Miscellaneous goods	5%

Table 3.6　U.S. consumption of SBR

Percent of Product	Total consumption/%
Passenger tires	50
Retread rubber	13
Truck tires	8
Special tires (aircraft, earth mover, etc)	4
Automotive mechanical goods	7
Miscellaneous use(domestic appliances medical equipment, construction)	18

Table 3.7 U. S. consumption of Polybutadiene

Percent of Product	Total consumption/%
Passenger tires	45
Retread rubber	4
Truck tires	28
Special tires (aircraft, earth mover, etc)	1
Mechanical goods	2
Miscellaneous applications (polymer blends, polymer modifiers with polystyrene or styrene acrylonitrile butadiene terpolymers)	20

Before reviewing elastomer characteristics required to meet any given set of tire performance parameters, it is appropriate to identify two means by which the materials scientist may describe a polymer: polymer macrostructure and polymer microstructure. The macrostructure of a polymer defines the molecular weight and also crosslink distribution, polymer chain branching, and crystallite formation. The arrangement of the monomers within a polymer chain constitutes its microstructure. Butadiene can adopt one of three configurations as illustrated in Fig. 3.1.

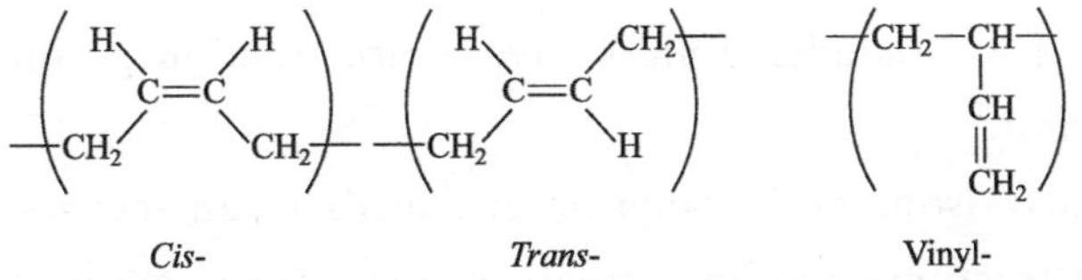

Fig. 3.1 Polymer microstructure: possible configurations for butadiene in SBR and BR.

These molecular configurations or stereo-chemistry can be described as follows:

① vinyl-(1,2)   The third and fourth carbon atoms are pendant; the first and second carbon atoms participate in the polymer backbone.

② *trans*-(1,4)   The hydrogen atoms attached to the carbon-carbon double bond on the polymer backbone are on opposite sides.

③ *cis*-(1,4)   The two hydrogen atoms attached to the carbon-carbon double bond in the polymer are on the same side of the double bond.

Table 3.8 illustrates the effect of the catalyst system on polymer microstructure.

Table 3.8 Polybutadiene microstructure

Catalyst	Isomer level to ±1%		
	Cis/%	*Trans*/%	Vinyl/%
Li	35	55	10
Ti	91~94	2~4	4
Co	96	2	2
Nd	98	1	1
Ni	96~98	0~1	2~4

The relative levels of each of the three isomers in a polymer such as BR can have a dramatic effect on the material's performance. For example lithium-catalyzed solution polymers, with approximately 36% *cis* content, tend to process easily, whereas high-*cis* Ti and Ni polymers (92% *cis*) are more difficult to process at factory processing temperatures but show better abrasion resistance. High-*trans* BR (93% *trans*) tends to be a tough, crystalline material at room temperature. High-vinyl butadiene BR polymers in tire treads tend to show good wet skid and wet traction performance.

Nordsiek documented a series of empirical guidelines which might be used in designing a polymer for a set of tire performance targets. By preparing various blends of BR and SBR, Nordsiek produced a series of compounds in which the T_g increased from -100 ℃ to -30 ℃. He noted the following points:

① As the T_g increases there is a near-linear drop in abrasion resistance.

② Wet grip or traction improves nearly linearly with the increase in compound T_g.

③ Addition of a catalyst modifier during the preparation of solution polymerized, lithium-catalyzed BR results in an increase in the 1,2-vinyl butadiene level in the polymer and causes an increase in T_g. There is a corresponding drop in abrasion resistance and an increase in wet traction.

④ Inclusion of styrene leads to an increase in traction performance and loss in abrasion resistance. There is a linear relationship between styrene and vinyl-1,2-butadiene. Approximately two vinyl-1,2-butadiene units gave a tire traction performance equivalent to that of one styrene unit.

⑤ Inclusion of 3,4-isoprene in polyisoprene leads to an increase in T_g and a corresponding increase in traction, and an increase in the percentage incorporation of 1,2- or 3,4-piperylene in polypiperylenes results in a T_g increase, causing a loss in abrasion resistance and an increase in grip.

This allowed the tanδ temperature curve of a tread compound run from -100 ℃ to $+100$ ℃ to be segmented into zones which would characterize that tire tread compound's performance (Table 3.9). Such property targets enable development of the concept of "integral rubber"; i.e., a polymer can be designed to meet rolling resistance, traction, and tread wear targets without a drop in overall tire performance.

Table 3.9 Characterization of an idealized tread compound: tanδ-temperature curve

Temperature zone/℃	Feature	Performance parameter
-60 to -40	T_g	Abrasion
-20	—	Low-temperature properties
$+20$	—	Wet traction
$+40$ to $+60$	—	Rolling resistance
$+80$ to $+100$	—	Heat buildup

Day and Futamura evaluated the impact of variation in 1,2-butadiene and styrene content in SBR on the properties of a compounded formulation. In short, (1) an increase in styrene produced an increase in tensile strength, (2) an increase in vinyl-1,2-butadiene resulted in a drop in both tear strength and ultimate elongation, and (3) at equal T_g, neither vinyl-1,2-butadiene nor styrene level affected the formulation's hysteretic properties.

Brantley and Day then conducted a study to compare the tire performance of emulsion- and solution-polymerized SBR. The authors noted that solution-polymerized polymers, which tend to have a narrower molecular weight distribution and lower T_g than equivalent emulsion-polymerized polymers, have lower hysteretic properties. They then showed that a solution SBR with the same bound styrene as an emulsion SBR will give lower rolling resistance, improved dry traction, and better tread wear. Emulsion SBR, however, tends to show better wet skid, wet traction, and wet handling performance.

Kern and Futamura later elaborated on this work by evaluating the impact of vinyl-1,2-butadiene level in a solution SBR and again comparing this with an emulsion SBR. Though this work was conducted with passenger tires, many of the principles should be applicable to the range of tires such as light truck and heavy-duty truck tires.

The authors collected the test data shown in Table 3.10. From these data it can be noted that the number average molecular weight, or M_n, of a commercial emulsion SBR such as IISRP 1500 or 1712 is typically 90000 to 175000.

Table 3.10 Comparison of emulsion and solution polymerized SBR

Property	Emulsion SBR	Solution SBR
Viscosity (M_L 1 + 4 at 100℃)	50	7
Time to optimum cure/min at 150℃	40	25
Tensile strength/MPa	26	21
Ultimate elongation/%	400	300
Rebound/%	48	61

The primary molecular weight of a solution-polymerized polymer produced with an anionic lithium catalyst can, in contrast, be increased toward 250 000 without gelation. In addition, emulsion polymerized SBR contains only about 92% rubber hydrocarbon as a result of the presence of residues from the production process; solution polymers tend to be near 100% hydrocarbon. As a consequence, the authors concluded that the number average molecular weight can be considered the key parameter of polymer macrostructure, particularly with respect to the hysteretic characteristics of a tread formulation. Hence the differences in macrostructure between emulsion and solution polymerized polymers will dictate many of their properties in a tire tread compound.

When considering only solution polymers, polymer microstructure has a greater effect on tire tread compound performance. Table 3.11 illustrates the impact on tire traction, rolling resistance, and tread wear of a polybutadiene tread on which the vinyl-1,2-butadiene level had been increased from 10% to 50%. The corresponding drop in wear and increase in tire rolling resistance are in agreement with the empirical rules presented by Nordsiek who attributed such tire property trends to the polymer T_g.

Table 3.11 Effect of polymer butadiene vinyl level on tire performance[a]

Vinyl level	10%	50%
Glass transition temperature	-90℃	-60℃
Tire properties		
Wet traction	100℃	120℃
Rolling resistance rating	100℃	95℃
Tread wear rating	100℃	90℃

[a] Higher rating is better.

Table 3.12 shows how polybutadiene microstructure and macrostructure, i.e., molecular weight, M_w, and M_n, polydispersity, and branching can affect the processability of a polymer. A study with both cobalt-and neodenium-catalyzed polybutadiene showed the relationship between polydispersity or molecular weight distribution and increases in stress relaxation. Increases in stress relaxation, as measured by the Mooney viscometer, will infer greater difficulty in compound processing, gauge control, "nerve," and extrudate calendered sheet shrinkage.

Table 3.12 Macrostructure and mooney viscometer

Catalyst	Polymer sample	M_w	M_n	M_w/M_n	M_L 1+4	Mooney stress relaxation
Cobalt	1	338	156	2.17	47	4.50
	2	318	131	2.43	45	7.50
	3	321	125	2.57	46	9.00
	4	303	108	2.81	44	14.00
Neodenium	1	353	186	2.10	50	5.00
	2	381	103	3.70	42	8.00
	3	347	87	3.99	44	9.00
	4	368	86	4.28	42	10.00

Halobutyl rubber (HIIR) is used primarily in tire inner liner and white sidewalls. These elastomers are best for tire air retention owing to lower air permeability as well as aging and fatigue resistance. The chlorinated (CIIR) and brominated (BIIR) versions of isobutylene isoprene rubber (IIR) can be blended with other elastomers to improve adhesion between HIIR compounds and those based on general purpose elastomers, and improve vulcanization kinetics.

Attempts at using halogenated isobutylene based polymers in tread compounds has been limited, even though such tread compounds display good performance in winter applications and have good traction performance. A new isobutylene polymer modified with p-methylstyrene and then brominated is also available that offers a fully saturated backbone to resist aging while improving compatibility with general purpose elastomers such as natural rubber and styrene-butadiene rubbers.

It is common to blend more than one type of rubber within a given tire tread compound. An example of this is the truck drive axle tire tread compounds that must not only possess high strength but must also have good fatigue resistance. In passenger car tires, as many as four different polymers may be used for the tread compound totaling 100phr; e.g., 25phr emulsion SBR, 25phr solution SBR, 30phr BR, and 20phr NR. If solution SBR categories can be considered as each one falling with a 10°C T_g range, there are at least nine groups of specialty SSBR polymers commercially available, in addition to the range of proprietary polymers chemical operations produce to support tire manufacturing.

Fig. 3.2 shows the effect of T_g on wet skid. If an increase in wet grip is required with minimum impact on rolling resistance, then a change in T_g is best accomplished via an increase in the vinyl-butadiene level rather than in the bound styrene content. Alternatively, if wear is of higher importance, T_g should be adjusted by a change in the bound styrene level. The optimum T_g could therefore be obtained by adjustment of either the vinyl-butadiene or styrene contents to obtain the required wet grip, rolling resistance, and wear performances. It has been demonstrated that an increase in wear performance would lead to a trade-off in traction performance (Fig. 3. 3).

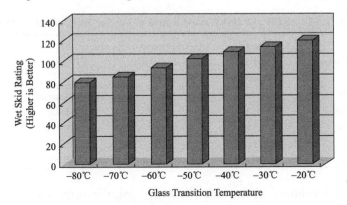

Fig. 3. 2   Effect of T_g on tire traction performance.

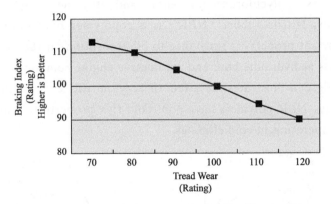

Fig. 3. 3   Relationship between wear and braking qualities.

High molecular weight commercial polymers are oil extended to facilitate processing and also to enable the production of polymers that will yield compounds with better mechanical properties than those with lower molecular weight polymers of corresponding structure. Table 3. 13 displays a selection of emulsion SBR grades. Aromatic oils can raise the glass transition temperature of the corresponding oil-free polymer. Naphthenic oils will tend to shift the transition temperature below the value of the oil-free rubber.

Table 3. 13   Oil extended emulsion SBR

IISRP polymer	Mooney nominal viscosity (M_L 1 + 4)	Styrene/%	Oil type	Oil level(PHR)
1707	50	23.5	Naphthenic	37.5
1712	50	23.5	Aromatic	37.5
1720	40	23.5	Naphthenic	50.0
1721	55	40.0	Aromatic	37.5

The primary function of oil in rubber is to facilitate improvement in processing, i. e., the ease of mixing in an internal mixer, to improve mixed compound uniformity such as viscosity, and to improve downstream processing such as in extrusion. The specific oils used in oil-extended elastomers have been categorized into essentially five groups, which are summarized in Table 3.14.

Table 3.14　ASTM and IISRP classification of oils for oil extended elastomers

Type	Asphaltenes	Polar compound content/%	Saturated Hydrocarbon content/%	Category	Viscosity gravity constant
101	0.75	25.0	20.0	Highly Aromatic	>0.900
102	0.50	12.0	20~35	Aromatic	0.90
103	0.30	6.0	35~65	Naphthenic	0.87
104A	0.10	1.0	65.0	Paraffinic	>0.820
104B	0.10	1.0	65.0	Paraffinic	0.820 Max

Though natural rubber, SBR, and BR represent the largest consumption of elastomers, several additional polymers merit a brief discussion because of their economic significance, i. e., nitriles, polychloroprene, butyl, and ethylene-propylene-diene monomer (EPDM) elastomers. Nitrile rubber (NBR) is a copolymer of acrylonitrile and butadiene. Its most important property is resistance to oil absorption; it therefore finds extensive use in such products as hydraulic hose and automotive engine components, where oil resistance is essential. Fig. 3.4 illustrates the effect of acrylonitrile level on oil absorption (IRM 903 oil). Conversely, NBR polymers have poor cold flex properties, which prohibits their use on equipment operating in cold climates.

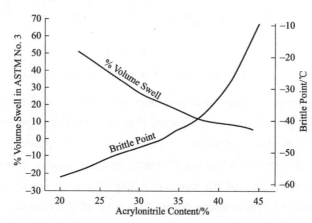

Fig. 3.4　Acrylonitrile content and NBR oil absorption.

NBR tends to break down readily on a mill or banbury. Peptizers are not normally required, though antigel agents are needed if mixing temperatures exceed 140℃. Because of the polymer's low green strength, sufficient shear during mixing is not achieved to enable use of SAF or ISAF carbon blacks. It can also result in poor processing qualities such as mill bagging. Antioxidants are essential in NBR compounds as NBR will oxidize readily in hot air. Polymerized 2,2,4-trimethyilihydroquinolene is the most effective antioxidant. Antiozonants and waxes are ineffective with NBR compounds.

Polychloroprene is made either from acetylene or butadiene:

$$HC\equiv CH \longrightarrow HC\equiv C-CH=CH_2 \xrightarrow{HCl} H_2C=\underset{Cl}{C}-CH=CH_2$$
Acetylene   Chloroprene

$$H_2C=\underset{Cl}{C}-CH=CH_2 \xrightarrow{Polymerization} +CH_2-CH=\underset{Cl}{C}-CH_2\,\underset{}{\,}_n \quad (3\text{-}1)$$
Polychloroprene

$$H_2C=CH-CH=CH_2 \xrightarrow{Cl_2} Cl-CH_2-CH=CH-CH_2-Cl + Cl-CH_2-\underset{Cl}{CH}-CH=CH_2$$
Butadiene

$$Cl-CH_2-CH=CH-CH_2-Cl \longrightarrow Cl-CH_2-\underset{Cl}{CH}-CH=CH_2 \quad (3\text{-}2)$$

$$Cl-CH_2-\underset{Cl}{CH}-CH=CH_2 \xrightarrow{NaOH} CH_2=\underset{Cl}{C}-CH=CH_2 + NaCl + H_2O$$
Chloroprene

Acetylene is reacted to form vinyl acetylene, which is then chlorinated to form chloroprene. This can then be polymerized to polychloroprene. Polychloroprene contains approximately 85% *trans-*, 10% *cis-*, and 5% vinyl-chloroprene. Because of its high *trans*-content, polychloroprene tends to crystallize readily.

Depending on the grade of polymer, polychloroprene can be vulcanized by zinc oxide or magnesium oxide. Tetramethylthiuram disulfide can serve as a retarder. Polychloroprene is inferior to NBR for oil resistance but is still significantly better than natural rubber, SBR, or BR. Like NBR it also finds extensive use in such products as oil seals, gaskets, hose linings, and automotive engine transmission belts where resistance to oil absorption is important.

Butyl rubbers are a copolymer of isobutylene and isoprene:

$$CH_2=\underset{CH_3}{\overset{CH_3}{C}} + CH_2=\underset{CH_3}{C}-CH=CH_2 \longrightarrow +CH_2-\underset{CH_3}{\overset{CH_3}{C}}\!\!\,\underset{}{\,}_n CH_2-\underset{CH_3}{C}=CH-CH_2- \quad (3\text{-}3)$$
Isobutylene   Isoprene   Butyl

Isobutylene and isoprene are in a ratio of approximately 50 : 1. Chlorobutyl rubber and bromobutyl rubber are produced by the halogenation of butyl rubber. Butyl rubber and halobutyl rubber are highly impermeable to air and show very low water absorption, and good heat and ozone resistance. As noted earlier, they therefore find extensive use in liners of radial tires, covers and insulation of high-voltage electric cables, and automobile engine and radiator hoses.

High-tensile-strength butyl compounds generally use FEF-or GPF-grade carbon blacks. Vulcanization systems tend to be based on thiazole accelerators such as mercaptobenzothiazole disulfide (MBTS) and thiuram accelerators such as tetramethylthiuram disulfide (TMTD). Low-tensile-strength compounds will use a clay or silica reinforcing filler in place of carbon black.

Copolymerization of ethylene and propylene produces an elastomeric polymer which is virtually inert because of the absence of carbon-carbon double bonds (EPM). Such polymers thus tend to be crosslinked with peroxides or by radiation. To improve the reactivity of ethylene-propylene copolymers, 1% to 10% of a third monomer can be added to give a terpolymer or Ethylene-propylene-diene monomer (EPDM). The primary diene monome used in EPDM are 1,4-hexadiene, dicyclopentadiene, and ethylidene norbornene. Introduction of an unsaturated monomer such as ethylidene norbornene will enable use of sulfur-based crosslinking systems.

EPDM tends to show good resistance to ozone attack, oxidation resistance, and moisture resistance. It is therefore used in applications which require good weather resistance and heat stability. Roofing materials, outer covers of high-voltage electric cables, and selected automotive hoses use EPDM. See Table 3.15 for the abbreviations of selected elastomers.

Table 3.15 Nomenclature for selected elastomers

Abbreviation	Full name
AU	Polyester urethane
BR	Polybutadiene
BIIR	Brominated isobutylene-isoprene (bromobutyl)
CIIR	Chlorinated isobutylene-isoprene (chlorobutyl)
CPE	Chlorinated polyethylene
CR	Chloroprene rubber
CSM	Chlorosulfonyl polyethylene
EAM	Ethylene-vinyl acetate copolymer
EPDM	Terpolymer of ethylene, propylene, and a diene with a residual unsaturated portion in the chain
EPM	Ethylene-propylene copolymer
EU	Polyether urethane
HNBR	Hydrogenated acrylonitrile-butadiene rubber (highly saturated nitrile rubber)
IIR	Isobutylene-isoprene rubber (butyl)
IR	Synthetic polyisoprene
NBR	Acrylonitrile-butadiene rubber
SBR	Styrene-butadiene rubber
E-SBR	Emulsion styrene-butadiene rubber
S-SBR	Solution styrene-butadiene rubber
X-NBR	Carboxylated nitrile-butadiene rubber
X-SBR	Carboxylated styrene-butadiene rubber
YSBR	Block copolymers of styrene and butadiene

3.3 Filler Systems

Fillers, or reinforcement aids, such as carbon black, clays, and silicas are added to rubber formulations to meet material property targets such as tensile strength and abrasion resistance. Carbon black technology is as complex as polymer science, and an extensive range of blacks are available, each imparting specific sets of properties to a compound. The correct choice of carbon black is therefore as important as the development of a formulation's polymer system in meeting a product performance specification. Table 3.16 displays the general classes of rubber-grade carbon blacks as defined in ASTM Standard D1765-04.

Table 3.16 Types of carbon blacks

Type	ASTM designation	Particle size/nm	General use
SRF	N 762	61~100	Nontread component
GPF	N 660	49~60	Nontread component
FEF	N 550	40~48	Nontread component
FF	N 475	31~39	Nontread component
HAF	N 330	26~30	Tread and other component
ISAF	N 220	20~25	Tread
SAF	N 110	11~19	Tread

3.3.1 Carbon Black Properties

Carbon black can be described qualitatively by a series of properties: particle size (and surface area); particle size distribution; structure (particle aggregates); surface activity (chemical functional groups such as carboxyl, and ketones). Key properties describing a carbon black are listed in Table 3.17.

Table 3.17 Carbon black properties

ASTM designation	Iodine number	DBP	Compressed DBP	NSA Multipoint	STSA	Tint strength
N 110	145	113	97	127	115	123
N 115	160	113	97	137	124	123
N 120	122	114	99	126	113	129
N 121	121	132	111	122	114	119
N 125	117	104	89	122	121	125
N 134	142	127	103	143	137	131
N 219	118	78	75			123
N 220	121	114	98	114	106	116
N 231	121	92	86	111	107	120
N 234	120	125	102	119	112	123
N 299	108	124	104	104	97	113
N 326	82	72	68	78	76	111
N 330	82	102	88	78	75	104
N 339	90	120	99	91	88	111
N 343	92	130	104	96	92	112
N 347	90	124	99	85	83	105
N 351	68	120	95	71	70	100
N 358	84	150	108	80	78	98
N 375	90	114	96	93	91	114
N 472	250	178	114	270	145	
N 550	43	121	85	40	39	
N 630	36	78	62	32	32	

Continued

ASTM designation	Iodine number	DBP	Compressed DBP	NSA Multipoint	STSA	Tint strength
N 650	36	122	84	36	35	
N 660	36	90	74	35	34	
N 762	27	65	59	29	28	
N 772	30	65	59	32	30	
N 990		43	37	8	8	
N 991		35	37	8	8	

Note: ① Iodine number: Measure of surface area (particle size). The higher the iodine number, the smaller the particle size.
② DBP: Measure of structure or size of carbon black aggregate. The higher the DBP number, the higher the structure.
③ Tint: Optical absorbance, which increases with smaller particles.

Carbon black terms are defined in Table 3.18. Further reference can also be made to ASTM Standards D1566-04 on general compounding terms and D3053-04 specifically for carbon blacks.

As an empirical guide, an increase in a carbon black aggregate size or structure will result in an improvement in cut growth and fatigue resistance. A decrease in particle size results in an increase in abrasion resistance and tear strength, a drop in resilience, and an increase in hysteresis and heat buildup. The impact of carbon black type and loading on tread compound performance has been studied by Hess and Klamp, who evaluated 16 types of carbon black in three tread formulations with varying oil levels. The authors documented a number of criteria relating carbon black to the hysteretic properties of rubber compounds. These included loading, aggregate size, surface area, aggregate size distribution, aggregate irregularity (structure), surface activity, dispersion, and phase distribution within a heterogeneous polymer system.

Table 3.18 Definition of carbon black terms

Carbon Black Terms	Definition
Furnace carbon black	Class of carbon blacks produced by injection of defined grades of petroleum feedstock into a high-velocity stream of combustion gases under a set of defined processing conditions, e.g., N 110 to N 762.
Thermal carbon black	Type of carbon black produced by thermal decomposition of hydrocarbon gases, e.g., N 990, N 991.
Microstructure	Carbon black microstructure describes the arrangement of carbon atoms within a carbon black particle.
Particle	Small spherical component of a carbon black aggregate produced by fracturing the aggregate. Particle size is measured by electron microscopy.
Aggregate	Distinct, colloidal mass of particles in its smallest dispersible unit.
Agglomerate	Arrangement or cluster of aggregates.
Structure	Measure of the deviation of the carbon black aggregate from a spherical form.
Iodine number	Weight in grams of iodine absorbed per kilogram of carbon black. Measure of particle surface area. The smaller the particle size, the greater the iodine number.
Carbon black DBP	Volume of dibutyl phthalate in cubic centimeters absorbed by 100g of carbon black. DBP number is a measure of the structure of the carbon black aggregate.

Carbon Black Terms	Definition
Tint	Tint is a ratio of the reflectance of a reference paste to that of a sample paste consisting of a mixture of zinc oxide, plasticizer, and carbon black.
CTAB	Measure of the specific surface area corrected for the effect of micropores. CTAB (cetyl trimethylene ammonium bromide) is excluded from the smaller interstices and thus better represents the portion of a particle surface area in contact with the polymer.
Nitrogen surface area	Measure of total particle surface area, due to nitrogen gas being able to cover the full surface including pores without interface from surface organic functional groups.
Compressed DBP	The DBP test, but where the sample undergoes a series of compressions (4 times to 12000kg) before testing. This enables a measure of changes the carbon black will undergo during compound processing.
Pellet	Mass of compressed carbon black formed to reduce dust levels, ease handling, and improve flow.
Fines	Quantity of dust present in a pelletized carbon black; should be at the minimum level possible.
Pellet hardness	Measure of the load in grams to crush a defined number of pellets. It is controlled by the quantity of pelletizing agent. For best pellet durability and compound mixing, pellet hardness range should have a narrow distribution. Examples of pelletizing agents are lignosulfonates and molasses.
Ash	Residue remaining after burning carbon black at 550℃ for 16 hours primarily a measure of the quality of plant cooling water.
Toluene discoloration	Hydrocarbons extractable in toluene from carbon black; can be used as a measure of the residence time in a furnace.
Hydrogen and oxygen content	Residual hydrogen and oxygen remaining after carbon black is produced; will be in the form of phenolic lactonic, carboxylic, quinonic, and hydroxyl functional groups. Such groups can have significant effects on vulcanization kinetics and reinforcement potential of the carbon black.

From tire testing of the selected carbon black types, the following points were noted:

① Reduction of carbon black loading lowers tire rolling resistance. At a constant black loading, an increase in oil level will increase rolling resistance but also improve traction (at low oil levels, an increase in oil level may decrease compound hysteresis by improving carbon black dispersion).

② Increasing black fineness raises both rolling resistance and traction.

③ An increase in the broad aggregate size distribution decreases the tire rolling resistance with constant surface area and DBP.

④ Tread-grade carbon blacks can be selected to meet defined performance parameters of rolling resistance, traction, wear, etc.

Fig. 3.5 illustrates the general trends for tread-grade carbon black loading and the effect on compound physical properties. As carbon black level increases, there are increases in compound heat buildup and hardness and, in tires, an increase in rolling resistance and wet skid properties. Tensile strength, compound processability, and abrasion resistance, however, go through an optimum after which these properties deteriorate.

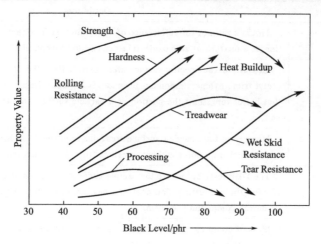

Fig. 3.5　Effect of carbon black level on compound properties.

To exploit the results of work such as that of Hess and Klamp in improving tire rolling resistance, Swor and coworkers have developed new-technology N 200 series carbon blacks which, they claim, will give a better balance of tire performance properties and the general principles will be applicable to a range of tire designs.

In attempting to predict the direction which future research in carbon black technology will follow, a review of the literature suggests that carbon black-elastomer interactions will provide the most potential to enhance compound performance. It has been demonstrated that carboxyl, phenolic, quinone, and other functional groups on the carbon black surface react with the polymer and provided evidence that chemical crosslinks exist between these materials in vulcanizates. Ayala and coworkers determined a rubber-filler interaction parameter directly from vulcanizate measurements. The authors identified the ratio σ/n, where σ is the slope of the stress-strain curve which relates to the black-polymer interaction, and n is the ratio of dynamic modulus E' at 1% and 25% strain amplitude and is a measure of filler-filler interaction. This interaction parameter emphasizes the contribution of carbon black-polymer interactions and reduces the influence of physical phenomena associated with networking. Use of this defined parameter enabled a number of conclusions to be made:

① The σ/n values obtained provided a good measure of black-polymer interaction for a range of polymers including SBR, IIR, NR, and NBR.

② Higher σ/n values were obtained for SBR and NBR, the aromatic structure in SBR and the polar —CN group in NBR clearly influence in black-filler interaction.

③ Analysis of dry carbon black surface indicated the presence of a range of hydrocarbon groups, which is in line with earlier work. These groups are capable of reacting with other functional groups.

Given the establishment of organic functional groups on the carbon black surface, Wolff and Gorl investigated the reactivity of organosilane such as bis(3-triethoxysilylpropyl) tetrasulfane with furnace blacks. The authors deduced that such groups as carboxyl, lactol, quinone, and ketone will react with the ethoxy group of bis(3-triethoxysilylpropyl) tetrasulfane and which then become pendant on the carbon black surface:

$$\left| \begin{array}{l} \text{-carboxyl} \\ \text{-quinone} \\ \text{-phenol} \\ \text{-ketone} \end{array} \right. \xrightarrow{X-Si(OR)_3} \left| \begin{array}{l} \text{carboxyl-} \\ \text{quinone-} \\ \text{phenol-} \\ \text{ketone-} \end{array} \right. \left| \begin{array}{c} -O-\overset{O-R}{\underset{O-R}{Si}}-X \\ -O\diagdown_{Si}\diagup^{O-R} \\ -O\diagup \diagdown X \end{array} \right. \qquad (3\text{-}4)$$

Carbon black surface Carbon black surface

On the basis of extract analysis and compound properties of organosilane-treated carbon black, Wolff and Gorl concluded:

① Carbon black is able to bind with a specific amount of trialkoxysilane.

② The quantity of bound organosilane correlates with carbon black particle surface area and level of oxygen-containing functional groups.

③ The triethoxysilyl group constitutes the reactive part of the silane, forming a covalent bond with the carbon black.

④ Reaction of bis (triethoxysilylpropyl) tetrasulfane with carbon black allows a reduction of compound hysteresis.

This work laid the foundation for many of the newer technology carbon blacks. These fall into two categories, postprocess modification where the surface of the carbon black is treated to improve its properties, and in-process modification where another material is introduced to again enhance the basic properties of the filler.

Examples of postprocess systems under evaluation include surface oxidation using ozone, hydrogen peroxide, or nitric acid. Such approaches are used in the production of conductive blacks. Reaction with diazonium salts, plasma treatment, and polymer grafting are also under investigation. In-process modification includes metal addition, development of inversion blacks or nanostructure blacks, and carbon black-silica dual phase fillers.

3.3.2 Silica and Silicates

Addition of silica to a rubber compound offers a number of advantages such as improvement in tear strength, reduction in heat buildup, and increase in compound adhesion in multicomponent products such as tires. Two fundamental properties of silica and silicates influence their use in rubber compounds: ultimate particle size and the extent of hydration. Other physical properties such as pH, chemical composition, and oil absorption are of secondary importance.

Silicas, when compared to carbon blacks of the same particle size, do not provide the same level of reinforcement, though the deficiency of silica largely disappears when coupling agents are used with silica. Wagner reported that addition of silica to a tread compound leads to a loss in tread wear, even though improvements in hysteresis and tear strength are obtained. The tread wear loss can be corrected by the use of silane coupling agents.

The chemistry of silica can be characterized as follows:

① Silica, which is amorphous, consists of silicon and oxygen arranged in a tetrahedral structure of a three-dimensional lattice. Particle size ranges from 1nm to 30nm and surface area from $20m^2/g$ to $300m^2/g$. There is no long-range crystal order, only short-range ordered domains in a random arrangement with neighboring domains.

② Surface silanol concentration (silanol groups —Si—O—H) influence the degree of surface hydration.

③ Silanol types fall into three categories—isolated, geminal (two —OH hydroxyl groups on the same silicon atom), and vicinal (on adjacent silicon atoms)—as illustrated in Fig. 3.6.

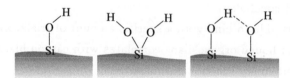

Fig 3.6 Typical silanol groups on silica: "Isolated", "geminated", and associated silanols.

④ Surface acidity is controlled by the hydroxyl groups on the surface of the silica and is intermediate between those of P-OH and B-OH. This intrinsic acidity can influence peroxide vulcanization, although in sulfur curing, there is no significant effect. Rubber-filler interaction is affected by these sites.

⑤ Surface hydration caused by water vapor absorption is affected by surface silanol concentration. High levels of hydration can adversely affect final compound physical properties. Silicas are hydroscopic and thus require dry storage conditions.

To illustrate the influence of surface hydroxyl groups and hydration levels on rubber properties, Wagner took a series of silicas of different surface areas, hydroxylated to different extents, and then added them to an SBR compound at 50phr (Table 3.19). The author concluded that a reduction in silanol level as a result of an increase in absorbed water will decrease cure time, tensile strength, and also abrasion resistance.

Table 3.19 Effect or surface hydration on silica properties

Properties	Silica A	Silica B
Surface area/(m^2/g)	152.0	152.0
Loss at 105℃/(%H_2O)	6.8	0.5
Mooney scorch	14.0	16.0
Rheometer t_{-90}	27.0	47.0
Tensile strength/MPa	18.3	21.7
Elongation at break	480	480
300% modulus/MPa	7.0	9.5
Pico abrasion	67.0	103.0

In general, silicas produce relatively greater reinforcement in more polar elastomers such as NBR and CR than in nonpolar polymers such as SBR and NR. The lack of reinforcement properties of silica in NR and SBR can be corrected through the use of silane coupling agents. An essential prerequisite for a coupling agent is that the molecule be bifunctional, i. e. , capable of reacting chemically with both the silica and either directly or indirectly with the polymer via participation in the vulcanization reaction or sulfur crosslinking process.

Use of silicas in rubber compounds offers two advantages: reduction in heat buildup when used as a part for part replacement of carbon black and improvement in tear strength, cut, chip, and chucking resistance. When loadings approach 20%, however, the drop in abrasion resistance of, for example, a tread compound renders the formulation no longer practical. Silane coupling agents offer the potential to overcome such drops in compound performance. Therefore, to compound silica effectively, a discussion of the properties and chemistry of coupling agents, and specifically silane coupling agents, is pertinent.

Silicas can be divided into three groups or classes. These include standard or conventional silicas, semi-highly dispersible (semi-HD) or easily-dispersible silica, and the latest group developed is termed highly dispersible silica or HDS (Table 3. 20). The silanol composition on the surface of three types of silicas remains to be elucidated, but it would be anticipated that the HDS silicas would have higher concentrations of geminal groups, whereas the conventional silica would have a greater amount of isolated silanols.

Table 3. 20 Silica groups

Surface Area/(m^2/gm)	90~130	130~180	180~220
Conventional	Tire casing Non-tire internal & external components	Tire treads Tire casings & non-tire products external components	Tire treads & non-tire external components for abrasion resistance
Semi-highly dispersible	Tire casing Non-tire products external components	Tire treads Tire casings	Tire treads
Highly dispersible	Tire casings Tire treads	Tire treads	High performance tire treads

3. 3. 3 Chemistry of Silane Coupling Agents

There are three silane coupling agents of commercial significance and these have similar properties: mercaptopropyltrimethoxysilane (A189), bis(triethoxysilylethyltolylene) polysulfide (Y9194), and bis(3-triethoxisilylpropyl) tetrasulfane (TESPT). Commercial designations are in parentheses. The coupling agent TESPT has been covered more extensively in the literature than other silane coupling agents; however, the following discussion on the use of silane coupling agents is applicable to all three materials.

TESPT, a bifunctional polysulfidic organosilane, was introduced as a coupling agent to improve the reinforcement properties of silicas in rubbers. Use of coupling agents offers the following advantages:

① Lowers heat buildup and hysteresis in silica-loaded compounds;
② Increases 300% modulus and tensile strength, again, in silica-loaded compounds;
③ Improves reinforcing effect of clays and whiting;
④ Serves as a reversion resistor in equilibrium cure systems;
⑤ Improves DIN abrasion resistance.

The mechanism of silane coupling agent reinforcement comprises two phases: ① the hydrophobation reaction in which coupling agent reacts with silica, and ② the formation of crosslinks between the modified silica and polymer. Silanization of the silica surface can occur quite readily, though with TESPT systems, the reaction is generally carried out in situ at between 150℃ and 160℃ in an internal mixer. Though an excess of silanol groups are present on the silica surface and reaction rates are fast, this high temperature is required because of the steric hindrance around the silylpropyl group in TESPT.

As noted earlier, three types of functional silanol groups exist on the silica surface: isolated hydroxyl groups, geminal groups (two —OH groups on one Si atom), and vicinal groups (see Fig. 3.6). The silanization reaction is illustrated by the followed:

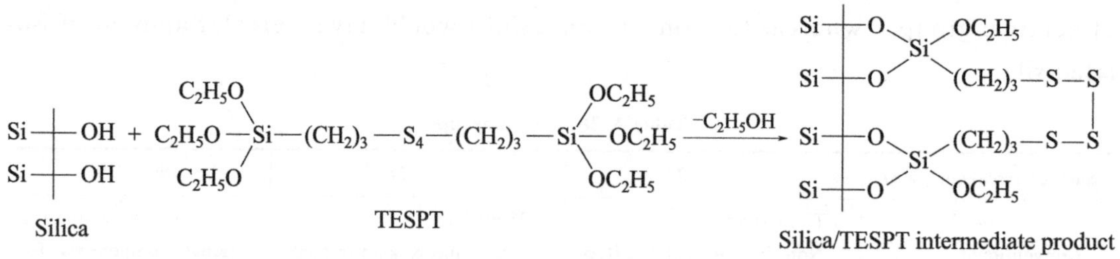

$$\begin{array}{c}\text{Si} - \text{OH} \\ \text{Si} - \text{OH}\end{array} + \begin{array}{c}\text{C}_2\text{H}_5\text{O} \\ \text{C}_2\text{H}_5\text{O} - \text{Si} - (\text{CH}_2)_3 - \text{S}_4 - (\text{CH}_2)_3 - \text{Si} \\ \text{C}_2\text{H}_5\text{O}\end{array} \begin{array}{c}\text{OC}_2\text{H}_5 \\ -\text{OC}_2\text{H}_5 \\ \text{OC}_2\text{H}_5\end{array} \xrightarrow{-\text{C}_2\text{H}_5\text{OH}} \text{Silica/TESPT intermediate product}$$

Silica　　　　　　　　TESPT

(3-5)

The filler/silane intermediate can now react with the allyl position of unsaturated sites on the polymer chain. The vulcanization of rubber is known to proceed via reaction of an accelerator, such as a sulfenamide, with sulfur, zinc oxide, and stearic acid, to generate a sulfurating agent. Eq. (3-6) gives a somewhat simplistic schematic of the vulcanization reaction. On completion of the reaction, the pendant accelerator will cleave off (i.e., Captax) after generation of a crosslink. This accelerator residue, Captax, is an accelerator in its own right and continues to participate in further crosslinking as vulcanization continues. In silica reinforcement systems containing TESPT, Wolff has suggested that the reaction is similar when the TESPT/silica intermediate is present instead of sulfur, in which case the crosslinking agent is the polysulfidic sulfur chain. Wolff showed that mercaptobenzothiazyl disulfide (MBTS) reacts with the tetrasulfane group, thus forming 2 moles of the polysulfide:

$$\text{(scheme 3-6)} \tag{3-6}$$

The silica particle is on one side and the mercaptobenzthiazolyl on the other. This polysulfidic pendant group on the silica surface will now undergo crosslink formation with the polymer in much the same way as occurs in rubber-bound intermediates that convert to crosslinks. Wolff suggested that the MBT entity reacts with the allyl position of a double bond of the rubber, thus releasing MBT and forming the rubber-silica bond.

$$\text{(scheme 3-7)} \tag{3-7}$$

Proper compounding of silica with coupling agents has permitted the use of such filler systems in applications including shoe soles; engine mounts in which coupling agent/silica NR compounds provide the necessary hysteretic properties; tire treads in which, again, hysteretic properties are important; and a range of other applications such as golf balls.

3.3.4 Other Filler Systems

A series of additional filler systems merit brief discussion, not because of their reinforcement qualities but because of their high consumption. These include kaolin clay (hydrous aluminum silicate), mica (potassium aluminum silicate), talc (magnesium silicate), limestone (calcium carbonate), and titanium dioxide.

As with silica, the properties of clay can be enhanced through treatment of the surface with silane coupling agents. Thioalkylsilanes can react with the surface to produce a pendant thiol group which may react with the polymer through either hydrogen bonding, van der Waal forces, or crosslinking with other reactive groups:

$$\text{Clay} \begin{bmatrix} O-(CH_2)_x-SH \\ O-(CH_2)_x-SH \end{bmatrix} \tag{3-8}$$

Such clays show improved tear strength, an increase in modulus, improved component-to-component adhesion in multicomponent products, and improved aging properties.

Calcium carbonate is used as a low-cost filler in rubber products for static applications such as carpet underlay. Titanium dioxide finds extensive use in white products such as white tire sidewalls where appearance is important.

3.4 Stabilizer Systems

The unsaturated nature of an elastomer accounts for its unique viscoelastic properties. However, the presence of carbon-carbon double bonds render elastomers susceptible to attack by oxygen, ozone, and also thermal degradation. A comprehensive review of elastomer oxidation and the role of anti-oxidants and antiozonants is available.

3.4.1 Degradation of Rubber

Oxidation of elastomers is accelerated by a number of factors including heat, heavy metal contamination, sulfur, light, moisture, swelling in oil and solvents, dynamic fatigue, oxygen, and ozone. Three variables in the compound formulation can be optimized to resist degradation: polymer type, cure system, and antidegradant system.

Thermooxidative stability is primarily a function of the vulcanization system. Peroxide vulcanization or cure systems tend to perform best for reversion resistance as a result of the absence of sulfur and use of carbon-carbon crosslinks. Efficient vulcanization (EV) systems that feature a low sulfur level (0.0~0.3phr), a high acceleration level, and a sulfur donor similarly show good heat stability and oxidation resistance. Such systems do, however, have poor resistance to fatigue because of the presence of predominantly monosulfidic crosslinks. Conventional cure systems that feature a high sulfur level and low accelerator concentration show poor heat and oxidation resistance because the polysulfidic crosslinks are thermally unstable and readily oxidized. Such vulcanization systems do, however, have better fatigue resistance. Semi-EV cure systems, which are intermediate between EV and conventional systems, are a compromise between resistance to oxidation and required product fatigue performance.

Oxidation proceeds by two fundamental mechanisms.

① Crosslinking: A predominantly di- or polysulfidic crosslink network breaks down into monosulfidic crosslinks. Compound hardness increases, fatigue resistance decreases, and the compound becomes much stiffer. SBR, EPDM, NBR, and polychloroprene tend to show this behavior.

② Chain scission: The polymer chain breaks, causing a softening of the compound and decreased abrasion resistance. Natural rubber tends to show such degradation.

The degradation of unsaturated elastomers is an autocatalytic, free radical chain reaction, which can he broken into three steps:

Initiation

$$RH(elastomer) \xrightarrow[heat]{O_2, light} R\cdot + HOO\cdot \text{ (free radical)} \qquad (3-9)$$

Propagation

$$R\cdot + O_2 \longrightarrow ROO\cdot$$
$$ROO\cdot + R'H \longrightarrow ROOH + R'\cdot$$
$$ROOH \longrightarrow RO\cdot + \cdot OH$$
$$RO\cdot + R''H \longrightarrow ROH + R''\cdot$$
$$\cdot OH + R'' \longrightarrow R''\cdot + H_2O \qquad (3-10)$$

Termination

$$R\cdot + \cdot R$$
$$R\cdot + ROO\cdot \longrightarrow \text{Nonradical products}$$
$$ROO\cdot + \cdot OOR \qquad (3-11)$$

Like any chemical process, the rate of reaction will increase with temperature. Increase in service temperature will thus accelerate the degradation of rubber, the rate of reaction with oxygen being governed by the Arrhenius equation.

Ultraviolet light initiates free radical oxidation at the exposed surface of an elastomeric product to generate a layer of oxidized rubber. Heat, moisture, or high humidity can then initiate crazing of the surface which subsequently can be abraded off. Such degradation of the surface is more severe with non-black stocks than with black compounds. Nonblack compounds such as white tire sidewalls thus require higher levels of nonstaining antioxidants than carbon black-loaded formulations.

Heavy transition metals ions such as iron, manganese, and copper catalyze oxidation of elastomers. Compounds of manganese or copper such as oleates and stearates are readily soluble in rubber, enabling rapid oxidation of the polymer. para-Phenylenediamine antidegradants are used to hinder the activity of such metal ions.

A major cause of failure in rubber products is surface crack development. The growth of such cracks under cyclic deformation results in fatigue failure. Fatigue-related cracks are initiated at high stress zones. Attack by ozone can induce crack initiation at the surface which then propagates as a result of flexing. Ozone-initiated cracking can be seen as crazing on the sidewalls of old tires. Ozone readily reacts with the carbon-carbon double bonds of unsaturated elastomers to form ozonides. Under strain, ozonides readily decompose, resulting in chain cleavage and a reduction in polymer molecular weight. Such polymer molecular weight reduction becomes apparent as surface crazing and cracking:

$$R-CH=CH-R \xrightarrow{O_3} R-HC\underset{O-O}{\overset{O}{\diagup\diagdown}}CH-R \xrightarrow{H_2O} \begin{array}{c} R-C{\diagup\diagdown}_H^O \\ + \\ R-C{\diagup\diagdown}_{OH}^O \end{array} \quad (3-12)$$

Unsaturated elastomer　　　Ozonide　　　Decomposition products

Polymer blends, in which the constituent polymers are incompatible, tend to improve fatigue resistance. For example, natural rubber and polybutadiene show good resistance to fatigue, crack initiation, and growth because of the formation of heterogeneous polymer phases; a crack growth in one polymer phase is arrested at the boundary with the adjacent polymer phase.

Natural rubber and polybutadiene blends tend to be used in tire side-walls which undergo flexing, and also in tire treads which have a lug pattern and contain high-stress zones at the base of the tread blocks. In summary, the addition of antidegradants becomes important in order to protect the elastomeric compound from this broad range of enviromental, chemical, and service related aging phenomena.

3.4.2 Antidegradant Use

The selection criteria governing the use of antidegradants can be summarized as follows:

① Discoloration and staining: In general, phenolic antioxidants tend to be nondiscoloring and amines are discoloring. Thus for elastomers containing carbon black, more active amine antioxidants are preferred as discoloration is not important.

② Volatility: As a rule, the higher the molecular weight of the antioxidant, the less volatile it will be, though hindered phenols tend to be highly volatile compared with amines of equivalent molecular weight. Thus, correct addition of antioxidants in the compound mix cycle is critical if loss of material is to be avoided.

③ Solubility: Low solubility of an antidegradant will cause the material to bloom to the surface, with consequent loss of protection of the product. Therefore, solubility of antidegradants, particularly antiozonants, controls their effectiveness. The materials must be soluble up to 2.0 phr, must be able to migrate to the surface, but must not be soluble in water or other solvents such as hydraulic fluid so as to prevent extraction of the protectant from the rubber.

④ Chemical stability: Antidegradant stability against heat, light, oxygen, and solvents is required for durability.

⑤ Concentration: Most antidegradants have an optimum concentration for maximum effectiveness after which the material solubility becomes a limiting factor. para-Phenylenediamines offer good oxidation resistance at a loading of 0.5 phr to 1.0 phr and antiozonant protection in the range 2.0 phr to 5.0 phr. Above 5.0 phr para-phenylenediamines tend to bloom.

⑥ Environment, health, and safety: For ease of handling and avoidance of dust and inhalation, antidegradants should be dust free while free flowing.

3.4.3 Antidegradant Types

(1) Nonstaining antioxidants

This class of antioxidants is subdivided into four groups: phosphites, hindered phenols, hindered bisphenols, and hydroquinones. Hindered bisphenols such as 4,4-thiobis(6-t-butyl-m-cresol) are the most persistent of the four classes of material. Because of their lower molecular weight, hindered phenols tend to be volatile. Phosphites tend to be used as synthetic rubber stabilizers, and hydroquinones such as 2,5-di-$tert$-amylhydroquinone are used in adhesives:

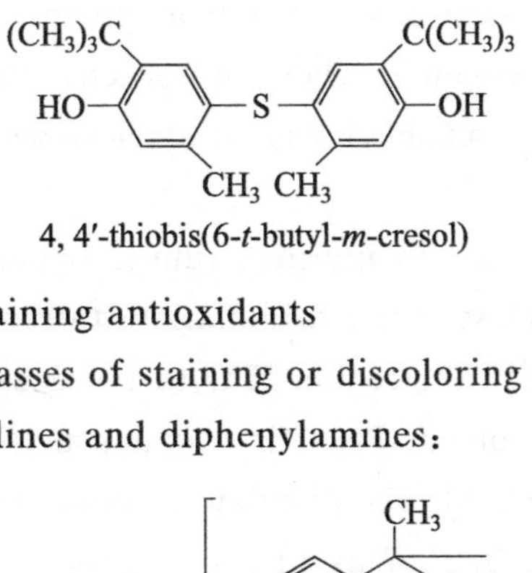

4,4'-thiobis(6-t-butyl-m-cresol)

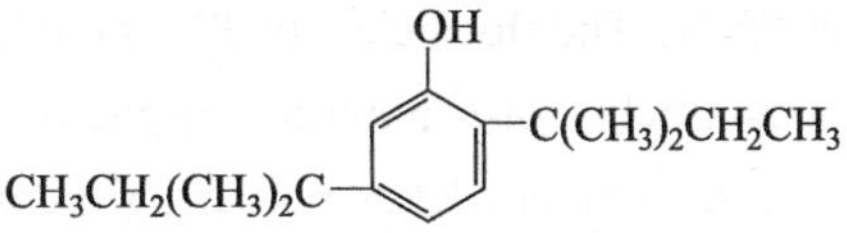

2,5-di-$tert$-amylhydroquinone

(2) Staining antioxidants

Two classes of staining or discoloring antioxidants find extensive use, polymerized dihydroquinolines and diphenylamines:

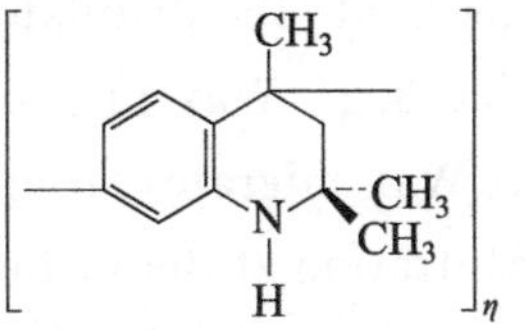

Polymerized 1,2-dihydro-2,2,4-trimethylquinoline

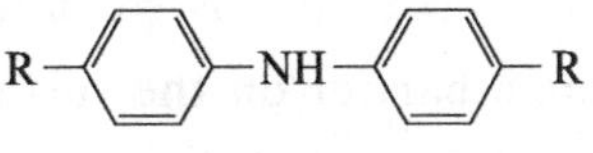

Diphenylamine class antioxidants

Dihydroquinolines differ in the degree of polymerization, thus influencing migratory and long-term durability properties. They are good general antioxidants and also are effective against heavy metal prooxidants such as nickel and copper ions. The polymeric nature of dihydroquinolines results in low volatility and migratory properties in a vulcanizate. Thus, there is minimum loss of protectant through extraction or diffusion, durability is improved, and high-temperature stability is improved. Diphenylamine antioxidants tend to show a directional improvement in compound fatigue resistance.

(3) Antiozonants

para-Phenylenediamines (PPDs) are the only class of antiozonants used in significant quantities. The general structure is:

$$R-NH-\langle\bigcirc\rangle-NH-R'$$

They not only serve to protect rubber products from ozone but also improve resistance to fatigue, oxygen, heat, and metal ions. There are three general categories of paraphenylenediamines and are listed as follows.

i. *Dialkyl PPDs*: The substituent R groups are both alkyls, as in diisopropyl-*p*-phenylene diamine. The R group can range from C_3 up to C_9. Dialkyl PPD antidegradants tend to induce higher levels of scorch in a compound than other classes of PPD antidegradants, and tend to migrate faster than other PPD because of their low molecular weight. They lack persistence.

ii. *Alkyl-aryl PPDs*: One R group is an aromatic ring; the other is an alkyl group. The most widely used PPD in this class is *N*-1, 3-dimethylbutyl-*N*-phenyl-*p*-phenylenediamine. This antiozonant offers good dynamic protection, good static protection when combined with wax, better compound processing safety and scorch safety, and, slower migratory properties, allowing it to be more persistent and suitable for long product life.

iii. *Diaryl PPDs*: The third class of PPDs contain two aromatic pendant groups, as in diphenyl-*p*-phenylenediamine or di-*β*-naphthyl-*p*-phenylenediamine. They are less active than alkyl-aryl PPDs and also tend to bloom, thus rendering them unsuitable for many applications.

(4) Waxes

Waxes are an additional class of materials used to improve rubber ozone protection primarily under static conditions. Waxes used in elastomeric formulations fall into two categories: Microcrystalline wax has a melting point in the region 55℃ to 100℃ and is extracted from residual heavy lube stock of refined petroleum. Paraffin wax has melting points in the range 35℃ to 75℃ and is obtained from the light lube distillate of crude oil.

The properties of waxes are listed in Table 3.21. Wax protects rubber against static ozonolysis by forming a barrier on the surface. Wax migrates from the bulk of the rubber continuously, maintaining an equilibrium concentration at the surface. Microcrystalline waxes migrate to the rubber surface at a slower rate than paraffins because of the higher molecular weight and branching. Furthermore, microcrystalline waxes tend to perform best at high service temperatures, whereas paraffin waxes protect best at low temperatures. This is related to the rate of migration of the wax to the product surface.

Table 3.21 Composition of paraffin and microcrystalline waxes

Characteristics	Microcrystalline	Paraffin
Molecular weight	500~800	340~430
Melting point/℃	55~100	35~75
Mean carbon chain length	C_{25}	C_{60}
Features	Branched molecules	Linear molecules

It should be noted that under dynamic conditions, the protective wax film breaks down, after which the antiozonant system in the rubber formulation will serve as the primary stabilizer or protection mechanism. Waxes are used to ensure protection against ozone for products in storage, such as tires in a warehouse.

In summary, a number of empirical guidelines can be used to develop an antidegradant system for an elastomeric formulation:

① Short-term static protection is achieved by use of paraffinic waxes;

② Microcrystalline waxes provide long-term ozone protection while the finished product is in storage;

③ A critical level of wax bloom is required to form a protective film for static ozone protection;

④ Optimized blends of waxes and PPDs provide long-term product protection under both static and dynamic applications and over a range of temperatures;

⑤ Excess levels of wax bloom can have a detrimental effect on fatigue resistance, because the thick layer of wax can crack under strain and the crack can propagate into the product.

3.5 Vulcanization System

Vulcanization, named after Vulcan, the Roman God of Fire, describes the process by which physically soft, compounded rubber materials are converted into high-quality engineering products. The vulcanization system constitutes the fourth component in an elastomeric formulation and functions by inserting crosslinks between adjacent polymer chains in the compound. A typical vulcanization system in a compound consists of three components:

① activators;

② vulcanizing agents, typically sulfur; and

③ accelerators.

The chemistry of vulcanization has been reviewed elsewhere in this text. It is appropriate, however, to review each of these components within the context of developing a compound for a defined service application.

3.5.1 Activators

The vulcanization activator system consisting of zinc oxide and stearic acid has received much less research effort than other components in the rubber compound. Stearic acid and zinc oxide levels of 2.0 phr and 5.0 phr, respectively, are accepted throughout the rubber industry as being adequate for achievement of optimum compound physical properties when in combination with a wide range of accelerator classes and types and also accelerator-to-sulfur ratios. To clarify why zinc oxide is selected over the other metal oxides, a comparative study was conducted with magnesium oxide, calcium oxide, titanium dioxide, lead oxide, and zinc oxide. All the metal oxides were evaluated in ASTM D3184; compound numbers 1A (gum stock) and 2A (which contains carbon black), are also referred to as American Chemical Society (ACS) compounds 1 and 2, respectively (Table 3.22). Test data are presented in Table 3.23 and Table 3.24.

Table 3.22 ASTM D3184 formulations 1A and 2A (ACS 1 and 2)

ACS 1		ACS 2	
Natural rubber	100	Natural rubber	100
Metal oxide	6	Metal oxide	5
Stearic acid	0.5	Stearic acid	2
Sulfur	3.5	Carbon black (IRB 5)	35
MBT	0.5	Sulfur	2.25
		TBBS	0.7

Table 3.23 Effect of metal oxide type on compound properties of ACS 1 base formulation

Compound	1	2	3	4	5	6
Metal oxide	MgO	CaO	TiO_2	FeO	ZnO	PbO
Cation Electronegativity	1.2	1.0	1.5	1.9	1.6	1.8
Free sulfur/%	2.69	2.51	2.60	2.74	1.43	1.03
Monsanto rheometer at 150℃, deta torque[M_H-M_L]/(dN·m)	14.0	17.0	19.0	24.0	29.0	43.0
Tensile strength/MPa	4.84	6.37	3.09	4.40	14.80	20.0
Elongation/%	731	695	817	530	667	634
300% modulus/MPa	0.90	1.14	0.58	1.85	2.96	2.2
Shore A hardness at 21℃	32	34	26	24	38	42
ASTM tear strength, die B/(kN/m)	38	53	11	21.5	68	67

Table 3.24 Effect of metal oxide type on compound properties of ACS 2 base formulation

Compound	1	2	3	4	5	6
Metal oxide	MgO	CaO	TiO_2	FeO	ZnO	PbO
Cation Electronegativity	1.2	1.0	1.5	1.9	1.6	1.8
Free sulfur/%	0.39	0.35	0.72	0.68	0.15	0.15
Monsanto rheometer at 150℃, deta torque [M_H-M_L]/(dN·m)	24.0	23.5	18.50	29.5	61.5	54.0
Tensile strength/MPa	14.21	20.28	12.03	15.41	25.63	26.38
Elongation/%	631	592	595	565	492	502
300% modulus/MPa	2.94	4.86	2.91	2.53	11.04	8.9
Shore A hardness at 21℃	39	46	40	39	57	57
ASTM tear strength, die B/(kN/m)	30	61	22	31.5	161	140

A plot of the electronegativity of the six metals of the oxides evaluated in the study versus rheometer torque (M_H-M_L) indicates that outside a given electronegativity range of 1.6 to 1.8, optimum vulcanizate properties will not be obtained (see Fig. 3.6~Fig. 3.8). Electronegativity is a measure of the metal atom affinity for electron attraction. Viewing Fig. 3.6~Fig. 3.8 it can be concluded that for metals of electronegativity less than 1.55, a consequent shift to ionic bonding with sulfur induces a reduction in electrophilicity in the penultimate sulfur atoms of complexes：

$$X—S_x—S^- \cdots M^{2+} \cdots S^-—S_x—X$$
$$R—H \cdots S_y—X$$

Conversely, with metals of electronegativity greater than 1.85, such as iron, the greater covalent character of the M+··S-linkage with reduced charge separation would adversely affect generation of amine or carboxylate ligands to the metal ion as in

which in turn will reduce the solubility of the sulfurating reagent, consequent drop in sulfurating agent activity, and resultant drop in vulcanizate properties.

In summary, zinc is most suited to participate in formation of the sulfurating complex. Coordination of external ligands (ROO—, R_2'NH) of the zinc atom causes the bonding between XS-S_x··· and ···S_y-SX groups to weaken, thereby increasing the contribution of the polar canonical form:

$$XS—S_x^-—Zn^{2+}—S_y^-—SX$$

This effect is induced by ligands satisfying vacant 4p orbitals and distributing positive charge from the metal. The result will be increased nucleophilicity of XSS_x^- but decreased electrophilicity of XS_y^+ in the sulfurating complex. The same is true for Cd^{2+} and Pb^{2+} complexes which have vacant p orbitals to accommodate coordination ligands. In the case of Mg^{2+} and Ca^{2+} complexes, coordination will not readily occur, the reduced ease of formation being further influenced by the inability of the metal to achieve an inert gas configuration as in more stable organometallics. Toxicity of CdO and PbO prohibits their use, and thus ZnO has found virtually universal use in the rubber industry, the ultimate loading in a compound being dependent on the product application.

As part of the metal oxide study, a comparative study of oleic acid and stearic acid, each at 1.0 phr, 2.0 phr, and 3.0 phr, was conducted on ASTM No. 2A (ACS 2) compound.

Table 3.25 Influence of fatty acid level in vulcanizates

Compound	1	2	3	4	5	6
Fatty acid	Stearic			Oleic		
phr	1.0	2.0	3.0	1.0	2.0	3.0
Crosslink density (rating)	100	94	106	75	80	89
Activation energy/(kJ/mol)	131.5	101.5	97.6	135.1	114.2	110.5
Tensile strength/MPa	27.50	26.8	26.9	28.5	28.0	26.4
Elongation/%	545	535	538	591	576	551
Shore A hardness	52	53	50	50	52	52
Tear strength ASTM die B/(kN/m)	72	112	103	72	94	80
Aged tensile strength/MPa	17.50	18.1	21.3	15.8	16.8	17.3

The data outlined in Table 3.25 illustrate a number of points:

① An increase in the fatty acid level reduces vulcanization activation energy, the effect being greater for stearic acid.

② Stearic acid/ZnO-activated compounds show higher crosslink densities compared with oleic acid systems.

③ Aging and tear strength properties of stearic acid/ZnO compounds are superior to those of oleic acid systems.

The effectiveness of stearic acid in activating vulcanization is a function of its solubility in the elastomer, molecular weight, and melting point.

3.5.2 Vulcanizing Agents

Three vulcanizing agents find extensive use in the rubber industry: sulfur, insoluble sulfur, and peroxides. The chemistry of peroxides has been reviewed elsewhere. Rhombic sulfur is the most common form of sulfur used in the rubber industry and, other than normal factory hygiene and operational procedures, does not require any special handling or storage. Sulfur is soluble in natural rubber at levels up to 2.0 phr. Above this concentration, insoluble sulfur must be used to prevent migration of sulfur to the compound surface, i.e., sulfur bloom.

3.5.3 Accelerators

Accelerators are products which increase both the rate of sulfur crosslinking in a rubber compound and crosslink density. Secondary accelerators, when added to primary accelerators, increase the rate of vulcanization and degree of crosslinking, with the terms primary and secondary being essentially arbitrary. A feature of such binary acceleration systems is the phenomenon of synergism. Where a combination of accelerators is synergistic, its effect is always more powerful than the added effects of the individual components.

Accelerators can be readily classified by one of two techniques:

① Rate of vulcanization: Ultra-accelerators include dithiocarbamates and xanthates. Semiultra-accelerators include thiurams and amines. Fast accelerators are thiazoles and sulfenamides. A medium-rate system is diphenylguanidine. A slow accelerator is thiocarbanilide.

② Chemical classifications: Most accelerators fall into one of the following eight groups.

| Aldehydeamines | Sulfenamides | Thioureas | Dithiocarbamates |
| Guanidines | Thiurams | Thiazoles | Xanthates |

Factors involved in the selection of vulcanization systems must include the type of elastomer, type and quantity of zinc oxide and fatty acid, rate of vulcanization, required resistance to fatigue, and service conditions. It is also recommended that use of nitrosamine-generating accelerators be avoided.

The type of elastomer will influence the rate of cure and also the resultant crosslink network. Natural rubber tends to cure faster than SBR. Cure systems containing thiuram accelerators such as tetramethylthiuram disulfide will show short induction times and fast cure rates compared with a system containing diphenylguanidine.

Sulfenamide accelerators represent the largest class of accelerators consumed on a global basis:

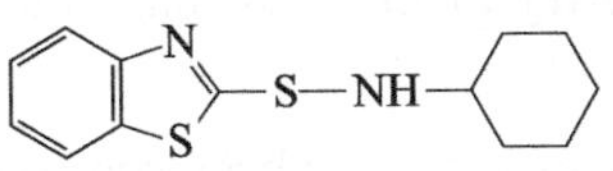

Cyclohexylbenzothiazole sulfenamide (CBS) 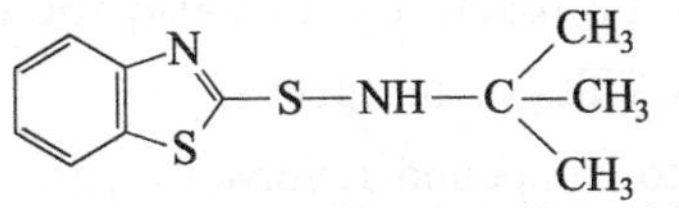   tert-Butyl-2-benzothiazole sulfenamide (TBBS)

The mechanism and chemistry of vulcanization have been reviewed earlier. It is therefore more appropriate to define the general principles governing the activity of an accelerator such as a sulfenamide. Three parameters merit elucidation:

① Bond strength of the sulfur-nitrogen bond: Sulfenamides are cleaved into mercapto benzothiazole and amine fragments during formation of the sulfurating complex, and the amine forms ligands with the zinc ion. Bond energy must be sufficiently low so as not to prevent generation of active accelerator species or sulfurating reagent.

② Stereochemistry of the amine fragment: The steric bulk of the amine ligand coordinated with the zinc ion, if too large, can hinder the formation of an active sulfurating agent. This is seen as an increase in induction times, change in vulcanization rate, and, ultimately, change in physical properties.

③ Basic strength of the amine fragment: An increase in the basicity of the amine fragment of the sulfenamide results in an increase in the rate of vulcanization. More basic amines also tend to induce poor scorch resistance (Table 3.26).

Table 3.26 Effect of sulfenamide amide fragment basicity (pK_b) on compound scorch activity

R radical	pK_a of free amine	Mooney scorch t_{10} at 135°C	Cure rate index
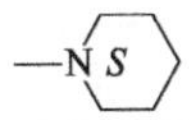 —N⟨S⟩ (piperidine-S ring)	3.3	29.8	1.00
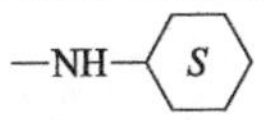 —NH—⟨S⟩	3.7	32.7	2.05
—NH—C(CH₃)₃	4.2	31.8	2.10
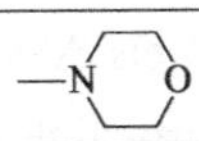 —N⟨O⟩ (morpholine)	6.2	42.4	2.32
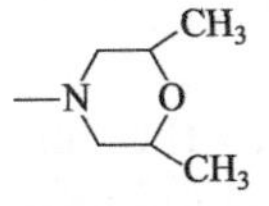 —N⟨O⟩(CH₃)₂ (dimethyl morpholine)	6.2	45.2	2.57

1. Compound: 100 phr SBR-1500, 50 HAF, 4.0 ZnO, 2.0 stearic acid, 10.0 oil, 1.5 antioxidant, 1.75 sulfur, equimolar accelerator levels.
2. Cure rate index is ($t_{35} - t_{10}$) at 135°C on a Mooney plastimeter.

3.5.4 Retarders and Antireversion Agents

The induction time or scorch resistance of a compound can be improved by addition of a retarder. *N*-Cyclohexylthiophthalimide (CTP) is by far the largest-tonnage retarder used in the rubber industry. The reader is referred to the review by Morita for discussion of the mechanism of CTP reactivity and also the chemistry of other special retarders such as the thiosulfonamide class of materials.

Resistance to compound reversion, particularly of natural rubber compounds, has received more attention because of the broad range of requirements including faster processing of compounds in production, processing at higher temperatures, and, perhaps more important, extension of product service life. Three antireversion agents have been used commercially:

① A semi-EV system is a compromise designed to produce, in structural terms, a vulcanizate containing a balance of monosulfidic and polysulfidic crosslinks at a defined optimum cure state. If polysulfidic crosslinks are to persist over extended periods, new ones must be created to replace those lost through reversion. With use of normal accelerations systems, there is limited opportunity for such events. Maintenance of a polysulfidic network through the curing process thus dictates utilization of a dual-cure system both of which are independent of each other. This is the principle of the equilibrium cure (EC) system. Here, bis(3-triethoxysilylpropyl)tetrasulfane (TESPT) is added as a slow sulfur donor (Fig. 3.7).

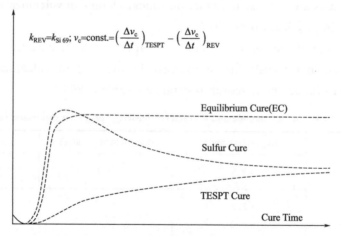

Fig. 3.7 Rheometer profile of the EC system.

② Bis(citraconimidomethyl)benzene, commercial name Perkalink 900, has been introduced which functions exclusively as a reversion resistor. It is understood to react via a Diels-Alder reaction to form a six membered ring on the polymer chain (Fig. 3.8). The ultimate crosslink is thermally stable and replaces sulfur crosslinks that disappear during reversion.

③ Sodium hexamethylene-1,6-bisthiosulfide dihydrate, when added to the vulcanization system, breaks down and inserts a hexamethylene-1,6-dithiyl group within a disulfide or polysulfide crosslink. This is termed a hybrid crosslink. During extended vulcanization periods or accumulated heat history due to product service, polysulfidic-hexamethylene

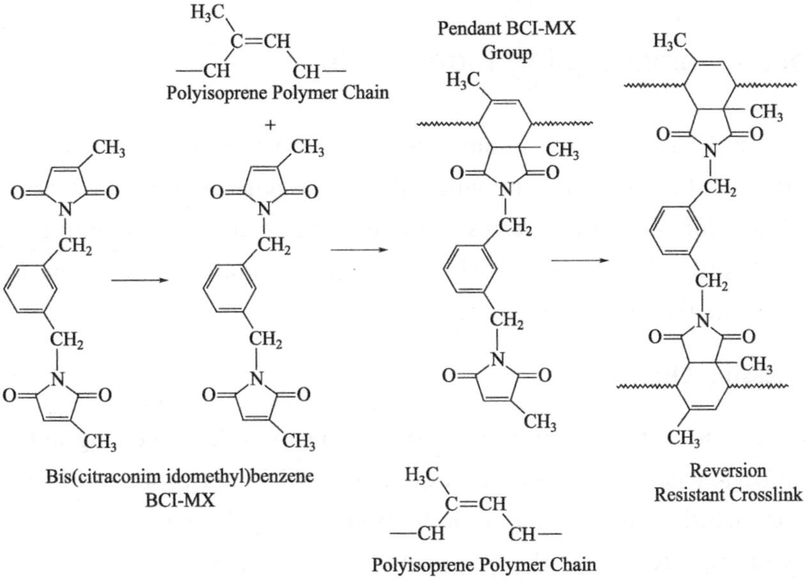

Fig. 3.8 Proposed reversion resistance mechanism of bis(citraconimidomethyl)benzene (BCIMX).

crosslinks shorten to produce thermally stable elastic monosulfidic crosslinks. At levels up to 2.0 phr, there is little effect on compound induction or scorch times, nor on other compound mechanical properties.

$$^+Na^-SO_3-S-(CH_2)_6-S-SO_3^-Na^+ \cdot 2H_2O$$

Hexamethylene-1,6-bis(thiosulfate)disodium salt,dihydrate

Table 3.27 shows industry recognized abbreviations for various accelerators.

Table 3.27 Comparison of the scorch activity of a range of selected commercial accelerators[a,b]

Accelerator	Abbreviation	Scorch time/min	
		ACS 1	ACS 2
Diisopropylbenzothiazole	DIBS	63	21
Cyclohexyl-2-benzothiazolesulfenamide	CBS	26	15
tert-Butyl-2-benzothiazolesulfenamide	TBBS	60	23
Oxydiethylene benzothiazole-2-sulfenamide	MBS	60	24
Dimethylmorpholine benzothiazole-2-sulfenamide	—	60	27
Benzothiazyl-N,N-diethylthiocarbamyl sulfide	—	26	15
Mercaptobenzothiazole	MBT	12	9
Benzothiazyl disulfide	MBTS	73	12
Tetramethylthiuram monosulfide	TMTM	25	12
Tetramethylthiuram disulfide	TMTD	13	6
Tetraethylthiuram disulfide	TETD	18	9
Zinc dibutyldithiocarbamate	ZDBC	5	4
Zinc dimethyldithiocarbamate	ZDMC	6	4
Diphenylguanidine	DPG	16	10

[a] Data obtained from the formulation ACS 2 (ASTM D3184-89, 2a) but containing 50phr HAF.

[b] Use of nitrosamine-generating accelerators should be avoided.

3.6 Special Compounding Ingredients

In addition to the four primary components in a rubber formulation, i. e., the polymer system, fillers, stabilizer system, and vulcanization system, there are a range of secondary materials such as processing aids, resins, and coloring agents (e. g., titanium dioxide used in tire white sidewalls). These are briefly discussed to establish a guideline for the use of the materials in practical rubber formulations.

3.6.1 Processing Oils

Process oils in a rubber formulation serve primarily as a processing aid. Oils fall into one of three primary categories: paraffinic, naphthenic, and aromatic. The proper selection of oils for inclusion in a formulation is important. If the oil is incompatible with the polymer, it will migrate out of the compound with consequent loss in required physical properties, loss in rubber component surface properties, and deterioration in component-to-component adhesion, as in a tire. The compatibility of an oil with a polymer system is a function of the properties of the oil such as viscosity, molecular weight, and molecular composition. Table 3.28 defines the physical properties of three typical classes of oils.

Table 3.28 Physical properties of three classes of oils used in the rubber industry

Physical property	ASTM methods	Paraffinic	Naphthanic	Aromatic
Specific gravity	D1250	0.85~0.89	0.91~0.94	0.95~1.0
Pour point/℃	D97	−18 to −13	−40 to −7	+4 to +33
Refractive index	D1747	1.48	1.51	1.55
Aniline point	D611	200~260	150~210	95.0~150.0
API gravity	D287	28.0~34.0	19.0~28.0	10.0~19.0
Molecular weight	D2502	320~650	300~460	300~700
Aromatic content/%		15.0	44.0	68.0

Aniline point is a measure of the aromaticity of an oil. It is the point at which the oil becomes miscible in aniline. Thus the lower the aniline point, the higher the aromatic content. All three classes of oils contain high levels of cyclic carbon structures; the differences are in the number of saturated and unsaturated rings. Oils can therefore be described qualitatively as follows:

① Aromatic oils contain high levels of unsaturated rings, unsaturated naphthanic rings, and pendant alkyl and unsaturated hydrocarbon chains. The predominant structure is aromatic.

② Naphthenic oils have high levels of saturated rings and little unsaturation.

③ Paraffinic oils contain high levels of naphthenic rings but also higher levels of alkyl pendant groups, unsaturated hydrocarbon pendant groups, and, most important, fewer naphthenic groups per molecule. Pure paraffins from refined petroleum condense out as wax.

The selection of an oil for a given polymer depends on the presence of polar groups in the polymer, such as —CN groups in NBR and —Cl in CR. Hydrogen bonding and van der Waals forces impact on the effectiveness of an oil in a compound. Table 3.29 presents a general guide for selection of an oil for a given polymer. This selection guide is necessarily brief and there are many exceptions. The key parameters to be noted though are the oil's tendency to discolor the product, the oil's tendency to stain adjacent component in a product, and the solubility of the oil in the polymer.

Table 3.29 Oil selection guide for range of commercial elastomers

Oil	Polymer	Examples of product applications
Naphthenic	Ethylene-propylene rubber	Sealants, caulking, Adhesives General rubber products
	EPDM	
	Polychloroprene	
	SBR	
	PBD	
Paraffinic	Natural rubber	Textile application Caulking Sealants
	Polyisoprene	
	Butyl	
	SBR	
	Polychloroprene	
Aromatic	Natural rubber	Tires Automotive components
	SBR	
	Polybutadiene	

3.6.2 Plasticizers

Though processing oils, waxes, and fatty acids can be considered as plasticizers, within the rubber industry the term plasticizer is used more frequently to describe the class of materials which includes esters, pine tars, and low-molecular-weight polyethylene.

Phthalates are the most frequently used esters. Dibutylphthalate (DBP) tends to give soft compounds with tack; dioctylphthalate (DOP) is less volatile and tends to produce harder compounds because of its higher molecular weight. Polymeric esters such as polypropylene adipate (PPA) are used when low volatility is required along with good heat resistance.

Though total consumption is tending to fall, pine tars are highly compatible with natural rubber, give good filler dispersion, and can enhance compound properties such as fatigue resistance and component-to-component adhesion which is important in tire durability. Other low-volume plasticizers include factice (sulfur-vulcanized vegetable oil); fatty acid salts such as zinc stearate, which can also act as a peptizer; rosin; low-molecular-weight polypropylene; and organosilanes such as dimethylpolysiloxane.

3.6.3 Chemical Peptizers

Peptizers serve as either oxidation catalysts or radical acceptors, which essentially remove free radicals formed during the initial mixing of the elastomer. This prevents polymer recombination, allowing a consequent drop in polymer molecular weight, and thus the reduction in compound viscosity. This softening polymer then enables incorporation of the range of compounding materials included in the formulation. Examples of peptizers are pentachlorothiophenol, phenylhydrazine, certain diphenylsulfides, and xylyl mercaptan. Each peptizer has an optimum loading in a compound for most efficiency.

Peptizers such as pentachlorothiophenol are generally used at levels between 0.1 phr and 0.25 phr. This enables significant improvement in compound processability, reduction in energy consumption during mixing, and improvement in compound uniformity. High levels can, however, adversely affect the compound properties, as excess peptizer continues to catalyze polymer break-down as the product is in service.

3.6.4 Resins

Resins fall into one of three functional categories: ①extending or processing resins, ②tackifying resins, and ③ curing resins. Resins have been classified in an almost arbitrary manner into hydrocarbons, petroleum resins, and phenolic resins.

Hydrocarbon resins tend to have high glass transition temperatures so that at processing temperatures they melt, thereby allowing improvement in compound viscosity mold flow. They will, however, harden at room temperature, thus maintaining compound hardness and modulus. Within the range of hydrocarbon resins, aromatic resins serve as reinforcing agents, aliphatic resins improve tack, and intermediate resins provide both characteristics. Coumarone-indene resin systems are examples of such systems. These resins provide:

(1) Improved tensile strength as a result of stiffening at room temperature;

(2) Increased fatigue resistance as a result of improved dispersion of the fillers and wetting of the filler surface;

(3) Retardation of cut growth by dissipation of stress at the crack tip (as a result of a decrease in compound viscosity).

Petroleum resins are a by-product of oil refining. Like hydrocarbon resins, a range of grades are produced. Aliphatic resins which contain oligomers of isoprene tend to be used as tackifiers, whereas aromatic resins, which also contain high levels of dicyclopentadiene, tend to be classed more as reinforcing systems.

Phenolic resins are of two types, reactive and nonreactive. Nonreactive resins tend to be oligomers of alkyl-phenyl formaldehyde, where the para-alkyl group ranges from to C_4 to C_9. Such resins tend to be used as tackifying resins. Reactive resins contain free methylol groups. In the presence of methylene donors such as hexamethylenetetramine, crosslink networks will be created, enabling the reactive resin to serve as a reinforcing resin and adhesion promoter.

3.6.5 Short Fibers

Short fibers may be added to compounds to further improve compound strength. They can be processed just as other compounding ingredients. Short fibers include nylon, polyester, fiberglass, aramid, and cellulose. The advantages of adding short fibers to reinforce a compound depend on the application for which the product is used; however, general advantages include improved tensile strength, improvement in fatigue resistance and cut growth resistance, increase in stiffness, increased component or product stiffness, improved cutting and chipping resistance as in tire treads.

3.7 Compound Development

The preceding discussion reviewed the range of materials which are combined in an elastomeric formulation to generate a defined set of mechanical properties. Elastomeric formulations can be developed by one of two techniques. Model formulations can be obtained from raw material supplier literature or other industry sources such as trade journals. Such formulations approximate the required physical properties to meet the product performance demands. Further optimization might then include, for example, acceleration level determination to meet a required compound cure induction time, and carbon black level evaluation to match a defined tensile strength target. Where more complex property targets must be met and no model formulations are available, a more efficient technique is to use either Taguchi analysis or multiple regression analysis.

A series of components in a formulation can be optimized simultaneously through use of a computer optimization. A number of models are suitable for use in designed experiments. Regardless of the technique or model selected, a series of simple steps are still pertinent before the experimental work is initiated:

① Definition of the objective of the work;
② Identification of the variables in the formulation to be analyzed;
③ Selection of the appropriate analysis for the accumulated experimental data;
④ Analysis of the data within the context of previously published data and knowledge of the activity and characteristics of the raw materials investigated;
⑤ Statistical significance of the data (data scatter, test error, etc.).

The designed experiment will then entail:

① Define the key property targets, such as tensile strength, fatigue resistance, and hysteretic properties;
② Select an appropriate design, for example a two-variable factorial or three-, four-, or five-variable multiple regression;
③ Calculate multiple regression coefficients from the accumulated experimental data. The coefficients can be computed from the regression equations which can be either a linear equation,

$$\text{property} = a X + b Y + C$$

in the case of a simple factorial design, or a second-order polynomial where interactions between components in a formulation can be viewed:

$$\text{property} = a X + b Y + c Z + d X^2 + e Y^2 + f Z^2 + g XY + h XZ + j YZ + C$$

Here, a property or the dependent variable might be modulus, and X、Y、and Z are independent variables such as oil level, carbon black level, and sulfur level. The terms a, b, c, d, etc., are coefficients for the respective dependent variables, and C is a constant for the particular model. Clearly, other equations are possible but depend on the objective of the study in question.

④ Construct appropriate contour plots to visualize trends in the data and highlight interaction between components in the formulation;

⑤ Compute an optimization of the ingredients;

⑥ If required, run a compound confirmation study to verify the computed compound optimization.

A wide range of experimental designs are available, and it is recommended that the attached reference be reviewed for further information.

Table 3.30～Table 3.35 display a series of model formulations on which further compound optimization can be based. Additional formulations are available in industry publications such as those from the Malaysian Rubber Producer's Research Association.

Table 3.30 Examples of tire compounds

Compounds	Tread	Sidewall	Wire coat	Ply coat	Inner liner
Polymer system	NR	NR	NR	NR	IIR
	BR	BR		BR	CIIR
	S-SBR				BIIR
	E-SBR				NR
Filler system	SAF	FF	HAF	HAF	GPF
	ISAF	FEF			
	HAF	GPF			
Vulcanization system	Semi-EV	Adapted to polymer system	Conventional system	Conventional system Semi-Ev	Adapted to polymer system
Miscellaneous components	Oils	Antidegradants	Adhesion promoters	Adhesion promoters	—
	Waxes	Waxes			

Table 3.31 Model truck tire tread formulation

Natural rubber	100.00	Stearic acid	2.00
Carbon black (N 220)	50.00	Zinc oxide	5.00
Peptizer	0.20	Sulfur	1.20
Paraffin wax	1.00	TBBS	0.95
Microcrystalline wax	1.00	DPG	0.35
Aromatic oil	3.00	Retarder (if required)	0.25
Polymerized dihydrotrimethylquinoline	1.00		

Table 3.32 Model sidewall formulation

Natural rubber	50.00	Dimethylbutylphenyl-p-phenylenediamine	3.00
Polybutadiene	50.00	Stearic acid	2.00
Carbon black (N 330)	50.00	Zinc oxide	4.00
Peptizer	0.20	Sulfur	1.75
Paraffin wax	1.00	TBBS	1.00
Microcrystalline wax	1.00	DPG	0.35
Aromatic oil	10.00	Retarder (if required)	0.25
Polymerized dihydrotrimethylquinoline	1.00		

Table 3.33 Model tire casing ply coat

Natural rubber	70.00	Polymerized dihydrotrimethylquinoline	1.00
Polybutadiene	30.00	Dimethylbutylphenyl-p-phenylenediamine	2.50
Carbon black (N 660)	55.00	Stearic acid	2.00
Peptizer	0.20	Zinc oxide	4.00
Paraffin wax	1.00	Sulfur	2.50
Microcrystalline wax	1.00	MBTS	0.80
Aromatic oil	7.00		

Table 3.34 Model conveyor belt cover

Natural rubber	100.00	Polymerized dihydrotrimethylquinoline	1.00
Carbon black (N 330)	50.00	Dimethylbutylphenyl-p-phenylenediamine	2.50
Peptizer	0.20	Stearic acid	2.00
Paraffin wax	1.00	Zinc oxide	4.00
Microcrystalline wax	1.00	Sulfur	2.50
Aromatic oil	5.00	TBBS	1.00

Table 3.35 Carpet underlay

SBR	100.00	Microcrystalline wax	1.00
Calcium carbonate	150.00	Aromatic oil	60.00
Reclaim (as filler)	20.00	Stearic acid	1.50
Clay	25.00	Zinc oxide	3.00
Iron oxide	4.00	Sulfur	3.00
Sodium bicarbonate	10.00	MBT	1.50
Blowing agent	1.00	TMTD	0.50
Peptizer	0.20	DPG	0.50
Paraffin wax	1.00		

3.8 Compound Preparation

In a modern tire or general products production facility, rubber compounds are prepared in internal mixers. Internal mixers consist of a chamber to which the compounding ingredients are added. In the chamber are two rotors that generate high shear forces, dispersing the fillers and other raw materials in the polymer. The generation of these shear forces results in the production of a uniform, quality compound. After a defined mixing period, the compound is dropped onto a mill or extruder where mixing is completed and the stock sheeted out for ease of handling. Alternatively, the compound can be passed into a pelletizer.

Depending on the complexity of the formulation, size of the internal mixer, and application for which the compound is intended, the mix cycle can be divided into a sequence of stages. For an all-natural-rubber compound containing 50 phr carbon black, 3 phr of aromatic oil, an antioxidant system, and a semi-EV vulcanization system, a typical Banbury mix cycle will be as follows:

Stage 1: Add all natural rubber; add peptizer if required. Drop into a mill at 165°C.

Stage 2: Drop in carbon black, oils, antioxidants, zinc oxide, stearic acid, and miscellaneous pigments such as flame retardants at 160°C.

Stage 3: If required to reduce compound viscosity, pass the compound once again through the internal mixer for up to 90 seconds or 130°C.

Stage 4: Add the cure system to the compound and mix it up to a temperature not exceeding 115°C.

Computer monitoring of the internal mixer variables such as power consumption, temperature gradients through the mixing chamber, and mix times enables modern mixers to produce consistent high-quality compounds in large volumes. The mixed compound is then transported to either extruders for production of extruded profiles, canlenders for sheeting, or injection molding.

Depending on the compound physical property requirements, compounds can be prepared on mills. Mill mixing takes longer, consumes larger amounts of energy, and gives smaller batch weights. The heat history of the compound is reduced, however, and this can be advantageous when processing compounds with high-performance fast acceleration systems. Two-roll mills function by shear created as the two rolls rotate at different speeds (friction ratio). This ratio of rolls speeds is variable and is set dependent on the particular type of compound. The higher the friction ratio, the greater the generated shear and intensity of mixing.

3.9 Environmental Requirements in Compounding

In addition to developing products to satisfy customers, the environmental implications of the technology must be taken into consideration. The environmental impact on compound development must be viewed in two parts:

① product use and long-term ecological implications;

② health and safety, in both product service and product manufacture.

An example of the impact of product usage and the environmental implications is tire rolling resistance and its effect on vehicle fuel consumption.

Reduction in tire rolling resistance results in a drop in vehicle fuel consumption. This has an immediate impact on the generation of exhaust gases such as carbon monoxide, carbon dioxide, and nitrous oxides. The crown area of the tire, which includes the tread and belts, accounts for approximately 75% of the radial passenger tire rolling resistance. Improvements in the hysteretic properties of the tread compound will therefore enable a reduction in tire rolling resistance and consequent improvements in vehicle fuel economy. The crown area and particularly the tread compound also affects the life cycle of the tire. Longer-wearing tires (including retreading) delay the point in time when used tires must enter the solid waste disposal system.

Critical to a tire's life cycle performance is the ability to maintain air pressure. Tire inner liners composed of halobutyl-based compounds exhibit very low air and moisture permeability. Therefore, tires built with the proper selection of compounds can reduce the rate of premature failure, again delaying entry into the scrap tire and solid waste streams.

Table 3.36 illustrates how the incorrect selection of a tire inner liner polymer will lead to more rapid deterioration in tire performance properties. Replacement of chlorobutyl with natural rubber or reclaim butyl will lead to a more rapid loss in tire air pressure and loss in overall tire performance.

Table 3.36 Relationship between HIIR/IIR content and permeability, ICP, and step-load endurance

Rubber hydrocarbon content in liner	HIIR/IIR, %Vol	Liner permeability ($\times 10^8$) at 65℃	Equilibrium intracarcass pressure (IOF) /MPa	FMVSS 109 step-load hours to failure
100 BIIR	65.2	3.0	0.032	61.5
75 BIIR/25 NR	48.3	4.2	0.063	56.9
65 BIIR/35 NR/20 IIR*	42.2	5.9	0.063	40.2
60 SBR/40 NR/20 IIR*	16.1	6.9	0.090	31.5

* IIR via whole tube reclaim, containing ~50% by weight IIR. Treated as filler.

Improved tire designs have enabled reduction in noise levels. This has become an important environmental consideration. Optimum footprint pressures reduce damage to highway pavements and bridges. All of these improvements in tire rolling resistance, life cycle duration, noise generation, and tire footprint pressure have been incorporated into the full range of tires, from small automobile to heavy truck to large earthmover equipment tires. Today's radial tires use 60% to 80% natural rubber as the polymer portion of compounds.

Because natural rubber is obtained from trees, it is an ideal renewable resource, and thus as a biotechnology material is preferred to petroleum-based synthetic polymers, when equivalent compound properties can be attained.

Tires are one of the most durable technological products manufactured today. They are a resilient, durable composite of fabric, steel, carbon black, natural rubber, and synthetic polymers. The qualities that make tires or other engineered rubber products a high-value item create a special challenge of disposal. Tires and other rubber products, such as conveyor belts and hydraulic hoses, are not biodegradable and cannot be recycled like glass, aluminum, or plastic. Four potential applications for such products entering the solid waste stream have been identified:

① The calorific energy of tires is higher (35MJ/kg) than that of coal (24MJ/kg). With properly designed equipment, tires can be burned to produce heat in cement kilns.

② Tires can be burned in furnaces at power-generating facilities to produce electrical energy.

③ Ground up scrap tires are beginning to find use in some special asphalt applications.

④ Tires with the proper installation technology can serve a variety of applications in the construction industry as marine reefs, energy-efficient house construction, highway bank reinforcement, and erosion control.

These four methods of disposal represent the best options for scrap tire and rubber products disposal. It is anticipated that a variety of new applications for disposal of scrap rubber products will emerge in the future.

In summary, materials scientists must consider the implications of their materials choices, from the quantity of energy to manufacture the product, to the performance during its useful life cycle, and finally to disposal methods.

The Environmental Protection Agency (EPA) also provides constraints that the materials scientist must consider in the design of compounds. As most rubber compounds contain approximately 6 to 20 different materials, not only the materials themselves must be clean and harmless, but any by-products that form during product tire manufacturing must also be harmless to humans and the environment.

The aromatic content of carbon blacks and oils was once considered hazardous. Data were generated that showed that carbon black was stabilized and did not represent a hazard to workers. Resins for cure or tack, antioxidants, antiozonants, and cure accelerators also must be investigated to ensure that the material and any impurities meet changing health and safety standards. Materials safety data sheets and chemical health and toxicity data must be maintained on all materials.

Nitrosamine-generating chemicals represent an area where suspect materials have been removed from rubber products, even though no governing legislation has yet been drafted. Nitrosamines can be formed when secondary amine accelerators are used to cure rubber. These accelerator changes have a very significant effect on the total rubber industry.

Solvent composition and volatility limits can have significant effects on synthetic rubber production and also tire manufacturing. Limits of exposure to some trace impurities defined in the U. S. Federal Clean Air Act are to be based on the hazard represented, not simply the best available measurement capability.

In conclusion, the materials scientist must continue to adjust to the changes in both the environment and health and safety standards.

3.10 Summary

This chapter has reviewed both the types and the properties of elastomers, compounding with a range of filler or reinforcement systems such as carbon black, and enhancement of filler performance by novel use of compounding ingredients such as silane coupling agents. Other issues such as antioxidant systems and vulcanization systems were also discussed. The role of the modern materials scientist in the tire and rubber industry is to use materials to improve current products and develop new products. Four key parameters govern this development process:

① Performance: The product must satisfy customer expectations.

② Quality: The product must be durable and have a good appearance, and appropriate inspection processes must ensure consistency and uniformity.

③ Environment: Products must be environmentally friendly in manufacturing, use, and disposal.

④ Cost: The systems must provide a value to the customer.

In meeting these goals rubber compounding has evolved from a 'black art', as it was at the start of the 20th century, to a complex science necessitating knowledge in advanced chemistry, physics, and mathematics.

References

[1] Worldwide Rubber Statistics, 2003, International Institute of Synthetic Rubber Producer Houston, TX, 2003.

[2] S. Datta, Rubber Compounding, Chemistry and Applications, B. Rodgers (Ed.), Marc Dekker Inc., New York, 2004.

[3] ASTM D2227-96: Standard specification for natural rubber technical grades, 2002.

[4] The Rubber Manufacturers Association, The International Standards of Quality and Pack aging for Natural Rubber Grades, The Green Book, The International Rubber Quality an Packaging Conference, Office of the Secretariat, Washington, D.C., January 1979.

[5] D. Bernard, C. S. L. Baker, and I. R. Wallace, NR Technology, Vol. 16, 1985.

[6] The Synthetic Rubber Manual, 14th ed., International Institute of Synthetic Rubber Producers, Houston, TX, 1999.

[7] S. M. Mezynski and M. B. Rodgers, Heavy Duty Truck Tire Materials and Performance, Kautschuk Gummi Kunststoffe, Frankfurt, Vol. 46, 1993, pp. 718-726.

[8] B. D. Simpson, The Vanderbilt Rubber Handbook, 12th ed., R. Babbit (Ed.), R. T. Vande bilt Co.,

Inc., Norwalk, CT, 1978.

[9] K. H. Nordsiek, The Integral Rubber Concept --An Approach To An Ideal Tire Tread Rubber, Kautschuk Gummi Kunstsoffe, Frankfurt, Vol. 38, 1985, pp. 178-18.

[10] M. B. Rodgers, W. H. Waddell, and W. Klingensmith, Rubber Compounding, in Kirk-Othmer Encyclopedia of Chemical Technology, 5th ed., John Wiley & Sons, New York, 2004.

[11] G. L. Day and S. Futamura, Paper 22 presented at a meeting of the Rubber Division, American Chemical Society, New York, 1986.

[12] H. L. Brantley and G. L. Day, Paper 33 presented at a meeting of the Rubber Division, American Chemical Society, New York, 1986.

Chapter 4
Reinforcement of Elastomers by Particulate Fillers

4.1 Introduction

The reinforcement of elastomers by particulate fillers has been extensively studied in the past, particularly in the 1960s and 1970s. The first reason is naturally the drastic changes in mechanical properties that induces fillers reinforcement: many of the usual applications of elastomers could not be envisaged without the use of particulate fillers. The other reason seems to us to be of a very different nature, and probably resides in the 'mystery' of the reinforcement mechanism that has fascinated many scientists and remains, despite their efforts, mainly not understood today.

It is necessary to define precisely what is reinforcement, because this word covers very different meanings when applied to thermoplastics, thermosets, or elastomers. Confusion is mainly due to the fact that reinforcement qualifies an increase in mechanical properties, but what is expected as mechanical properties is very different considering the different matrices and applications.

For plastics, reinforcement results in an increase in modulus and hardness. The effect of particulate fillers is quite clear, they replace a part of the matrix, so modulus becomes higher, but deformation at break decreases in the same time.

The situation is very different for elastomers: the use of reinforcing fillers induces a simultaneous increase modulus and deformation at break. Curiously, the replacement of a part of the deformable matrix by solid objects doesn't reduce its deformability. The increase of these two antagonistic properties characterizes elastomer reinforcement. This fascinating paradox, despite not being fully understood, explains the ability of reinforced elastomers to provide unique material properties and applications and justify their success in different technological fields.

4.2 Preparation of Fillers

4.2.1 Nonreinforcing Fillers

As it will be discussed later, the size of the filler is probably one of the most important properties for reinforcement. So, particulate fillers obtained by grinding of minerals or by coarse precipitation are usually nonreinforcing fillers because of their size: they are too big. Such fillers can even be used in elastomers but just confer them a very slight increase in modulus and a very significant drop in break properties occurs.

4.2.2 Reinforcing Fillers

4.2.2.1 Carbon Black

(1) Historical Processes

Carbon black, any of a group of intensely black, finely divided forms of amorphous carbon, usually obtained as soot from partial combustion of hydrocarbons, used principally as reinforcing agents in automobile tires and other rubber products. Carbon black is considered a petrochemical because it is made from natural gas or petroleum residues. There are several processes involving either incomplete combustion (burning off the hydrogen of a hydrocarbon, such as methane, and leaving the carbon) or by externally applied heat in a furnace, splitting the hydrocarbon into hydrogen and carbon. The most important of all the uses of carbon black is in compounding rubber to be used in tires. An average tire of a passenger automobile contains about two kilograms of carbon black.

(2) Furnace Process

Carbon black particles are usually spherical in shape and less regularly crystalline than graphite. Carbon black changes into graphite if heated at 3000℃ for a prolonged period. Among the most finely divided materials known, carbon blacks vary widely in particle size depending on the process by which they are made. Channel or impingement black is made by the impingement of smoky flames from tiny jets on iron channels; the deposited black is scraped off by moving the channels over stationary scrapers. Furnace blacks are made in refractory chambers by incomplete combustion of any of various types of gaseous or liquid hydrocarbons. Thermal blacks are produced in the absence of air when hydrocarbons are decomposed by contact with heated refractories. Lampblack, the oldest known black pigment, is produced by burning oil, usually coal-tar creosote, in shallow pans, in a furnace with the draft regulated to give a heavy smoke cloud. Acetylene black is produced in refractory chambers in the absence of air by the decomposition of acetylene gas preheated to 800℃. It is used in applications requiring high electrical conductivity, such as dry cells.

(3) Post Treatments, Surface Modification

The great majority of carbon black post-treatment studies have been conducted to increase strength/quality of reinforcement. So the chemical modifications that have been tested are strongly linked to the different theories envisaged for reinforcement.

In the 1960s, carbon black-elastomer interaction was considered as the result of a chemical bonding between acidic surface functions and natural rubber alkaline moieties. So many studies have been conducted to increase carbon black activity by surface oxidation: oxygen at high temperatures, H_2O_2, ozone, nitric acid. The type of oxidation used determines the number and the type of functions obtained; it is interesting to underline that such chemical modifications are used at industrial scale for specialty carbon blacks (inks, pigments).

In the early 1980s, Danneberg proposed the mechanism of molecular slippage and post treatments turned to chemical grafting of polymeric chains onto carbon black surface. More recently, the need for low hysteresis compounds has reactivated chemical modification studies. Many modification processes have been proposed: functionalization, surface coating of carbon black by silica, and alumina.

4.2.2.2 Silicas

The use of silica in rubber mixes can not be considered as new at all, because this filler has been used in rubber formulations since the beginning of the 20th century. Silicas are not reinforcing fillers in the proper sense, because silica-reinforced mixes exhibit much lower mechanical properties, particularly considering modulus at break and abrasion resistance. So silicas weren't used as reinforcing fillers but mainly in association with carbon black. Two major breakthroughs have transformed this facility product into a reinforcing filler that can achieve all carbon black mixes properties with, in addition, a decreased hysteresis of major interest for tire applications. The first step was made in the 1970s by Wolff, who proposed a specific silane coupling agent, the TESPT. The second step arises in the 1990s with an R. Rauline's patent, which introduces the use of specific precipitation silica, elastomers, and mixing conditions to achieve reinforcement.

(1) Precipitated Silicas

Silicas used as reinforcing fillers are mainly obtained by precipitation. The process basically consists in the preparation of a silica glass by alkaline fusion of pure sand and an alkaline salt. Then this glass is solubilized in water at high temperature and acid precipitated. The silica suspension obtained is then filtered, washed, and dried.

In order to obtain reinforcing silicas, much care must be taken in precipitation recipes (to obtain small rigid objects) and drying conditions (to maintain high dispersibility). It is also interesting to underline that silicas can be very easily chemically modified by doping or grafting of species during or at the end of their preparation.

(2) Fumed Silicas

Fumed silicas are obtained by high temperature oxydecomposition of SiH_4, or other methyl hydride precursors ($SiHMe_3$, $SiH_2Me_2 \cdots$):

$$n SiH_4 + 2n O_2 \longrightarrow n SiO_2 + 2n H_2 O$$

Coming out of the furnace, fumed silicas are obtained in a fluffy form, and because of their high temperature of formation, they present a very stable morphology and few surface silanols compared to precipitation silicas. This confers a high dispersibility and reactivity to fumed silicas but, because of their higher price, they are rarely used in the rubber industry.

4.2.2.3 New Reinforcing Fillers

Very recently, many studies have been conducted to identify new reinforcing systems. These systems are similar to silica compounds and characterized by the use of a coupling agent to chemically bond elastomer chains to filler surface. Many reinforcing systems have been patented: alumina oxyhydroxide and oxide, titanium oxides, and silicon nitride/carbide.

4.3 Morphological and Physicochemical Characterization of Fillers

As will be demonstrated later, morphology and physicochemical properties of reinforcing fillers are of crucial importance because they directly define their reinforcement ability. Their characterization formerly was based essentially on morphological properties (surface area and structure), but because of the use of silicas as reinforcing filler, there is now a strong need for dispersibility and surface chemistry characterization.

4.3.1 Filler Morphology Characterization

4.3.1.1 Filler Morphology

It is important to emphasize that the actual morphology of carbon blacks has remained unknown for decades, even if it was commonly used in the rubber industry. This is due to the very small size of its constituting objects; they are smaller than 0.1mm and can only be resolved by transmission electron microscopy.

(1) MET

As observed by transmission electron microscopy, carbon blacks appear as irregular chainlike, branched aggregates of partially fused spheres. Aggregates constitute the smallest dispersible unit of carbon black and are virtually unbreakable in usual conditions of use; therefore, aggregates must be considered as the actual reinforcing objects.

The chainlike, branched structure of aggregates makes them very bulky, and their effective volume is much higher than the volume of the aggregate itself. This observation is of primary importance because the effective volume of the aggregate will be more or less its volume in the mixes and define which part of rubber can be deformed and not. This bulkiness is usually called structure and generally measured by other methods (see later section on structure); some very interesting studies have been conducted in the past to classify carbon black aggregates in different shape classes (bulk, ellipsoid, linear, etc.).

Even if they are usually called primary particles, spheres that constitute aggregate are partially fused together and never exist by themselves. Anyway, their size is of great importance because it defines the actual surface of interaction between carbon black and elastomeric phase: the lower the size of primary particles, the higher the interface extension. Primary particle size distribution has been estimated by TEM image analysis, but carbon black surface area is usually more efficiently obtained by adsorption methods (see later section on surface area).

At very high magnification, it is possible to observe directly the internal structure of carbon black primary particles. They are constituted by overlapping graphitic layers that locally present a quasi-crystalline turbostratic structure with an approximately 0.35nm interlayer spacing, close to pure graphite (~0.332nm).

(2) STM, AFM

Scanning tunnelling microscopy (STM) has been applied to carbon black characterization. As suggested by TEM, carbon black surface morphology consists in the overlapping of graphitic sheets in an onionlike structure (Fig. 4.1).

Carbon black surfaces appear surprisingly ordered, and graphitic edges should be identified with chemically reactive zones that were previously assigned to 'amorphous' zones.

Atomic force microscopy (AFM) has also been used for three-dimensional characterization of carbon black aggregates.

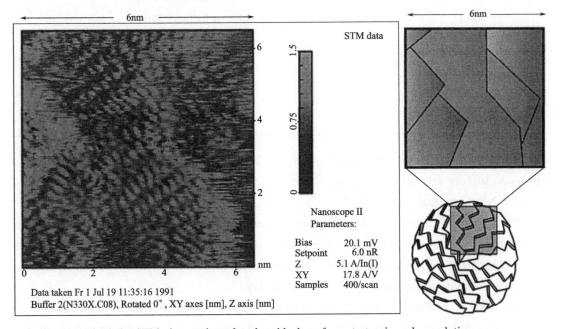

Fig. 4.1 STM observation of carbon black surface at atomic scale resolution.

(3) Silicas

Electron transmission microscopy of silicas is much more difficult, particularly for precipitation silicas, which tend to real glomerate during preparation and are fairly often unstable under high magnification. Reinforcing silicas observed by TEM present a morphology very close to carbon black; they are constituted by small, chainlike, branched aggregates. However, the identification of silicas' primary particles is much more difficult because they are significantly more fused.

4.3.1.2 Surface Area

(1) Introduction

Surface area is probably the most important morphological characterization of reinforcing fillers because it corresponds to the extension of the interface, i.e., the interaction between elastomer and filler surface. As evidenced by transmission electron microscopy, surface area is directly linked to the size of primary particles so that American Society of

Testing Materials (ASTM) has chosen this parameter for carbon black nomenclature. More precisely, ASTM nomenclature includes four digits, the first one relates to vulcanization speed (N as normal or S as slow), then tree numbers, which first correspond to the primary particle diameter.

ASTM numbers	Primary particle diameter/nm	Previous nomenclature
900-999	201~500	MT: Medium Therm
800-899	101~200	FT: Fine Therm
700-799	61~100	SRF: Semi-Reinforcing Furnace
600-699	49~60	GPF: General Purpose Furnace
500-599	40~48	FEF: Fine Extrusion Furnace
400-499	31~39	FF: Fine Furnace
300-399	26~30	HAF: High Abrasive Furnace
200-299	20~25	ISAF: Intermediate Super Abrasive Furnace
100-199	11~19	SAF: Super Abrasive Furnace
000-099	1~10	—

Particle size diameter can only be done by TEM characterization and is difficult and costly, so surface area of fillers is usually obtained by different adsorption methods.

(2) Nitrogen Adsorption/BET

Adsorption of nitrogen and surface area determination by BET method is probably the most widely used method for surface area characterization of reinforcing fillers. This method is very sensitive, reliable, and can be applied to all reinforcing fillers because it is either not or very weakly influenced by surface chemistry.

The main drawback of BET characterization is that the surface area obtained includes micropores whose surface can not be reached by elastomeric chains, which are much bigger than nitrogen. So, many are now using the t-plot method that allows the determination of the net surface excluding micropores.

(3) CTAB Adsorption

Cetyl triethyl ammonium bromide (CTAB) adsorption is widely used in carbon black industry. This method consists of making a suspension of a known mass of carbon black into a water CTAB solution of known concentration. Carbon black is then filtered and the quantity of adsorbed CTAB is determined by titration of the remaining CTAB in the filtrate. Surface area of carbon black is then deducted from the amount of CTAB adsorbed, using a previously determined calibration with a reference carbon black.

CTAB surface area characterization is not influenced by micropores because of the size of CTAB molecule, but it can be influenced by surface chemistry and impurities. Hence, it tends to disappear and to be replaced by BET/t-plot, which is much more reliable.

(4) Iodine Adsorption

Iodine surface characterization proceeds in exactly the same way as instead of

CTAB. This method CTAB, except that the adsorbed species is I_2 can only be applied to carbon black characterization. Because iodine probably partly adsorbs and partly reacts with double bonds on the surface, the iodine method is extremely sensitive to any surface functions, modifications, or contaminations. Hence, this method, formerly very widely used in carbon black industry, is now replaced by CTAB or BET/t-plot.

4.3.1.3 Structure

If structure is rather easy to define on the basis of MET images, it is much more difficult to measure it quantitatively. Nevertheless, the determination of structure is of primary importance because structure defines the actual volume of the filler in the mix and therefore the level of strain amplification of the deformable phase.

(1) TEM Measurements

Some attempts have been made to use TEM measurements to determine structure of fillers. But in spite of the well-constructed studies, the qualification of TEM images that are two-dimensional do not lead to a three-dimensional image.

(2) DBP Absorption

Practically, structure determination of fillers is obtained by a very simple and easy method: dibutyl phtalate absorption (DBPA). This method consists of filling up the voids in and between aggregates with DBP, which is a viscous liquid. Historically, the DPB volume was determined by making a solid pellet of carbon black and DBP with a spoon. Now, this delicate measurement is made automatically with a couple-monitored Banbury, in which a well-known quantity of dry carbon black is placed. DBP is then added drop by drop, and the structure value is obtained when torque reaches a given value.

DBP measurements are surprisingly reliable and can be easily and quickly obtained. Nevertheless, DBP measurement can be sensitive to surface chemistry, and values obtained with fillers of different nature can't be directly compared. For silicas, dioctyl phtalate can also be used instead of DBP.

DBP values also depend on pelletization/granulation of the filler. This makes sense because DBP measures the total void volume: intra-aggregate voids, which are the pertinent parameter, but also inter-aggregate voids, which essentially reflect pelletization/granulation. Thus, fluffy or loosely pelletized black will have higher DBP value; in the same way, spray-dried silicas will have greater DBP value than granulated ones, even if their aggregates are exactly the same.

(3) CDBP Absorption

Very frequently, DBP adsorption is made with carbon black previously submitted to very high pressure into a cylinder, for example, 165MPa four times. Crushed DBP absorption or CDBPA are equal to DBP (for low structure carbon blacks) or significantly lower than DBP (for high structure carbon blacks).

The high pressure crush procedure is supposed to reproduce aggregate breakage during mixing, but because DBP measures either intra-aggregates and inter-aggregates voids, it is difficult to settle if the decrease in DBP absorption is due to the compaction of the filler induced by the pressure or to an actual breakage of aggregates.

(4) Mercury Porosimetry

Filler structure can also easily be determined by mercury porosimetry. Filler is placed in a small chamber and mercury is forced into the void by increasing pressure. The intrusion curve gives the volume of mercury intruded in pores for each applied pressure. Usually, intrusion curves present a well defined step at high pressure, which corresponds to the filling of the smallest pores (intra-aggregate and inter-primary-particles voids). Mercury porosimetry gives a structure index which excludes inter-aggregate voids and so is more representative of the intrinsic structure of the filler. In addition, mercury porosimetry allows a direct determination of filler surface area, excluding micropores, and can be applied to any particulate filler.

4.3.1.4 Aggregate Size Distribution

Aggregate size distribution is the last morphological characterization of reinforcing fillers. This measurement is very rarely used, despite the great interest in knowing the size of aggregates, which directly influences distances between reinforcing objects in the mix, and therefore the strain amplification. This surprising situation is mainly due to the fact that this determination is particularly difficult to make.

(1) TEM/AI Measurement

Transmission electron microscopy/image analysis (TEM/AI) has been used for a long time to determine aggregate size distribution of carbon black and silicas. Such studies are very costly because they need at least a few thousand aggregate size measurements to determine precisely the size distribution. Nevertheless, using TEM/AI aggregates are measured as two-dimensional projections, which probably maximizes their sizes.

(2) Disk Centrifugation

One better method to access aggregate size distribution is probably to use disk centrifugal photo or X-ray sedimentometers. This apparatus consists of a transparent void disk that can be rotated at very high speed. A sedimentation medium is first injected into the rotating disk, and then a very small quantity of filler suspension is injected at its surface. Sedimentation is registered by light or X-ray transmission.

Aggregate size distribution of carbon blacks and silicas can be easily obtained by this method. Its main drawback is that, unlike with TEM/AI, aggregate sizes are probably underestimated because they settle following their lowest project area.

(3) Tint

Coming from the ink and pigment industry, the tint measurement is a very simple way to evaluate the mean aggregate size of carbon black. This characterization consists of making a paste of known amounts of carbon black, white solid powder (titanium or zinc oxide), and oil (DBP, for example). Then the reflectivity of the paste is measured; it has been demonstrated that tint values will roughly correlate with mean aggregate size and can be considered as an indicator of aggregate size distribution broadness.

4.3.2 Dispersibility

Even if the relationship between filler dispersion and abrasion resistance is well established, relatively few studies have been done on the characterization of filler dispersibility. This is mainly due to the fact that carbon black dispersibility was commonly judged satisfactory, partly because it is indeed high, but more probably because all mixing apparatuses were designed for dispersing carbon blacks.

The use of silica has given a new light to this domain, because, contrary to carbon black, dispersibility is one of the key properties for achieving silica reinforced mixes. Obviously, filler dispersibility is mainly influenced by interactions between agglomerates and/or aggregates, in other words, the force/energy needed in order to separate two objects. For carbon black, these interactions are mainly due to van der Waals forces, which are very low compared to the hydrogen bonding existing between silica objects.

In addition, pelletization process has a great influence on dispersibility: any action leading to a higher compaction of the filler increases interaction between filler objects and so decreases its dispersibility.

(1) Reflectivity

The most commonly used technique to qualify filler dispersibility is to study light reflectivity of clean-cut mixes. Some apparatuses have been developed to evaluate filler dispersion using a calibrated set of reference mixes (Dispergrader). However, such characterization mainly detects dispersion defects of a few tens of microns, and direct comparison of carbon black and silica mixes has to be done cautiously. In any case, it is necessary to make a mix, which means choosing a formula, a mixer, and mixing conditions; thus the result can not be considered as an intrinsic dispersibility measurement of the filler, but just reflect the dispersibility of the filler in one mix with a set of mixing conditions.

(2) Laser Granulometry

Recently, a method has been patented to determine filler dispersibility. It consists of measuring continuously the size of the filler by laser granulometry during an ultrasonic desagglomeration. This characterization can be applied to any filler and is an intrinsic property; however, the use of water as a desagglomeration medium can be a problem because of its high polarity compared to elastomers.

4.3.3 Filler Physicochemistry

Compared to morphology, filler chemistry has been only slightly studied, partly because of the difficulty of such characterisations and more probably because since the 1970s reinforcement is broadly considered as a physical interaction between elastomer and filler. So carbon black chemical characterizations mainly date from the 1960s, and few new technical methods have been applied to carbon black surface characterization since this time. The situation is somewhat different for silicas, because silica reinforcement is the consequence of a chemical reaction of silane with silica surface. Few studies have been published in the elastomer reinforcement area, probably because silica surface was already well characterized for other applications. Concerning physicochemical characterization, the studies are limited to surface energy distribution determination, which will be discussed first.

4.3.3.1 Surface Energy

Elastomer reinforcement by carbon black is generally considered as the consequence of the adsorption of polymeric chains onto carbon black surface. Therefore carbon black surface energy knowledge is of primary importance in carbon black characterization. However, very few carbon black surface energy measurements have been published; this can be easily understood considering the difficulty of such measurements on a highly heterogeneous and tortuous surface.

Immersion calorimetry allows carbon black or silica surface energy determination. This technique can be used with different liquids or solutions of low mass elastomers, but it presents the main drawback of giving a mean surface energy value (about $50 \sim 70 mJ/m^2$), when surface adsorption of polymeric trains probably preferentially occurs on the highest surface energy sites. This justifies the use of inverse gas chromatography (IGC) for filler surface energy characterization. This technique can be used in two very different modes: 'infinite' or 'finite' dilution.

In 'infinite' dilution, very small quantities of alkanes of growing number of carbons are adsorbed onto carbon black or silica surface; from their retention times, it is possible to calculate the dispersive component of surface energy $\gamma_{s,d}$. Because of the very low surface coverage during characterization, this value corresponds to the highest energetic sites. According to this technique, surface energy grows with carbon black surface area; because high surface area carbon blacks are highly reinforcing, this result should be considered encouraging. Surface energy values obtained can reach values of $300 \sim 500 mJ/m^2$, which can't be considered as realistic for carbonaceous surfaces. In addition, this determination seems to be also sensitive to chemical modification of carbon black surface even if the probes used should only characterize the dispersive part of surface energy.

'Finite' dilution is a more powerful technique in that it is possible to obtain the complete energetic site distribution for carbon black or silica. In this technique, surface is fully covered by the probe and the distribution is calculated by a specific post treatment of desorption signal.

Using this technique, carbon blacks present approximately the same surface energy distribution differing only in the number of adsorption sites. The energetic site distribution is particularly broad, with sites of high ($\sim 100 mJ/m^2$) to low ($\sim 10 mJ/m^2$) energy. Mean values are consistent with these obtained by immersion calorimetry.

Finally, it should be mentioned that a procedure similar to IGC, a 'finite' dilution, can be applied to nitrogen adsorption isotherm and allow surface nanoroughness characterization of any filler.

4.3.3.2 Surface Chemistry

(1) Carbon Black

① Impurities   Because of its manufacturing process, carbon black surface includes some organic and mineral impurities.

Organic impurities are mainly poly aromatic hydrocarbons (PAH). They correspond to partially unconverted fuel that has been readsorbed onto carbon black. These PAHs are present at a very low content and, because of their firm adsorption on carbon black, the extraction must be conducted in a Soxhlet apparatus with a strong solvent (toluene) and at high temperature (80℃). However, it has been demonstrated that organic impurities have no significant effect on carbon black reinforcement.

Mineral impurities come from quench and pelletization steps in the carbon black production process. As presented before, the decrease in temperature of carbon black and exhaust gases is mainly obtained by injection of a great mass of water. Additional water is also added to carbon black during pelletization. Even if this water is purified, the remaining mineral salts precipitate onto the carbon black surface and, because of the high temperature, are reduced to basic salts. Mineral impurities of carbon blacks can easily be extracted by solubilization in water, as in the so-called 'pH of carbon black' in which carbon black is suspended in water and the pH of the filtered water measured. Mineral impurities don't seem to alter carbon black reinforcement properties but they have a significant effect on vulcanization speed, which increases with the pH value of carbon black.

② Oxygenated Functions   Oxygenated functions on carbon black surface were observed in the early 1950s and completely characterized by H. P. Boehm in the 1960s. At this time, interaction between carbon black and natural rubber was considered the consequence of chemical reactions between the carbon black surface's acidic groups and basic moieties present in the natural rubber structure.

The carbon black surface function characterization consists of suspending a given amount of carbon black in solutions of known normality of basis of different strength: $NaHCO_3$, Na_2CO_3, NaOH in water, and EtONa in ethanol. Then carbon black is filtered and the number of reacted acidic groups obtained by titrating the remaining basis in filtrate (Fig. 4.2).

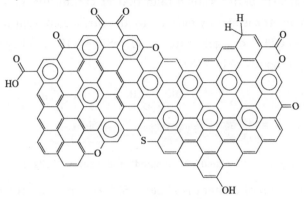

Fig. 4.2   Chemical functions on carbon black surface.

In the 1960s, carbon blacks were mainly prepared by channel processes, and their acidic functions were present at about 10^{-3} eq/g, which allows relatively easy determinations; but now, with furnace blacks, the surface acidic functions are generally of about 10^{-5} eq/g, and specific techniques or drastic reaction conditions must be used. Obviously, such delicate determination must be conducted on previously extracted black, in order to eliminate basic mineral impurities that would hinder any characterization of the rare acidic groups present on carbon black.

This observation allows one to believe that acidic groups on carbon blacks are mainly produced by surface oxidation in the production process, probably during drying following pelletization. Therefore, acidic groups could be considered an alteration of carbon black surface. This point of view is supported by the fact that oxidized blacks which have an acidic surface group content of about 10^{-2} eq/g exhibit a very low reinforcement ability, even if their slow vulcanization is corrected.

③ Double Carbon Bonds, Hydrogen Content   All chemical studies done in the past on carbon black have focused on chemical impurities or on functions produced by partial surface oxidation of carbon black, and not on its own surface reactivity.

Now carbon blacks can not be considered as chemically inert surfaces. Their reaction with iodine or oxygen, their structure evidenced by STM, demonstrates the presence of a great number of reactive double bonds on their surface.

Such double bonds could react with sulfur, olefins, and radicals to provide chemical bonding between carbon black surface and polymer, but their direct quantitative determination has never been obtained. There is a strong correlation between hydrogen content of carbon blacks and their reinforcement ability. The content and reactivity of hydrogen present on graphitic edges have been determined by isotopic exchange and correlated to carbon black reinforcement ability. This characterization is particularly difficult and usually hydrogen mass content is used; these values also are surprisingly well correlated to reinforcement ability of carbon blacks.

(2) Silica

① Silanols   Because of its numerous and longstanding uses in other applications, silica surface chemistry is clearly better known than that of carbon black.

Silica surface chemistry is mainly defined by the surface content in Si-OH, silanols can be 'isolated', $O=Si-OH$, or 'geminated', $O=Si=(OH)_2$. They are generally highly associated by hydrogen bonds (Fig. 4.3).

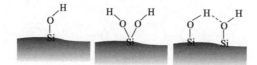

Fig. 4.3 "Isolated" "geminated" and associated silanols.

For fumed silicas, silanol content is about 2 Si-OH/nm^2, but for precipitation silicas, silanol content can reach values as high as 6 to more than 10 Si-OH/nm^2. It is essential to emphasize that such high silanol content cannot be considered as true per se: considering bond length and Silicon-oxygen arrangement, a content of 2~3 Si-OH/nm^2 corresponds to a full coverage of the surface with silanols. Such high values are generally attributed to the existence of poly silicic acid chains, $-[Si(OH)_2]_n-O-Si(OH)_3$, and to the fact that BET surface area doesn't take into account pore of very small size in which some silanols are located.

Silanol surface content can not reasonably be considered an indicator for silica reinforcement ability; indeed, because of its size, the coupling agent used to bond silica surface and elastomer can not react with more than two or three silanols of the surface. Moreover, a high number of silanols will induce more associations by hydrogen bonding and so decrease their chemical reactivity.

② Other Surface Functions Silica surface chemistry cannot be used to determine the silanol surface content. Particularly, for silicas synthesized by precipitation, some other chemical species can modify silica's surface reactivity.

For example, because of an incomplete hydrolysis of silicate, $\equiv$Si-O-Na can be observed; because of an incomplete washing of silica after filtering, sulfate can be absorpted onto silica surface. In addition, some salts used for silica processing can also be added and modulate its reactivity.

Obviously, as discussed earlier in the silica production process, silica surface can also be modified by other chemical species added during and after its preparation. In any case, this modifier only changes silanol reactivity by enhancing or decreasing its acidity.

4.4 The Mix: a Nanocomposite of Elastomer and Filler

Even if the term nanocomposite is usually not used in reinforcement by particulate fillers, it would be particularly adapted: mixing of reinforcing solids and elastomers is not limited to the arithmetic 'sum' of the properties of both taken independently but gives a synergetic alliance that achieves new properties.

Moreover, the term filler is more or less inadequate, because the particulate solid is not used to fill a void that is to diminish the cost of the elastomeric product. Indeed, elastomer and reinforcing filler should be considered as two inseparable parts of equal merit in the composite. As expressed nicely by Papirer, "carbon black and polymer is the wedding of the century".

This introduction is much more than a semantical debate; it underlines the importance of a global approach including reinforcing filler and elastomeric matrix relationships.

4.4.1 Dispersion, Aggregate Sizes, and Distances

4.4.1.1 Dispersion

(1) Dispersion of Filler in Rubber Matrix

The quality of particulate filler dispersion in the elastomer matrix is of primary importance for compound mechanical and use properties. In very recent years, filler dispersion characterization has been brought again into light because of the difficulties encountered to disperse silica in rubber.

Usually, filler dispersion is achieved in a Banbury mixer and is presented to proceed in two different steps. In the first step, filler is distributed in pure polymer in the form of pellets or subpellets (i. e., agglomerates) and then, in the second step, subpellets are eroded into aggregates by an 'onion peeling' process. This second phase, which eliminates agglomerates and determines inter-actual-aggregate distances is much longer and necessitates higher mechanical energetic input.

As discussed earlier, dispersibility and so dispersion is mainly controlled by the strength of interaction between agglomerates and/or aggregates, which is a direct consequence of their surface energy. The influence of surface energy on dispersion has been clearly demonstrated by the use of different matrices; when matrix surface energy increases and becomes closer to filler surface energy, dispersion is facilitated.

(2) Characterization

Filler dispersion characterization is particularly difficult because it must be conducted on a very broad range of scale: from microscopic undispersed agglomerates, which are defects and will decrease a product's life, to nanoscopic distances between aggregate, which will greatly influence reinforcement level. Indeed, filler dispersion characterization has been conducted with a large number of analytical techniques: optical, electronic, and atomic force microscopy, but also X-rays and neutron diffraction. The main difficulty is to recompose a global image from very different data, because any of these techniques gives directly a complete description of dispersion.

(3) Influence of Filler's Properties

In addition to filler's surface energy, which is of major importance, dispersion can also be influenced by filler's morphological properties. Dispersion is highly influenced by filler surface area: the higher the surface area, the lower the dispersion. This result is probably due to the fact that high surface area usually has smaller aggregates, which will develop more interactions with their neighbors in the dry state.

Filler structure also has a neat influence on dispersion: the higher the structure, the higher the dispersion. This result is well established and likened to the fact that more 'open' aggregate structures develop a lower number contact with their neighbors in the dry state.

4.4.1.2 Object Sizes in the Mix

As discussed earlier, filler occurs as a distribution of different aggregate sizes. This characterization is interesting as a potential, but this final aggregate size distribution could not be achieved in mixes. Indeed, dispersion can be incomplete and some agglomerates can be left in the mix, shifting the actual size of reinforcing objects to a higher value. On the other hand, because of the high shear strengths developed during mixing, some aggregate breakage can also occur and produce an actual size of reinforcing objects lower than expected on the basis of filler characterization. Therefore, considering the possibility of filler in the mix, we will use the term object, which can refer to either aggregates or agglomerates.

Obviously, aggregate size distribution characterization in the mix is very delicate. Some transmission electron microscopy observations have been conducted on microtome thin cuts, but such characterizations are restricted to a small number of aggregates and can only lead to qualitative conclusions.

Direct characterization of object distribution in the mix has also been conducted using X-ray or neutron diffraction, but such approaches are strongly limited by the high concentration of filler objects and their refraction index, which is relatively close to that of rubber. One other way to characterize object size distribution is to extract the filler from the mix by thermal or catalyzed polymer decomposition; these procedures probably greatly affect object size, because of possible reagglomerations.

The characterization of filler object size distributions in the mix mainly remains a domain to develop. In any case, it is generally accepted that aggregate size of carbon black decreases during mixing, even if any of the methods used can eliminate possible artifact. About this, it is interesting to recall that crushed DBP, as previously discussed, has been developed to take into account possible aggregate breakage. Considering silica, explicit data has been published considering this problem, but the same possible aggregate breakage seems also possible.

4.4.1.3 Distances

Characterization of inter object distances in the mix is the reciprocal problem of object size determination and involves the same difficulties and the same lack of experimental data.

However, it should be noted that usual filler loadings used in elastomers, (around 20% in volume), are very close to the maximal fraction that can be incorporated into the elastomeric matrix. This fraction is very low compared to compact spheres arrangement, but, as discussed earlier, filler's aggregate present a highly open structure; moreover, maximum loading is highly dependent on the filler's structure.

Based on simple models, it as been demonstrated that interobject distances are in the range of a few tens of nanometers; this result is consistent with the fact that carbon black mixes are electrically conducting, which implies that inter object distances are low enough to allow tunnel conduction.

Moreover, it has been demonstrated that optimum filler loading corresponds to very similar inter aggregate distances, without regard to aggregate size or structure. This very interesting result underlines the decisive influence of interaggregate distances in elastomer reinforcement.

4.4.2 Filler-Elastomer Interaction

4.4.2.1 Carbon Black

(1) Elastomer Adsorption.

① 'Filler Network' Because of carbon black high surface energy, elastomeric chains are strongly adsorbed onto its surface. This adsorption, even if it is limited to a small part of the elastomeric chains, called "trains", drastically slows down their mobility.

As a simplified-but slightly incorrect-picture, it can be considered the 'trains' have a lowered transition temperature. The exact thickness of this layer remains disputed but values of 1~5nm are usually considered. Anyway, it is very noticeable that such thickness corresponds at least to 3% to 15% of total elastomeric phase. Taking into account carbon black structure/tortuosity, values as high as 30% have been proposed.

A more accurate approach considers that elastomeric chains present a gradient of mobility coming from carbon black surface to bulk. The high surface areas and loadings of carbon blacks used in elastomer reinforcement induce such small distances between reinforcing objects that almost any elastomeric chain contacts at least one aggregate. In addition, because the statistical size of polymeric chains is in the range of interaggregate distances, close neighboring objects are probably bounded together by chains adsorbed onto both aggregates. The bonding of carbon black aggregates constitutes the filler network.

② Bound Rubber The filler network is clearly evidenced by bound rubber Measurements. Bound rubber is a very specific measurement done on green mixes; it consists of determining the part of rubber that can not be extracted by a good solvent. A small part of rubber, previously weighted, is put in toluene and submitted to extraction at a room temperature. Samples swell but usually not delitate; the surrounding solvent is regularly renewed to ensure an optimal extraction, and samples are weighted to follow extraction progression. After 1 or 2 weeks, extraction is completed and samples are dried and weighted. The weight difference corresponds to the soluble part of elastomer, that is, to the chains that were weakly adsorbed on the carbon black surface. When the extraction temperature increases, the bound rubber decreases and, above about 80℃, samples completely delitate, indicating the disappearance of the continuous networking of carbon black aggregates by elastomer chains.

Because bound rubber measures elastomer adsorption onto filler surface, it is highly dependent on filler loading, specific surface, and structure, which are parameters that can be measured independently. However, at given loading and carbon black surface area and structure, it has been demonstrated that bound rubber is also dependent on carbon black surface energy.

(2) Chemical Surface Bonding

Before the 1970s, carbon black reinforcement of elastomers was generally considered chemical by nature. It was supposed that carbon black surface acidic groups were reacting with natural rubber basic moieties conducting to a strong covalent bond that was responsible for carbon black reinforcement ability.

In the 1970s, furnace gradually replaced channel technology. But even if furnace carbon blacks present ten times less surface acidic groups, their reinforcing ability remains unchanged or increased. On the other hand, synthetic elastomers, which obviously have more basic moieties than natural rubber, were also perfectly reinforced by carbon black. In addition, the preparation of surface-oxidized carbon or grafted blacks leads to a decreased reinforcement ability.

It was then obvious that chemical reaction of carbon black surface acidic groups with natural rubber basic moieties was not responsible for reinforcement. So the newly discovered mechanism of "molecular slippage", proposed by Dannenberg and based on molecular adsorption, was quickly and fully adopted.

These observations do not allow the full refutation of chemical reinforcement theory. Chemical bonding by acidobasic reaction is clearly rejected, but, other chemical reactions could occur in carbon black mixes. For example, elastomeric chain breaking during mixing can lead to radicals that could react with carbon black surface; sulfur-direct bonding of elastomer and carbon black could also be envisaged.

4.4.2.2 Silica

(1) Silica-Silane Reaction

In contrast to carbon black, it is necessary to use a coupling agent to achieve silica elastomers reinforcement. TESPT, triethoxysilylpropyltetrasulfide, is the most widely used coupling agent.

$$\text{(EtO)}_3\text{Si}-\text{CH}_2\text{CH}_2\text{CH}_2-\text{S}_4-\text{CH}_2\text{CH}_2\text{CH}_2-\text{Si(OEt)}_3$$

TESPT is a bifunctional molecule with a triethoxysilyl moiety reactive toward Silica's silanols and a polysulfidic moiety that reacts with elastomeric chains Because reaction temperatures are somewhat different, the reaction of TESPT with silica surface mainly occurs during mixing when the reaction of polysulfidic moiety takes place during curing (Fig. 4.4).

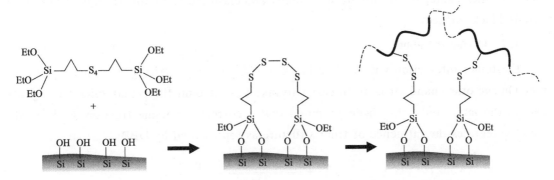

Fig. 4.4 TESPT loadings are about one $Si(OEt)_3$ per nm^2, which roughly corresponds to a complete coverage of silica surface. So, for precipitation silicas with content 6 to 10 Si-OH/nm^2, many surface silanols will remain unreacted.

(2) Polymer Adsorption

When a coupling agent is used to generate a covalent bond between silica surface and elastomeric chains, it also limits polymer adsorption because of its shielding effect. So in silica-silane-elastomer compounds, the "filler network" will be much lower than in carbon black-elastomer systems. However, it remains qualitatively the same, and elastomer chain mobility is also limited in the close neighboring of silica surface. Obviously, if any coupling agent is used, polymer adsorption will naturally occur; in addition, because of the high polarity of silica, some direct interaction between silica aggregates will also take place and constitute an additional filler-filler network. These effects will not happen in silica reinforced systems when an appropriate amount of coupling agent is used. Bound rubber determination is also applied to silica compounds, even if the numerous possible interactions naturally limit the interpretation of the values.

4.5 Mechanical Properties of Filled Rubbers

4.5.1 Mechanical Properties in Green State

The increased life time expected from reinforcement by particulate fillers naturally refers to cured pieces. The incorporation of reinforcing fillers greatly changes the viscosity of green compounds, conducting to a mainly plastic behavior that allows their processing.

4.5.1.1 Viscosity

It is generally reported that elastomers filled with a volume fraction φ present a viscosity following:

$$\eta = \eta_0 (1 + 2.5\varphi + 14.1\varphi^2) \tag{4-1}$$

It is interesting to remember that this relationship has been established for solid spherical objects having any interaction among them and/or with the surrounding medium. As widely discussed earlier in this chapter, rein-forcing systems are very far from this: carbon black or silica aggregate are highly structured, and elastomeric chains strongly adsorb onto carbon black surfaces.

4.5.1.2 Occluded Rubber

Elastomer interaction with carbon black or silica is very difficult to estimate and correct. On the other hand, it is much easier to take into account the actual volume of aggregate in the mix, and it has been proposed that a corrected volume fraction ϕ_c be used, which integrates the influence of filler structure as represented by DBP.

$$\phi_c = \frac{\phi}{2} \times \left[1 + \frac{1 + 0.02139 \times \text{DBP}}{1.46}\right] \text{ et}$$
$$\eta = \eta_0 (1 + 2.5\phi_c + 14.1\phi_c^2) \tag{4-2}$$

The ϕ_c value represents the "actual" size of filler aggregates in the mix; The ϕ_c includes, naturally, the filler object itself plus a significant volume of polymer that is shielded from deformation by aggregate tortuousity. This part of the polymer that will not be deformed is usually called occluded rubber. Nevertheless, occluded rubber must not be confused with the polymer part whose molecular mobility is changed by adsorption. Occluded rubber, which is mainly trapped in aggregate fractal sites, only represents a part of the volume of elastomer whose molecular motion is slowed down. Occluded rubber and viscosity increases with filler structure and loading; on return, specific surface area of the filler has an influence on green mix viscosity.

4.5.1.3 Shear Dependence of Viscosity, Non-Newtonian Behavior

The presence of reinforcing fillers also increases the non-Newtonian behavior of elastomers. This effect is mainly due to the fact that the incorporation of fillers in elastomers decreases the volume of the deformable phase. As discussed in the following text, this decrease is not limited to the actual volume of the filler, but must also include the existence of occluded rubber. So, when filled mixes are submitted to shear forces, because of the lower deformable volume, the actual deformation and speed of deformation are much higher than in unfilled mixes. This phenomenon is usually called Strain amplification effect; obviously strain amplification is not specific to reinforced systems but to any filled polymer.

The influence of filler is not limited to this enhancement of the non-Newtonian behavior of elastomers. At very small shear rates, filled green compounds also exhibit an additional increase of viscosity that can not be explained by strain amplification. This effect is usually attributed to the existence of the filler network: the direct bonding of reinforcing objects by adsorbed chains implies a increased force to be broken. Obviously this influence can only be observed at very low strain, because a very small increase of interaggregate distances immediately implies a desorption of the bridging elastomeric chains.

4.5.2 Mechanical Properties in Vulcanized State

4.5.2.1 Introduction

As for pure elastomers, the vulcanization step provides sulfur bridges between elastomeric chains and connects them into an infinite network. Vulcanization is supposed to be mainly unaffected by the presence of the reinforcing fillers and transforms the roughly plastic green mixes into viscoelastic vulcanizates. For silica mixes, the high temperature of the vulcanization step allows the reaction of the polysulfidic moiety of the coupling agent with elastomer, ensuring the chemical covalent bonding of polymer to silica surface.

As in the green state, the strain amplification, due to the limited volume of the actually deformable phase, remains the first order result of filler incorporation. For a given macroscopic deformation, the actual deformation of the polymeric matrix will always be much higher, obviously depending on the filler volume and its structure, which defines occluded rubber volume. The viscosity equation is usually generalized to Young or shear modulus G^*:

$$G^* = G_0(1 + 2.5\phi_c - 14.1\phi_c^2) \tag{4-3}$$

where G^* is the shear modulus of the unfilled vulcanized matrix at the same shear strain. Obviously, it is possible to vary the modulus by changing vulcanization conditions, providing more or fewer sulfur bridges and so higher or lower G^* values. Anyway, as evidenced by the equation, vulcanization changes G_0^* but roughly does not affect reinforcement by itself.

4.5.2.2 Small-Strain Properties, Dynamic Viscoelastic Measurements

(1) Payne Effect

Reinforced vulcanized samples generally present a marked viscoelastic behavior that is usually studied by dynamic viscoelastic measurements. In this experiment, a sample is subjected to periodic sinusoidal shear strain γ (at defined frequency ω and temperature T). Its dynamic shear modulus G^* is complex and can be written as the sum of the storage modulus G', and the loss modulus G''.

$$G^*(\gamma) = G'(\gamma) + iG''(\gamma) \tag{4-4}$$

The dynamic storage modulus G' presents an interesting variation when γ increases: at very low shear strain, G' is constant (phase ❶ in Fig. 4.5), then strongly decreases (phase ❷ in Fig. 4.5) and reaches a plateau value (phase ❸ in Fig. 4.5). This evolution of G' is usually described as the Payne effect.

This change in G' also corresponds to an important variation of G'' that passes through a maximum value. The phase angle δ between stress and strain, is given by:

$$\mathrm{Tan}\delta(\gamma) = \frac{G''(\gamma)}{G'(\gamma)} \tag{4-5}$$

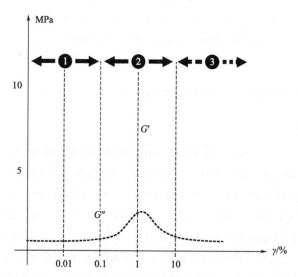

Fig. 4.5 Schematic illustration G' and G'' variation.

The evolution of G' and G'' in the range of 0.1 to 0.5 strain amplitude is of major importance because this domain corresponds to the most common solicitations of filled rubber compounds, for example in tire tread applications.

The loss modulus G'' must not be confused with hysteretic losses of which the expression naturally depends on solicitation mode: hysteretic losses are proportional to G'' (constant strain), or $G''/G^{*2} \sim \tan(\delta)/G'$ (constant stress) or to $\tan\delta$ (constant energy).

It is also important to stress that filled elastomers are a very complex thermorheological system; particularly, G' and G'' variations do not follow the same laws in frequency and temperature.

(2) Mechanism

The Payne effect is widely accepted as the mechanical consequence of the progressive destruction of the "filler network" under shear strain.

The attribution of the Payne effect to the filler network is strongly supported by the fact that carbon black pastes, made with carbon black and low molecular weight oils, present very similar G^* levels at very low shear strain. Obviously, when strain increases, G^* drops drastically for carbon black pastes and much more slowly for filled rubber compounds, because of the progressive desorption of elastomeric chains.

Dannenberg's molecular slippage model, which will be discussed in detail in the next section, gives a good schematic view of the molecular mechanism that is responsible for the Payne effect. Numbers ❶, ❷, and ❸ refer to Fig. 4.5.

① At equilibrium, elastomer chains are adsorbed onto filler surface (state ⓪). When strain increases, it induces a progressive extension of elastomer chain segments that bridges filler particles (state ❶). Obviously, this extension is much greater than macroscopic deformation because of strain amplification. At very low strain, the macroscopic deformation energy is stored in elongated chains as elastic energy and so can be fully recovered when strain decreases; G'' is low and constant.

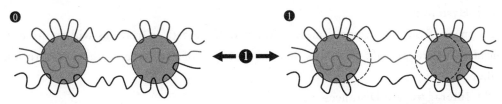

② At higher deformation, it is necessary to decompose the solicitation cycle. During the first extension at higher rate, stored elastic energy overpasses adsorption energy, and elastomer chains progressively desorb from filler surface (state ❷, ★sites desorbed).

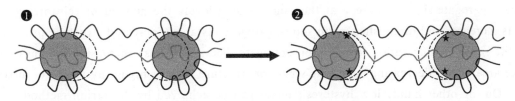

This desorption lengthens the bridging polymer segments that have to direct impacts. First, G', which is roughly inversely dependent on the length of bridging chains, decreases; second, a part of the initially elastic energy stored in deformed chains is converted into molecular mobility and mechanically lost, which corresponds to an increase in G''.

The decrease of deformation that immediately follows the first extension at the higher rate does not exactly lead to initial state; indeed, in the short time of the dynamic deformation cycle, adsorption can not reach equilibrium state and remains imperfect, as illustrated by state ❸.

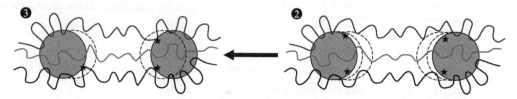

Thus during phase ❷, bridging elastomeric chains undergo adsorption-desorption cycles between 'pseudo' equilibrium state ❷ and state ❸:

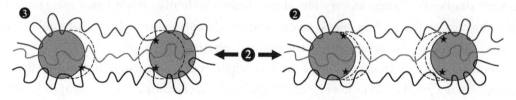

Obviously, elastomer desorption occurs gradually, because of the very broad interaggregate distance distribution that induces an also broad distribution of bridging elastomer segments. This explains the smooth decrease of G' for mixes and its step shape for shorter molecules such as oils.

③ Progressively, desorption induces a homogenization of bridging elastomer segment lengths. This homogenization and the stabilization of the modulus will be discussed in detail in the next section.

(3) Hysteresis

It appears from the previous mechanism that G' and G'' variations will be directly influenced by interaggregate distances and the strength of elastomeric chain adsorption on the filler surface.

(4) Interaggregate Distances

Any change in mix composition or processing that influences inter-aggregate distance distribution will change hysteresis: the lower the average distance, the higher the hysteresis and vice versa. Filler loading evidently decreases interaggregate distances. When filler mean aggregate size decreases, at the same loading level, the number of reinforcing objects increases and diminishes mean inter-aggregate distance. Increased structure provides, at the same loading ratio, lower inter-aggregate distances. Thus, increasing loading, surface area (i.e., decreasing aggregate size), or structure induces higher hysteresis of mixes.

On the other hand, low hysteresis mixes can be achieved by dispersion methods that increase inter-aggregate distances; prolonged or two-stage mixing or master-batch techniques have been used in order to decrease hysteresis.

(5) Adsorption Strength

Coupling agents that mask filler surface strongly reduce elastomer adsorption and thus hysteresis of mixes. This partially explains the low hysteresis value of silica mixes. It is also possible to use functionalized elastomers to reduce mixes hysteresis; such elastomers have specific chemical moieties that react with the filler surface. These reactions lower hysteresis because of the decrease of polymer dangling chains and by shielding of the filler surface.

4.5.2.3 Large Strain Properties

(1) Observations

At large strain, dynamic viscoelastic measurements can not be made accurately because of the important self-heating of the sample during the experiment. Therefore large-strain properties are usually determined by uniaxial extension (Fig. 4.6).

As was observed for G^* in dynamic shear-strain measurements, a clear decrease of Young's modulus $E + \sigma/\varepsilon$ is observed at low strain ($\varepsilon < 1$). This corresponds to the previously described phases ❶ and ❷, even if phase ❶ is obviously not observable. (see Fig. 4.5).

At higher extension rates ($\varepsilon > 1$), Young's modulus increases and reaches a "pseudo maximum" just before the sample break.

A very significant observation is the stress softening effect, also called the Mullins effect. In this experiment, a compound sample is stretched to ε_1 and returned to zero strain, then stretched again. For strain below ε_1, its stress-strain curve is significantly below the first one but rejoins it at ε_1. Stress softening is dependent on the initial strain level; it can be partially reduced by thermal treatment but not be totally effaced (Fig. 4.7).

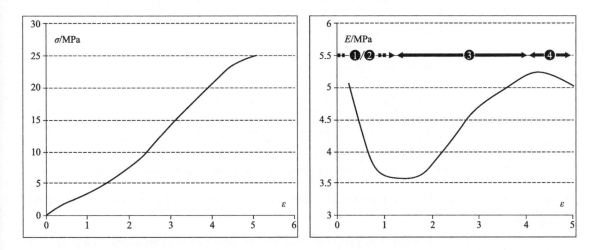

Fig. 4.6 Stress and Young's modulus of reinforced compound.

(2) Interpretation

The paradox of reinforcement by particulate fillers is that there is a simultaneous increase of modulus and elongation at break. This fact is clearly illustrated by the comparison of stress-strain curves of pure and carbon black-filled elastomer (Fig. 4.8).

The modulus increase is the logical consequence of strain amplification due to the replacement of a part of elastomeric deformable phase by a particulate rigid filler: for a macroscopic deformation ε, the local deformation of bridging chains is much higher. Strain amplification should also induce a neat decrease in elongation at break, which is not observed: here is the paradox. Even if this point clearly remains mainly undecided, one way

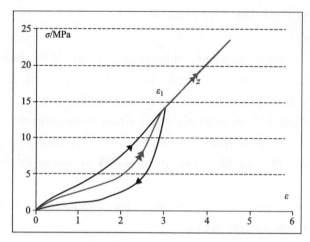

Fig. 4.7 Schematical illustration of stress softening effect.

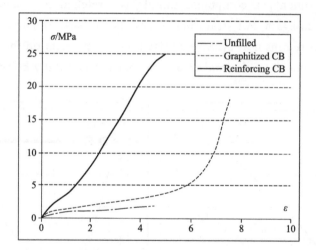

Fig. 4.8 Stress-strain curve of unfilled, graphitized, and reinforcing carbon black samples.

to surpass this paradox is to consider that fillers allow a locally more cooperative sharing of the stress.

In the early 1960, Bueche was probably the first to consider carbon black as a part of a polyfunctional network. In his model, carbon black aggregates were chemically linked by chains and constitute what Medalia and Kraus describe as a "giant multifunctional crosslink" network.

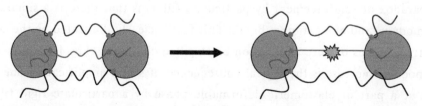

Even if Bueche's model tried to give a molecular origin of reinforcement it remains difficult to consider that it ensures a massive local sharing of the stress: when the shortest chain reaches its finite extensibility, it is really not evident that a large part of the stress is shared by other bridging chains. Moreover, Bueche's model supposes chemical bonding of elastomeric chains which is, aminimo, debatable (see previous text). In any case, it must be mentioned that chain breaks during extension have been demonstrated using ESR.

At the end of the 1960s, Dannenberg completely renewed reinforcement understanding by proposing the "molecular slippage" model that we have previously used to illustrate Payne's effect. In contrast to Bueche, Dannenberg suggested that interaction between elastomer and carbon black was mainly caused by adsorption and not by chemical bonding. Because of its low energy and reversibility, adsorption permits elastomer-filler contacts to change continuously and so allows the homogenization of bridging segment lengths, which ensures the local sharing of stress.

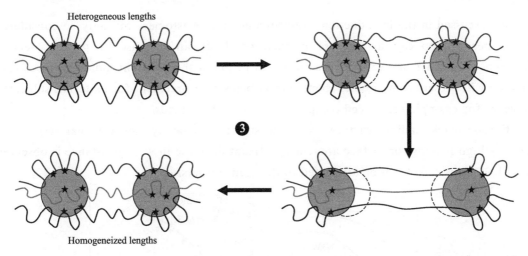

It is noteworthy that the concept of "giant multifunctional crosslink" network associated with the "molecular slippage" model proposes a possible mechanism for length homogenization of bridging chains and gives a rather satisfactory answer to the reinforcement paradox. So the increase of modulus in phase ❸ is due to this stress sharing between bridging chains.

Obviously, the obtained homogenization directly depends on the maximum strain at which the compound has been accommodated. In other words, chain segment lengths are homogenized for any strain below the maximum strain; at higher strains, a new homogenization should occur. This naturally corresponds to the so-called stress softening effect previously described.

The stress sharing by segment homogenization is naturally limited by the number of bridging segments between reinforcing objects; when all connecting chains reach approximately the same length, modulus is at its maximum.

A further increase of strain produces total chain slippage of the shortest chain; then the number of bridging segments decreases (phase ❹). Each remaining bridging chain must support an increased force that produces their massive dewetting and macroscopic break. The filler dewetting under strain has been evidenced by TEM direct observation.

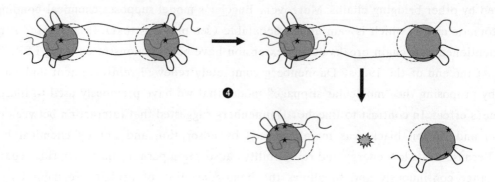

4.5.3 Applications

As discussed in the introduction, reinforcement of elastomers can only be considered for a specific application, because it corresponds to an increase of product service life. Hence, in order to give some practical illustration of the different topics discussed earlier, we will present some results about carbon black reinforcement for materials subjected to wear, for example reinforced compounds used for tire treads.

Carbon black loading, surface area, and structure basically increase wear resistance; for very high loadings or surface area, a significant decrease in wear resistance is observed (Fig. 4.9); this effect can be attributed to deficient dispersion.

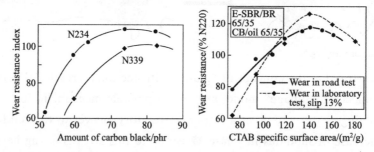

Fig. 4.9 Wear resistance index versus carbon black loading and surface area.

Carbon black surface activity, as revealed by hydrogen content, has also a significant influence on wear resistance (Fig. 4.10).

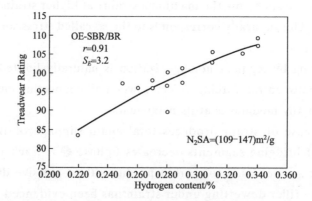

Fig. 4.10 Treadwear versus carbon black hydrogen content.

Carbon black dispersion also influences abrasion resistance, and the maximum observed for high carbon black loadings and surface area can be significantly shifted by using specific dispersion techniques like masterbatching (Fig. 4.11).

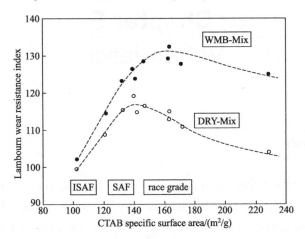

Fig. 4.11 Comparision of abrasion resistance for dry mixes and master batches.

References

[1] F. Bueche, J. Appl. Polym. Sci. 5, 271 (1961).

[2] F. Bueche, J. Appl. Polym. Sci. 4, 107 (1960).

[3] J. B. Donnet and G. Heinrich, Bull. Soc. Chim. Fr. p. 1609 (1960).

[4] J. Le Bras and E. Papirer, J. Appl. Polym. Sci. 22, 525 (1983).

[5] E. M. Dannenberg, RubberChem. Technol. 48, 410 (1975).

[6] J. B. Donnet, E. Papirer, and A. Vidal, Grafting of Macromolecules onto Carbon Blacks, "Chem. and Physics of Carbon," P. L. Walker, Jr. and P. A. Thrower (Eds.), Marcel Dekke New York, 1975.

Chapter 5
Vulcanization

5.1 Introduction

Most useful rubber articles, such as tires and mechanical goods, cannot be made without vulcanization. Unvulcanized rubber is generally not very strong, does not maintain its shape after a large deformation, and can be very sticky. In short, unvulcanized rubber can have about the same consistency as chewing gum.

The first commercial method for vulcanization has been attributed to Charles Goodyear. His process (heating natural rubber with sulfur) was first used in Springfield, Massachusetts, in 1841. Thomas Hancock used essentially the same process about a year later in England. Since those early days, there has been continued progress toward the improvement of the process and in the resulting vulcanized rubber articles. In addition to natural rubber, over the years, many synthetic rubbers have been introduced. Also, in addition to sulfur, many other substances have been introduced as components of curing (vulcanization) systems. This chapter is an overview of the science and technology of vulcanization. Emphasis is placed on the vulcanization of general-purpose "high-diene" rubbers, such as natural rubber (NR), styrene-butadiene rubber (SBR), and butadiene rubber (BR), by sulfur in the presence of organic accelerators.

The accelerated-sulfur vulcanization of these rubbers along with the vulcanization of other rubbers which are vulcanized by closely related technology [e.g., ethylene-propylene-diene monomer rubber (EPDM), butyl rubber (IIR), halobutyl rubbers, and nitrile rubber (NBR)] comprises more than 90% of all vulcanization. Nevertheless, we give some consideration to vulcanization by the action of other vulcanization agents such as organic peroxides, phenolic curatives, and quinoid curatives.

Dynamic vulcanization (DV) is also considered. DV is the crosslinking of one polymer in a blend of polymers during its mixing therein, all polymers of the blend being in the molten state. The process is used in the preparation of thermoplastic elastomeric compositions from rubber-plastic blends.

5.2 Definition of Vulcanization

Vulcanization is a process generally applied to rubbery or elastomeric materials. These materials forcibly retract to their approximately original shape after a rather large mechanically imposed deformation. Vulcanization can be defined as a process which increases the retractile force and reduces the amount of permanent deformation remaining after removal of the deforming force. Thus vulcanization increases elasticity while it decreases plasticity. It is generally accomplished by the formation of a crosslinked molecular network (Fig. 5.1).

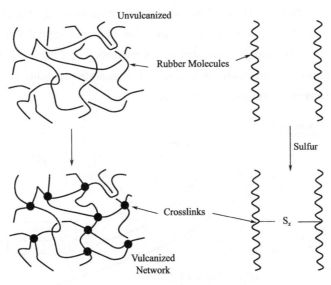

Fig. 5.1 Network formation.

According to the theory of rubber elasticity, the retractile force to resist a deformation is proportional to the number of network supporting polymer chains per unit volume of elastomer. A supporting polymer chain is a linear polymer molecular segment between network junctures. An increase in the number of junctures or crosslinks gives an increase in the number of supporting chains. In an unvulcanized linear polymer (above its melting point), only molecular chain entanglements constitute junctures.

Vulcanization, thus, is a process of chemically producing network junctures by the insertion of crosslinks between polymer chains. A crosslink may be a group of sulfur atoms in a short chain, a single sulfur atom, a carbon to carbon bond, a polyvalent organic radical, an ionic cluster, or a polyvalent metal ion. The process is usually carried out by heating the rubber, mixed with vulcanizing agents, in a mold under pressure.

5.3 Effects of Vulcanization on Vulcanizate Properties

Vulcanization causes profound changes at the molecular level. The long rubber molecules (molecular mass usually between 100000 and 500000 daltons) become linked together with junctures (crosslinks) spaced along the polymeric chains, with the average distance between junctures corresponding to a molecular mass between crosslinks of about 4000 to 10000 daltons. As a result of this network formation, the rubber becomes essentially insoluble in any solvent, and it cannot be processed by any means which requires it to flow, e. g. , in a mixer, in an extruder, on a mill, on a calender, or during shaping, forming, or molding. Thus, it is essential that vulcanization occur only after the rubber article is in its final form.

Major effects of vulcanization on use-related properties are illustrated by the idealization of Fig. 5. 2. It should be noted that static modulus increases with vulcanization to a greater extent than does the dynamic modulus. Here, static modulus is more correctly the equilibrium modulus, approximated by a low strain, slow-strain-rate modulus. Dynamic modulus is generally measured with the imposition of a sinusoidal, small strain at a frequency of 1~100Hz. The dynamic modulus is a composite of viscous and elastic behavior, whereas static modulus is largely a measure of only the elastic component of rheological behavior.

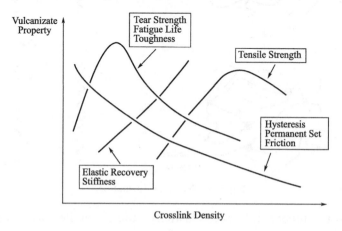

Fig. 5. 2 Vulcanizate properties as a function of the extent of vulcanization.

Hysteresis is reduced with increasing crosslink formation. Hysteresis is the ratio of the rate-dependent or viscous component to the elastic component of deformation resistance. It is also a measure of deformation energy that is not stored (or borne by the elastic network) but that is converted to heat. Vulcanization then causes a trade-off of elasticity for viscous or plastic behavior. Tear strength, fatigue life, and toughness are related to the breaking energy. Values of these properties increase with small amounts of crosslinking, but they are reduced by further crosslink formation. Properties related to the energy-to-break increase with increases in both the number of network chains and hysteresis. Since hysteresis decreases as more network chains are developed, the energy-to-break related properties are maximized at some intermediate crosslink density. It should be noted that the properties given in Fig. 5. 2 are not only functions of crosslink density. They are also affected by the type of crosslink, the type of polymer, and type and amount of filler, etc.

Reversion is a term generally applied to the loss of network structures by nonoxidative thermal aging. It is usually associated with isoprene rubbers vulcanized by sulfur. It can be the result of too long of a vulcanization time (overcure) or of hot aging of thick sections. It is most severe at temperatures above about 155℃. It occurs in vulcanizates containing a large number of polysulfidic crosslinks. Though its mechanism is complex, a good deal about the chemical changes that occur during the reversion of natural rubber has been deduced.

Sometimes the term "reversion" is applied to other types of nonoxidative degradation, especially with respect to rubbers not based on isoprene. For example, thermal aging of SBR (styrene-butadiene rubber), which can cause increased crosslink density and hardening, has been called reversion since it can be the result of overcure.

5.4 Characterization of the Vulcanization Process

Important characteristics related to the vulcanization process are the time elapsed before crosslinking starts, the rate of crosslink formation once it starts, and the extent of crosslinking at the end of the process. There must be sufficient delay or scorch resistance (resistance to premature vulcanization) to permit mixing, shaping, forming, and flowing in the mold before vulcanization. Then the formation of crosslinks should be rapid and the extent of crosslinking must be controlled (Fig. 5.3 and Fig. 5.4).

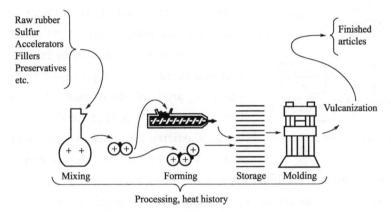

Fig. 5.3　The effect of processing on heat history.

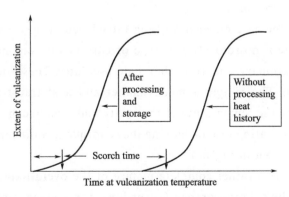

Fig. 5.4　The effect of heat history (processing) on scorch safety.

Scorch resistance is usually measured by the time at a given temperature required for the onset of crosslink formation as indicated by an abrupt increase in viscosity. The Mooney viscometer is usually used. During this test, fully mixed but unvulcanized rubber is contained in a heated cavity. Imbedded in the rubber is a rotating disc. Viscosity is continuously measured (by the torque required to keep the rotor rotating at a constant rate) as a function of time. The temperature is selected to be characteristic of rather severe processing (extrusion, calendering, etc.).

Both the rate of vulcanization after the scorch period and the final extent of vulcanization are measured by devices called cure meters. Many workers contributed to this development. Widely used cure meters are oscillating disc rheometers of the type introduced by the Monsanto Company in about 1965. The development of the oscillating disc rheometer, largely through the efforts of R. W. Wise, was the beginning of modern vulcometry, which has become standard practice in the industry. Before the development of the cure meter, it was necessary to measure mechanical properties of many specimens of a rubber sample, each vulcanized for a different length of time at a given temperature.

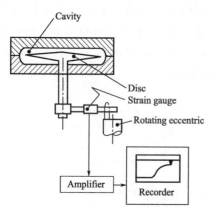

Fig. 5.5 Oscillating disc rheometer.

In order to measure the vulcanization characteristics, the rubber is enclosed in a heated cavity (Fig. 5.5). Imbedded in the rubber is a metal disc that oscillates sinusoidally in its plane about its axis. Vulcanization is measured by increase in the torque required to maintain a given amplitude (e. g., degrees of arc) of oscillation at a given temperature. The torque is proportional to a low strain modulus of elasticity. Since this torque is measured at the elevated temperature of vulcanization, the portion of it due to viscous effects is minimal. Thus it has been assumed that the increase in torque during vulcanization is proportional to the number of crosslinks formed per unit volume of rubber. The torque is automatically plotted against time to give a so-called rheometer chart, rheograph, or cure curve.

Later versions of the cure meter have been introduced (e. g., Fig. 5.6). The cavity is much smaller and there is no rotor. In this type of cure meter, one-half of the die (e. g., the upper half) is stationary and the other half oscillates. These instruments are called moving-die rheometers. The sample is much smaller and heat transfer is faster. Also, because there is no rotor, the temperature of the cavity and sample can be changed more rapidly. In either case (oscillating disc or moving die), torque is automatically plotted against time. Such a chart is shown in Fig. 5.7.

The cure curve gives a rather complete picture of the overall kinetics of crosslink formation and even crosslink disappearance (reversion) for a given rubber mix. In some cases, instead of reversion, a long plateau or marching cure can occur. The cure meter is, therefore, extensively used to control the quality and uniformity of rubber stocks (also called rubber compounds).

Vulcometry started as a research tool to study vulcanization. It was then used to con-

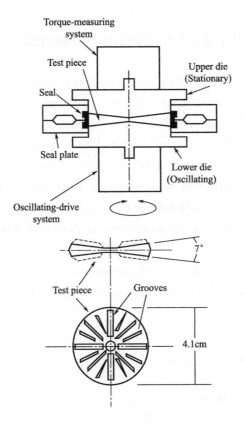

Fig. 5.6 Moving-die rheometer.

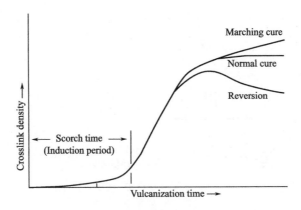

Fig. 5.7 Rheometer cure curve.

trol uniformity of rubber mixed in the factory. Also programmed temperature-profile vulcometry has been used to develop recipes for industrial use. The cure temperature-time profile of an industrial mold can be imposed on the curing cavity of the cure meter. The test sample can then be vulcanized in the cure meter under the same conditions as those encountered in the factory. Both the extent of cure and temperature can be simultaneously displayed as functions of time.

5.5 Vulcanization by Sulfur without Accelerator

Initially, vulcanization was accomplished by using elemental sulfur at a concentration of 8 parts per 100 parts of rubber (phr). It required 5 hours at 140℃. The addition of zinc oxide reduced the time to 3 hours. The use of accelerators in concentrations as low as 0.5phr has since reduced the time to as short as 1 to 3 minutes. As a result, elastomer vulcanization by sulfur without accelerator is no longer of much commercial significance. An exception to this is the use of about 30 or more phr of sulfur, with little or no accelerator, to produce molded products of hard rubber or "ebonite". Even though unaccelerated sulfur vulcanization is not of commercial significance, its chemistry has been the object of much research and study.

The chemistry of unaccelerated vulcanization is controversial. Many slow reactions occur over the long period of vulcanization. Some investigators have minded that the mechanisms involved free radicals:

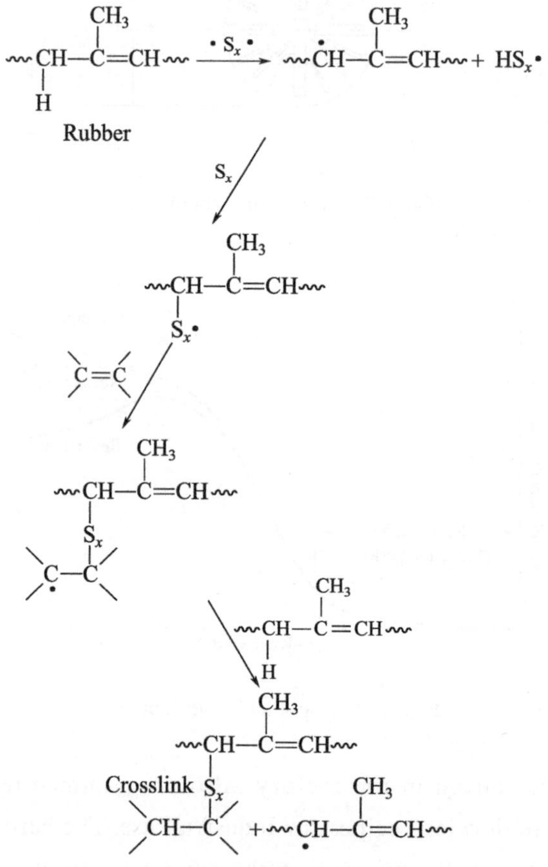

$$\text{\textasciitilde CH-C(CH}_3\text{)=CH\textasciitilde} \xrightarrow{\cdot S_x \cdot} \text{\textasciitilde CH-C(CH}_3\text{)=CH\textasciitilde} + HS_x\cdot$$

Rubber

Scheme 1

Other investigators have promoted ionic mechanisms:

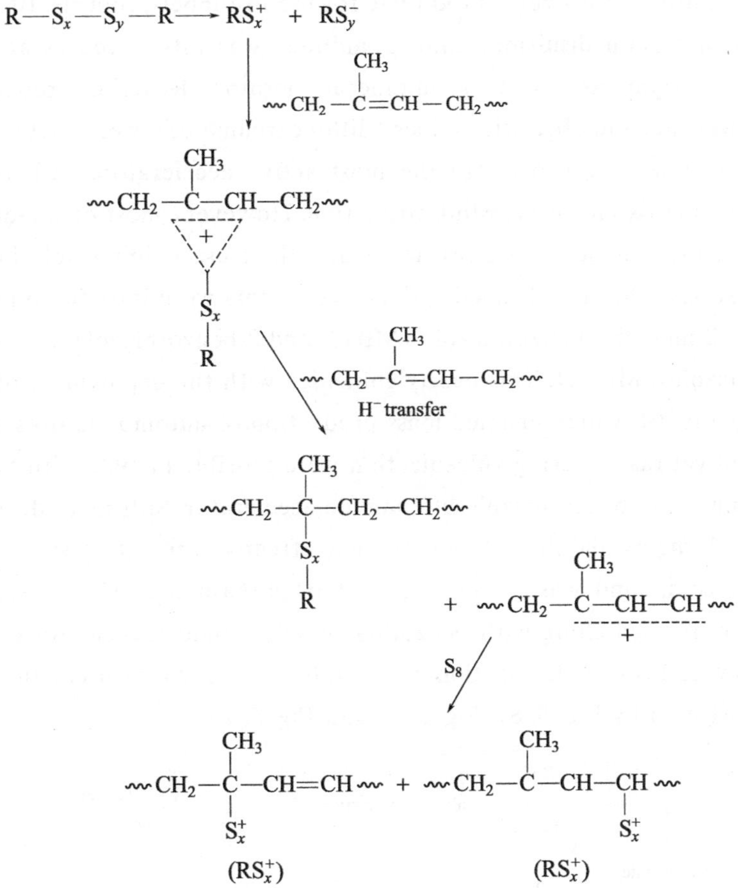

$$R-S_x-S_y-R \longrightarrow RS_x^+ + RS_y^-$$

Scheme 2

The intermediates,

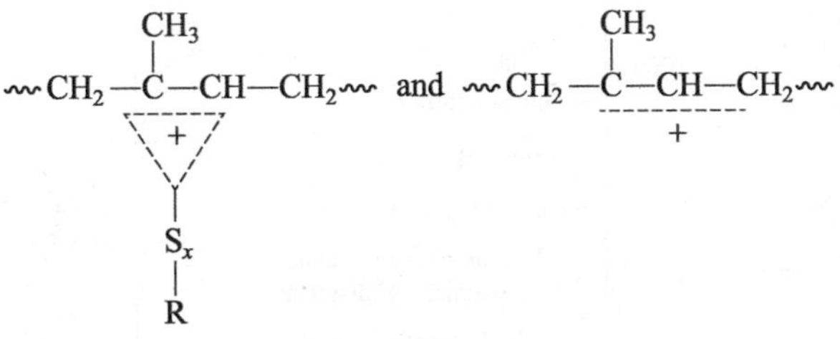

were proposed in **Scheme 2** to explain the fact that model compound reactions gave both unsaturated and saturated products, sulfur atoms being connected to both secondary and tertiary carbon atoms.

5.6 Accelerated-Sulfur Vulcanization

Organic chemical accelerators were not used until 1906 [65 years after the Goodyear-Hancock development of unaccelerated vulcanization (Fig. 5.8)] when the effect of aniline on sulfur vulcanization was discovered by Oenslager. This could have been, at least partially, in response to the development of pneumatic tires and automobiles near the turn of the century. Aniline, however, is too toxic for use in rubber products. Its less toxic reaction product with carbon disulfide, thiocarbanilide, was introduced as an accelerator in 1907. Further developments lead to guanidine accelerators. Reaction products formed between carbon disulfide and aliphatic amines (dithiocarbamates) were first used as accelerators in 1919. These were and are still the most active accelerators with respect to both crosslinking rate and extent of crosslink formation. However, most of the dithiocarbamate accelerators give little or no scorch resistance and their use is impossible in many factory processing situations. The first delayed-action accelerators were introduced in 1925 with the development of 2-mercaptobenzothiazole (MBT) and 2-benzothiazole disulfide (or 2,2-dithiobisbenzothiazole, MBTS). This nearly coincided with the deployment of cord-ply construction (1920~1930) which enabled mass production of automobile tires. Even more delayed action and yet faster curing vulcanization were possible in 1937 with the introduction of the first commercial benzothiazolesulfenamide accelerator. Still more delay became possible in 1968 with the availability of an extremely effective premature vulcanization inhibitor (PVI). This compound was N-(cyclohexylthio)phthalimide (CTP), small concentrations of which were used along with benzothiazolesulfenamide accelerators. The history of the progress toward faster vulcanization but with better control of premature vulcanization or scorch is illustrated by Fig. 5. 8, Fig. 5. 9, and Fig. 5. 10.

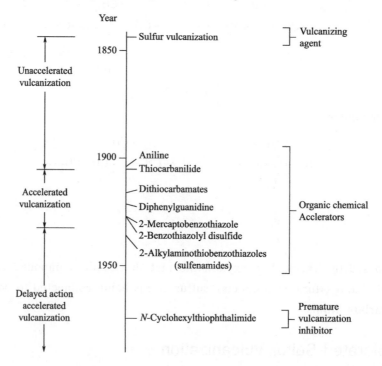

Fig. 5. 8　The history of vulcanization by sulfur.

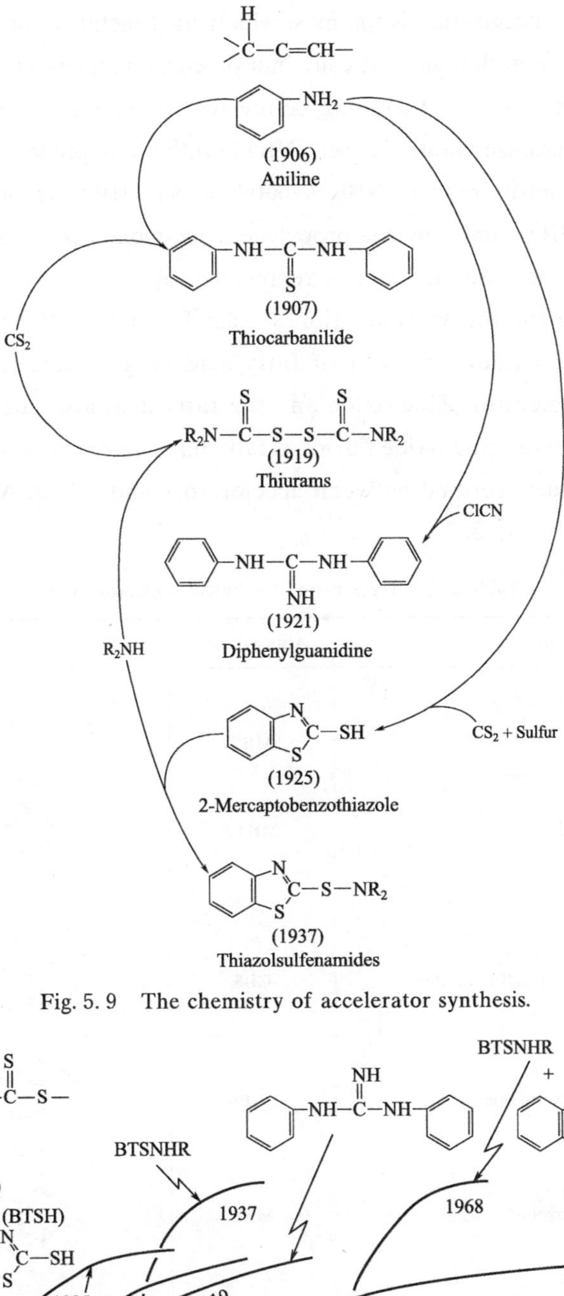

Fig. 5.9 The chemistry of accelerator synthesis.

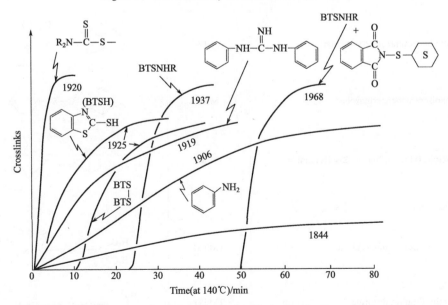

Fig. 5.10 Improvements in the accelerated-sulfur vulcanization of natural rubber.

Accelerated-sulfur vulcanization is the most widely used method. For many applications, it is the only rapid crosslinking technique that can, in a practical manner, give the delayed action required for processing, shaping, and forming before the formation of the intractable vulcanized network. It is used to vulcanize natural rubber (NR), synthetic isoprene rubber (IR), styrene-butadiene rubber (SBR), nitrile rubber (NBR), butyl rubber (IIR), chlorobutyl rubber (CIIR), bromobutyl rubber (BIIR), and ethylene-propylene-diene-monomer rubber (EPDM). The reactive moiety for all of these elastomers can be represented by.

Typically a recipe for the vulcanization system for one of the above elastomers contains 2~10 phr of zinc oxide, 1~4 phr of fatty acid (e.g., stearic), 0.5~4 phr sulfur, and 0.5~2 phr of accelerator. Zinc oxide and the fatty acid are vulcanization system activators. The fatty acid with zinc oxide forms a salt that can form complexes with accelerators and reaction products formed between accelerators and sulfur. Accelerators are classified and illustrated in Table 5.1.

Table 5.1 Accelerators for Sulfur Vulcanization

Compound	Abbreviation	Structure
Benzothiazoles		
2-Mercaptobenzothiazole	MBT	
2,2′-Dithiobisbenzothiazole	MBTS	
Benzothiazolesulfenamides		
N-Cyclohexylbenzothiazole-2-sulfenamide	CBS	
N-t-butyobenzothiazole-2-sulfenamide	TBBS	
2-Morpholinothiobenzothiazole	MBS	
N-Dicyclohexylbenzothiazole-2-sulfenamide	DCBS	
Dithiocarbamates		
Tetramethylthiuram monosulfide	TMTM	
Tetramethylthiuram disulfide	TMTD	

Continued

Compound	Abbreviation	Structure
Zinc diethyldithiocarbamate	ZDEC	$\left(\begin{array}{c}C_2H_5\\C_2H_5\end{array}\!\!>\!\!N\!-\!\overset{\overset{S}{\|}}{C}\!-\!S\right)_{\!2}\!\!Zn$
Amines		
Diphenylguanidine	DPG	Ph-NH-C(=NH)-NH-Ph
Di-*o*-tolylguanidine	DOTG	(o-tolyl)-NH-C(=NH)-NH-(o-tolyl)

Frequently, mixtures of accelerators are used. Typically, a benzothiazole type is used with smaller amounts of a dithiocarbamate (thiuram) or an amine type. An effect of using a mixture of two different types of accelerator can be that each activates the other and better-than-expected crosslinking rates can be obtained. Mixing accelerators of the same type gives intermediate or average results.

We should note here that there is urgency to reduce the use of accelerators based on secondary amines, which can react with nitrogen oxides to form suspected carcinogenic nitrosamines. This is especially a problem with dithiocarbamate-type accelerators. Proposed accelerators, which do not give carcinogenic nitrosamine derivatives, include dibenzylamine-derived dithio-carbamates and those based on sterically hindered amines.

Different types of accelerators impart vulcanization characteristics which differ with respect to both scorch resistance and crosslinking rate. Fig. 5.11 is a map of accelerator system characteristics. Within groups or types, differences can be obtained by choosing the individual accelerators. In the group of benzothiazolesulfenamides, the scorch resistance and vulcanization time increase in the order: TBBS or CBS, MBS, DCBS.

The effect of the addition of small concentrations of the premature vulcanization inhibitor (PVI), N-(cyclohexylthio) phthalimide, is also given by Fig. 5.11. This retarder is frequently used to independently control scorch resistance with little effect on the rate of crosslinking. Before the development of N-(cyclohexylthio) phthalimide as a PVI, acidic retarders, e. g., salicylic acid, acetylsalicylic acid, phthalic anhydride, and benzoic acid, were used.

These additives improved scorch resistance but also gave greatly reduced rates of crosslink formation after the delay. Another retarder of the past was N-nitrosodiphenylamine, which is less active and not now used because of toxicological concerns.

5.6.1 The Chemistry of Accelerated-Sulfur Vulcanization

The general reaction path of accelerated-sulfur vulcanization is thought to be as follows: Accelerator reacts with sulfur to give monomeric polysulfides of the structure Ac-S_x-Ac where Ac is an organic radical derive from the accelerator (e. g., benzothiazyl-). The monomeric polysulfides interact with rubber to form polymeric polysulfides, e. g., rubber-S_x-

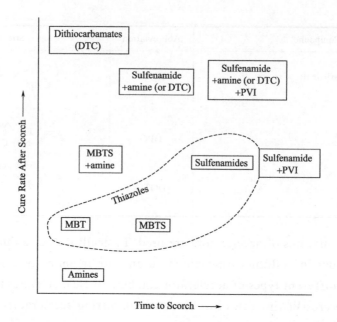

Fig. 5.11 Vulcanization characteristics given by various accelerators and combinations.

Ac. During this reaction, 2-mercaptobenzothiazole (MBT) is formed if the accelerator is a benzothiazole derivative and if the elastomer is natural rubber. In SBR the MBT becomes bound to the elastomer molecular chain probably as the thioether rubber-S_x-Ac. When MBT itself is the accelerator in natural rubber, it first disappears then reforms with the formation of BT-S-S_x-S-BT and rubber-S_x-Ac. Finally, the rubber polysulfides react, either directly or through an intermediate, to give crosslinks, rubber-S_x-rubber. A reaction scheme can be written as follows:

There are obvious differences between accelerated vulcanization and unaccelerated vulcanization. Greater crosslinking efficiencies and greater crosslinking rates are obtained with accelerated vulcanization. But there are more subtle differences. Results from model reactions with curing ingredients indicate that sulfur becomes attached to the rubber hydrocarbon almost exclusively at allylic positions. This is not the case with unaccelerated sulfur vulcanization, thus:

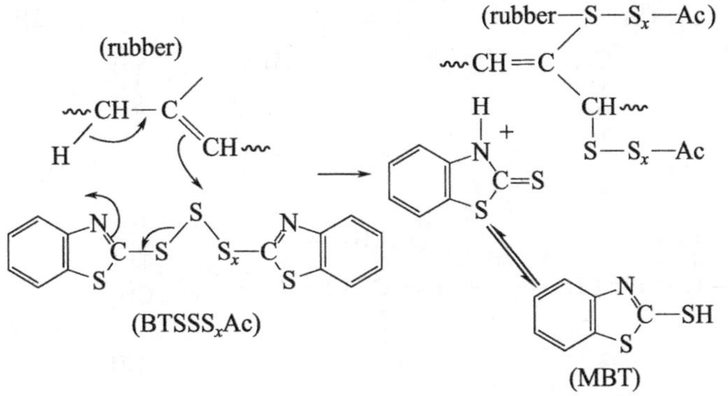

Rather than the eight-member-ring intermediate shown above, one could propose a six-member-ring sulfurization intermediate:

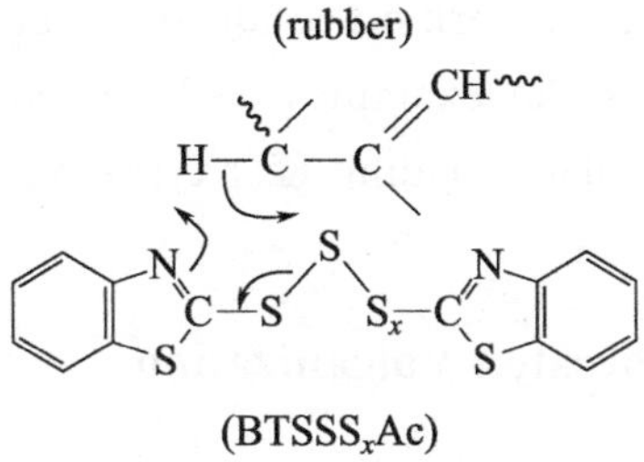

This is similar to dithiocarbamate acceleration:

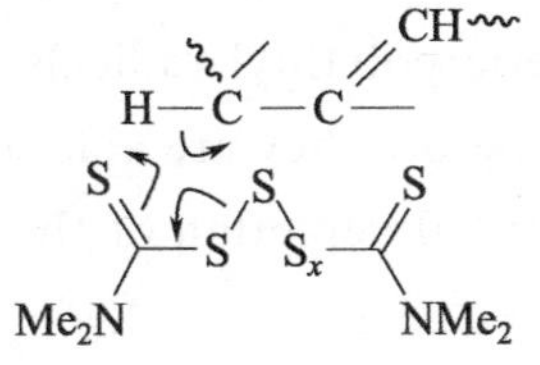

It is proposed that zinc must be present for sulfuration. At any rate, crosslinks could then form, in a number of ways, e. g.:

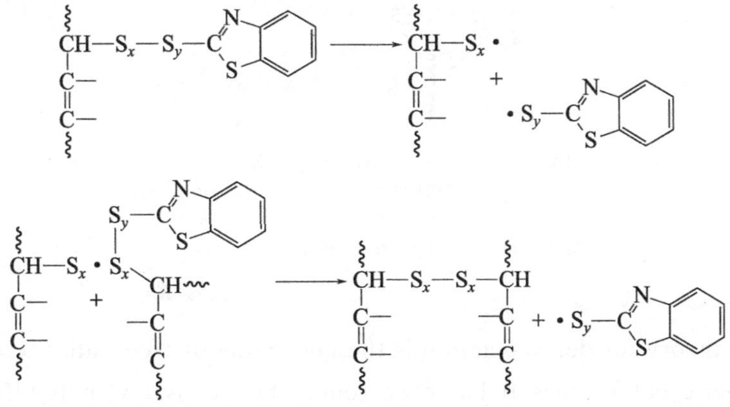

By using solid-state C^{13} NMR spectroscopy, Koenig and his group have added much detail to the chemical structure of the sulfurated network back-bone. The following diagrams show the types of structures that have been assigned to the attachment points of the sulfur atoms to the rubber molecular backbone:

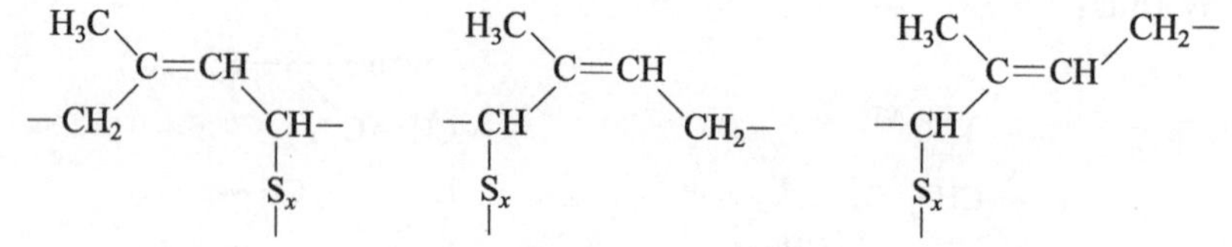

$$
\begin{array}{ccc}
\text{A1c} & \text{B1c} & \text{B1t} \\
\text{Unsaturated A2} & \text{Saturated A2} & \text{C1c}
\end{array}
$$

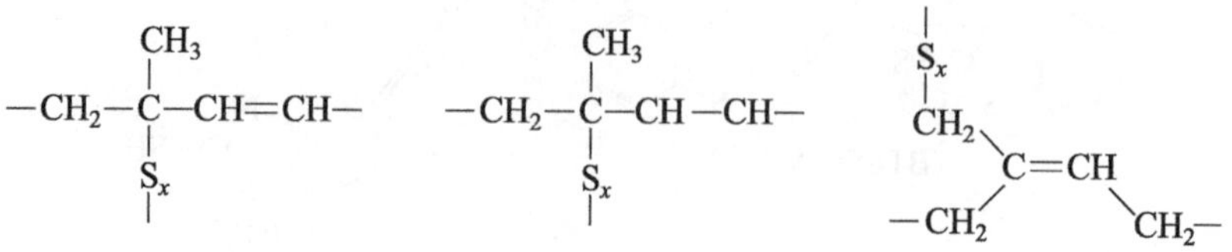

Koenig's group has done much work on conditions that change the relative amounts of the various types of attachments. For example, both the B1c and B1t type polysulfides increase with the level of carbon black loading (for types N110, N220, N326, N330, N550, and N765).

5.6.2 Delayed-Action Accelerated Vulcanization

If crosslink formation is by a free radical mechanism, delayed action could be the result of a quenching action by the monomeric polysulfides formed by reactions between accelerator and sulfur. If the polymeric polythiyl radicals (crosslink precursors) are rapidly quenched by an exchange reaction before they are able to form crosslinks, crosslink formation would be impeded until substantial depletion of the monomeric polysulfides. This is illustrated as follows:

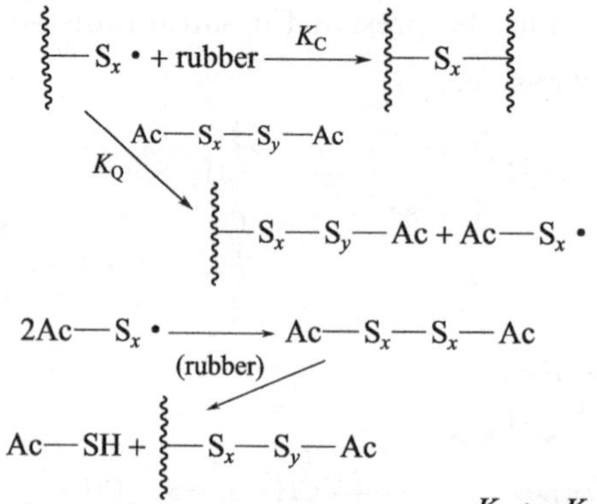

$$-S_x \cdot + \text{rubber} \xrightarrow{K_C} -S_x-$$

$$\text{Ac}-S_x-S_y-\text{Ac} \xrightarrow{K_Q} -S_x-S_y-\text{Ac} + \text{Ac}-S_x \cdot$$

$$2\text{Ac}-S_x \cdot \xrightarrow{\text{(rubber)}} \text{Ac}-S_x-S_x-\text{Ac}$$

$$\text{Ac}-\text{SH} + -S_x-S_y-\text{Ac}$$

$$K_Q \gg K_C$$

Thus, one theory for delayed action is the quenching of free radical crosslink precursors by monomeric polysulfides. It has been found that if bisalkyl polysulfides are mixed with uncured rubber stocks, more delay results. It is also been shown that the early reaction products formed by the interaction between accelerator and sulfur (Ac-S_x-Ac) are inhibitors of crosslink formation. The very substances which give rise to the formation of the crosslink precursor (rubber-S_x-Ac) inhibit the formation of the crosslinks. We note that other mechanisms for delayed action have been proposed. In the case of acceleration by benzothiazolesulfenamides, the accelerator is depleted in an autocatalytic fashion with the formation of 2-mercaptobenzothiazole (MBT). The rate of this depletion is about proportional to the amount of MBT present. There is strong evidence which indicates that the following reactions occur in sulfenamide accelerated systems:

$$\text{BT-S-NHR} + \text{BT-SH} \longrightarrow \text{BT-S-S-BT} + \text{RNH}_2$$
sulfenamide   MBT   MBTS
accelerator

If MBT could be taken out of the system as fast as it forms, substantial increases in processing safety would result. Such is the case when the premature vulcanization inhibitor, N-(cyclohexylthio)phthalimide (CTP) is present. This compound and others like it react rapidly with MBT to form 2-(alkyldithio)benzothiazoles, R-S-S-BT, which are active accelerators which do not interact rapidly with the sulfenamide accelerator:

$$\text{L-S-R} + \text{BT-SH} \longrightarrow \text{BT-S-S-R} + \text{L-H}$$
premature   MBT   2-(alkyldithiyl)-   inactive
vulcanization   benzothiazole   compound
inhibitor

where L is a "leaving group" of the premature vulcanization inhibitor (e.g., phthalimido- for CTP). The importance of scorch control cannot be overemphasized. Present-day tire plants could not compete without good control of scorch resistance or processing safety as it is commonly called. Such safety is necessary in order to rapidly process rubber mixes at high temperatures (through extrusion, calendering, etc.) into preforms for molding (e.g., tire components). Delayed action mechanisms and reaction kinetics have been discussed and reviewed elsewhere.

5.6.3 The Role of Zinc in Benzothiazole-Accelerated Vulcanization

An increase in the concentration of fatty acid and hence increases in the concentration of available Zn^{2+} causes an increased overall rate in the early reactions (during the delay period), which lead to the formation of rubber-S_x-Ac. However, it gives rise to a decrease in the rate of crosslink formation but an increase in the extent of crosslinking. The increase in the rates of the early reactions has been explained by the interaction:

$$\text{BT-S-S-BT} + Zn^{2+} \rightleftharpoons \text{BT-S-S-BT} \cdots Zn^{2+}$$

where the chelated form of the accelerator is more reactive than the free accelerator during the early reactions:

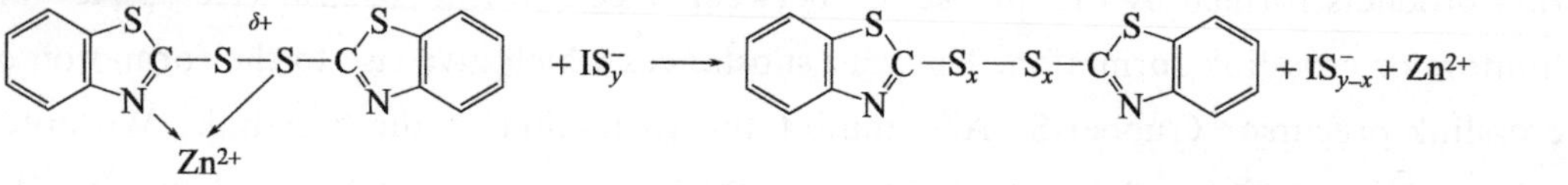

[Reaction scheme showing benzothiazole-S-S-benzothiazole complex with Zn^{2+} reacting with IS_y^- to form product with $IS_{y-x}^- + Zn^{2+}$]

Here, IS_y^- is an ionized form of linear sulfur. It could be rapidly formed in reaction between sulfur and any of a number of initiating species. Others have proposed that the presence of Zn^{2+} can increase the rate of sulfurization through the formation of complexes of the type:

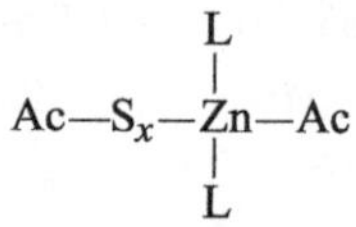

$$Ac-S_x-\underset{\underset{L}{|}}{\overset{\overset{L}{|}}{Zn}}-Ac$$

where L is a ligand such as an amine molecule.

The decreased specific rate of crosslink formation, and the increased extent of crosslinking due to the presence of Zn^{2+} in benzothiazole accelerated vulcanization, have been explained by the following scheme:

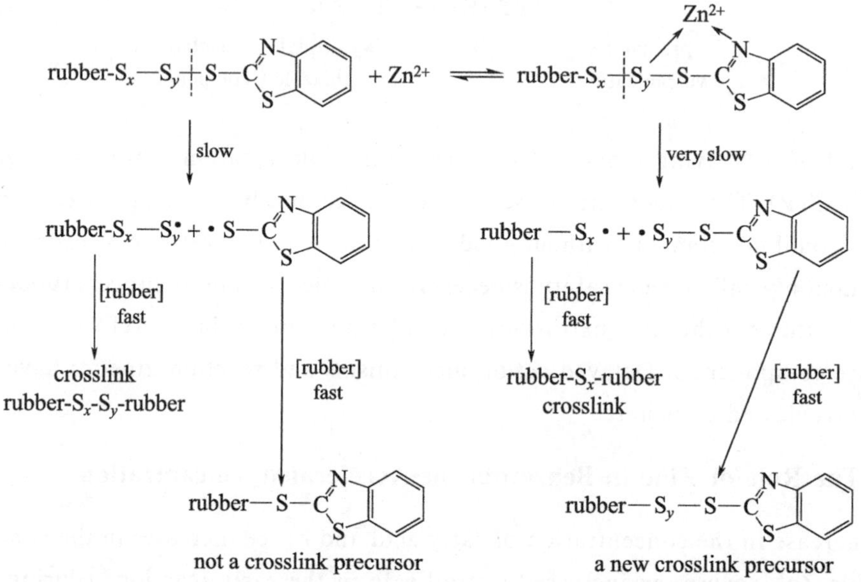

[Reaction scheme showing two pathways: left pathway (slow) from rubber-S_x-S_y-S-C(benzothiazole) producing rubber-S_x-$S_y^\bullet$ + $\bullet$S-C(benzothiazole), leading via [rubber] fast to crosslink rubber-S_x-S_y-rubber and via [rubber] fast to rubber-S-C(benzothiazole) "not a crosslink precursor"; right pathway (very slow) with Zn^{2+} coordination, producing rubber-$S_x\bullet$ + $\bullet S_y$-S-C(benzothiazole), leading via [rubber] fast to rubber-S_x-rubber crosslink and via [rubber] fast to rubber-S_y-S-C(benzothiazole) "a new crosslink precursor"]

Zinc chelation changes the position of the S—S bond most likely to break. Since a stronger bond must break, the rate is slower. Though the rate of crosslinking is slower, the extent of crosslink formation is increased since less sulfur is used in each crosslink. That is, the crosslinks are of lower sulfidic rank.

The presence of zinc compounds can also promote the reduction the sulfur rank of crosslinks during high-temperature ageing of the vulcanizate, e. g. , during reversion. In some cases zinc compounds actually promote the decomposition of crosslinks.

5.6.4 Achieving Specified Vulcanization Characteristics

For many years, it was difficult to independently control the two main vulcanization characteristics, scorch resistance (processing safety) and rate of crosslink formation. In the case of natural rubber (NR), if one had chosen a fast accelerator system in order to obtain short curing times in the press, then process safety would have suffered greatly. If one had chosen a delayed action acceleration system, then the rate of vulcanization in the press would have been limited. The development of the highly efficient premature vulcanization inhibitor N-(cyclohexylthio)phthalimide (CTP) changed all of that, since great improvements in scorch resistance with little or no change in crosslinking rate became possible. Thus the rate of crosslink formation can be adjusted by the selection of accelerators. For example, the moderately fast delayed action accelerator t-butylbenzothiazolesulfenamide (TBBS) can be partially replaced by a small amount of a coaccelerater such as 0.1 to 0.2 phr of tetramethylthiuram disulfide (TMTD) or tetramethyl thiumram monosulfide (TMTM) to obtain a greatly increased cure rate; however, the scorch resistance will be significantly reduced. In such a case, the scorch resistance can be regained by the addition of 0.05 to 0.25 phr of CTP, without a noticeable decrease in the rate of crosslinking.

It is true that merely increasing the concentration of TBBS will give an increase in cure rate with only a small change in scorch resistance. However, the increase in accelerator concentration will generally be rather large and the concentration of sulfur will be adjusted downward to keep the hardness and stiffness constant (maintaining constant crosslink density). The relatively large change in the concentrations of sulfur and accelerator will cause changes in vulcanizate performance properties.

In rubber mixes containing only a synthetic rubber, such as styrene-butadiene rubber (SBR) or butadiene rubber (BR), the effects of cure system changes may not be as pronounced as they are in the case of NR. However, if even a relatively small amount of NR is present, the effects of cure system changes on the vulcanization process parameters resemble those obtained with NR alone. One of the curing characteristics that one would like to control is reversion that can occur in compounds containing natural rubber. There is more than one approach to reducing the amount of reversion. One can use sulfur donors or increase the ratio of accelerator concentration to sulfur concentration. One could carry out the vulcanization at a reduced temperature for a longer period. However, these approaches give rise to effects that will have to be compensated. Another approach is use of additives such as certain bisimides, e. g. , N,N-m-phenylene-biscitraconimide and N,N-m-phenylene-bismaleimide or trialkoxysilylalkylpolysulfides such as bis-(3-triethoxisilyl propyl)-tetrasulfide.

5.6.5 Effects on Adhesion to Brass-Plated Steel

The adhesion between rubber and brass-plated steel (e. g. , steel tire cords for belted radial tires) has been the subject of much study and speculation. Brass plating is presently the major method of obtaining adhesion between natural rubber and the steel of tire cords. Over the years there has been much speculation about its mechanism, but there is agreement on one aspect of the adhesion of natural rubber to brass-plated steel: the actual adhesive between the natural rubber and the brass-plated cord, formed in situ during the vulcanization process, is an interfacial layer of sulfides and oxides of copper.

The adhesive layer between the rubber and cord is generally considered to be formed by the interaction between the copper and the vulcanization system. As a result of this, optimization of the vulcanization system with respect to adhesion is critical. Also, a change in the composition of the brass coating on the steel wires, or a change in the thickness, can require a change in the vulcanization system in order to maintain the optimum level of adhesion.

Reviews on the subject of brass-plated steel cord-natural rubber adhesion have been written by van Ooij who has done much of the work in the field. Van Ooij has given a model for rubber brass adhesion, in which a copper sulfide layer forms on the brass before the onset of crosslink formation. The thin film of copper sulfide has good adhesion and cohesion. In addition, the film is so porous that rubber molecules can become entangled with it. It is not required that the film forms simultaneously with the formation of crosslinks during vulcanization; but, rather, it is required that the copper sulfide film be completely formed before crosslinking starts. Indeed, adhesion between brass-plated steel and natural rubber can frequently be improved by the use of the retarder, CTP or by using a more delayed action accelerator such as N-dicyclohexylbenzothiazole-2-sulfenamide (DCBS).

Failure rarely occurs between the rubber and the copper sulfide film. It generally occurs cohesively within the sulfide film or adhesively in a layer below the sulfide film. Sulfidation of the brass surface is not due to its interaction with elemental sulfur, but it is the result of the interaction between the brass surface and accelerator-sulfur reaction products which can be represented by the general structure, Ac-S_y-Ac and Ac-S_y-H, where Ac is an accelerator-derived moiety (e.g., benzothiazolyl group). The value of the subscript, y, increases with the ratio of the concentration of sulfur to the concentration of accelerator used in the curing system. Generally, high sulfur levels and high ratios of sulfur concentration to accelerator concentration favor good rubber-to-brass adhesion.

The choice of accelerator also has an effect on the quality of adhesion between cord and rubber. The accelerator should not form a stable copper complex which dissolves in the rubber. This would be quite corrosive to the brass plating. In this respect, benzothiazoles and their sulfenamides are much better than dithiocarbamates. DCBS is a particularly good sulfenamide accelerator for rubber-to-brass adhesion.

5.6.6 The Effect on Vulcanizate Properties

Increases in sulfur and accelerator concentrations give higher crosslink densities and, therefore, higher moduli, stiffness, hardness, etc. However, as the ratio of the concentration of accelerator to the concentration of sulfur increases, the proportion of monosulfidic crosslinks increases in natural rubber stocks (also called rubber compounds). Greater amounts of accelerator (with respect to sulfur) also give an abundance of pendent groups of the type, -S_x-Ac, which are attached to and "dangle" from the rubber molecular chain. Higher ratios of sulfur concentration to accelerator concentration give both more polysulfide crosslinks and more sulfur combined with the rubber chains to form sulfur-containing six-membered heterocyclic rings along the rubber molecular chains. In addition, conjugated olefinic double bonds appear along the polymer backbone chain. These features are indicated by Fig. 5.12. Such changes in the vulcanizate network structure, no doubt, are responsible for changes which occur in vulcanizate properties as a result of changes made in the curing-system recipe.

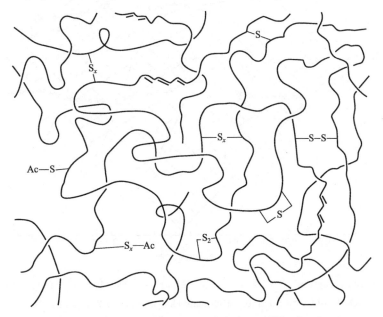

Fig. 5.12 Crosslink types and chain modifications.

Effects of changes in the concentrations of accelerator and sulfur on vulcanizate properties have been studied by using the following recipe (parts by weight): natural rubber, 100; N330 carbon black, 50; N-isopropyl-N-phenyl-p-phenylenediamine (IPPD antidegradant), 2; zinc oxide, 5; stearic acid, 3; plasticizer, 3; sulfur, variable; N-cylohexylbenzothiazolesulfenamide (CBS), variable.

The effects of changes in the accelerator concentration on 300% modulus (jargon for stress at 300% tensile strain), thermal-oxidative aging, and fatigue life (DeMattia flex crack) are given in Fig. 5.13. The effects on 300% modulus are indicated by the diagonal contours of negative slope. They are parallel and show that the stress at 300% strain increases with an increase in either sulfur or accelerator concentration. The contours for % retention of ultimate elongation after hot air aging (at 100°C for 2 days) indicate that oxidative aging in the presence of IPPD, depends only on the concentration of sulfur. Higher concentrations of sulfur give poor aging characteristics in correlation with the higher number of points of chain sulfuration. This suggests that sulfur substitution along the chain can activate chain scission by reactions with oxygen.

Another view is that sulfur interferes with the antidegradant activity (in this case with IPPD). The contours for flex fatigue life are complex. The test is run such that the speci-

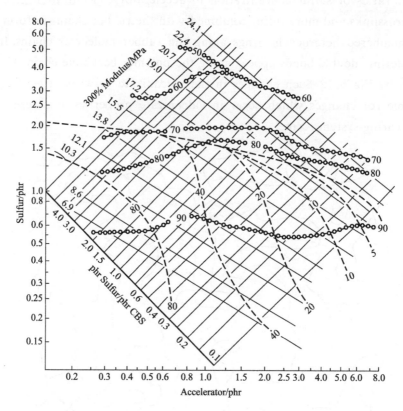

Fig. 5. 13 Vulcanizate properties: —, 300% modulus (Mpa); ---, De Mattia flex fatigue life (kHz×10^{-1}); -O-O-O-O-, % retention of ultimate elongation after 2 days at 100℃.

mens are about equally strained; however, there is some question as to whether the tests should be run at equal strain or at equal strain energy. For some cases, where strain is restricted by fabric reinforcement, fatigue test data should be compared at equal strain amplitude. For other applications, where the strain is not limited, the tests should be run at equal strain energy. The contours as presented here can be interpreted in terms of either constant strain or constant strain energy: All points on the chart can be compared at an approximately equal strain per cycle; however, if we interpolate between the flex-life contours but only along a constant modulus contour, we can extract values corresponding to approximately equal strain energy per cycle. By choosing higher modulus contours, we are considering higher strain energies.

Considering the group of flex-life contours as a whole, or at approximately constant strain energy per cycle, we may conclude the following: a high level of either sulfur or accelerator gives poor flex life. However, by the selection of the proper ratio of sulfur concentration to accelerator concentration, higher modulus vulcanizates can be obtained with at least some optimization of fatigue life. The flex life at approximately equal strain energy per cycle can be illustrated by extracting values along the 13.8Mpa modulus contour line. Table 5.2 can then be constructed.

Table 5.2 Fatigue life at constant 300% modulus

phr Sulfur/phr sulfenamide	Flex life/kc	phr Sulfur/phr sulfenamide	Flex life/kc
6	400	1	400
5	500	0.6	350
4	530	0.4	270
3	550	0.3	250
2	550	0.2	190

For the strain energy corresponding to a 300% modulus of 13.8Mpa, the maximum flex life (as measured by the DeMattia flex test) is obtained when 2.5 times as much sulfur as accelerator is used. Other optimum ratios for various 300% moduli can be obtained from the contours. Some of these are given in Table 5.3.

Table 5.3 Optimized fatigue life

Stress at 300% strain, 300% Modulus/MPa	Optimum sulfur conc./accelerator conc. ratio	Optimized flex life/kc
6.9	3.50	800
10.4	3.00	800
13.8	2.50	550
15.5	1.00	300
17.2	0.45	120
19.0	0.27	70

These optimum ratios and fatigue life data should not be considered to be universal values. Different recipes, different types of antidegradants, different types of fillers, different concentrations of antidegradants and fillers, different base polymers, different types of fatigue tests, etc., give rise to different optimum sulfur concentration-accelerator concentration ratios and different optimum fatigue life values. Nevertheless, the trends given here have been generally noted.

The low values for fatigue life at low levels of sulfur, but high levels of accelerator, have been attributed to high concentrations of accelerator-terminated appended groups and high concentrations of monosulfidic crosslinks. Monosulfidic crosslinks are not able to exchange, rearrange, or break to relieve stresses without the breakage of main chains.

On the other hand, polysulfidic crosslinks are able to rearrange under stress. The rearrangement of a crosslink occurs in two steps: (1) breaking, (2) reforming. Only the breaking of the weak polysulfide crosslinks is required for the strengthening of the vulcanizate network. It is only required that enough of the crosslinks be weak (in comparison to backbone bonds) for the rubber to be strong. At any rate, when moderately high concentrations of sulfur (with respect to accelerator) are used, flex life improves, presumably due to the presence of enough weak or rearrangeable polysulfidic crosslinks.

When even higher concentrations of sulfur are used (with the maintenance of constant modulus), flex life decreases. It is possible that this is due to the large amount of cyclic chain modification associated with high levels of sulfur.

Natural rubber compositions, vulcanized by high levels of accelerator and low levels of sulfur, have been called EV and semi-EV vulcanizates. Here, "EV" means efficient vulcanization, since sulfur is used efficiently in the production of crosslinks. On the average, the crosslinks are shorter than in the case of conventional vulcanization; they contain more monosulfidic crosslinks and less polysulfidic ones, or their average sulfur rank is lower. Though EV vulcanizates suffer with respect to fatigue resistance, they are frequently used because of their excellent reversion resistance (resistance to nonoxidative thermal aging or overcure) and good resistance to thermal-oxidative aging. The resistance to reversion or thermally induced loss of crosslinks is thought to be the result of the greater intrinsic stability of the lower rank (disulfidic and especially monosulfidic) crosslinks. Semi-EV vulcanizates (wherein the sulfur concentration to accelerator-moiety concentration ratio is at an intermediate value) are an advantageous compromise in which fairly good unaged fatigue life is obtained, but maintained after aging. Rather than using high levels of accelerator to obtain EV and semi-EV vulcanizates, it is sometimes advantageous to replace some of the sulfur with a so-called sulfur donor. Examples of these are tetramethylthiuram disulfide (TMTD) and 4,4-dithiodimorpholine (DTDM).

This type of vulcanization system design was reported by McCall. He found that by judiciously balancing the levels of accelerator, sulfur, and DTDM, he could obtain good vulcanization characteristics, good thermal stability, good flex life, and superior retention of flex life.

5.6.7 Accelerated-Sulfur Vulcanization of Various Unsaturated Rubbers

Over the years, much of the research on accelerated sulfur vulcanization was done by using natural rubber as a model substrate. Natural rubber was the first elastomer and therefore the search for understanding of vulcanization originated with work on natural rubber. Even in recent years most of the work published on the study of vulcanization has been related to natural rubber. This was because of the tradition of doing research on natural rubber and because of the fact that the largest knowledge base to build upon was with respect to natural rubber. It should be mentioned that a large factor in the establishment of the tradition of research on the vulcanization of natural rubber was the British Rubber Producers Research Association or BRPRA (now called the Malaysian Rubber Producers Research Association or MRPRA). Of course, this institution is essentially devoted to natural rubber.

Most of the work cited in the previous sections is related to natural rubber. However, some studies have been directed to the vulcanization of butadiene 1,4-polymers. Other basic work on the vulcanization of ethylene-propylene-diene-monomer rubber (EPDM) has been carried out. The chemistry of the accelerated vulcanization of BR, SBR, and EPDM appears to have much in common with the vulcanization of natural rubber: Before the formation of crosslinks, the rubber is first sulfurated by accelerator-derived polysulfides (Ac-S_x-Ac) to give macromolecular, polysulfidic intermediates (rubber-S_x-Ac). However, whereas in the case of MBTS or benzothiazolesulfenamide-accelerated sulfur vulcanization of natural rubber, MBT is given off during the formation of rubber-S-BT from the attack of rubber by BT-S_x-BT, in the case of BR and SBR, MBT is not eliminated and remains unextractable presumably because it becomes bound as the macromolecular thioether rubber-S_x-BT. (BT is a 2-benzothiazolyl group). As in the case of natural rubber, the average length of a crosslink (its sulfidic rank, the value of x in the crosslink, rubber-S_x-rubber) increases with the ratio of sulfur concentration to accelerator concentration (S/Ac) used in the compounded rubber mix. However, in the case of BR or SBR, the crosslink sulfidic rank is not nearly as sensitive to S/Ac as it is in the case of natural rubber. Model compound studies of the vulcanization of EPDM (e.g., wherein ethylidenenorbornane was used as a model for EPDM) indicate that the polysulfidic rank of the EPDM crosslinks probably responds to changes in S/Ac in a natural rubberlike fashion. A difference here is that evidence for model monosulfidic crosslinks was lacking while model disulfidic crosslinks were more apparent than in the case of natural rubber vulcanization.

Reversion (when defined as the loss of crosslinks during nonoxidative thermal vulcanizate aging) is a problem associated mainly with natural rubber or synthetic isoprene polymers. It can occur only under severe conditions in butadiene rubber; in SBR, instead of the softening associated with the non-oxidative aging of natural rubber, one can observe hardening (the so-called marching modulus) during extensive overcure. In natural rubber and synthetic isoprene-polymer rubbers, the crosslinks tend to be more polysulfidic than in the case of BR or SBR. The highly polysulfidic crosslinks are more heat-labile than their lower rank cousins in BR and SBR; they are more likely to break and then form cyclic chain modifications. But the reason for the formation of the crosslinks of higher polysulfidic rank in isoprene rubbers than in butadiene polymers is grandly elusive, though it almost has to be related to the methyl groups that are substituents along the isoprene-polymer chains but that are absent from butadiene-polymer chains.

The effect of zinc is much greater in the vulcanization of isoprene rubbers than it is in the vulcanization of BR and SBR. Again, the reason for the difference is not known, but a strong speculation is that this difference is also related to the presence of methyl groups only in the case of the isoprene rubbers.

5.6.8 Selected Accelerated-Sulfur System Recipes

Examples of recipes are given in Table 5.4. These recipes are not intended as ultimate solutions to compounding problems. Variations will undoubtedly be necessary to meet particular requirements.

Table 5.4 Recipes for accelerated sulfur vulcanization systems[a]

	NR	SBR	Nitrile(NBR)		Butyl (HR)	EPDM
			1	2		
Zinc oxide	5.00	5.00	3.00	2.00	3.00	5.00
Stearic acid	2.00	2.00	0.50	0.50	2.00	1.00
Sulfur	2.50	1.80	0.50	0.25	2.00	1.50
DTDM[b]	—	—	—	1.00	—	—
TBBS[b]	0.60	1.20	—	—	—	—
MBTS[b]	—	—	2.00	—	0.50	—
MBT[b]	—	—	—	—	—	0.50
TMTD[b]	—	—	1.00	1.00	1.00	1.50
Vulcanization conditions[c]						
Temperature/℃	148	153	140	140	153	160
Time/min	25	30	60	60	20	20

[a] Concentrations in phr.

[b] DTDM, 4,4-dithiodimorpholine; TBBS, N-t-butylbenzothiazole-2-sulfenamide; MBTS, 2,2-dithiobisbenzothiazole (2-benzothiazole disulfide); MBT, 2-mercaptobenzothiazole; TMTD, tetramethylthiuram disulfide.

[c] Conditions change depending on other aspects of the compositions.

5.6.9 Vulcanization by Phenolic Curatives, Benzoquinone Derivatives, or Bismaleimides

Diene rubbers such as natural rubber, SBR, and BR can be vulcanized by the action of phenolic compounds, which are (usually di-) substitute by —CH_2—X groups where X is an —OH group or a halogen atom substituent. A high-diene rubber can also be vulcanized by the action of a dinitrosobenzene that forms in situ by the oxidation of a quinonedioxime that had been incorporated into the rubber along with the oxidizing agent, lead peroxide.

The attack upon rubber molecules by the vulcanization system can be visualized in a way similar to that which was postulated for the sulfurization of the rubber molecules by the action of accelerated-sulfur vulcanization systems. Reaction schemes for these two types of vulcanization can be written as follows:

(a) Vulcanization by phenolic curatives.

$$X-CH_2-\underset{R}{\underset{|}{\bigcirc}}(OH)-CH_2X \xrightarrow[\text{Acid catalyzed}]{-HX} X-CH_2-\underset{R}{\underset{|}{\bigcirc}}(=O)=CH_2$$

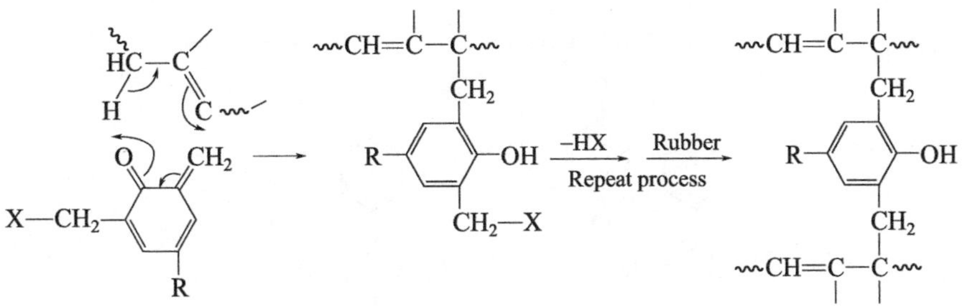

(b) Vulcanization by benzoquinonedioxime.

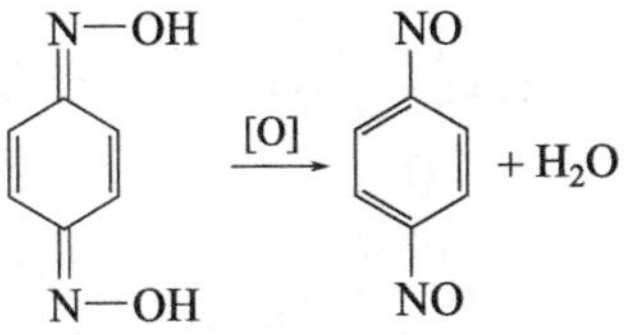

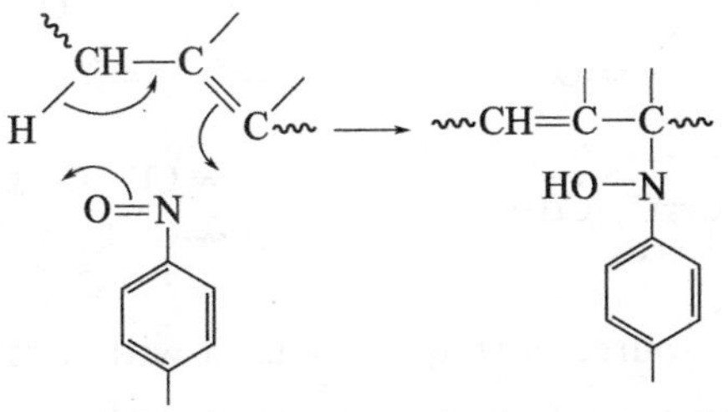

As shown above, the chemical structural requirements for these types of vulcanization are that the elastomer molecules contain allylic hydrogen atoms. The attacking species from the vulcanization system must contain sites for proton acceptance and electron acceptance in proper steric relationship. This will then permit the following rearrangement, where A is the proton acceptor site and B is the electron acceptor site:

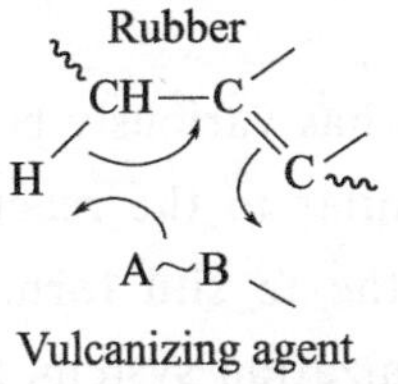

This is an explanation for the fact that this type of vulcanization is not enabled by double bonds per se, without allylic hydrogens in the elastomer molecules. It should be pointed out that the phenolic curative can also act by a slightly different mechanism to give crosslinks which contain chromane structural moieties, the allylic hydrogens still being required.

Another vulcanizing agent for high-diene rubbers is m-phenylenebismaleimide. A catalytic free radical source such as dicumyl peroxide or benzothiazyl disulfide (MBTS) is usually used to initiate the reaction. A reaction scheme for this type of vulcanization is as follows:

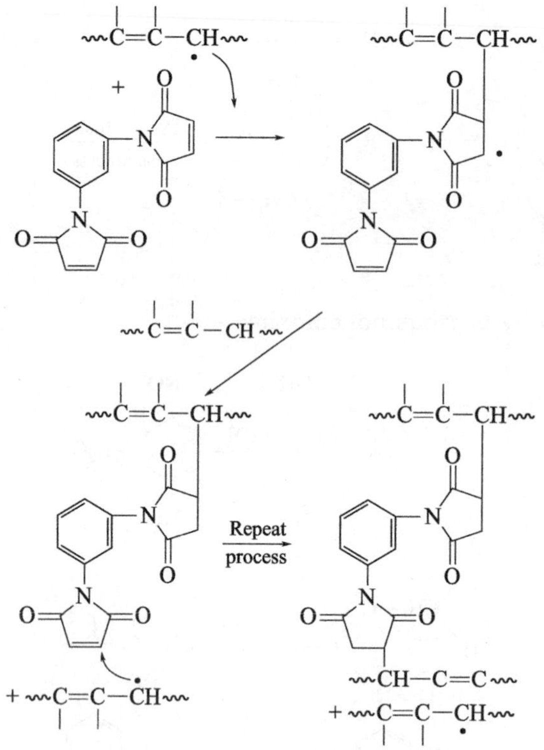

Although a free radical source is frequently used with a maleimide vulcanizing agent, at high enough vulcanization temperatures, the maleimides react with the rubber without the need for a free radical source. This could occur as shown here:

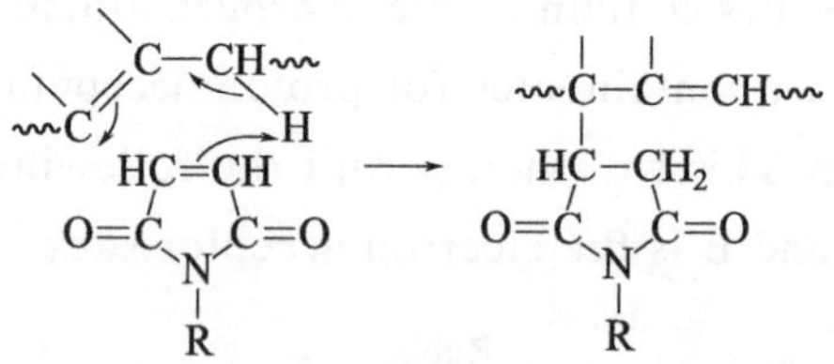

This is another example of what has variously been called a pseudo-Diels-Alder, ene, or "no-mechanism" reaction. It is similar to the reaction written for the attack of rubber molecules by phenolic curatives or the in situ formed nitroso derivative of the quinoid (e. g., benzoquinonedioxime) vulcanization system. It is also closely related to the sulfurization scheme written for accelerated sulfur vulcanization. Comparisons between accelerated sulfur, phenolic, quinoid, and maleimide vulcanization can then be visualized as follows:

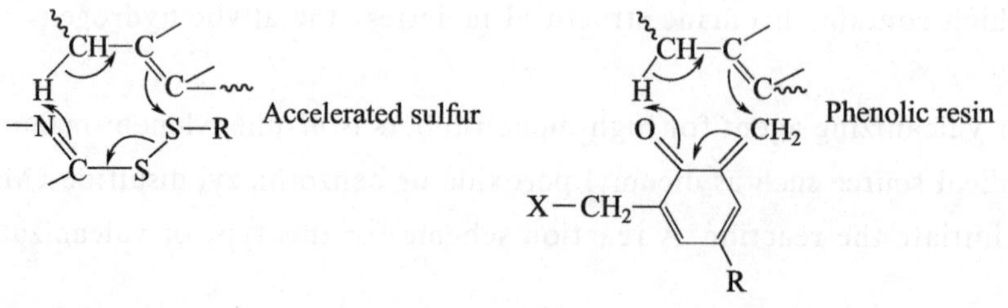

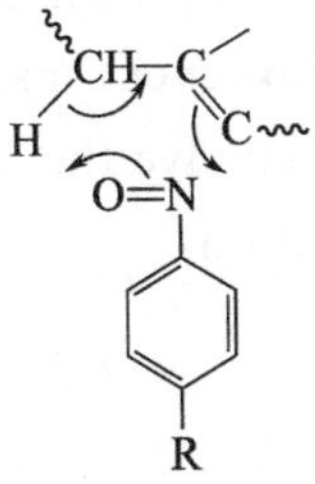

Quinone derivatives

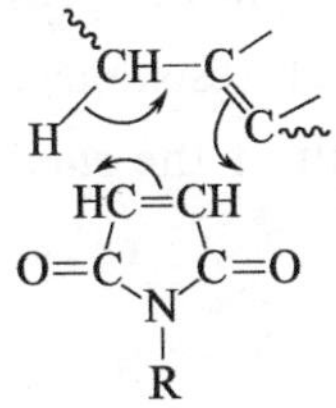

maleimides

Selected recipes for vulcanization by phenolic curatives, benzoquinone dioxime, or m-phenylenebismaleimide are given by Table 5.5. Vulcanizates based on these types of curatives are particularly useful in cases where thermal stability is required.

Table 5.5 Recipes for vulcanization by phenolic curatives, quinone derivatives, or maleimides[a]

	IIR		SBR		NBR
	1	2	1	2	
Zinc oxide	5.00	5.00	—	—	—
Lead peroxide (Pb_3O_4)	—	10.00	—	—	—
Stearic acid	1.00	—	—	—	—
Phenolic curative (SP-1056)[b]	12.00	—	—	—	—
Benzoquinonedioxime (GMF)	—	2.00	—	—	—
m-Phenylenebismaleamide (HVA-2)[c]	—	—	0.85	0.85	3.00
2-Benzothiazyl disulfide (MBTS)	—	—	2.00	—	—
Dicumyl peroxide	—	—	—	0.30	0.30
Vulcanization condition[d]					
Temperature/°C	180	153	153	153	153
Time/min	30	20	25	25	30

[a] Concentrations in phr.
[b] Schenectady chemicals.
[c] Du Pont.
[d] Conditions change depending on other aspects of the compositions.

5.6.10 Vulcanization by the Action of Metal Oxides

Chlorobutadiene or chloroprene rubbers (CR), also called neoprene rubbers, are generally vulcanized by the action of metal oxides. CR can be represented by the structure:

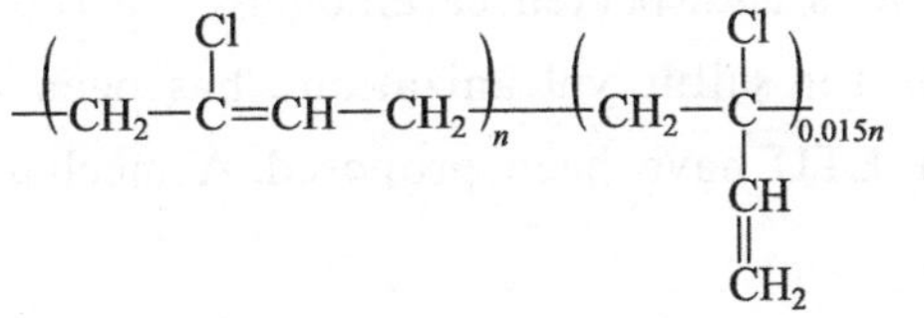

$$+CH_2-\underset{Cl}{C}=CH-CH_2+_n \ (CH_2-\underset{\underset{CH_2}{\overset{CH}{\|}}}{\overset{Cl}{C}})_{0.015n}$$

The crosslinking agent is usually zinc oxide, which is used along with magnesium oxide. CR can be vulcanized in the presence of zinc oxide alone; however, magnesium oxide is necessary to give scorch resistance. The reaction is thought to involve the allylic chlorine atom, which is the result of the small amount of 1,2-polymerization:

$$\begin{array}{c} \text{Cl} \\ | \\ -\text{CH}_2-\text{CH}- \\ | \\ \text{CH} \\ \| \\ \text{CH}_2 \end{array}$$

A mechanism that has been written for the vulcanization of CR by the action of zinc oxide and magnesium oxide is as follows:

$$\begin{array}{c} \text{Cl} \\ | \\ -\text{CH}_2-\text{C}- \\ | \\ \text{CH} \\ \| \\ \text{CH}_2 \end{array} \rightleftharpoons \begin{array}{c} -\text{CH}_2-\text{C}- \\ \| \\ \text{CH} \\ | \\ \text{CH}_2 \\ | \\ \text{Cl} \end{array} \xrightarrow{\text{ZnO}} \begin{array}{c} -\text{CH}_2-\text{C}- \\ \| \\ \text{CH} \\ | \\ \text{CH}_2 \\ | \\ \text{OZnCl} \end{array}$$

$$\begin{array}{c} -\text{CH}_2-\text{C}- \\ \| \\ \text{CH} \\ | \\ \text{CH}_2 \\ | \\ \text{OZnCl} \end{array} + \begin{array}{c} \text{Cl} \\ | \\ \text{CH}_2 \\ | \\ \text{CH} \\ \| \\ -\text{CH}_2-\text{C}- \end{array} \longrightarrow \begin{array}{c} -\text{CH}_2-\text{C}- \\ \| \\ \text{CH} \\ | \\ \text{CH}_2 \\ | \\ \text{O} \\ | \\ \text{CH}_2 \\ | \\ \text{CH} \\ \| \\ -\text{CH}_2-\text{C}- \end{array} + \text{ZnCl}_2$$

$$\text{ZnCl}_2 + \text{MgO} \longrightarrow \text{ZnO} + \text{MgCl}_2$$

Most accelerators used in the accelerated-sulfur vulcanization of other high-diene rubbers are not applicable to the metal oxide vulcanization of neoprene rubbers. An exception to this is in the use of the so-called mixed curing system for CR, in which metal oxide vulcanization is combined with accelerated-sulfur vulcanization. Along with the metal oxides, tetramethylthiuram disulfide (TMTD), N,N-di-o-tolylguanidine (DOTG), and sulfur are used. This may be desirable for high resilience or for good dimensional stability. The accelerator which has been most widely used with metal oxide cures is ethylenethiourea (ETU) or 2-mercaptoimidazoline. Further extensive use of ETU in the vulcanization of CR is somewhat in doubt since it is a suspected carcinogen. The related compound, thiocarbanalide, an old accelerator for sulfur vulcanization, has been revived for CR vulcanization. Other substitutes for ETU have been proposed. A mechanism for ETU acceleration has been given by Pariser:

[Reaction scheme showing chloroprene unit reacting with ETU (ethylenethiourea), followed by ZnO/−ZnCl⊕ step, then further rearrangements producing ethyleneurea and a sulfide intermediate, and finally crosslinking between two chloroprene chains with loss of ZnCl₂.]

Examples of recipes for metal oxide vulcanization are given in Table 5.6. It should be noted that in one case, calcium stearate was used instead of magnesium oxide to obtain better aging characteristics.

Table 5.6 Vulcanization systems for chloroprene rubber[a]

ZnO	5.00	5.00	5.00
MgO	4.00	—	4.00
Calcium stearate	—	5.50	—
Stearic acid	—	—	1.00
TMTM	—	—	1.00
DOTG	—	—	1.00
ETU	0.5	0.5	—
Sulfur	—	—	1.00
Vulcanization condition[b]			
Temperature/℃	153	153	153
Time/min	15	15	15

[a] Concentrations in parts by weight per 100 parts of neoprene W.

[b] Conditions change depending on other aspects of the compositions.

5.6.11 Vulcanization by the Action of Organic Peroxides

Most elastomers can be vulcanized by the action of organic peroxides. Diacyl peroxides, dialkyl peroxides, and peresters have been used. Dialkyl peroxides and *t*-butyl perbenzoate give efficient crosslinking. Di-*t*-butyl peroxide and dicumyl peroxide give good vulcanizates, but the former is too volatile for general use. Dicumyl peroxide is widely used, however its vulcanizates have the odor of acetophenone, which is a byproduct of the vulcanization process. Other nonvolatile peroxides of the same class, which give vulcanizates free of the odor of acetophenone, are 1,1-bis(*t*-butylperoxy)-3,3,5-trimethylcyclohexane and 2,5-dimethyl-2,5-bis(*t*-butylperoxy)hexane. This latter compound is particularly good for vulcanization at higher temperatures (as high as 180℃) since it is more thermally stable than the others.

It should be noted that acidic compounding ingredients (fatty acids, certain carbon blacks, and acidic silicas) can catalyze on radical-generating, wasteful decomposition of peroxides. Other compounding ingredients such as antidegradants can reduce crosslinking efficiency by quenching or altering the free radicals before they can react with the polymeric substrate.

Peroxides are vulcanizing agents for elastomers that contain no sites for attack by other types of vulcanizing agents. They are useful for ethylene-propylene rubber (EPR), ethylene-vinylacetate copolymers (EAM), certain millable urethane rubbers, and silicone rubbers. They are not generally useful for vulcanizing butyl rubber [poly(isobutylene-*co*-isoprene)] because of a tendency toward chain scission, rather than crosslinking, when the polymer is subjected to the action of a peroxide. Elastomers derived from isoprene and butadiene are readily crosslinked by peroxides; but many of the vulcanizate properties are inferior to those of accelerated-sulfur vulcanizates. However, peroxide vulcanizates of these diene rubbers may be desirable in applications where improved thermal ageing and compression set resistance are required.

5.6.11.1 Peroxide Vulcanization of Unsaturated Hydrocarbon Elastomers

The initiation step in peroxide-induced vulcanization is the decomposition of the peroxide to give free radicals, thus

$$\text{Peroxide} \longrightarrow 2R \cdot$$

where R is an alkoxyl, alkyl, or acyloxyl radical, depending on the type of peroxide used.

Dibenzoyl peroxide gives benzoyloxyl radicals but dicumyl peroxide gives cumyloxyl and methyl radicals. If the elastomer is derived from butadiene or isoprene, the next step is either the abstraction of a hydrogen atom from an allylic position on the polymer molecule or the addition of the peroxide-derived radical to a double bond of the polymer molecule.

$$R \cdot + \sim CH_2-\underset{|}{C}=\underset{|}{C}\sim \longrightarrow \sim \overset{\cdot}{C}H-\underset{|}{C}=\underset{|}{C}\sim + RH$$

$$R \cdot + \sim CH_2-\underset{|}{C}=\underset{|}{C}\sim \longrightarrow \sim CH_2-\underset{|}{\overset{R}{C}}-\underset{|}{\overset{\cdot}{C}}\sim$$

For isoprene rubber, the abstraction route predominates over radical addition. Two polymeric free radicals then unite to give a crosslink.

$$2\sim\overset{\bullet}{C}H-\underset{|}{C}=\underset{|}{C}\sim \longrightarrow \begin{array}{c} \sim CH-\underset{|}{C}=\underset{|}{C}\sim \\ | \\ \sim CH-\underset{|}{C}=\underset{|}{C}\sim \end{array}$$

Crosslinks could also form by a chain reaction that involves the addition of polymeric free radicals to double bonds.

$$\begin{array}{c} \sim\overset{\bullet}{C}H-\underset{|}{C}=\underset{|}{C}\sim \\ + \\ \sim CH_2-\underset{|}{C}=\underset{|}{C}\sim \end{array} \longrightarrow \begin{array}{c} \sim CH-\underset{|}{C}=\underset{|}{C}\sim \\ | \\ \sim C-\overset{\bullet}{C}=CH_2\sim \\ | \end{array}$$

$$\begin{array}{c} \sim CH-\underset{|}{C}=\underset{|}{C}\sim \\ | \\ \sim C-\overset{\bullet}{C}-CH_2\sim \\ | \\ + \\ \sim CH_2-\underset{|}{C}=\underset{|}{C}\sim \end{array} \longrightarrow \begin{array}{c} \sim CH-\underset{|}{C}=\underset{|}{C}\sim \\ | \\ \sim C-CH-CH_2\sim \\ | \\ + \\ \sim\overset{\bullet}{C}H-\underset{|}{C}=\underset{|}{C}\sim \end{array}$$

In this case crosslinking occurs without the loss of a free radical so that the process can be repeated until termination by radical coupling. Coupling can be between two polymeric radicals to form a crosslink or by an unproductive processes: a polymeric radical can unite with a radical derived from the peroxide. If a polymeric radical decomposes to give a vinyl group and a new polymeric radical, a scission of the polymer chain is the result.

Few monomeric radicals are lost by coupling with polymeric radicals when dialkyl peroxides are used as the curative. Also, if the elastomer is properly chosen, the scission reaction is not excessive. For dicumyl peroxide in natural rubber, the crosslinking efficiency has been estimated at about 1.0. One mole of crosslinks is formed for each mole of peroxide; crosslinking is mainly by the coupling of two polymeric radicals. One peroxide moiety gives two monomeric free radicals, which react with rubber to give two polymeric radicals that couple to form one crosslink. In the case of BR or SBR, the efficiency can be much greater than 1.0, especially if all antioxidant materials are removed. A chain reaction is indicated here. It might be explained by steric considerations. In butadiene-based rubbers, double bonds are quite accessible. Radical addition to double bonds could give highly reactive radicals that would be likely to add to other polymer double bonds. A chain of additions might be more likely in butadiene rubber than in the presence of hindering methyl groups in isoprene rubbers.

One might expect that nitrile rubber would also be vulcanized with efficiencies greater than 1.0; however, though the double bonds in nitrile rubber are highly accessible, the crosslinking efficiency is somewhat less than 1.0.

5.6.11.2 Peroxide Vulcanization of Saturated Hydrocarbon Elastomers

Saturated hydrocarbon polymers are also crosslinked by the action of organic peroxides, though the efficiency is reduced by branching. Polyethylene is crosslinked by dicumyl peroxide at an efficiency of about 1.0, saturated EPR gives an efficiency of about 0.4, while butyl rubber cannot be cured at all. For polyethylene, the reaction scheme is similar to that of the unsaturated elastomers.

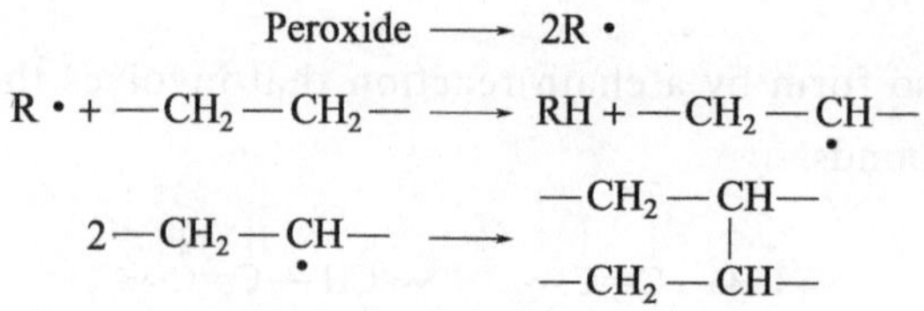

$$\text{Peroxide} \longrightarrow 2R\cdot$$

$$R\cdot + -CH_2-CH_2- \longrightarrow RH + -CH_2-\overset{\cdot}{C}H-$$

$$2-CH_2-\overset{\cdot}{C}H- \longrightarrow \begin{array}{c} -CH_2-CH- \\ | \\ -CH_2-CH- \end{array}$$

However, branched polymers undergo other reactions.

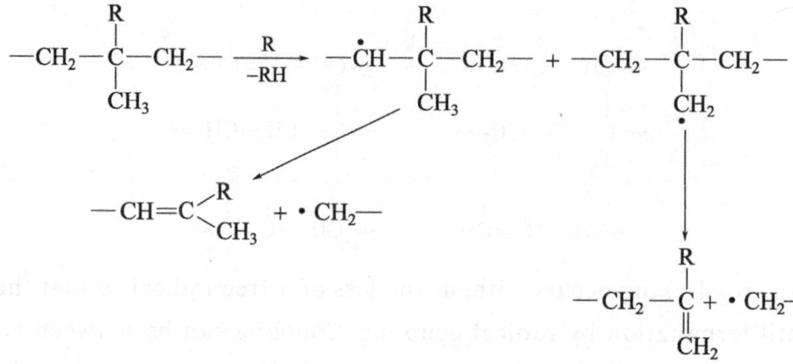

$$-CH_2-\underset{\underset{CH_3}{|}}{\overset{\overset{R}{|}}{C}}-CH_2- \xrightarrow[-RH]{R\cdot} -\overset{\cdot}{C}H-\underset{\underset{CH_3}{|}}{\overset{\overset{R}{|}}{C}}-CH_2- \;+\; -CH_2-\underset{\underset{\overset{\cdot}{C}H_2}{|}}{\overset{\overset{R}{|}}{C}}-CH_2-$$

$$-CH=C\underset{CH_3}{\overset{R}{\diagdown}} + \cdot CH_2-$$

$$-CH_2-\underset{\underset{CH_2}{\|}}{\overset{\overset{R}{|}}{C}} + \cdot CH_2-$$

Here, though the peroxide has been depleted, no crosslinks have been formed between polymer chains, and the average molecular weight of the polymer has even been reduced by scission.

Sulfur, or the so-called coagents, can be used to suppress scission. Examples of coagents are m-phenylenebismaleimide, high-1,2-(high-vinyl)polybutadiene, triallyl cyanurate, diallyl phthalate, ethylene diacrylate, etc. Their mechanism of action may be as follows:

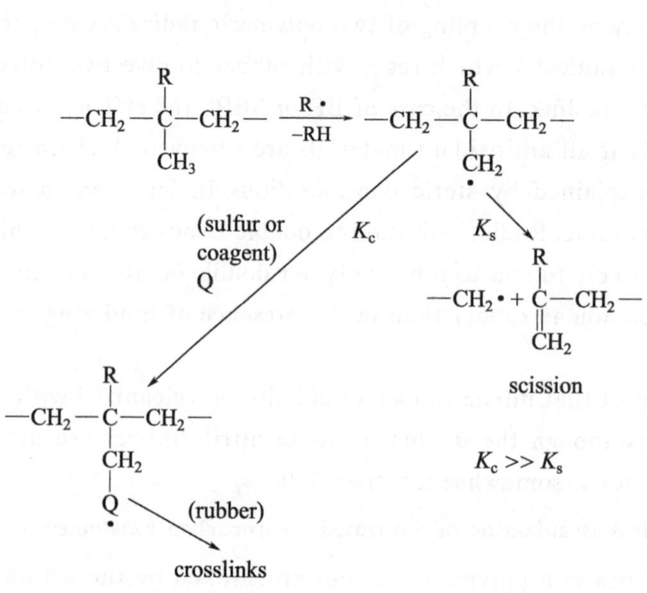

$$-CH_2-\underset{\underset{CH_3}{|}}{\overset{\overset{R}{|}}{C}}-CH_2- \xrightarrow[-RH]{R\cdot} -CH_2-\underset{\underset{\overset{\cdot}{C}H_2}{|}}{\overset{\overset{R}{|}}{C}}-CH_2-$$

(sulfur or coagent) Q K_c K_s

$$-CH_2\cdot + \underset{\underset{CH_2}{\|}}{\overset{\overset{R}{|}}{C}}-CH_2-$$

scission

$$-CH_2-\underset{\underset{\underset{\overset{\cdot}{Q}}{|}}{CH_2}}{\overset{\overset{R}{|}}{C}}-CH_2-$$

(rubber) ↘ crosslinks

$K_c \gg K_s$

5.6.11.3 Peroxide Vulcanization of Silicone Rubbers

Silicone rubbers can be represented by

$$\left(\begin{array}{c} R \\ | \\ Si-O \\ | \\ CH_3 \end{array}\right)_n$$

where R can be methyl, phenyl, vinyl, trifluoropropyl, or 2-cyanoethyl. Silicone rubbers that contain vinyl groups can be cured by dialkyl peroxides such as dicumyl peroxide. Saturated silicone rubbers require diacyl peroxides such as bis-(2,4-dichlorobenzoyl) peroxide. In the case of saturated siloxane rubbers, the mechanism is hydrogen atom abstraction followed by polymeric radical coupling to give crosslinks. Nonproductive use of peroxide results from the coupling of the polymeric radicals with the lower-molecular-weight free radicals formed by the decomposition of the peroxide curative. The incorporation of vinyl groups improves the crosslinking efficiency.

Vulcanization is frequently done in two steps. After a preliminary vulcanization in a mold, a high-temperature (e.g., 180°C) postcure is carried out in air. The high-temperature postcure removes acidic materials, which can catalyze hydrolytic decomposition of the vulcanizate. Also, the high temperature enables the formation of additional crosslinks of the following type:

$$\begin{array}{ccc} & CH_3 & CH_3 \\ & | & | \\ \sim\sim Si-O- & Si-O\sim\sim \\ & | & | \\ & CH_3 & O \\ & CH_3 & | \\ & | & \\ \sim\sim Si-O- & Si-O\sim\sim \\ & | & | \\ & CH_3 & CH_3 \end{array}$$

5.6.11.4 Peroxide Vulcanization of Urethane Elastomers

Urethane elastomers suitable for peroxide vulcanization are typically prepared from an hydroxyl-group-terminated oligomeric adipate polyester and 4,4-methylenediphenylisocyanate (MDI). A typical structural representation is as follows:

$$\left[\begin{array}{c} O \\ \| \\ C-NH-\phi-CH_2-\phi-NH-\overset{O}{\underset{\|}{C}}-O\left(R-O-\overset{O}{\underset{\|}{C}}-C_4H_8-\overset{O}{\underset{\|}{C}}-O\right)_x R-O \end{array}\right]_y$$

Hydrogen atoms can be abstracted from arylated methylene groups, but hydrogen atoms may also be abstracted from alpha-methylene groups of the adipate moieties. Though they are usually sufficient, vulcanization efficiencies can be increased by the incorporation of urea structures into the polymer chain.

5.6.11.5 Recipes for Peroxide Vulcanization

Examples of starting-point recipes are given in Table 5.7. Outstanding characteristics of peroxide vulcanizates are low permanent set and high thermal stability of the network.

Table 5.7 Recipes for peroxide vulcanization[a]

	NR	SBR	EPR	Silicone rubber	Millable urethane
Dicumyl peroxide	1.0	1.0	2.7	—	2
Bis(2,4-dichlorobenzoyl)peroxide	—	—	—	1.0	—
Triallyl cyanurate	—	—	1.5	—	—
Vulcanization conditions[b]	—	—	—	—	—
Temperature/℃	150	150	160	115, 250[c]	153
Time/min	45	45	30	141, 440[c]	45

[a] Concentrations in phr.

[b] Conditions change depending on other aspects of the compositions.

[c] Temperature and time of postcure in air.

5.6.12 Dynamic Vulcanization

Dynamic vulcanization (DV) is the vulcanizing or crosslinking of one polymer during its molten-state mixing with another polymer or with other polymers. The polymers are first thoroughly mixed and then, during further mixing, one of the polymers is obliged to become crosslinked, whereas the remaining other polymeric material remains uncrosslinked. The process produces a dispersion of crosslinked polymer in a matrix or continuous phase of uncrosslinked polymer. If the dispersed crosslinked material is elastomeric and the continuous or matrix material is of a melt-processable plastic, then the composition can be used as an impact-resistant thermoplastic resin, or if there is a large enough proportion of rubber in the composition, it might be suitably used as a thermoplastic elastomer (TPE).

Fischer used the DV process to prepare compositions containing partially vulcanized rubber. It has since been found that improved, very strong elastomeric compositions of EPDM and polypropylene could be prepared by dynamic vulcanization provided that the rubber was completely vulcanized.

The DV process for thermoplastic elastomers can be described as follows: after sufficient melt-mixing of plastic and rubber, vulcanizing agents are added. Vulcanization of the rubber phase occurs as mixing continues. After removal from the mixer, the cooled blend can be chopped, extruded, pelletized, injection molded, etc. Such a composition is described as a dispersion of very small particles of vulcanized rubber in a thermoplastic resin matrix. Such compositions are prepared commercially by a continuous process by using a twin-screw extruder.

Dynamic vulcanization gives the following improvements, in comparison with blends that have not been dynamically vulcanized: reduced set, improved ultimate properties, improved fatigue resistance, improved resistance to attack by hot oils, greater stability of melt-phase morphology, greater melt strength, etc.

5.6.12.1 EPDM-Polyolefin Composition

The dynamic vulcanization of blends of EPDM rubber with polyolefins (PP or PE) has been described. The rubber-plastic proportions and the extents of vulcanization were varied. In a few instances the rubber was first press cured and then ground to various particle sizes. The ground rubber particles were then mixed with molten polypropylene. It was found that the ultimate properties (UE and UTS) varied inversely with rubber particle size. Since the smallest particle sizes of vulcanized rubber were obtained by dynamic vulcanization (not by grinding of cured rubber), the more durable compositions were obtained by dynamic vulcanization.

Only a small amount of crosslink formation is required for a large improvement in tension set. However, tensile strength improves rather continuously as the crosslink density of the rubber phase is increased. Compositions can be vulcanized by accelerated sulfur, methylolphenolic materials (e. g. , catalyzed by $SnCl_2$), or other curatives.

As the concentration of the polyolefin resin increases, the compositions become less like rubber and more like plastic. Modulus, hardness, tension set, and strength increase.

5. 6. 12. 2 NBR-Nylon Composition

Excellent elastomeric NBR-Nylon compositions have also been prepared by dynamic vulcanization during the melt-mixing of intimate blends of NBR with various nylons. In this case, the effect of curatives was complicated by the fact that some nitrile rubbers tend to self-cure at temperatures of mixing.

Sulfur, phenolic, maleimide, or peroxide curatives can be used. The thermoplastic elastomeric compositions prepared by the dynamic vulcanization of NBR-Nylon blends are highly resistant to hot oil. As in the case of the EPDM-Polyolefin blends, increases in the amount of rubber in the composition reduce stiffness but increase resistance to permanent set.

5. 6. 12. 3 Other Elastomeric Compositions Prepared by Dynamic Vulcanization

In addition to EPDM-Polyolefin and NBR-Nylons combinations, a large number of other rubber-plastic combinations have been used to prepare thermoplastic vulcanizates by dynamic vulcanization.

The best compositions are prepared when the surface energies of the rubber and plastic material are matched, when the entanglement molecular length of the rubber molecule is small, and when the plastic material is crystalline. It is also necessary that neither the plastic nor the rubber decompose in the presence of the other at temperatures required for melt-mixing. Also, in each case, a curing system appropriate for the rubber under the conditions of melt-mixing is required.

5. 6. 12. 4 Technological Applications

The lower cost of thermoplastic processing is the motivation for the development of thermoplastic elastomers. However, failure in the achievement of truly rubberlike properties has impeded the acceptance of thermoplastic-elastomer technology. Nevertheless, relatively recently commercialized compositions based on polypropylene and completely vulcanized EPDM have many of the excellent properties of the polyurethane and copolyester-type thermoplastic elastomers and even improved set and fatigue resistance.

Applications of these materials can be listed as follows: caster wheels, convoluted bellows, diaphragms, gaskets, seals, tubing, mounts, bumpers, glazing seals, shields, suction cups, torque couplings, vibration isolators, plugs, connectors, rollers, oil-well injection lines, handles, grips, hose covers, vacuum tubing, bushings, grommets, protective sleeves, shock isolators, ducts, various hoses (e. g. , hydraulic, agricultural spray, paint spray, plant air-water, mine hose, etc.), wire and cable insulation and strain relief, jacketing, etc.

References

[1] P. J. Flory, "Principles of Polymer Chemistry," Cornell Univ. Press, Ithaca, NY, 1953, Cha11.
[2] L. Bateman, C. G. Moore, M. Porter, B. Saville, in "The Chemistry and Physics of Rubber Like Substances," Bateman (Ed.), John Wiley & Sons, Inc. , New York, 1963, Chap. 19.
[3] W. Hofmann, "Vulcanization and Vulcanizing Agents," Maclaren and Sons Ltd. , London 1967.
[4] A. Y. Coran, in "Science and Technology of Rubber," F. R. Eirich (Ed.), Academic Press, New York, 1978, Chap. 7.
[5] N. J. Morrison, M. Porter, Rubber Chem. Technol. 57, 63 (1984).
[6] G. E. Decker, R. W. Wise, D. Guerry, Rubber Chem. Technol. 36, 451 (1963); A. I. Juve, P. W. Karper, L. O. Schroyer, A. G. Veith, Rubber Chem. Technol. 37, 434 (1964).
[7] E. H. Farmer, F. W. Shipley, J. Polym. Sci. 1, 293 (1946).
[8] E. H. Farmer, J. Chem. Soc. p. 1519 (1947).
[9] E. H. Farmer, J. Soc. Chem. Ind. 66, 86 (1947).
[10] L. Bateman, C. G. Moore, M. Porter, J. Chem. Soc. p. 2866 (1958).
[11] G. Oenslager, Ind. Eng. Chem. 23, 232 (1933).
[12] M. Weiss, U. S. Patent 1, 411, 231 (1922).
[13] S. Malony, U. S. Patent 1, 343, 222 (1920).
[14] C. Bedford, U. S. Patent 1, 371, 922 (R) C4 (1921)
[15] L. Sebrell, C. Bedford, U. S. Patent 1, 522, 687 (1925).
[16] G. Bruni, E. Romani, India Rubber J. 62, 63 (1921).
[17] E. Zaucker, M. Bogemann, and L. Orthner, U. S. Patent 1, 942, 790 (1934).
[18] M. W. Harmon, U. S. Patent 2, 100, 692 (1937).
[19] A. Y. Coran, J. E. Kerwood, U. S. Patent 3, 546, 185 (1970).
[20] R. H. Campbell, R. W. Wise, Rubber Chem. Technol. 37, 635 (1964).
[21] R. H. Campbell, R. W. Wise, Rubber Chem. Technol. 37, 650 (1964).
[22] P. L. Hu, W. Scheele, Kautsch. Gummi 15, 440 (1962).
[23] N. J. Morrison, M. Porter, Rubber Chem. Technol. 57, 63 (1984).
[24] P. Ghosh, S. Katare, P. Patkar, J. M. Caritjers, V. Venkatasubramanian, K. A. Walker, Rubber Chem. Technol. 76, 592 (2003).
[25] T. D. Skinner, Rubber Chem. Technol. 45, 182 (1972).
[26] P. J. Nieuwenhuizen, J. Reedijk, M. Van Duin, and W. J. McGill, Rubber Chem. Technol. 70, 368 (1997).

Chapter 6
Processing of Unvulcanized Elastomer

Generally elastomer processing involves two major steps. First one is the designing of a mixing formulation for a specific end-use and the second one is the production process by which rubber compound is transformed into final product. When designing a mixing formulation the compounder must take account not only of those vulcanisate properties essential to satisfy service requirements but also cost of the raw materials and the production process. There should always be a compromise between cost of production and quality of the product. This chapter is an attempt to deal with different processing techniques normally used in the rubber industry.

6.1 Introduction

The processing of a rubber formulation is a very important aspect of rubber compounding. The raw polymer can be softened either by mechanical work termed mastication or by chemicals known as peptisers. Under processing conditions various rubber chemicals, fillers and other additives can be added and mixed into the rubber to form an uncured rubber compound. These compounding ingredients are generally added to the rubber through one of the two basic type of mixers; two roll mill or internal mixers.

6.2 Two Roll Mill

The first use of the two roll mill was in the 1830s in USA. Hancock's Pickle was patented in 1837, although models had actually been in use from the early 1820s. The first machine that appears suitable for rubber was a twin rotor design patented by Paul Pfleiderer in 1878/1879. Two roll mill consists of two horizontal, parallel heavy metal rolls which can be jacketed with steam and water to control the temperature. These rolls are connected to the motor through gears to adjust the speed. Rolls turn towards each other with a pre set adjustable gap or nip to allow the rubber to pass through to achieve high shear mixing (Fig. 6.1).

Fig. 6.1 Two-roll rubber mixing mill.

6.2.1 Friction Ratio

The speed of the two rolls is often different. The back roll usually turns at a faster speed than the front roll, this difference increases the shear force. The difference in roll speeds is called friction ratio, which is dependent upon the mill's use. For natural rubber mixing a ratio of 1 : 1.25 for the front to back roll is common.

6.2.2 Cooling

Cooling is employed either through cored rolls or through peripherally drilled rolls. The principal one employs cored rolls i.e., water is sprayed onto the outside of an axially drilled central core.

6.2.3 Other Attachments

Mills are fitted with a metal tray under the rolls to collect droppings from the mill. Guides are plates which are fitted to the ends of the rolls to prevent the rubber from contamination with grease etc. Safety measures are also attached to the mill for protecting the operator as well as the mill.

6.2.4 Mixing Process

There are five stages in the mixing process. They are:
① Banding the rubber on the first roll;
② Viscosity reduction by mastication or peptisation;
③ Incorporation of ingredients;
④ Distribution;
⑤ Dispersion.

When a highly elastic rubber of high molecular mass is fed into the mixer, it must be converted to a state in which it will accept particulate additives. This stage is called viscosity reduction. It is achieved either by a physical mechanism called mastication or by chemical means called peptisation. Now the rubber is ready to flow around the additives, incorporating and enclosing them in a matrix of rubber. Incorporated additives are then available for distribution. For better incorporation and distribution, with the help of a cutting knife give suitable cuts from either sides of the front roll.

During distributive mixing the rubber flows around the filler particle agglomerates and penetrate the interstices between particles in the agglomerate and the rubber mix becomes less compressible and its density increases. The rubber which has penetrated the interstices becomes immobilised and is no longer available for flow. This immobilisation reduces the effective rubber content of the mixture. The incompressibility of the mixture allows high forces to be applied to the particle agglomerates, causing them to fracture. This action is called dispersive mixing, which serves the purpose of separating the fragments of agglomerates once they have been fractured. The addition of plasticizers facilitates easy incorporation of the fillers. Curatives are added at the end of the mixing cycle. After thorough incorporation of all the ingredients the mix is homogenised and the batch is then sheeted out. For best mixing procedure the temperature is kept at 75~80℃ by careful adjustment of flow of cooling water through the rolls. The sequence of mill mixing is as follows:

① Band the rubber;
② Mastication/peptisation;
③ Addition of cure activators;
④ Half of the filler and oil;
⑤ Rest of the fillers;
⑥ Curatives;
⑦ Homogenisation;
⑧ Sheeting out the compound.

It is better to keep the rubber compound at ambient temperature for one day, for better consistency in properties.

6.3 Internal Mixers

The internal mixers were initially developed by Fernley H. Banbury from 1916 onwards. Both two roll mills and internal mixers are batch mixers, mill mixing is relatively a slow process, and the batch size is limited. Internal mixers overcome these problems by ensuring rapid mixing and large output. An internal mixer consists of two horizontal rotors with wings or protrusions, encased by a jacket (Fig. 6.2).

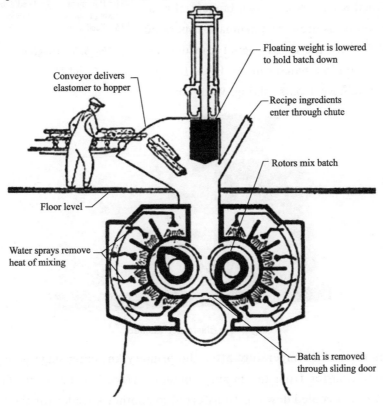

Fig. 6.2 Diagrammatic section of Banbury mixer.

6.3.1 The Intermix

The concept of Intermix was developed in the UK during the early 1930s by an unknown engineer of the ITS Rubber Company. Construction and detailed design of the Intermix was contracted to Francis Shaw a company of Manchester, who eventually acquired and patented the design (Fig. 6.3).

6.3.2 Rubber Kneaders

There are two different types of internal mixers used in the industry at large. The first type is more commonly known as a "Banbury" type intensive mixer and the second type is known as a "Kneader". The primary difference between the two types of mixers is rotor, throat, chamber and floating weight design. The former also discharges the batch through a bottom door where as the kneader tilts to discharge the batch (Fig. 6.4).

Conventional Kneaders have two tangential non-intermeshing rotors as well as pneumatic operated floating weights. With the conventional kneader design the temperature in a batch can not be sufficiently controlled to achieve 1 pass mixing. With conven-

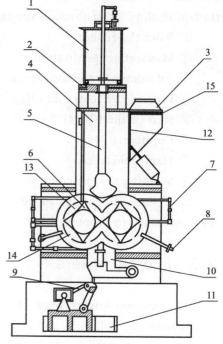

1—Cylinder; 2—Sealing; 3—Feeding port;
4—Pressure chamber; 5—Stung treng bolt;
6—Rotor; 7—Pipe line; 8—Thermal couple;
9—Discharge bolt; 10—Discharge door;
11—Basement; 12—Feed door; 13—Upper mixing chamber; 14—Lower mixing chamber;
15—Feed lock

Fig. 6.3 Diagrammatic section of Shaw Intermix.

Fig. 6.4 Elastomer kneader.

tional kneaders the batch temperature after the primary kneading stage is high because of poor temperature transfer from the mixing contact surfaces to the batch. Therefore, the batch has to be either cooled down or transferred to another kneader for the final kneading stage. This additional step is cost prohibitive as well as time consuming.

The MXI-Intermeshing Kneader imparts superior dispersion by reducing filler particle size during the kneading process. The reduction of particle size is achieved by the intermeshing rotor design. Traditional kneaders have two counter rotating rotors with each mixing rotor having two wings affixed on it. The two wing rotors typically rotate at two different speeds through connecting gears. The wings move material from one portion of the chamber to the other while also providing material movement along the rotor axis. These kneader do not have intermeshing rotors and therefore can have differential rotor speeds.

Conventional Kneaders have a one piece rotor design which includes a rotor shaft with two wings welded on the shaft. Water cooling is provided through a passage in the rotor shaft and small jackets in each wing. This cooling method is not sufficient for single pass mixing. The MXI-Kneader consists of a two piece rotor design. An over-sized rotor shaft and a cast blade shell portion. The cast blade shell is provided with a spiral water passage which is close to the material contact surface. The assembled rotor has a much larger outside diameter than conventional kneaders. This allows for more cooling surface as well as larger mixing surfaces.

Conventional kneader rotors have a shaft and two wings one wing is typically shorter than the other, to have adequate material movement inside the mixing chamber. The MXI-Kneader has a rotor shaft with one long wing (blade) and two nogs (small blades) for mixing. The conventional kneader's blades are typically long high and narrow. The new MXI-Kneader has much wider land width and stubby in shape. The much wider rotor tip (land width) greatly enhances the dispersion effect. The materials are subjected to a larger smearing action of the batch against the rotor tip to chamber wall as well as the rotor tip to rotor shaft. In the non-intermeshing type kneader no mixing occurs between the rotor tip and rotor shaft due to the non intermeshing design.

Conventional kneaders use pneumatic pressure to push the batch down into the rotors and mixing chamber with a floating weight (ram). This pneumatic system is unreliable and inconsistent. The pneumatic ram moves completely uncontrolled and the ram position is controlled by the rotor dragging force as well as the size of rubber pieces it is trying to force into the rotors. The hydraulic ram exerts positive pressure on the batch and can be accurately controlled in the desirable position which leads to better batch to batch consistency.

6.3.3 Intermix and Banbury

In Tire factories and large rubber factories the internal mixers has practically replaced the two roll mill for the preparation of compounds. Both these machines are used in rubber industry. Compared to tangential system in Banbury the Intermeshing system in Intermix have more effective temperature control, drive power is around 10%～20% higher. But optimum fill level is 5% lower because of the narrow intermeshing zone. The basic difference between the two machines lies in the rotor design. The intermix is an example for the interlocking type rotors and the Banbury is of noninterlocking type. In both cases the rotors run at even speed and the nogs or wings are designed to produce a friction ratio between the rotors. In the Banbury, the mixing process is carried out between the rotors and the jacket. In the Intermix the work is done between the rotors. Both the machines are fitted with a ram to ensure that the rubbers and powders are in contact.

6.3.3.1 Machine Sizes

A range of sizes of machines are available. The most popular size of machine is one with a batch load of about 200kg of compound.

6.3.3.2 Rotor Speeds

Internal mixers of 200kg size can be obtained with rotor speeds in the range of about 20~66rev/min. To carry out a mixing, certain number of rotor revolutions are needed, then the mixing time is directly proportional to the rotor speed.

6.3.3.3 Ram Thrust and Fill Factor

The thrust applied to the ram affects the output of the internal mixer. Increase in the ram thrust reduces the voids in the machine. For efficient mixing the fill factor is also important. For a given rotor speed and ram pressure, there is a correct volume of the compound to give efficient mixing. If this is divided by the volume of the chamber the fill factor is obtained. For rubber compounds normally it will be 70%~80%. The remaining corresponds to the voids in the mixture. If the fill factor is accurate, large output of better quality will result. Increase in the ram thrust increases the rate of increase of temperature, and reduces the mixing cycle, and gives more rapid ingredient absorption, and gives greater reproducibility in mixing.

6.3.3.4 Cooling Arrangement

Drilled sides are now common for cooling arrangements. The drilled sides comprise cooling passages drilled under the surface of the body. If cooling is higher, slippage can occur between the rubber and the rotors. As this gives inefficient mixing, warm water is circulated through the machine. The temperature of the water can be regulated by cooling water when heat is being generated and by electrical heaters when the machine is too cold. When a mix has been completed in the internal mixer, cool it as quickly as possible. The batch of compound is either dropped into an extruder, or to a two roll mill. The compound is then treated with anti-tack prior to cooling and storing. Various degrees of automation are possible for these systems.

6.3.3.5 Mixing Procedure

The following facts should be noted for better mixing:

① Generally the efficiency of mixing depends on the sequence of material input to the mixer. The steps for ingredient addition should be minimum. For each addition the ram should be raised, with the ram up, there is no pressure on the mix, which leads to little effective mixing.

② Particulate fillers should be added at the earliest stage of mixing. This helps to achieve good dispersion as a result of the high viscosity at the initial low temperature. A higher viscosity will lead to an increased shear stress at a given rotor speed. For the same reason plasticizers should be added at the later stage. Oils may coat the rotors and chamber wall and cause slippage and reduce mixing efficiency. They are therefore added together with fillers to reduce this action.

③ The curative package should not be added at he elevated temperature stage. Batches are usually dumped from an internal mixer on to a mill where they may be further worked while being cooled. Curatives are added at this point.

6.3.3.6   Upside Down Mixing

This method involves adding all the dry ingredients other than the elastomer to the mixer first, then all the liquids, and finally the elastomer.

6.3.3.7   Advantages

① It is faster and simplest.

② It is employed when the polymer content is less than 25%, and also for polymer having poor self-adhesion.

③ It is effective for compound having large volume of liquid plasticisers and large particle size fillers.

6.3.3.8   Disadvantages

① Small particle size carbon blacks cannot be effectively mixed by this technique, as it does not provide a high level of dispersion.

② If the polymer is of high viscosity upside down mixing will result in the development of temperature and lead to poor dispersion.

③ Clays which are difficult to wet due to low surface energy do not incorporate well in this method.

6.3.3.9   Take-Off Systems

After the mixing the batch has to be cooled and converted into strips or sheets suitable for feeding to the next process. This is done by dumping the batch through the drop door at the bottom of the mixer on to a cool mill capable of handling the entire batch. A three or four roll calender is used when uncured rubber is applied to a textile fabric or steel cord as a coating. Extruders are used when the uncured stock is to be shaped into a tyre tread, a belt cover, or hose tube for example. Extruders with lower screw length to screw diameter (L/D) ratios are considered hot feed extruders while one with high ratio are called cold feed extruders.

6.4   Continuous Mixers

The continuous mixing of rubber compounds is very much in its infancy. The earliest machines used for continuous processing of true curable materials were the Extruding, Venting and Kneading (EVK) machine made by Wernar and Pfliederer and Mixing, Venting, Extruding (MVX) machine made by Farrell Bridge. EVK primarily used in EPDM extrusion compound area using powdered polymer, MVX for cable compounding and in the production of tire compounds using granulated Polymer. Recent work on continuous mixing of rubbers has centred on modified twin screw compounders that have been used for some considerable number of years for in the plastic compounding industry. Because of the numerous compounding ingredients used in the tire industry in their varying physical form, an economic and sufficiently accurate proportioning of the compounds in a continuous mixer is barely possible. Now-a-days continuous mixers are used in rubber compounding only for partial operations such as making batches either consisting of elastomer and filler or another one containing chemicals.

Both mills and internal mixers are batch mixers. In order to replace the batch mixing process, attempts were made since world war II to develop a continuous mixing technique. Examples are Double R mixers from Francis Shaw in the late 1940s, the continuous mixer from Farrel's Corporation in the 1960s, and the Transfer mix from U.S Rubber Company in the late 1960s. Continuous mixing normally demands that solids are fed in a particulate form. In order to convert rubber bales into pellets; disintegrators are needed. All continuous mixing operations face the problem of how to weigh continuously the multiplicity of very variable weights of rubbers and their compounding ingredients with the accuracy required. Disintegrating rubber will consume power and powdered rubbers cost more than baled rubber.

6.5 Trouble Shooting the Mixing Process

Problems in the mixing process are usually due to inadequate dispersion, contamination, poor processability on the dump mill, scorchy compound and batch to batch variation. Once the problem has been identified, corrective action is often simple. The list below suggests possible causes of such problems.

(1) Inadequate dispersion or distribution
- Insufficient work input, mixing time;
- Order of ingredient addition not proper;
- Batch size too large or too small;
- Insufficient ram pressure, wrong rotor speed, wear of rotors and chamber wall;
- Cold polymer (this applies especially to natural rubber, EPDM and butyl);
- Excessive moisture in fillers;
- Oils added at temperature below pour point.

(2) Scorchy compound
- Too high a heat history after addition of curatives;
- Accelerator added too soon;
- Inadequate distribution of curatives;
- Too high a rotor speed;
- Materials added at too high a temperature;
- Inadequate cooling of compound after take-off.

(3) Contamination
- Physical contamination of one or more ingredients;

- Insufficient clean-out between batches of different base polymers;
- Oil-seal leak.

(4) Poor handling on dump mill
- Incorrect roll temperatures, speed and friction ratio;
- Too high a loading of clay fillers, viscous plasticizer;
- Poor distribution or dispersion;
- Scorchy compound;
- Compound left on mill too long.

(5) Batch to batch variation
- Variation in initial loading temperatures, ram pressure, cooling water flow, or temperature;
- Variation in compounding ingredients;
- Variation in dump time, temperature, or energy input;
- Variation in milling time or mill settings;
- Variation in amount of cross-blending on mill.

6.6 Extrusion Process

6.6.1 Introduction

In elastomer processing, extruder is mainly used for shaping the elastomer compound into the desired profile before it is finally processed. There are two type of extruder—ram extruder and screw extruder. Ram extruder has high operating cost and lower output. Now, the screw extruder is used mainly for the production of tubing, channel, tire treads and for the insulation of the wire and cables (Fig. 6.5).

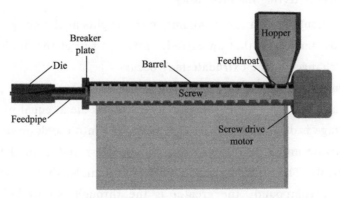

Fig. 6.5 Schematic representation of an elastomer extruder.

6.6.2 Screw Extruder

The extruder consists of a feed hopper, cylindrical barrel, rotating screw, head attachment and a die. The screw is driven by the an eletric motor through appropriate reduc-

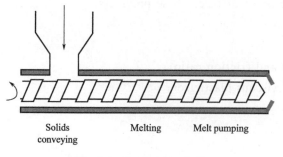

Fig. 6.6 Screw extruder.

tion gear system. The compound to be shaped is fed into the machine through the feed hopper. The width of the compound strip fed is slightly less than the width of the feed hopper and the thickness of the strip should be slightly less than or equal to the depth of the flight of the screw. The barrel is usually made of hardened steel and is jacketed for the circulation of steam or cold water. Heating of the barrel is necessary in the early stages, when it is started temperature is developed inside in this stage, steam supply is cut off and cold water is circulated to maintain the temperature (Fig. 6.6).

The head attachment of the extruder varies in shape according to the purpose for which it is used. The design of the head is very important to get free movement of the compound at equal pressures and speeds from all side of the head into the die. Any point within the head where the compound doesn't move is known as dead spot. Provision for heating and cooling should be provided at the head attachment for better control of temperature. The die of the extruder shape the compound into the desired profile. For better shape and finish of the extrudate, the design of the dies is very important. The die is the hottest part of the extruder and is usually heated initially by a gas flame. The cross sectional area of the die should never be lower than 5% less or greater than 30% more of the cross sectional area of the extruder. The extrudate coming out from the die is usually carried to the next stage of processing, through conveyor system. Cooling of the extrudate is done by immersion in water or by a spray of cold water. Talc is applied to the extrudate.

6.6.2.1 Parameters Affecting the Processing

① The screw should have a lower volume in the flights at the out going than at the in going end. It is most important that an extruder screw is full at the discharge end, otherwise dimensional changes in the extrudate may occur.

② The design of the head is very important, the head equalizes the pressure from the screw and barrel and the compound moves smoothly to the die at equal pressure and speed.

③ The last stage is die, which forms the compound into the desired shape. Die should be designed to operate under conditions of minimum stress and at predetermined running speed and temperature. The extrudate should be produced under these conditions. The lower the viscosity of the compound, the greater is the through put to be expected in unit time. Dimensional variations is at minimum when the compound has the minimum entrapped stress.

6.6.2.2 Hot Feed, Cold Feed and Vacuum Extruder

Depending on the design of the screw and barrel, extrudate can be three types namely, hot feed, cold feed and vacuum extruders. The screw of the vacuum zone there is provision at the barrel for connecting it to a vacuum device. In the vacuum zones, the screw is either deeper or more widely cut or the cylinder in that zone is slightly bigger than in the other zones. Vacuum extruder helps to remove any traces of moisture or entrapped air from the compound. Hence it is used in shaping articles for open steam cure, hot air cure, molten salt cure and fluidized bed cure. Since moisture and air trapped in the compound is removed during vacuum extrusion, the product will be free from porosity. Depending on the design of the screw, the extruder may be used for hot or cold feeding of the compound. The hot feed extruder has got a short barrel, the length to diameter ratio of the screw is low, in the range of 5 : 1 and the compression ratio is nearly equal to unity. In the case of cold feed extruder the length to diameter ratio of the screw is high, in the range of 20 : 1 and the compression ratio is greater than unity. In hot feed extruders, pre-milled rubber is fed into the extruder, output will be uniform and the production equilibrium can be attained within a short time. As its name refers cold elastomer compound can be fed into cold elastomer feed extruders. Out put depends on the nature of the compound and it takes longer time to attain production equilibrium. Hot feed extruder has high operating cost and higher output than cold feed extruder. Product of consistent quality can be obtained only if the compound fed into the machine is uniform viscosity, temperature and volume. Maintain the temperature of the barrel, screw, head and die constant.

6.6.2.3 Defects in Extrusion

(1) Die swell

As the compound comes out from the die, it shrinks along its length resulting in slight increase in overall dimension of the extrudate. Nature of the compound, uniformity of the speed stock and speed of the screw and take off conveyer systems can affect die swell.

(2) Rough surface

Poor finish of the extrudate may be due to poor dispersion of the compounding ingredients, very high money viscosity of the polymer, very low temperature and pressure of extrusion. By proper adjustment of this good finish of the product can be obtained.

(3) Porosity

This is due to the presence of excess moisture in the compounding ingredients, use of high volatile compounding ingredients and presence of entrapped air. Proper drying of fillers before using, addition of material like calcium oxide in the compound and use of vacuum extruder can reduce porosity in the extrudate.

(4) Collapse of the material

Collapse of the extrudate occurs when the quality of the polymer used is poor, viscosity of the compound is very low and when the processed material is recycled several times.

6.6.2.4 Troubleshooting the Extrusion Process

Despite the many feedback microprocessor control system available on the market, it is still often the skill, experience, and understanding of the extruder operator that determines the success or failure of an extrusion operation. Success or failure has to be measured in economic terms, that is the hourly production rate of the process. This depends on minimizing scrap and downtime both in start-up.

6.7 Calendering

6.7.1 Introduction

Calenders are used in the elastomer industry primarily to produce elastomer compounds and sheets of various thicknesses, coating textiles or other supporting materials with this elastomer sheet or frictioning fabrics with elastomer compound.

6.7.2 Machinery

A variety of products like sheeting for lining, hospital bed sheets, films, frictioning of tire fabrics for Bicycle/motor cycle/Auto tires, hoses beltings, profiling, cushion gum, single coating, laminating doubling etc. Special Calenders for profiles and Inner liner can be done. The Elastomer Calender Machines are made in a wide range of sizes, from small laboratory unit up to the largest production calender machines (Fig. 6.7).

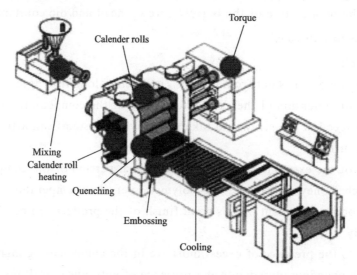

Fig. 6.7　Calendering unit.

6.7.3 Calendering Unit

Rubber calenders are differentiated by the number of rolls, their arrangements, and their size (diameter and width). They can have two, three, or four rolls in a variety of configuration. For the production of tire stock, belting, and sheeting the 3 roll vertical calender with 2400 diameter, 6800 width rolls, and four-roll Z and L calenders with 2897800 rolls are standard. Four-roll calenders are used for applying compound to both sides of tire cord fabrics in one operation.

6.7.3.1 Types

There are three main types of calender: the "I" type, "L" type and "Z" type (Fig. 6.8~ Fig. 6.10).

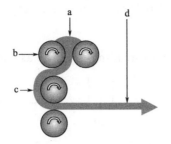

Fig. 6.8 Roller setup in a typical "I" type calender.

Fig. 6.9 Roller setup in a typical inverted "L" type calender.

Fig. 6.10 Roller setup in a typical "Z" type calender.

The "I" type, as seen in Fig. 6.8, was for many years the standard calender used. It can also be built with one more roller in the stack. This design was not ideal though because at each nip there is an outward force that pushes the rollers away from the nip (Fig. 6.9). The "L" type is the same as seen in Fig. 6.9 but mirrored vertically. Both these setups have become popular and because some rollers are at 90 to others their roll separating forces have less effect on subsequent rollers. "L" type calenders are often used for processing rigid vinyls and inverted "L" type calenders are normally used for flexible vinyls (Fig. 6.10). The "Z" type calender places each pair of rollers at right angles to the next pair in the chain. This means that the roll separating forces that are on each roller individually will not effect any other rollers. Another feature of the "Z" type calender is that they lose less heat in the sheet because as can be seen in Fig. 6.10 the sheet travels only a quarter of the roller circumference to get between rollers. In most other types this is about half the circumference of the roller.

6.7.3.2 Feeding

To ensure steady operations of the calender, and to control shrinkage, the compound has to be preheated to around 93℃, and thoroughly fluxed and plasticized before being fed to the calender. Calenders use rolls with axial drilling about 50mm under the surface, through which water at a pre-set temperature is continuously circulated. The rolls with the axial hole through the centre are known as cored rolls and the latter are periphery drilled rolls. Periphery drilled rolls without controlled—temperature water going through them are unsatisfactory in operation, since hot and cold water used alternately increase and decrease the roll temperature too rapidly. Periphery drilled rolls are normally heated and cooled at a speed of 1℃/min i.e., it give much better control.

6.7.3.3 Sheeting

This process is carried out on a three—roll calender with thickness control by a feed back system from the product. The rolls are crowned to compensate for deflection under load, and so to maintain a constant roll gap across the width of the sheet.

6.7.3.4 Frictioning

This is impregnating a textile or metallic fabric between two rolls running at different speeds so that the elastomer compound is forced into the interstices of the substrate.

6.7.3.5 Spreading

The main working part of a typical spreading machine as shown in (Fig. 6.11).

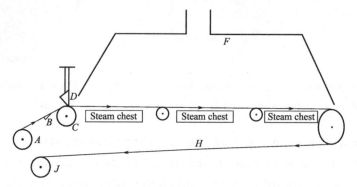

Fig. 6.11 Diagrammatic sketch of spreading machine.

A roll of dried or pre-treated fabric is fitted on to location A and the leader cloth is fed through the rest of the machine until finally taken up on roller J from A, the cloth passes over a spreader bar B to ensure that all creases are removed from the fabric and to keep it under the correct lateral tension. The smooth tensioned fabric is then fed over the bearer roller C and under the doctor blade D, which is pre-set to give the correct build-up of dough on the fabric surface. The angle between the blade and the fabric and the distance between them control the coat thickness and the degree of "strike through" (degree of penetration) of the dough. The greater the angle at which the blade meets the moving fabric, the greater the degree of penetration. The fabric then enters the steam chest area, where the solvent is driven off and removed by means of the extraction unit F; the speed of travel of the fabric is dependent on the rate of solvent removal. On emerging from the end of the steam chest, spread fabric requires cooling before it is rolled up on roller J. This is achieved by means of the festooning device, placed at H, which may consist of a single or several rollers; if the dough is of a sticky nature, it may also necessary to use a liner cloth or dust the surface with talc to prevent blocking together of the elastomer-fabric laminate during storage. Once a machine has been set up to run, it is necessary to carry out the spreading of the first few yards at the low speed to check the coating thickness against specification, either by means of a vernier gauge or electronically. Normal running speed of the order of 10m/min.

6.7.3.6 Skim Coating or Topping by Means of the Calendering

The operation of applying a substantial thickness of elastomer to fabric on a calender is termed skim coating or topping. In this method, compound is fed around a calender roll from a calender nip, and the sheeted compound is applied to the fabric at a second nip. The elastomer sheeting must be travelling at the same speed as the fabric at the point where it is laid on to the fabric; however, sheeting can be produced from roll which run at the same speed or with a friction ratio in the nip (Fig. 6.12).

6.7.3.7 Temperature Effects

The temperature of the fluid melt has been found to be highest at the rollers. This happens for two reasons:

① The shear is highest at the sides in laminar flow and therefore friction and heat is also highest there.

② The heat is added to the system through the rollers, and the fluid doesn't conduct it very well. The

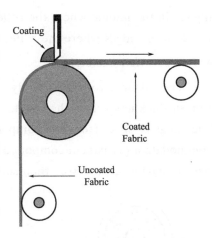

Fig. 6.12 Topping by means of a calender.

effects of this tend to grow in magnitude with more and more viscus the fluid is. If one were to raise the rolling temperature there would be changes in the above fluid mechanics. It would decrease the viscosity; consequently decreasing the power input, pressure and roll separating forces in the fluid. It would also lower the chances of a fracture in the fluid and make the surface finish better, but this all comes at the price and increases the chances of thermal degradation.

6.7.3.8 Velocity Effects on Final Product

The calender is able to produce the polymer sheeting at a fast rate. It can produce sheeting at a rate between 0.1m/s and 2.0m/s. By increasing the speed the heat has even less time to spread throughout the fluid from the rollers causing an even greater temperature variation. It also causes an increase in shear forces in the fluid at the rollers, which increases the chances of surface defects like fractures. The speed clearly needs to be chosen very carefully in order to produce a quality product.

6.7.3.9 Roll Bending

In calendering the rollers are under great pressures, which can reach up to 41MPa in the final nip. The pressures are highest in the middle of the width of the roller and due to this the rollers get deflected. This deflection causes the sheet being made to be thicker in its center than it is at its sides. There are three methods that have been developed to compensate for this bending:

① Roll crowning;
② Roll bending;
③ Roll crossing.

Roll crowning uses a roller that has a bigger diameter in its center to compensate for the deflection of the roller. Roll bending involves applying moments to both ends of the rollers to counteract the forces in the melt on the roller. With roll crossing the rollers are put at a slight angle to each other and because of this the force of the rollers on the melt is higher in the middle where the rollers are on top of each other more, and less force is applied on the edges where the rollers are not directly over top of each other.

6.7.3.10 Roll Cambering

The calender rolls are usually cambered and are not parallel to compensate for variation in thickness across the sheet. This is an ideal solution for a calender which produces one gauge of sheet from one compound. Calenders can be recambered in a few hours to accommodate a permanent compound change or to take up the wear of the roll; if more than one compound is processed then some other device for resetting the crown will be needed.

6.7.3.11 Calendering Technology

Rubber compound behave as viscous non-Newtonian liquids. If a uniform gauge of sheeting is to be produced, then the viscosity of the compound must be constant. In order to achieve uniform viscosity, the temperature of both the compound and the calender must be controlled, as the viscosity of elastomer compound is affected very considerably by temperature (Fig. 6.13). When an unsupported sheet is taken from a calender nip, it shrinks along its length and increases in the thickness and the width. This results in elastomer sheets having a crown, i.e., they are thicker in the centre than at the edges.

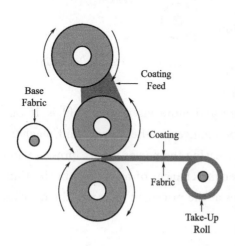

Fig. 6.13 Elastomer-fabric coating by calendering.

6.7.3.12 Unsupported Sheeting

Two-, three-, or four-roll calenders are used for the production of sheeting containing no textile fabric reinforcement.

For precision gauge control, a sheet is produced in a first nip, and this is then fed round a roll to a second nip. The second nip gives either less blistering or a thicker sheet for the same amount of blistering. The quality of sheeting is more dependent upon the quality of feed than on any factor other than the temperature of the calender rolls. Occasionally, three nips are used for calendering, i.e., a four roll.

6.7.3.13 Application of Elastomer from Solvent Dispersion or Dough

When it is necessary to apply a coating of elastomer to a fabric which is too delicate for the calendering process, or when the compound is not suitable, the technique of spreading is used. This process consists of the application of the compound dispersed in solvent at high concentration in the form of dough, as it is termed. A stationary blade (commonly called a doctor) regulates the thickness applied to the fabric as it is passed underneath the blade. The fabric then drawn over a heated chest where the solvent is evaporated and usually recovered for re-use by adsorption on an active carbon.

6.7.3.14 Troubleshooting Problems in Calendering

(1) Scorch
- Poor temperature control;
- Running speed too fast, leading to excessive shear heating;
- Stock warmed upon mill too long.

(2) Blistering
- Roll temperature too high;
- Feed bank too large, resulting in entrapped air;
- Sheet thickness too high.

(3) Rough or Holed sheet
- Inadequate stock warm-up;
- Amount of material in bank too small, or too large, to form rolling bank;
- Varying stock temperature.

(4) Tack
- Temperature of rolls too high;
- Incorrect stock feed temperature.

(5) Bloom
- Low solubility of some ingredients in formulation.

6.8 Continuous Vulcanization System

As a elastomer compound containing a curative system is held at the curing temperature the production of cross-links causes it to change from a viscoelastic fluid to an elastic solid. As it leaves the die the compound is still a fluid, and as the stresses built up in the passage through the head and the die relax, the dimension of the profile, originally those of the die, change.

6.8.1 Pressurized Steam Systems

These are commonly used for products having a core or other reinforcement and profile that are easy to seal, such as wire, cable, and hose. The time required for heat to penetrate to the centre of the cross-section depends on the diameter. The weight of elastomer per unit length is proportional to the square of the diameter.

6.8.2 Hot Air Curing Systems

These consists, basically, of an insulated tunnel, a metal mesh conveyor to support and move the profile through, and a counter current of air, heated to up to 300°C. Heat transfer is poor (coefficient $70kJ/m^2/h/°C$) and so lines of 100ft (30.48m) are required to complete the curing process. With compounds (e.g., EPDM), which are not readily susceptible to oxidation at ultra high temperature, shorter ovens can be used.

6.8.3 Microwave System

Microwave system provide quick and uniform heating throughout the profile, which is especially useful for thick profiles, profiles of varying thickness, and for sponge. The compound has to be microwave receptive (i.e., polar), which most polymer are not. However, many carbon black are, and if necessary, it is also possible to add other chemicals specifically to increase microwave receptiveness. Usually, a short microwave section is used immediately after the die to boost the extrudate temperature to curing temperature, followed by a hot air tunnel to maintain temperature until the profile is cured. There is much less heat loss with a microwave system than with the system described previously because the heat is generated in the elastomer itself. This high energy-efficiency makes the use of electricity, a more expensive source, economically feasible.

6.9 Moulding

Moulding is the operation of shaping and vulcanising the plastic elastomer compound by means of heat and pressure in a mould of appropriate form. There are three general moulding techniques:
① Compression;
② Transfer;
③ Injection moulding.

6.9.1 Compression Moulding

In compression moulding a pre-weighed, pre-formed piece is placed in the mould. The mould is closed, with the sample under pressure as it vulcanises. For the satisfactory large scale production of components, it is necessary to use carefully designed and well constructed steel moulds, suitable hardened and finished depending upon the surface quality required for the product. Cavity pressure is maintained by slightly overfilling the mould and holding it closed in a hydraulic press. Heat is provided by electricity, hot fluid or steam. Compression moulding is the oldest and most universally used technique and for many products the cheapest process because of its suitability for short runs and because of the low mould cost. The press used for conventional compressional moulding has two or more platens, which are heated either electrically or by saturated steam under pressure. The platens are brought together by pressure applied hydraulically to give a loading from 75 to 150 kgf/cm^2 of projected mould cavity area (Fig. 6.14).

6.9.2 Advantages

① Moulds have low investments cost.
② Due to the simplicity in the mould design, it is suitable for curing thick elastomer

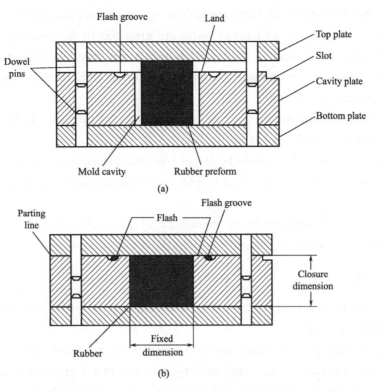

Fig. 6.14 Compression mould containing a elastomer preform:
(a) Before closing; and (b) after closing.

articles. Hence large elastomer articles (tyres, belts etc.,) and small articles (gaskets, washers etc.,) maybe cured by this process.

6.9.3 Transfer Moulding

Transfer moulding involves the transfer of a elastomer compound from a heated reservoir or transfer pot, through a narrow gate called sprue or runner into the closed cavity of a mould by a piston where the compound is cured at pre-determined temperature and pressure (Fig. 6.15).

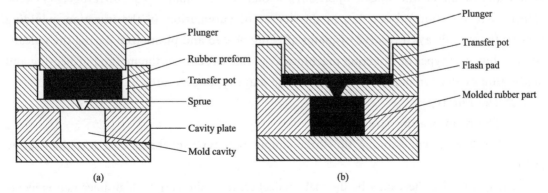

Fig. 6.15 Transfer mould showing: (a) preform transfer pot before closing; (b) after closing.

Transfer mould consists of three parts. The upper part called piston, the lower part the mould itself. Both upper and lower parts are attached to hydraulic press. The centre part which contains the cylinder, the injection nozzle is removable. Though the moulds are more expensive, this process permits better heat transfer. Important points to be remembered during transfer moulding.

① Clearance between the transfer pot and plunger (piston) should be optimised.

② The plunger should not tilt in the pot.

③ The plunger face area should be larger than the projected cavity area.

④ Compound should be optimised to reduce the vulcanised scrap in the mould cavity and the sprue.

⑤ Changes in pressure should be monitored by pressure transducers, since pressure variations during the transfer process can open the mould cavity to flash.

6.9.4 Advantages

① As the mould is closed before the elastomer charge is forced into it, closer dimensional control is achievable.

② In the transfer process fresh elastomer surfaces are produced. This allows the development of a strong elastomer to metal bonding with any insert in the mould.

③ Production cost is lower due to shorter cure times as a result of heating the elastomer due to flow through sprue, runner and gate and shorter downtime between runs as only one charge blank is necessary even if a multi-cavity mould is used.

6.9.5 Injection Moulding

Injection moulding is now a well established process in elastomer industry. The operation of an injection moulding machine requires feeding, fluxing and injection of a measured volume of a compound, at a temperature close to the vulcanisation temperature into a closed and heated mould. The process also requires a curing period, demoulding and if necessary mould cleaning or metal insertion before the cycle starts again. For maximum efficiency almost all of the above operations should be automatic. The difference between transfer and injection process lies in the degree of automation. In the injection moulding process there is always a reserve of material being heated and plasticized during the vulcanisation step. The types of injection machines available depend on the method of heating and peptisation of the compounded elastomer. The main types used are:

① The ram type.

② The reciprocating screw type.

③ Screw-ram type. Among the three simple ram type machines cost less than screw machines.

The mix receives heat only by thermal conduction from the barrel, high injection temperature and thermal homogeneity are difficult to achieve and they are not widely used. In the reciprocating screw type the screw acts both as an extruder and a ram. In this type of machine, the mix is heated and plasticised as it progresses along a retractable screw. When the necessary shot volume has accumulated in front of the screw. it is injected by forward ramming action of the screw. With this system more uniformly controlled feeding of the material can be achieved together with more rapid heating of the stalk, due to mechanical shearing, and a greater degree of thermal homogeneity. However, during the induction stage as the screw act as ram there is some leakage back past the flights of the screw and this limits the injection pressure. This type of machine is only possible for low shot volumes, or for very soft compounds. The preferred basic design for the elastomer injection moulding is the screw-ram type because they combine the advantages of both screw and ram type (Fig. 6. 16).

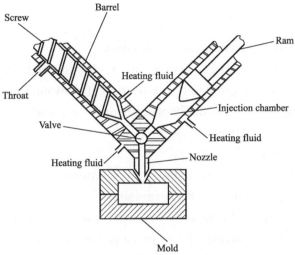

Fig. 6.16 Injection moulding machine with separate ram and screw.

In the standard "V" configuration the plasticated compound is fed through a check valve into an accumulation chamber. One disadvantage is that the first elastomer fed through the check valve is the last one to be injected. This can lead to adhesion and build up of elastomer on the face of injection piston, which can cure, break off and cause rejects and moulding problems. So modifications have developed to overcome this. In the first in-first out system the screw and ram though separate are inline. Initially the injection ram is in the forward position and the injection chamber is empty. As compound enters through the ram it is forced back by incoming material until a limit switch controlling short volume is activated. Injection then takes place through a special ball type torpedo which completes the plastication and thermal homogenisation. As the material does not reach the final injection temperature until it reaches the nozzle temperature earlier in the system can be relatively low (approximately 70 ℃). Such inline systems have gained popularity in recent years. Multi stage rotary press units can be fitted to many of the injection moulding system, thus enabling continuous moulding to take place. In this injection unit that feeds a number of moulds carried on a rotating carrousel. This system is economic and practical. Such machine may have automatic ejection of parts and runner system, cleaning and spraying of the moulds and automatic loading of metal inserts.

6.9.6 Compounds for Injection Moulding

Compounds for injection moulding must have sufficient scorch safety to flow through the nozzle, runners and gates without scorching but still cures rapidly in the mould. For maximum productivity, the compound has to injected rapidly into the mould at near vulcanisation temperature. This area has to be optimised within which a particular material will flow well and cure effectively without the danger of scorching. The three major areas in which data on a compound are required are rheological behavior, rate of vulcanisation and heat flow into and through the compound.

6.9.7 Advantages

① A high output of production can be obtained.
② Automation is the process can lead to cost saving and high quality.
③ Cure time may be reduced due to pre-heating of elastomer.
④ Uniform curing of variable thickness component is possible.
⑤ Flash trimming is eliminated.
⑥ There is no bumping.
⑦ The finished products can be removed more rapidly.
⑧ Complete filling of mould cavity is ensured.
⑨ Feeding is much easier in the strip form and is more economical.

RIM involves the rapid mixing of two or more highly reactive low molecular weight compounds before the injection of the mixture in a closed mould. The reaction may be polymerisation or molecular network formation in a very short time, approximately 30s The total cycle time is 1~2min. RIM is important for urethene elastomers because the process is very energy efficient.

6.9.8 Trouble Shooting the Moulding Process

6.9.8.1 Scorchiness of the Compound

Compounds must have sufficient processing safety to flow through the nozzle, runners and gates without scorching but still cure rapidly in the mould. For this the balance of viscoelastic and the curing characteristics of the compounds are extremely important. Most elastomer compounds that will compression mould can be satisfactorily injection moulded, provided that they flow well enough and are not too scorch sensitive. For maximum productivity the compound has to be injected rapidly into the mould at near vulcanisation temperature.

6.9.8.2 Shrinkage

On cooling both the mould cavity and the moulded part contract usually by a differential amount because the metal and elastomer have different coefficients of thermal contraction. Shrinkage is defined as the difference between the dimensions of the mould cavity and those of the moulded part, when both are measured at room temperature. The amount of shrinkage has to be allowed for in mould design. It will vary depending on the polymer, cure temperature, time and pressure.

6.9.8.3 Adhesion

Adhesion has two aspects, adhesion to the mould surface, which is not wanted, and adhesion to a metal insert in the part, which is wanted. Mould release agents are used to prevent the one, and adhesion promoters to ensure the other.

6.9.8.4 Backrinding

This is the term applied to the torn look that occurs at the mould parting line of compression moulded parts and at the gates of transfer and injection moulds. It is caused by thermal expansion of the elastomer after cross-linking, which can force the cross-linked elastomer into the space at the parting line or gate, causing it to rupture. The best way to minimise this is to minimise the shot weight commensurate with filling the cavity. Increasing scorch time can also help because it ensures that the mould is filled before curing begins as the injection temperature can be raised.

6.9.8.5 Mould Fouling

The build-up of material in a mould especially in the corners is a major problem. The cause is usually deposition of chemicals and their subsequent oxidation or degradation. These agents may originate in the elastomer, in fillers or from release agents. Thus, there are a wide variety of deposits whose severity varies from compound to compound and also depends on injection rate and mould temperature. Mould cleaning is often done by blasting with some abrasive particulate material such as glass beads, plastic or metal beads.

6.9.8.6 Orange Peeling

This is usually caused by the initial layer of elastomer in contact with the heated mould surface having cross-linked before succeeding layers have filled the mould. Usually occurs in injection moulding and the remedy is to increase the scorch time of the compound.

6.9.8.7 Porosity

This is due to under cure and the presence of volatiles, especially moisture in the compound. Higher injection and mould temperature or longer mould closed time should resolve this.

6.9.8.8 Blisters

Air entrapped in the elastomer compound is the usual cause. This can be eliminated by a higher back pressure, slower injection rate, or effective venting of the mould.

References

[1] J. S. Dick, Rubber Technology-Compounding and Testing for Performance, p. 17-18. Carl Hanser publishers, Munich (2001)

[2] http://tirenews4u.wordpress.com

[3]   C. M. Blow, C. Hepburn. Rubber Technology and Manufacture, p. 263-290. Published for the Plastics and Rubber Institute, London (1982)

[4]   A. K. Bhowmick, Rubber Products Manufacturing Technology, p. 347. Marcel Dekker Inc, New York (1994)

[5]   P. K. Freekley, Rubber Processing and Production Organisation, p. 46. Plenum Press, New York (1985)

[6]   P. R. Wood, Rubber Mixing. Rapra Technology Ltd. , Shawbery (1996)

[7]   P. S. Kim, J. L. White, Rubber Chem. Technol. 67, 880-891 (1994)

[8]   http: www. kneadermachinery. com

[9]   P. S. Johnson, Rubber processing, Hanser publishers, Munich (2001)

[10] S. R. Salma, Operation and maintenance of mixing equipment. In: Grossman, R. F. (ed.) The Mixing of Rubber, Chapman and Hall, London (1997)

[11] J. L. White, Rubber Processing, Technology, Materials, Principles, p. 484-485. Carl Hanser, Munich (1995)

[12] http: www. appropedia. org/polymer-calendering

[13] N. Nakajima, In: The science and practice of rubber mixing, p. 356. Rapra technology Ltd, UK (2000)

[14] J. G. Sommer, Rubber Chem. Techno. 58, 672 (1985)

[15] M. A. Wheelans, Injection moulding of rubber. Hastead press, London (1974)

[16] W. Hofmann, Rubber Technology Handbook. Carl Hanser Publishers, Munich (1989)

[17] A. Peterson, Rubber Mixing Technology Course. Center for Continuing Engineering Education, Univ. Of Wisconsin, Milwaukee (1999)

[18] L. N. Valsamis, et al. : Evaluating the performances of internal mixers. In: Grossman, R. F. (ed.) The Mixing of Rubber, Chapman and Hall, London (1997)

[19] N. Tokita, J. L. White, Appl Polym. Sci. 10, 1011 (1966)

[20] N. Tokita, Rubber Chem. Technol. 52, 387 (1979)

[21] Banbury Mixer. Bulletin No. 224-C Farrel Corporation, Ansonia, CT

[22] Shaw Intermix Mark 5 Series Farrell Corporation, Ansonia, CT

[23] T. Asai, et al. , Presented at Proceedings of International Rubber Conference, Paris (1983)

[24] N. O. Nortey, Rubber World. 49 (1999)

[25] E. Sheehan, L. Pomini, Rubber World. 50 (1997)

[26] Rubber machinery sourcebook Rapra Technology Ltd. Shawbury, UK (2000)

[27] T. Smith, Tire Technol. Int. 53 (1996)

[28] J. G. A. Lovegrove, Extrusion of Rubber Rapra Technologies Technical Report (1989)

[29] R. L. Christy, Rubber World 180, 100 (1979)

[30] J. Lambright, In: Long, H. (ed.) Basic Compounding and Processing of Rubber, ACS Rubber Division, Akron (1985)

[31] G. Cappelle, Rubber Products Manufacturing Technology Marcel Dekker Inc. New York (1994)

[32] A. K. Bhowmick, D. Mangaraj, Electron beam processing of rubber. In: Bhowmic et al. (ed.) Rubber Products Manufacturing Technology, Marcel Dekker Inc, New York (1994) Division, Akron (1985)

[33] P. R. Wood, Tire Technol. Int. 134 (2000)

[34] P. T. Dolezal, P. S. Johnson, Rubber Chem. Technol. 53, 253 (1980)

[35] P. K. Freekley, S. R. Patel, Rubber Chem. Technol. 58, 751 (1985)

Chapter 7
Rheological Behavior in Elastomer Processing

The fabrication of elastomer parts generally involves the mixing and processing of bulk unvulcanized compounds and sometimes solutions and emulsions through complex equipment. The ease or difficulty of fabrication depends on how these elastomer systems respond to applied stresses and deformations, their rheological properties. It is the purpose of this chapter to describe both rheological properties and the processing of unvulcanized elastomers and their compounds. We also consider some of the implications of rheology for processing.

7.1 Introduction

The study of the rheological properties and processing of elastomers and their solutions and compounds dates to the origins of the industry in the 1820s. The patent literature, memoirs, and reviews of the early 19th century contain numerous discussions of the flow and fabrication of natural rubber and gutta percha. The fundamental properties and methods of processing elastomer are associated with Thomas Hancock, Charles Macintosh, Edwin Chaffee, Charles Goodyear, Richard Brooman, Charles Hancock, and others, many long forgotten. It was not, however, until the development of three-dimensional linear viscoelasticity by Ludwig Boltzmann at the University of Vienna in 1874 that the understanding of the rheological properties of elastomery materials became sophisticated enough to allow rational study. Furthermore, it was another half-century before Bruno Marzetti of Pirelli SpA in Italy and (in the 1930s) J. R. Scott of the British elastomer Manufacturers Research Association (BRMRA), John H. Dillon of the Firestone Tire and elastomer Company, and Melvin Mooney of the U. S. elastomer Company (the last two in the United States) undertook the study of the deformation and flow of unvulcanized elastomer. In each case the motivation seems to have been the development of quality control instrumentation to ensure satisfactory processibility. Fortunately, each of the four was a careful, observant, and thoughtful scientist. Marzetti interpreted the extrusion of elastomer through a cylindrical die in terms of the flow of a fluid with a shear rate-dependent viscosity. This view was confirmed by Scott, Dillon, and Mooney. Scott was the first to realize that elastomer compounded with large quantities of small particles exhibited a yield stress below which there was no flow. This view was supported by Dillon and Johnston. Mooney obtained both the first quantitative viscosity-shear rate data on elastomer and the first measurements of elastic recoil, and Dillon and Cooper reported the first investigations of stress transients at the beginning of flow.

Post-World War II studies of the rheological properties of unvulcanized elastomers have been dominated by the idea that these materials are viscoelastic. Research along these lines was initiated by Leaderman, at the Textile Research Institute of Princeton University, who rediscovered the work of Boltzmann cited earlier. In the late 1940s, Tobolsky and his coworkers at Princeton University made extensive stress relaxation measurements on polyisobutylene. In succeeding years, linear viscoelastic measurements were performed on a wide variety of polymers in temperature regions in which they exhibited elastomery behavior. Tobolsky devised a program to relate the viscoelastic behavior to molecular parameters, such as molecular mass and glass transition temperature. This work carried out in the late 1940s and 1950s was reviewed in a 1959 monograph by Tobolsky.

In the 1960s, Bernstein, Kearsley, and Zapas made extensive studies of large strain nonlinear viscoelastic properties of polyisobutylene and proposed a three-dimensional constitutive equation to represent its behavior. In later years, non-linear transient experiments on polyisobutylene and natural rubber were similarly interpreted.

The hypothesis of Scott that highly filled elastomers exhibited yield values was confirmed by Zakharenko and his coworkers as well as the Basic Concepts of Mechanics by Vinogradov et al. (both in Moscow) and later by others. Yield values were also found to occur by subsequent researchers for thermoplastics filled with a wide variety of particles including talc, titanium dioxide, calcium carbonate, as well as carbon black. Mullins and Whorlow of the BRMRA found highly filled elastomer-carbon black compounds to exhibit strong time-dependent (thixotropic) characteristics. Their results were confirmed and extended by others. Since the late 1970s, there have been efforts to develop three-dimensional constitutive equations representing the yield value, thixotropic and viscoelastic characteristics of these compounds.

7.1.1 Quality Control Instrumentation

The first 40 years of the 20th century saw an enormous increase in the production of elastomer products especially in the tire industry. The horrendous nonuniformity of the wild elastomer used in the 19th century was reduced by the introduction of plantation elastomer from Malaya in 1910 (see the discussion of Litchfield). From the 1920s, efforts were made by various industrial and plantation-related scientists to develop improved quality control. Here we may cite the efforts notably of Marzetti of Pirelli, Williams of Firestone, Griffiths of the Dunlop elastomer Company, van Rossem and van der Meijden of the Netherlands Government elastomer Institute (NGRI), Karrer of the B. F. Goodrich Company, Dillon of Firestone, Mooney of U. S. elastomer, Hoekstra also of the NGRI, and Baader of Continental Gummiwerke. These efforts involved using capillary instruments, compressional flow between parallel disks, and shearing disk rotational rheometers.

In the late 1930s, the I. G. Farbenindustrie began commercial production of emulsion-polymerized butadiene-styrene copolymer (SBR) synthetic rubber under the designation Buna S. They adopted, after a comparison of many instruments, the Defo compressional flow instrument devised by Baader of Continental Gummiwerke. This instrument was used from the 1930s until the end of World War II to test and qualify the German Buna S. In the American government synthetic rubber program to develop SBR, termed by them GR-S, the Mooney shearing disk viscometer was adopted.

In the post-World War II period, the Mooney viscometer has tended to maintain its dominance, though it has been increasingly challenged by new generations of instruments which seek to measure viscoelastic characteristics. Mooney himself had urged the use of elastic recovery measurements following flow in his shearing disk rheometer and Baader had included recovery measurements in his original Defo test. Little attention was given to viscoelastic effects in processing until the 1960s. Researchers with U. S. elastomer/Uniroyal and B. F. Goodrich independently came to realize the importance of stress relaxation behavior. The B. F. Goodrich Company in the 1970s introduced a practical male-female biconical shear stress relaxation instrument known as the DSR. Subsequently the rubber and Plastics Research Association (RAPRA, successor to the BRMRA) developed a modified compression plastometer for this purpose.

In the 1980s, Bayer AG, successor to I. G. Farbenindustrie synthetic rubber activities, devised two new instruments in a program led by Koopmann. These were both a Mooney viscometer with stress relaxation measurement capability and an improved Baader Defo instrument.

The Bayer researchers preferred the improved Defo. The Mooney viscometer with stress relaxation is now manufactured by Alpha Technologies. The new Defo was made by Haake under Bayer license. Alpha Technologies manufactures an instrument that measures linear viscoelastic dynamic mechanical properties.

7.1.2 Processing

The earliest successful elastomer processing technology was that of Charles Macintosh, devised in the 1820s which prepared solutions of elastomer in a volatile solvent and coated it onto textile fabrics for the purpose of waterproofing. Macintosh's firm in Manchester and Thomas Hancock's in London were the two major early manufactures. In the 1830s, the Americans Edwin Chaffee and Charles Goodyear developed the mill, calender, and vulcanization molding technologies. This brought companies such as Farrel Foundry and Machine and the Birmingham Iron Foundry (from 1927 combined into Farrel Birmingham later Farrel Corp.) into the business of manufacturing mills and calenders. This technology was exported to Europe. By the 1870s, the Harburger Eisen and Bronzewerke (from the 1960s part of Krupp) began manufacturing calender rolls. The introduction of the screw extruder for gutta percha and elastomer in the 1870s led to creation of new machinery firms, notably Francis Shaw and Company in Manchester, England, and John Royle in Paterson, New Jersey. Germany's elastomer industry became concentrated in the city of Hannover, with the major firm (from 1870) being Continental Gummiwerke. Two important firms concentrating on manufacture of elastomer processing equipment including extrusion and calendering developed in Hannover. These were Paul Troester Maschinenfabrik (in 1892) and Hermann Maschinenbau (in 1896), both created in the last decade of the 19th century.

The 20th century saw the rise of the tire industry and the development of organic accelerators. These were coupled with the demands of the automotive industry and led to the development of the internal mixer to replace the two-roll mill. Originally, Werner and Pfleiderer was the leading firm, but their unwillingness to prosecute the patent of Fernley H. Banbury led to the development of his internal mixer technology by the Birmingham Iron Foundry and, after 1927, by Farrel-Birmingham (now Farrel Corp.).

The factory system based on internal mixers, screw extruders, calenders, and vulcanization presses has remained basically unchanged in the past half-century. Internal mixers have had major improvements, e. g. , intermeshing rotors proposed by Francis Shaw and Company and Werner and Pfleiderer, and variable intermeshing clearance rotors proposed by Pomini-Farrel SpA. Sophisticated computer control systems have been introduced. The early single hot-feed extruders have been replaced by cold-feed extruders with increasingly sophisticated design including pin barrel extruders as well as complex control systems.

7.1.3 Flow Simulation of Processing

The origins of flow simulation of processing should probably he traced to the establishment of the Navier-Stokes equations and its early solutions. Reynolds' 1886 simulation of the flow of lubricating oil in bearings has had enormous influence on succeeding flow simulations of viscous fluids moving through small clearances. Specific studies relating to the flow of elastomer in processing operations and the implications of non-Newtonian flow behavior date to the 1930s with the work of Mooney and Dillon and Johnston or flow-through tubes. These papers though are closely related to viscometry and a better beginning may be Ardichvili's 1938 analyses of flow between and bending of calender rolls, followed by Gaskell's 1950 non-Newtonian flow modeling of the former problem.

In the 1950s considerable attention was given to simulation of the screw extrusion process by researchers with the Goodyear Tire and elastomer Company, DuPont, and Bayer AG. By the 1960s, rather sophisticated models of nonisothermal non-Newtonian flow in metering regions of extruder screws were published by Griffith and Zamodits and Pearson. Simulation of polymer processing in the 1960s was dominated by J. R. A. Pearson of Cambridge University whose early activities in screw extrusion we have just cited. Pearson showed in a series of papers published in the early 1960s how hydrodynamic lubrication theory may be applied to die design. He subsequently showed how membrane theory could be applied to simulate processing operations such as tubular film extrusion and blow molding.

In the 1970s, attention turned to simulation of injection molding of thermoplastics, with the first papers concerned with either isothermal mold filling or nonisothermal filling very simple molds. From the late 1970s, commercial computer software for simulating non-isothermal injection molding was developed by both C. Austin and his firm Moldfow Australia and by K. K. Wang and his coworkers at Cornell University.

The 1970s also saw the first application of finite-element computational techniques to polymer processing operations, with Tanner and his coworkers playing a key early role with applications to extudate swell and wire coating. Subsequently, commercial computer software based on finite-element analysis of Newtonian/non-Newtonian fluid mechanics was developed by various entrepreneurs, notable among which was M. Crochet of the University Louvain-le-Neuve and his Polyflow. Crochet reviewed the progress of finite-element analysis in solving viscoelastic fluid problems.

Until the 1980s, most simulations of polymer operations related to thermoplastics problems. The special processing machinery of the elastomer industry had receive little attention. It is only in the late 1980s that realistic simulations appeared for internal mixers and pin barrel extruders. These have used primarily lubrication theory-based simulations.

7.2　Basic Concepts of Mechanics

In this section we develop the basic ideas of classical rheological thought. We presume elastomers to deform as continuous media and to be subject to the formalism of continuum mechanics. We begin by developing the idea of the nature of applied forces and the stress tensor.

The idea of the stress tensor in a material arises from the necessity of representing the influence of applied forces on deformation. The applied forces F acting on a body may be represented as the sum of contact forces acting on the surface and body forces f, such as gravitation, which act directly on the elements of mass. We may write (Fig. 7.1)

$$\boldsymbol{F} = \sum_i \boldsymbol{t}_i \Delta a_i + \sum_t \boldsymbol{f}_j \Delta m_j = \oint \boldsymbol{t} \mathrm{d} a + \oint \rho \boldsymbol{f} \mathrm{d} V \qquad (7\text{-}1)$$

where t is the force per unit area (stress vector) acting on the surface area elements Δa_i, and the $\oint$ indicates the integration exists over the entire surface.

The idea of the stress tensor comes from relating t to the unit normal vector $\boldsymbol{n}$ to the surface through

$$\boldsymbol{t} = \boldsymbol{\sigma} \times \boldsymbol{n} \qquad (7\text{-}2)$$

where $\boldsymbol{\sigma}$ is an array of nine quantities

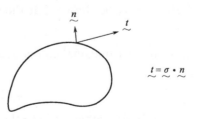

Fig. 7.1　Stress vector.

$$\boldsymbol{\sigma} = \begin{vmatrix} \sigma_{11} & \sigma_{12} & \sigma_{13} \\ \sigma_{21} & \sigma_{22} & \sigma_{23} \\ \sigma_{31} & \sigma_{32} & \sigma_{33} \end{vmatrix} \qquad (7\text{-}3)$$

known as the stress tensor or matrix. $\boldsymbol{\sigma}$ may be considered as a second-order tensor or a Gibbs dyadic

$$\boldsymbol{\sigma} = \sum_i \sum_j \sigma_{ij}\, \mathbf{e}_i \mathbf{e}_j \qquad (7\text{-}4)$$

The concepts of the stress vector and stress tensor were developed during the 1820s by Cauchy. The direction of the stress component is i. When the direction of the stress component i is perpendicular to the plane ($\sigma_{ij}\, i = j$), the stress is called the normal stress. When the direction i is tangent to the plane j ($\sigma_{ij}\, i \neq j$), the stress is called the shear stress.

Applying the divergence theorem to Eq. (7-1) gives

$$\oint t\, \mathrm{d}a = \oint \boldsymbol{\sigma} \cdot \mathbf{n}\, \mathrm{d}a = \int \nabla \cdot \boldsymbol{\sigma}\, \mathrm{d}V \qquad (7\text{-}5)$$

$$\boldsymbol{F} = \int [\nabla \cdot \boldsymbol{\sigma} + \rho \boldsymbol{f}]\, \mathrm{d}V \qquad (7\text{-}6)$$

where ∇ is the del operator

$$\nabla = \mathbf{e}_1 \frac{\partial}{\partial x_1} + \mathbf{e}_2 \frac{\partial}{\partial x_2} + \mathbf{e}_3 \frac{\partial}{\partial x_3} \qquad (7\text{-}7)$$

The complete dynamics of a deforming body requires including the contact forces, the body (gravitational) forces $\rho \boldsymbol{f}$ with inertial forces. For a macroscopic mass M

$$\boldsymbol{F} = \frac{\mathrm{d}}{\mathrm{d}t} \int \rho \boldsymbol{V}\, \mathrm{d}V = \int [\nabla \cdot \boldsymbol{\sigma} + \rho \boldsymbol{f}]\, \mathrm{d}V \qquad (7\text{-}8)$$

while for a macroscopic fixed-space "control volume" through which the mass may move

$$\int \frac{\partial}{\partial t}(\rho \boldsymbol{V})\, \mathrm{d}V + \oint \rho \boldsymbol{V}(\boldsymbol{V} \cdot \boldsymbol{n})\, \mathrm{d}a = \int [\nabla \cdot \boldsymbol{\sigma} + \rho \boldsymbol{f}]\, \mathrm{d}V \qquad (7\text{-}9)$$

It follows that at a point within the body

$$\rho \left[\frac{\partial \boldsymbol{V}}{\partial t} + (\boldsymbol{V} \cdot \nabla)\boldsymbol{V} \right] = \nabla \cdot \boldsymbol{\sigma} + \rho \boldsymbol{f} \qquad (7\text{-}10)$$

Eq. (7-10), which establishes the balance of forces at a point within the body, is known as Cauchy's law of motion. The components of the stress tensor are not independent of each other. By a balance of torques and angular moments similar to that leading to Eq. (7-10) it may be shown that the stress tensor is symmetric; i. e.,

$$\boldsymbol{\sigma} = \boldsymbol{\sigma}^T \quad \text{or} \quad \sigma_{ij} = \sigma_{ji} \qquad (7\text{-}11)$$

Thus, the off-diagonal components of Eq. (7-4), which are the shear stresses, are related through

$$\sigma_{12} = \sigma_{21}, \quad \sigma_{13} = \sigma_{31}, \quad \sigma_{23} = \sigma_{32}$$

The basic problem of rheology is the development of expressions for σ in terms of the deformation and kinematics of materials. The deformation behavior of continuous materials may then be determined through solutions of Eq. (7-10).

If a body is not subjected to applied forces, the stress components reduce to equal normal hydrostatic pressure components

$$\boldsymbol{\sigma} = -p\boldsymbol{I}, \quad \boldsymbol{I} = \begin{vmatrix} 1 & 0 & 0 \\ 0 & 1 & 0 \\ 0 & 0 & 1 \end{vmatrix} \tag{7-12}$$

where p is the pressure. More generally, when forces are applied, we may express the stress tensor in terms of the pressure and an extra stress tensor $\boldsymbol{P}$ through the relation

$$\boldsymbol{\sigma} = -p\boldsymbol{I} + \boldsymbol{P} \tag{7-13a}$$

Specifically for normal stresses

$$\sigma_{-ii} = -p + P_{-ii} \tag{7-13b}$$

and for shear stresses

$$\sigma_{ij} = P_{ij} \quad (i \neq j) \tag{7-13c}$$

and we need not distinguish between σ_{ij} and P_{ij} when $i \neq j$.

7.3 Rheological Properties

7.3.1 Gums

In this section we begin by reviewing experimental studies of the rheological behavior of unvulcanized elastomers and related materials. We then seek to correlate this behavior in terms of the theory of viscoelasticity. First, the linear theory of viscoelasticity in which there is broad consensus of agreement in the rheological community is discussed. We then describe the nonlinear theory, where the level of consensus is much less.

The experimental literature largely divides between studies of behavior in small strain and studies in steady shear flow. Much of the emphasis has been to relate such behavior to molecular structure.

7.3.1.1 Small-Strain Studies

During the 1940s and early 1950s, Leaderman, Tobolsky, Ferry, and others made extensive studies of the small-strain behavior of elastomers and related polymers. These studies involved creep (deformation under applied stress), stress relaxation following applied stresses, and imposed oscillatory strains. These and other experimental techniques used have been described in special detail in the monograph of Ferry. These studies showed that all of these deformations could be represented in terms of the superposition principle of Boltzmann [see Eq. (7-14)].

Tobolsky and his coworkers made extensive efforts to characterize the stress relaxation characteristics of elastomers, notably polyisobutylene. The stress would decay over time to zero at a rate dependent on temperature and molecular mass (Fig. 7.2). They expressed the relaxation through a series of exponentials or a spectrum of relaxation times. Consider the shear stress decay $\sigma(t)$ following a shear imposed strain γ_0. This may be used to define a shear relaxation modulus $G(t)$ through

$$\frac{\sigma(t)}{\gamma_0} = G(t) = \sum_{i=1}^{m} G_i e^{-t/\tau_i}$$

$$G(t) = \int_0^\infty H(\tau) e^{-t/\tau} \, d\ln\tau$$

(7-14)

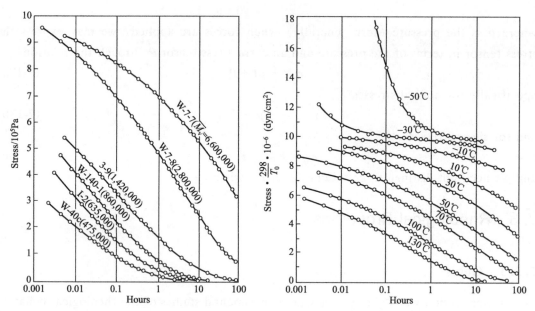

Fig 7.2 Stress relaxation from the measurements of Andrews et al.

Table 7.1 summarizes the G_i, τ_i relaxation times obtained for natural smoked sheet at 100℃.

Table 7.1 Linear Viscoelastic Representation for G_i of SMR5-NR (100℃)

i	G_i/Pa	τ_i/sec
m	2199	230
$m-1$	9166	37.7
$m-2$	33290	7.31
$m-3$	59200	0.925
$m-4$	59000	0.095

From Montes and White.

Tobolsky and his coworkers found they could study $G(t)$ over a very wide range of times by constructing "master curves" from data obtained at different temperatures. Specifically they found that:

$$G(t,T) = \frac{\rho T}{\rho_s T_s} G(t/\alpha_T)$$

(7-15)

which interrelates stress relaxation data at different temperatures T where α_T is a temperature-dependent shift factor applied to the time. Here ρ is density, and ρ_s and T_s are values at a standard temperature unique to each polymer.

The shift factor α_T may be represented as a universal function of $T - T_s$. The value of T_s is related to the glass transition temperature. This was established by Tobolsky and Ferry and their respective coworkers, but is associated primarily with Williams, Landel, and Ferry.

Tobolsky and his coworkers found that the $H(\tau)$ relaxation spectrum function for polyisobutylene may be represented by the combination of a wedge and box, i. e., as

$$H(\tau) = \frac{A}{\sqrt{\tau}} \quad \tau < \tau_1 \tag{7-16a}$$

$$H(\tau) = H_0 \quad \tau_2 < \tau < \tau_m \tag{7-16b}$$

H_0 and A were independent of molecular mass, but τ_m exhibited a strong dependence. Specifically

$$\tau_m = CM_w^{3.4} \tag{7-17}$$

Oscillatory strain experiments were also studied in this period, notably by Ferry and his coworkers. These were found to produce both "in-phase" and "out-of-phase" stress components. These were interpreted in terms of the expression

$$\sigma(t) = G'(\omega)\gamma(t) + \eta'(\omega)\frac{d\gamma}{dt} \tag{7-18}$$

where the term $\eta'(\omega)$ is known as the dynamic viscosity. We may also define a complex viscosity $\eta^*(\omega)$:

$$\eta^*(\omega) = \sqrt{[G'/\omega]^2 + (\eta')^2} \tag{7-19}$$

In later years, the $G(t)$ and $H(\tau)$ functions have been further investigated by various researchers including Ninomiya, Bogue et al., and notably Masuda et al. with respect to the influence of molecular mass distribution. The effect of the presence of high-molecular-weight polymer in a blend was to make $G(t)$ relax more slowly. The effect on $H(\tau)$ was to make $H(\tau)$ decrease more slowly at large τ. They found that at large times for narrow molecular mass distribution samples, the box of Eq. (7-16b) is replaced by a more complex function which includes a spike. Ninomiya and Bogue et al. seek to develop functional forms to represent the effect of the breadth of the molecular mass distribution of $H(\tau)$.

7.3.1.2 Shear Viscosity and Normal Stresses

Shear viscosity measurements on elastomers and polymer melts began with Mooney in 1936 and continued in the 1940s. Early measurements focused on molecular mass dependence and the dependence of the shear viscosity on shear rate. The shear viscosity was found to be a decreasing function of shear rate (Fig. 7.3) but to have an asymptotic viscosity, η_0, at lower shear rates. Fox and Flory found the zero shear viscosity of η_0 of linear polymer chains to depend on the 3.4 power of molecular mass. This was subsequently confirmed by various researchers. It was at first found that the zero shear viscosity for branched polymers was lower than that for linear polymers of the same molecular mass; however, subsequently Kraus and Gruver showed that above a molecular mass, characteristic of the molecular topology, η_0 for carefully prepared cruciform and Y-shaped polymers actually increases more rapidly perhaps with the sixth power.

The viscosity-shear rate dependence at higher shear weight also has a dependence on molecular mass distribution. In the mid 1960s, Vinogradov and Malkin found that plots of η/η_0 versus $\eta_0\dot{\gamma}$. were independent of temperature for all polymers. They also argued that this was a universal plot valid for all polymers. This was, however, found not to be the case. In succeeding years, various laboratories have found that such plots are strongly dependent on the molecular mass distribution. η/η_0 falls off more rapidly as molecular mass distribution narrows. Yamane and White correlated data on a wide range of polymers including polystyrene, polypropylene, and polybutene-1 to produce a master plot, which allows prediction of breadth of molecular mass distribution for viscosity-shear rate data.

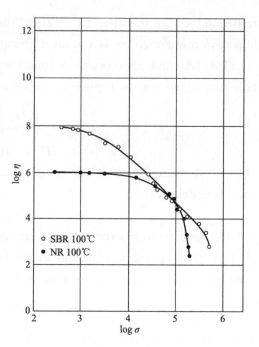

Fig. 7.3 Shear viscosity of elastomer gums: SMR natural rubber and SBR (Bridgestone-Firestone Duradene 706) as a function of shear stress.

Correlations of the type described in the preceding paragraph are limited to linear hompolymers and have no validity for long-chain branched polymers such as low-density polyethylene. Cox and Merz have made the remarkable experimental observation that the non-Newtonian shear viscosity function of flexible chain polymers has the same form as the complex viscosity-frequency function, i. e. ,

$$\eta(\dot{\gamma}) = \eta^*(\omega) \tag{7-20}$$

Normal stresses originally recognized by Weissenberg through observation of rod climbing effects in soap-hydrocarbon liquid suspensions and polymer solution systems began to be measured on thermoplastics in the 1960s. White and Tokita noted their occurrence in gum elastomers sheared in a Mooney disk rheometer. These were recognized by Weissenberg from the beginning to be associated with melt elasticity and to be sensitive to breadth of molecular mass distribution. White and Kondo addressed the dependence of the low-shear-rate principal normal stress difference coefficient Y1 and its dependence on molecular mass. They argued that the dependence should be of order 6.5 to 7.0 which was consistent with their own experiments and other results in the literature. In a 1978 paper, Oda, White, and Clark correlated normal stresses for polystyrenes with varying molecular mass distribution with shear stress. It was found that the principal normal stress difference N_1, when treated as a function of shear stress, was independent of temperature:

$$N_1 = A\sigma_{12}^a \tag{7-21}$$

where N_1 was larger at the same σ_{12} for broad molecular mass distributions but increased less rapidly with shear stress (i. e. , A was large and a was smaller).

7.3.1.3 Theory of Linear Viscoelasticity

By the late 1940s, experimental researchers led by Leaderman and Tobolsky found that the small-strain properties of elastomers and thermoplastic melts were described by the theory of linear viscoelasticity developed by Boltzmann in the 1870s. This view is made clear in the monographs of Leaderman and Mark and Tobolsky, which were published in this period. In the next two decades, this subject was developed in monographs by Tobolsky and, later, Ferry.

Boltzmann's formulation is based on linear materials with incomplete memory. Consider a material which when subjected to strain ε at time zero exhibits a stress.

$$\sigma = G(t)\gamma \tag{7-22}$$

where $G(t)$ is a modulus function that decays with time. If the response of this material is linear, the stress σ developed for a series of sequential deformations $\gamma_1, \gamma_2, \gamma_3, \cdots$ at times $t_1, t_2, t_3, \cdots$ will be at time t

$$\sigma = G(t-t_1)\gamma_1 + G(t-t_2)\gamma_2 + G(t-t_3)\gamma_3 + \cdots \tag{7-23}$$

$$= \int_{-\infty}^{t} G(t-s)\,d\gamma(s) = \int_{-\infty}^{t} G(t-s)\frac{d\gamma}{ds}ds \tag{7-24}$$

where the integral, refers to the limit for which the applied deformation in continuous. Eq. (7-21) is known as Boltzmann's superposition integral. If a constant shear rate is imposed at time zero, the stress will be given by

$$\sigma = \left[\int_0^t G(s)\,ds\right]\dot{\gamma} \tag{7-25}$$

For long times the integral corresponds to the zero shear viscosity η_0, i.e.,

$$\eta_0 = \int_0^\infty G(s)\,ds \tag{7-26}$$

Another deformation of interest is to impose a sinusoidal shear strain $\gamma_0 \sin\omega t$. The stress σ is now found to be expressible as

$$\sigma = \omega\left[\int_0^\infty G(s)\sin\omega s\,ds\right]\gamma_0\sin\omega t + \omega\left[\int_0^\infty G(s)\cos\omega s\,ds\right]\gamma_0\cos\omega t \tag{7-27}$$

i.e., there are in-phase and out-of-phase components [compare Eq. (7-18)]. Eq. (7-26) is often rewritten

$$\sigma(t) = G'(\omega)\gamma_0\sin\omega t + G''(\omega)\gamma_0\cos\omega t \tag{7-28}$$

where

$$G'(\omega) = \omega\int_0^\infty G(s)\sin\omega s\,ds \tag{7-29}$$

$$G''(\omega) = \omega\int_0^\infty G(s)\cos\omega s\,ds \tag{7-30}$$

$G'(\omega)$ is known as the storage modulus and $G''(\omega)$ as the loss modulus. Eq. (7-27) is often rewritten in the form

$$\sigma(t) = G^*(\omega)\sin(\omega t + \delta) \tag{7-31}$$

where δ is a phase angle and $G^*(\omega)$ is known as the complex modulus. Clearly,

$$G'(\omega) = G^*(\omega)\cos\delta \tag{7-32a}$$

$$G''(\omega) = G^*(\omega)\sin\delta \tag{7-32b}$$

which leads to

$$G^* = \sqrt{(G')^2 + (G'')^2} \tag{7-33}$$

$$\tan\delta = G''/G' \tag{7-34}$$

The quantity $\tan\delta$ is known as the loss tangent.

The "loss modulus" and "loss tangent" expressions for the terms $G''(\omega)$ and $\tan\delta$ suggest that the $G''(\omega)$ term in Eq. (7-27) is associated with a dissipative process. This should be apparent from the input $\gamma_0 \sin\omega t$ and the output stress component in $\cos\omega t$ of Eq. (7-18) where a dynamic viscosity $\eta'(\omega)$ is introduced

$$\eta' = \frac{G''(\omega)}{\omega} = \int_0^\infty G(s)\cos\omega s\,ds \tag{7-35}$$

It has become customary to represent $G(t)$ as a series or continuous spectrum of exponential terms as expressed in Eq. (7-14). Historically the reason for representations using exponentials is that a single exponential term represents the form of a model that had been proposed by Maxwell in the 1860s prior to the publication of Boltzmann. This model has the form of a differential equation which is equivalent to

$$\frac{d\sigma}{dt} = G\frac{d\gamma}{dt} - \frac{1}{\tau}\sigma \tag{7-36}$$

The solution of Eq. (7-36) is

$$\sigma = \int_0^\infty G e^{-(t-s)/\tau} \gamma(s)\,ds \tag{7-37}$$

This suggests that $G(t)$ could he expressed as a sum of exponentials, i. e., as presented in Eq. (7-14). The use of Eq. (7-14) for $G(t)$ leads to a zero shear viscosity of

$$\eta_0 = \int_0^\infty G(s)\,ds = \Sigma G_i \tau_i = \int_0^\infty H(\tau)\,d\tau \tag{7-38}$$

For sinusoidal oscillations, the storage and loss moduli are found to be

$$G'(\omega) = \Sigma \frac{\omega^2 \tau_i^2 G_i}{1+\omega^2 \tau_i^2} \tag{7-39a}$$

$$G''(\omega) = \Sigma \frac{\omega \tau_i G_i}{1+\omega^2 \tau_i^2} \tag{7-39b}$$

and the dynamic viscosity $\eta'(\omega)$ is

$$\eta'(\omega) = \Sigma \frac{\tau_i G_i}{1+\omega^2 \tau_i^2} \tag{7-39c}$$

We also define $\eta^*(\omega)$, the complex viscosity.

$$\eta^*(\omega) = \frac{1}{\omega}[(G'(\omega))^2 + (G''(\omega))^2]^{1/2} \tag{7-39d}$$

When $\omega \to 0$, $\eta'(\omega)$, and $\eta^*(\omega)$ have the same form as the steady shear viscosity and

$$\eta_0 = \eta(0) = \eta'(0) = \eta^*(0) \tag{7-40}$$

The formulation described above is one dimensional and expressed in terms of a shear stress. It is possible to obtain a three-dimensional representation. Indeed this was done in the original paper of Boltzmann. Eq. (7-21) may be written in terms of a stress tensor σ and pressure p as

$$\boldsymbol{\sigma} = -p\boldsymbol{I} + 2\int_{-\infty}^{t} G(t-s)\,\boldsymbol{d}(s)\mathrm{d}s \tag{7-41}$$

with

$$\boldsymbol{d} = \frac{1}{2}[\nabla \boldsymbol{v} + (\nabla \boldsymbol{v})^T] \tag{7-42}$$

$\boldsymbol{d}$ is known as the rate of deformation tensor and $\boldsymbol{I}$ a unit tensor. This may be shown by "integration by parts" to be equivalent to

$$\boldsymbol{\sigma} = -p\boldsymbol{I} + 2\int_{-\infty}^{t} \Phi(t-s)\gamma(s)\mathrm{d}s \tag{7-43}$$

where

$$\Phi(t) = -\frac{\mathrm{d}G(t)}{\mathrm{d}t} \quad \text{and} \quad G(\infty) = 0 \tag{7-44a}$$

We may write $\Phi(t)$ as

$$\Phi(t) = \sum \frac{G_i}{\tau_i} e^{-t/\tau_i} \tag{7-44b}$$

and γ is the infinitesimal strain tensor measure from the instantaneous state, i. e.,

$$\boldsymbol{\sigma} = -p\boldsymbol{I} + 2\int_{0}^{\infty} \Phi(z)\gamma(z)\mathrm{d}z \tag{7-45}$$

7.3.1.4 Theory of Nonlinear Viscoelasticity

A proper formulation of a special theory of nonlinear viscoelasticity was developed by Zaremba in 1903, roughly 30 years after the classic paper of Boltzmann; however the modern period, when there was general acceptance of such formulations, begins with the work of Oldroyd in 1950. In the 1950s, various special theories of nonlinear viscoelastic behavior were developed by Oldroyd, deWitt, Rivlin and Ericksen, and Noll. Lodge and Yamamoto sought to develop non-linear theory using molecular arguments. A totally general formulation of large strain viscoelastic behavior was presented by Green and Rivlin and Noll at the end of the decade. This general formulation was then applied to model various special types such as viscometric laminar shear flows and wave propagation.

Although consensus had rapidly developed around the Boltzmann formulation of linear viscoelasticity, this was not to be the case for the specific forms of the nonlinear theory to use. The general formulation of nonlinear theory devised notably by Green, Rivlin, Noll, Coleman, Ericksen, and Truesdell has not been further developed by newer generations of researchers. Rather, since the mid 1960s, it has declined in influence. From 1960 on, we have increasing numbers of new theories being published. Some of these theories are phenomenological, others based to varying extents on molecular models.

From the early 1970s, powerful numerical tools and high-capacity computers have become increasingly available. In Newtonian fluid dynamics, this led to great increases in our knowledge. For nonlinear viscoelastic fluids where there has been no consensus on constitutive equations, we have instead simulations being carried out for many different models, with a maze of differing predictions of flow behavior and numerical instabilities associated with particular models. By the 1990s, nonlinear viscoelasticity theory and fluid mechanics had taken on many aspects of farce as opposed to science.

It is, however, necessary that we try to weave some rational web that allows us to describe a rational nonlinear theory. The general formulation of Green and Rivlin took the position that the stress tensor is a general hereditary function of the strain history. The stress tensor was expressed as infinite series of integrals:

$$\boldsymbol{\sigma} = -p\boldsymbol{I} + \int_0^\infty \Phi(z)\boldsymbol{e}(z)\mathrm{d}z + \int_0^\infty \int_0^\infty [\Psi(z_1,z_2)\boldsymbol{e}(z_1)\boldsymbol{e}(z_2) \\ + \Sigma(z_1,z_2)[\mathrm{tr}\,\boldsymbol{e}(z_1)]\boldsymbol{e}(z_2)]\mathrm{d}z_1\mathrm{d}z_2 + \cdots \quad (7\text{-}46)$$

where $e(z)$ is a nonlinear strain measure and $\Phi(z)$, $\Psi(z_1, z_2)$, and $\Sigma(z_1, z_2)$ are memory functions. Here z is a time measured backward from the present. The strains e are also measured backward in time. $\Phi(z)$ is the linear viscoelasticity relaxation function of Eq. (7-45). The experiments of Zapas and Craft, among others, indicate that there is only one stress relaxation function. White and Tokita show that this is equivalent to reducing Eq. (7-46) to

$$\boldsymbol{\sigma} = -p\boldsymbol{I} + \int_0^\infty m(z)[\boldsymbol{c}^{-1} - \kappa\boldsymbol{c}]\mathrm{d}z \quad (7\text{-}47)$$

where $m(z)$ depends on the invariants of $\boldsymbol{c}^{-1}$, the Finger deformation, tensor, or $\boldsymbol{c}$, the Cauchy deformation tensor. This type of constitutive equation was arrived at by various authors including Bernstein, Kearsley, and Zapas and Wagner, among others, using different reasoning. Other authors arrived at similar constitutive forms but used different invariants such as those of the rate of deformation tensor. The general form of Eq. (7-47) is often referred to as a BKZ fluid (for Bernstein, Kearsley, and Zapas, who were the earliest proponents).

One particular form of Eq. (7-47) which has been successfully used for representing the nonlinear viscoelastic behavior of elastomers is due to Bogue and coworkers. Both Middleman and Montes and White have used the form

$$m(z) = \sum_i \frac{G_i}{\tau_{i\mathrm{eff}}} e^{-z/\tau_{i\mathrm{eff}}} \quad (7\text{-}48)$$

with

$$\tau_{i\mathrm{eff}} = \frac{\tau_i}{1 + a\tau_i \overline{\Pi_d}^{1/2}} \quad (7\text{-}49\mathrm{a})$$

$$\overline{\Pi_d^{1/2}} = \frac{1}{t} \int_0^t \Pi_d^{1/2} \mathrm{d}z \quad (7\text{-}49\mathrm{b})$$

$$\Pi_d = 2\,\mathrm{tr}\,d^2 = 2d_{ij}d_{ij} \quad (7\text{-}49\mathrm{c})$$

d is the rate of deformation tensor of Eq. (7-43) for gum elastomers. In the work of the latter authors the linear viscoelastic behavior through $G(t)$ for an SMR-5 natural rubber sample at 100°C was fit with the parameters given earlier in Table 7.1. The parameter a of Eq. (7-49a) was taken as 0.7. A comparison of Eq. (7-47)—Eq. (7-49) with steady-state shear viscosity data, transient shear viscosity data, transient shear viscosity at the startup of flow, and stress relaxation following flow is given in Fig. 7.4. The agreement is quite good. Criminale, Ericksen, and Filbey (see also White) have shown that for long duration shearing flows the general constitutive equations of form Eqs. (7-46) and (7-47) reduce to the form

$$\boldsymbol{\sigma} = -p\boldsymbol{I} + 2\eta\boldsymbol{d} + 4\Psi_2\boldsymbol{d}^2 - \Psi_1 \frac{\delta\boldsymbol{d}}{\delta t} \qquad (7\text{-}50\text{a})$$

where

$$\frac{\delta\boldsymbol{d}}{dt} = \frac{\partial}{\partial t}\boldsymbol{d} + (\boldsymbol{V}\cdot\nabla)\boldsymbol{d} - \nabla\boldsymbol{v}\cdot\boldsymbol{d} - \boldsymbol{d}\cdot\nabla\boldsymbol{V} \qquad (7\text{-}50\text{b})$$

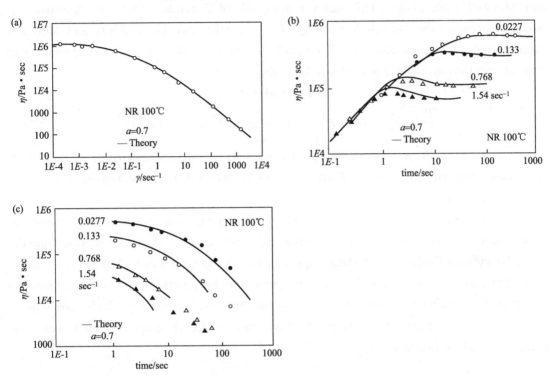

Fig. 7.4 Comparison of rheological model of Eq. (7-47)~Eq. (7-49) with experiment for natural rubber: (a) Steady-state shear viscosity; (b) Transient shear viscosity at beginning of flow; (c) Stress, relaxation following now.

Here η is the shear viscosity, and Ψ_1 and Ψ_2 are the first and second normal stress difference coefficients. These are functions of tr $\boldsymbol{d}^2$, the trace or sum of the diagonal components of $\boldsymbol{d}\cdot\boldsymbol{d}$ [see Eq. (7-49c)],

$$\eta = \eta(\operatorname{tr}\boldsymbol{d}^2) \; \Psi_1 = \Psi_1(\operatorname{tr}\boldsymbol{d}^2) \; \Psi_2 = \Psi_2(\operatorname{tr}\boldsymbol{d}^2) \qquad (7\text{-}51)$$

At low values of $\boldsymbol{d}$ and tr $\boldsymbol{d}^2$, the shear viscosity of Eq. (7-51) goes to the linear viscoelastic value, i.e.,

$$\eta \to \int_0^\infty G(s)\,ds \qquad (7\text{-}52\text{a})$$

and the quantity Ψ_1 goes to its first moment:

$$\Psi_1 \to 2\int_0^\infty sG(s)\,ds \qquad (7\text{-}52\text{b})$$

For a simple shear flow, i.e.,

$$v_1 = \dot{\gamma}x_2\boldsymbol{e}_1 + 0\boldsymbol{e}_2 + 0\boldsymbol{e}_3 \qquad (7\text{-}53)$$

Eq. (7-50) predicts the stress field, i.e.,

$$\sigma_{12}=\eta\dot{\gamma} \quad \sigma_{11}-\sigma_{22}=\Psi_1\dot{\gamma}^2, \quad \sigma_{22}-\sigma_{33}=\Psi_2\dot{\gamma}^2 \tag{7-54}$$

where η, Ψ_1, and Ψ_2 are dependent on $\dot{\gamma}^2$. Thus both shearing stresses and normal stresses are predicted. The results of this paragraph are important because they represent the basis of interpretation of shear flow viscometry of viscoelastic polymer melts and elastomers.

There is a major literature associated with a second class of theories based on generalizing Maxwell's Eq. (7-36). This was the approach of Zaremba, Oldroyd, deWitt, and Noll and predates the formulation of Eq. (7-46). Further papers by White and Metzner, Johnson and Segalman, and Phan-Tien and Tanner have developed this approach. This class of constitutive equations is generally of the form

$$\boldsymbol{\sigma}=-p\boldsymbol{I}+\boldsymbol{P}$$

$$\frac{\delta \boldsymbol{P}}{\delta t}+\sum_j a_j(\nabla \boldsymbol{v},\boldsymbol{P})+b(\boldsymbol{P})=2\eta(\operatorname{tr}\boldsymbol{d}^2)\boldsymbol{d}+c(\operatorname{tr}\boldsymbol{d}^2)\frac{\delta \boldsymbol{d}}{\delta t} \tag{7-55}$$

$\boldsymbol{P}$ is an extra stress tensor and a_j, b, h, and c are functions of $\operatorname{tr}\boldsymbol{d}^2$ as indicated. These formulations are much more arbitrary in character than the approach of Eq. (7-46) and Eq. (7-47).

The formulations of Eq. (7-47) and Eq. (7-55) have been criticized by Leonov, among others, as not being tested for consistency with the second law of thermodynamics. For Newtonian fluids, such testing requires a positive shear viscosity. For a linear viscoelastic material, one may show that the relaxation modulus function must be always positive to satisfy the second law. The requirements for Eq. (7-47) and Eq. (7-55) are not so clear. Leonov has sought to develop nonlinear viscoelastic rheological models based on thermodynamic arguments.

7.3.2 Compounds

Although experimental studies of the flow of elastomer compounds date to the 1930s, it was only in the late 1980s that there were extensive experimental studies. Reasonable rheological models also date to the 1980s; however, there have been relatively few investigations in this area and little consensus has been reached. In this section we first review the experimental literature. We then turn to rheological models that have been used. The central thrust of the rheological data is the existence of yield values and thixotropic characteristics in elastomer compounds.

7.3.2.1 Experimental

We begin this section by first describing steady-state shear flow behavior. We then turn to the discussion of time-dependent characteristics. In 1931, Scott on the basis of compression plastomer studies suggested that elastomer-particulate compounds at high loadings exhibited yield values; further, he suggested the expression

$$\sigma=Y+K\dot{\gamma}^n \tag{7-56}$$

to represent the shear stress-shear rate behavior at shear stresses higher than yield value

Y. This view was supported by a subsequent study of Dillon and Johnston using a capillary rheometer. Dillon and Johnston also used Eq. (7-56). Eq. (7-56) is usually named for Herschel and Bulkley, who used (or rather misused) this formulation earlier for polymer solutions. Scott-Herschel Bulkley is a better name for Eq. (7-56).

There was little discussion of this perspective during the next 25 years, and only in the 1960s was there renewed attention. In 1962, Zakharenko et al. in Moscow reported shear flow measurements of elastomer-carbon black compounds. In 1972, Vinogradov et al., also in Moscow, reported similar results for other elastomer-carbon black compounds and indicated the occurence of yield values. At the same time similar behavior was reported for talc-polypropylene compounds by Chapman and Lee of Shell and for titanium dioxide-polyethylene compounds by Minagawa and White.

From about 1980, there have been extensive investigations of the shear viscosity of elastomer-carbon black compounds and related filled polymer melts. Yield values in polystyrene-carbon black compounds in shear flow were found by Lobe and White in 1979 and by Tanaka and White in 1980 for polystyrene with calcium carbonate and titanium dioxide as well as carbon black. From 1982, White and coworkers found yield values in compounds containing butadiene-styrene copolymer, polyisoprene, polychloroprene, and ethylene-propylene terpolymer. Typical shear viscosity-shear stress data for elastomer-carbon black compounds are shown in Fig. 7.5(a) and (b). White et al. fit these data with both Eq. (7-56) and the expression

$$\sigma = Y + \frac{A}{1 + B\dot{\gamma}^{1-n}}\dot{\gamma} \tag{7-57}$$

Generally it was found that 0.15 volume fraction small particles seem to be required to produce yield values. The yield value Y and viscosity level increase with both increasing particle loading [Fig. 7.5(a), (b)] and decreasing particle size [Fig. 7.5(c)]. Note: Y is largest for the ISAF compounds and lower for the FEF. It is lowest for the SRF compounds. This is the same order as the particle size (ISAF has the smallest particles). Larger yield and viscosities are also found for calcium carbonate compounds. For elastomer compounds the yield value is of order 50 to 100 kPa and the power law exponent n about 0.2. In Table 7.2, we present values Y, A, B, K, and n of Eq. (7-56) and Eq. (7-57).

Osanaiye et al. made extensive measurements of creep in elastomer-carbon black compounds at very low stresses. It was found that there were stresses below which there was no flow. The yield values determined by these authors were somewhat lower than those reported earlier. The results of Osanaiye et al. were subsequently confirmed for various polymer melt/elastomer-particle compounds.

Studies of transient behavior of elastomer-carbon black compounds were first reported by Mullins and Whorlow in 1950. They found that there were strong time-dependent thixotropic effects in elastomer-carbon black compounds that were not found in gum elastomers. If one (1) sheared a gum elastomer, time (2) then halted the flow and (3) started up the flow again, within a short period the material would remember its earlier deformation

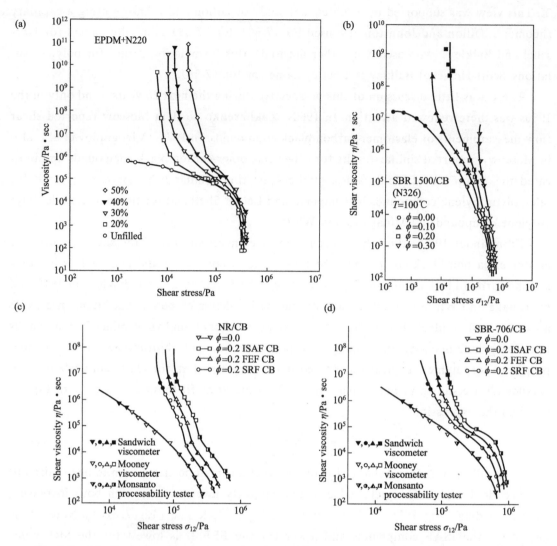

Fig. 7.5 (a) Shear viscosity-stress rate data for EPDM elastomer compound. Percentages are for volume percent. (b) Shear viscosity-shear rate for SBR elastomer compound. ϕ is volume fraction. (c) and (d) Shear viscosity-shear stress rate showing effect of particle size.

Table 7.2 Parameters Y, A, B, K, and n of the Scott-Herschel-Bulkley [Eq. (7-56)] and White et al. [Eq. (7-57)] Models for Elastomer Compounds[a]

Material	Carbon black type	Volume fraction	Y/kPa	$K=A/B$ /kPa·sec^n	A/kPa·sec	B/sec^{1-n}	n
NR	N 220	0.2	119	134.5	73	0.54	0.21
	N 550	0.2	79	105	97	0.93	0.20
	N 762	0.2	62	89	1.268	1.43	0.20
SBR	N 220	0.2	130	200	1.770	8.85	0.225
	N 550	0.2	82	187	7.130	3.82	0.21
	N 762	0.2	62	140	12.970	92.65	0.255

Continued

Material	Carbon black type	Volume fraction	Y/kPa	$K=A/B$ /kPa·$\sec^n$	A/kPa·sec	B/$\sec^{1-n}$	n
NR	N 330	0.2	63	103	587	5.68	0.175
	N 330	0.3	138	104	330	3.17	0.18
	N 326	0.2	65	108	4.500	41.5	0.2
SBR-1500	N 330	0.2	60.3	132	704	5.32	0.22
	N 330	0.3	158	141	1.030	7.31	0.19
	N 326	0.2	13	142	586	41.33	0.195
CR	N 330	0.2	56	113	178	15.8	0.22
EPDM	N 330	0.2	57.5	176	500	31.3	0.17

^{a} From shin et al.

history and respond with stresses lower than it had when it was sheared in the virgin state. If the period of rest was more than 1 minute, all previous history would be forgotten, and the stress response of the gum would be as a virgin material. This was not found in carbon black compounds. Following flow, at first, as with gums, their stress responses would be milder. Again, but now requiring a much longer time, the compounds would gradually gain back their apparent virgin response. However, unlike gums, the compounds would continue to build up a "structure" indefinitely. The longer the storage, the higher the stress transients required to break up the structure and induce steady shear flow. This same problem has been investigated by Montes et al. Typical examples of storage effects on stress transients at the beginning of flow are shown in Fig. 7.6. As the storage time increases, the stress overshoot transients become larger and larger. With sufficient storage times they exceed the initial sample.

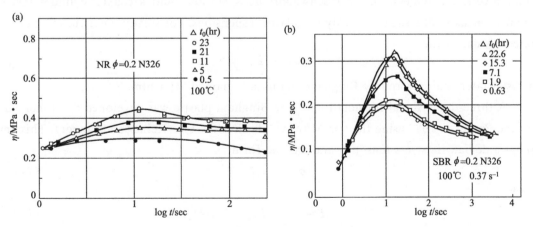

Fig. 7.6 Storage effects on stress transients at beginning of flow. (a) Transient shear viscosity of NR with 0.2 volume fraction N 326 black, $T=100℃$. (b) Transient shear viscosity of SBR with 0.2 volume fraction N 326 black, $T=300℃$.

Lobe and White studied stress relaxation following imposed strains in polystyrene-carbon black compounds and found that the stresses did not decay to zero, but to a finite value of stress roughly equal to the yield value of Eq. (7-56) and Eq. (7-57). Montes et al. have found similar effects in elastomer-carbon black compounds. This is shown in Fig. 7.7. Montes et al. found similar effects in stress relaxation following shear flow.

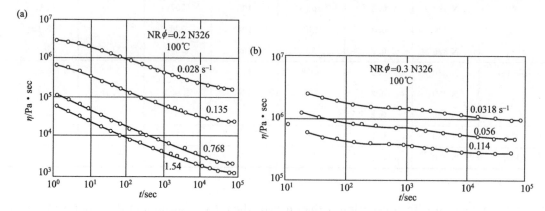

Fig. 7.7 Stress relaxation in natural rubber-carbon black compound following flow. (a) NR $\phi = 0.2$. (b) NR $\phi = 0.3$.

There have been many studies of the dynamic viscosity $\eta'(\omega)$ and complex viscosity $\eta^*(\omega)$ of elastomer-carbon black compounds and other filled systems. Nakajima et al. and later researchers found that the Cox-Merz rule mentioned earlier for flexible chain polymer melts is not valid for filled polymer melts or elastomer. Generally, one has the inequality $\eta^*(\omega) > \eta(\dot{\gamma})$. The complex viscosity is considerably larger.

7.3.2.2 Theory of Plastic Viscous Fluids

The theory of flow of materials with yield values dates to the work of Schwedoff of the University of Odessa in 1890. Schwedoff designed and built a coaxial cylinder instrument to measure the shear viscosity of gelatin suspensions. For steady shear flows he correlated his data with

$$\sigma = Y + \eta_B \dot{\gamma} \tag{7-58}$$

a result later rediscovered by Bingham and usually named after him.

A three-dimensional formulation of the Bingham plastic was developed by Hohenemser and Prager in 1932 using the von Mises yield criterion (see Reiner and Prager). This employs a deviatoric stress tensor T and has the form

$$\boldsymbol{\sigma} = -p_1 \boldsymbol{I} + \boldsymbol{T} \tag{7-59a}$$

with

$$\operatorname{tr} \boldsymbol{T} = T_{11} + T_{22} + T_{33} = 0 \tag{7-59b}$$

Accepting the von Mises yield surface

$$\operatorname{tr} \boldsymbol{T}^2 = 2Y^2 \tag{7-60}$$

T has the form above the yield stress defined by Eq. (7-60):

$$\boldsymbol{T} = \frac{Y}{\sqrt{\frac{1}{2} \operatorname{tr} \boldsymbol{T}^2}} \boldsymbol{T} + 2\eta_B \boldsymbol{d} \tag{7-61}$$

tr T^2 is the trace of the scalar product of $T \cdot T$, and d is the rate of deformation tensor of Eq. (7-29b). Oldroyd has shown that Eq. (7-49) may be written in the simpler form

$$T = \frac{2Y}{\sqrt{2 \text{ tr } d^2}} d + 2\eta_B d \tag{7-62}$$

Oldroyd first envisaged viscous fluids with yield values and non-Newtonian viscosities. He wrote

$$T = \frac{2Y}{\sqrt{2 \text{ tr } d^2}} d + 2\eta_B (\text{tr } d^2) d \tag{7-63}$$

where η_B depends on tr d^2. The Scott-Herschel-Bulkley equation, Eq. (7-56), may be expressed

$$T = \frac{2Y}{\sqrt{2 \text{ tr } d^2}} d + 2K [2 \text{ tr } d^2]^{(n-1)/2} d \tag{7-64}$$

and Eq. (7-57) as

$$T = \frac{2Y}{\sqrt{2 \text{ tr } d^2}} d + \left[\frac{A}{1 + B [2 \text{ tr } d^2]^{(1-n)/2}} \right] d \tag{7-65}$$

Parameters for the constitutive equations of Eq. (7-64) and Eq. (7-65) for elastomer compounds are contained in Table 7.2.

7.3.2.3 Plastic Viscoelastic Fluid Models

The concept of a material with a yield value that behaved in a differentially viscoelastic manner above the yield value was also due to Schwedoff. He reported transient experiments and concluded that the gelatin material responded as a viscoelastic material above its yield value. This led him to modify the one-dimensional Maxwell model into the form

$$\frac{d\sigma}{dt} = G \frac{d\gamma}{dt} - \frac{t}{\tau} (\sigma - Y) \tag{7-66}$$

and proposed it would represent the rheological properties of gelatin suspensions.

There was no follow-up to the work of Schwedoff as there had been to Maxwell. During most of the century following Schwedoff's paper, his work was remembered only through the monographs of Reiner. It is only with the publication of Hohenemser and Prager and more especially White and in 1979 that the approach initiated by Schwedoff was revived. Specifically, it was pointed out that Eq. (7-66) is a differential equation whose solution is

$$\sigma = Y + \int_0^\infty G e^{-(t-s)/\tau} \frac{d\gamma}{ds}(s) ds \tag{7-67}$$

This could be considered a special case of a more general form:

$$\sigma = Y + \int_0^\infty G(t-s) \frac{d\gamma}{ds} ds \tag{7-68}$$

Using the analog of Eq. (7-61), White proposed that for a material responding as a viscoelastic material above the yield value,

$$T = \frac{Y}{\sqrt{\frac{1}{2} \operatorname{tr} T^2}} T + H \qquad (7\text{-}69)$$

where H is a memory functional. This was shown to be equivalent to

$$T = \frac{2Y}{\sqrt{2 \operatorname{tr} H^2}} H + H \qquad (7\text{-}70)$$

Three-dimensional formulations for T including nonlinear viscoelastic integral forms are contained in the papers of White, White and Tanaka, and White and Lobe. These all have the general form of Eq. (7-70) with [compare Eq. (7-47)]

$$H = \int_0^\infty m(z) \left[c - \frac{1}{3}(\operatorname{tr} c^{-1}) I \right] dz \qquad (7\text{-}71)$$

where $m(t)$ is a memory function. Isayev and Fan have also proposed a formulation of the type of Eq. (7-70), but using the Leonov model as the basis of H.

7.3.2.4 Thixotropic Plastic Viscoelastic Fluid

The problem of constitutive equations involving thixotropy is an old one. In the 1930s, Freundlich and Jones, among others, associated thixotropy in suspensions of small particles with the occurrence of yield values.

The formulations described in the previous section show yield values coupled with viscous and viscoelastic flow, but not with thixotropy. It is possible to introduce thixotropy into plastic fluid behavior by allowing Y to depend on the history of $\operatorname{tr} d^2$. This approach was applied to fluids with viscoelastic response above the yield value by Suetsugu and White and later by Montes and White.

Alternate approaches to represent thixotropy rates of structural break-down and reformation may be developed involving the stresses, deformation rate, and internal structure. Linear theories of this type date to Goodeve and Whitfield.

The study of Montes and White, compares its predictions with the rheological behavior of elastomer compounds and we cite it here. Montes and White write the yield function [compare Eq. (7-49c)]

$$Y(\pi_d, t) = Y_f + (\beta_1 \Pi_d^{1/2} + \beta_2 \Pi_d) \int_{-\infty}^t \alpha \Pi_d^{1/2} e^{-a \Pi_d^{1/2}} (t-s) ds \qquad (7\text{-}72)$$

$m(z)$ is taken to have same form as Eqs. (7-48a) and (7-49). The modulus G_i was taken to have the same value as G_i^0, the value for the gum, and τ_i was taken to be $f(\phi) \tau_i^0$, where τ_i^0 is the value for the gum and $f(\phi)$ the magnitude of the increases of τ_i^0 for a non-interactive filler.

The formulation of Eq. (7-70) $\sim$ Eq. (7-72) was compared with experiment for an SMR-5 nature elastomer with 0.2 volume fraction N 326 carbon black. The steady-state viscosity is compared with experiment in Fig. 7.8(a), and the transient shear stress buildup at the start of flow is considered in Fig. 7.8(b). The shear flow-rest-shear flow data are contained in Fig. 7.8(c). The agreement is, in general, good.

An alternate formulation rather different from that of this section has been given by

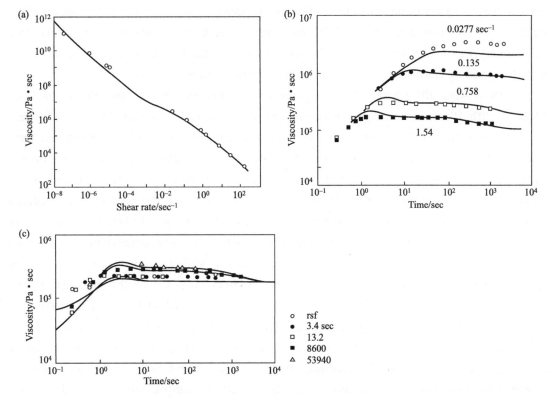

Fig. 7.8 Comparison of rheological model of Eq. (7-70)~Eq. (7-72) with experiment on elastomer-carbon black compound: (a) Steady shear viscosity; (b) Transient; (c) Shear-rest-shear flow behavior.

Leonov. It also combines thixotropic behavior and a yield value and is based on thermodynamic arguments.

7.4 Boundary Conditions

It is now well accepted that Newtonian liquids adhere to solid surfaces as pointed out by Goldstein. It was a very controversial subject as late as the 1840s, as may be seen by reading the writings of Stokes in this period.

There is a long history of concerns about the boundary conditions of elastomer and elastomer compounds on steel and other metal surfaces. Melvin Mooney described in his Goodyear Medal Address that the potential slippage of elastomer on steel surfaces was his first concern when he was assigned by the U. S. elastomer Company to work on the processability of unvulcanized elastomer in 1928. Indeed, it concerned him throughout his career and was also the subject of a paper published in the year that he retired, 1959. Mooney's experiences with elastomer caused him to be the first scientist to seek to develop experimental techniques to measure slippage of materials during flow. In a paper published in 1931, he described methods of determining slippage in both capillary and coaxing cylinder rheometers. We describe Mooney's often-quoted capillary analysis in Section 6.4. In a 1936 paper devoted to determining the shear viscosity of natural rubber in a coaxial cylinder viscometer, Mooney carefully put grooves into both the stationary cylinder and rotor to prevent slippage on either surface. This was also done in his 1934 shearing disk instrument intended for quality control. Mooney was not the only one in this early period concerned with slippage. Similar concerns were expressed by Garner in regard to the Williams compression plastomer.

In the 1940s, little attention was given to this subject, but in the 1950s, we find Mooney with Black discussing slippage in extrusion through slit dies. In a 1954 paper, Decker and Roth described several experiments in a Mooney shearing disk viscometer. They found that the measured shear viscosity depended on the depth of the serrations on the shearing disk and on the type of metal which was coated onto the surface. Mooney made extensive studies of slippage in the 1950s generally by comparing the results of serrated and smooth steel rotors in a Mooney viscometer. He found that slip was induced by adding soap to elastomer, especially emulsion polymerized butadiene-styrene copolymer. He reported experiments on polytetrafluroethylene-coated rotors, which showed greatly reduced torques apparently associated with slippage. In a 1958 review paper on the rheology of elastomers, Mooney presents "slippage" as if it is a well-established fact and describes the work cited earlier in which he seeks to quantify his results in terms of slip velocities.

In the two decades following Mooney's retirement, there was little in the literature referring to this phenomenon; however, in 1980, Turner and Moore of Avon elastomer returned to this subject. These authors constructed a controlled-pressure rotational rheometer to study the rheological and slip flow characteristics of elastomer and its compounds. It was thus possible to compare behavior at the same pressures. They first compared a gum butyl elastomer in this apparatus and then showed that the torque rotor speed data are the same for all three rotors; however, when a fully compounded nitrile elastomer (NBR) with high levels of plasticizers and fillers was investigated, it was found that the torques on the ground and polished rotors were significantly lower than that on the ground rotors. This was taken as evidence of slip. Slip was similarly found in an ethylene-propylene terpolymer highly filled with carbon black and oil. It was found that the slip velocity increased with temperature and was suppressed by applied pressure.

Ahn and White have shown that carbooxylic acids and amides induce slip in polyethylene and polypropylene melts presumably by exuding to the surface and forming a low viscosity layer. This was also found to be the case with carbon black compounds, but the situation with compounds containing polar particles was more complex.

In 1988, Montes and his coworkers at the University of Akron described experiments on a rebuilt multispeed Mooney viscometer which had been pressurized in the manner described by Turner and Moore.

Torque-rotor speed sweeps were made at various applied pressure for both gums and compounds. It was found that at applied pressures below 0.5 MPa (500 kPa), the torque was greatly reduced, indicating the occurrence of slippage (Fig. 7.9).

If a low pressure were initially applied and then a substantial pressure imposed, the torque would suddenly increase as shown in Fig. 7.10. This indicated the rapid occurrence

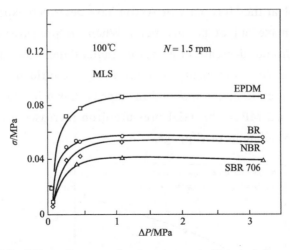

Fig. 7. 9   Effect of pressure level on torque in biconical rheometer.

of adhesion of polymer on the rotor. If the pressure were removed, the torque would slowly decay away (see Fig. 7. 10). The onset of slippage could be observed from the elastomer caps removed from the rotors; it would begin on the outer radius and propagate inward. These studies have been extended by Han et al. and White et al. from the same laboratories. These authors noted that the critical pressure for the onset of slippage, which seemed to be about 0. 2 MPa, was independent of the metal of construction with the exception of polytetrafluorethylene (PTFE) in which it was most serious. When, however, one operated at low applied pressures, brass and copper gave rise to the highest shear stresses, followed by aluminum, then different steels, and finally PTFE. Generally, slippage occurs at about the same critical pressure with smooth and serrated rotors, but in the region of slip, there are higher stresses with serrated rotors.

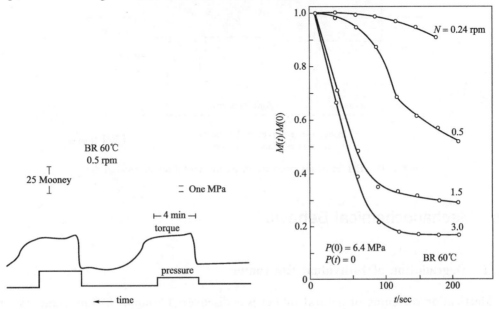

Fig. 7. 10   Influence of pressure history on torque history in a rotational rheometer.

Brzoskowski et al. at the University of Akron have described experiments involving extrusion through dies made out of porous metal. When air pressures are at a level of 0.2 MPa, there is an enormous decrease in pressure gradient along the die axis. Such decreases can be more than 90%. An experimental study showing the influence of decreasing the imposed air pressure on a elastomer compound in steps is shown in Fig. 7.11. When the pressure decreases below 0.2 MPa, the axial pressure drop increases drastically. These results strongly imply the development of slippage.

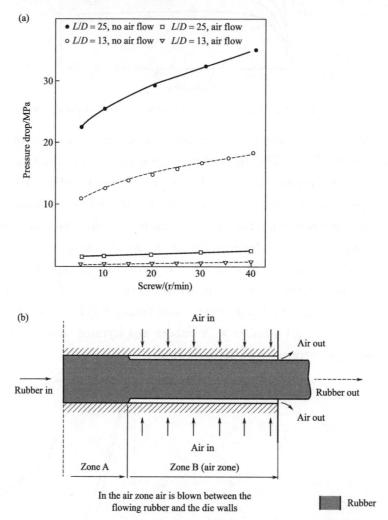

Fig. 7.11 Effect of air pressure in porous metal air-lubricated die.

7.5 Mechanochemical Behavior

7.5.1 Degradation of Individual Elastomers

Mastication softening of natural rubber was discovered around 1820 by Hancock, who also found that at any concentration level, solutions of masticated elastomer would possess a lower viscosity than the initial polymer. It was not, however, until the acceptance of the macromolecular hypothesis and the work of Cotton and Busse and Cunningham in the 1930s that the importance of oxygen (as well as the mechanism involved) in the mastication process came to light. The results of these researchers were subsequently verified by Pike and Watson. Essentially it is thought that the natural rubber degrades to form free radicals under the action of stresses and that the radicals are stabilized through reaction with oxygen:

$$R-R' \xrightarrow{stress} R\cdot + R'\cdot$$
$$R\cdot + O_2 \longrightarrow RO_2\cdot$$

One of the major points made, notably by Busse, was that the temperature has a peculiar influence on breakdown. At low temperatures, degradation is induced by applied stresses, and if the temperature is increased while the kinematics of mastication are unchanged, the rate of polymer degradation will decrease. If the temperature continues to be raised, the rate of degradation will pass through a minimum and begin to increase again. This is due to the interaction of the simultaneous decrease in viscosity and increase in rate of oxidative attack with temperature.

Generally it is found that there is little reduction in molecular mass when mastication is carried out in nitrogen. Busse and Cunningham and Pike and Watson have found that addition of free radical acceptors to natural rubber masticated in nitrogen leads to molecular weight reduction. Pike and Watson suggest that the free radical acceptors play a role similar to that of oxygen. Quantitative verification of this has been obtained by Ayrey et al. who related spectral and radiochemical determination of end groups to molecular mass reduction.

The influence of mastication on molecular mass distribution has been studied by Angier et al. and later investigators. It was found that high-molecular-weight species are preferentially degraded and narrower-molecular-weight-distribution materials are formed.

Although most investigations of stress-induced degradation have been carried out on natural rubber, studies on other elastomers (synthetic cis-1,4-polyisoprene, cis-1,4-polybutadience, butadiene-styrene copolymer) have appeared. Folt has studied the influence of mastication-induced degradation on the rheological properties of elastomers, especially polybuta-dienes and polyisoprenes. The greatest breakdown rates are found in cis-1,4-polyisoprene, a fact attributed to (1) stress-induced crystallization occurring during the deformation process, which results in especially high stresses, and (2) resonance stabilization of the radicals produced.

7.5.2 elastomer Blends

Angier and Watson found that graft copolymers could be prepared by masticating elastomer blends in the absence of oxygen. The systems studied included natural rubber/polychloroprene, natural rubber/butadiene-styrene copolymer, and polychloroprene/butadience-styrene co-polymer. The mechanism hypothesized was

$$A-A' \xrightarrow{stress} A\cdot + A'\cdot$$
$$B-B' \xrightarrow{stress} B\cdot + B'\cdot$$
$$A\cdot + B\cdot \longrightarrow A-B$$

7.5.3 Elastomers Swollen with Monomers

It has been shown by Angier and Watson that if an elastomer is swollen with a vinyl monomer (styrene, chlorostyrene, acrylic acid, methyl acrylate, methacrylic acid, methyl methacrylate, vinyl pyridine, methyl vinyl ketone, etc.), mastication in the absence of oxygen can lead to the formation of block copolymers. This would seem to occur through the mechanism

$$R-R' \xrightarrow{stress} R\cdot + R'\cdot$$
$$R\cdot + M \longrightarrow RM\cdot$$
$$RM\cdot + M \longrightarrow RMM\cdot$$
$$RMM\cdot + M \longrightarrow RM_{n+1}M\cdot$$

Angier et al. have shown that this process may also be applied to glassy vinyl plastics if they are processed in a elastomery state.

7.6 Rheological Measurements

7.6.1 Parallel Plate and Sandwich Rheometer

Perhaps the conceptually simplest type of rheometer can be constructed by sandwiching a material to be tested between two or three parallel plates that are separated by a distance H, and moving one plate parallel to the others at a velocity V (Fig. 7.12). The shear rate g. is V/H. For normal liquids this is not practical, but for elastomers and compounds it is very much so. Apparatus of this type have been designed and used by Zakharenko et al., Middleman, Goldstein, Furuta et al., Lobe and White, Toki and White, Montes et al., Osanaiye et al., and K. J. Kim and White. The apparatus (with constant-temperature chamber) may be placed in a tensile tester and operated in a mode with a fixed velocity V giving a constant shear rate. It may, on the other hand, be used in a creep mode with hanging weights. This provides constant stress experiments. At low stress levels one needs to compensate for the weight of the central member which exerts a gravitational stress. At very low stresses one may accurately determine the yield value of elastomer-carbon black compounds.

Osanaiye et al. have made measurements at shear stresses below the yield value. The shear rate $\dot{\gamma}$ is the ratio of the velocity of the moving member V to the perpendicular distance between the plates, i.e.,

$$\dot{\gamma} = V/H \tag{7-73}$$

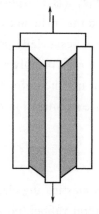

Fig. 7.12 Sandwich rheometer.

The shear stress σ_{12} is

$$\sigma_{12} = \frac{F}{2A} = \frac{F}{2(A_0 - VWt)} \tag{7-74}$$

The term VWt accounts for the decreasing active surface of the sandwich face. W is the width of the sandwich and t is experimental time.

7.6.2 Biconical Rheometer

The concepts of the cone-plate and biconical rheometers developed in the 1940s (Fig. 7.13). The cone-plate instrument is due to Freeman and Weissenherg and intended for modest-viscosity fluids. It has the basis of his "rheogoniometer" which also measured normal stresses. The biconical rheometer was developed in the same period by Piper and Scott of the BRMRA and was from the beginning intended for elastomer. Similar instruments are discussed by Turner and Moore and Montes et al.. In the latter instruments, the pressure is controlled by charging the elastomer into the rheometer by an attached pressure-driven device.

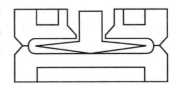

Fig. 7.13 Biconical rheometer.

The great advantage of cone-plate geometries occurring in both instruments is that for small-cone angles, the shear rate is constant throughout the gap between the cone and the plate. It is given by

$$\dot{\gamma} = \frac{\Omega}{h(r)} = \frac{\Omega}{r\tan\alpha} = \frac{\Omega}{\tan\alpha} \tag{7-75}$$

where $h(r)$ is the vertical distance from the plate to the cone and a is the angle between the cone and plate.

The shear stress in a biconical instrument may be computed from the torque M through the expression

$$M = 2\int_0^r 2\pi r^2 \sigma_{12} dr = \frac{4\pi r^3}{3}\sigma_{12} \tag{7-76}$$

The instruments of Turner and Moore and Montes et al. are pressurized by an external reservoir. This allows shear flow measurements to be carried out at constant shear rates and conditions for the development of slippage in individual compounds to be determined.

7.6.3 Shearing Disk Viscometer

The shearing disk viscometer introduced in 1934 by Mooney has played a very important role in the elastomer industry because of its utilization for monitoring and control of elastomer samples (Fig. 7.14).

It is related to the parallel disk type shear flow instrument that has been widely used for polymer solutions and melts.

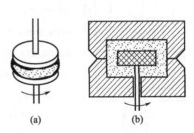

Fig. 7.14 Shear disk viscometers: (a) parallel disk, (b) Mooney.

The shear rate between the two disks at radial position r is
$$\dot{\gamma}(r) = r\Omega/h \tag{7-77}$$
where h is the distance between the disks and Ω is the rotation rate of the moving disk. The torque M in this geometry is given by
$$M = \int_0^r 2\pi r^2 \sigma_{12}\, dr \tag{7-78}$$
where σ_{12} varies with radius r. As has been shown by several researchers, Eq. (7-77) may be substituted into Eq. (7-78) and the integral differentiated with respect to the shear rate, $\dot{\gamma}(R)$, at the outer perimeter of the disk to yield
$$\sigma_{12}(R) = \left(\frac{S+3}{4}\right)\left(\frac{2M}{\pi R^3}\right) \tag{7-79}$$
where $S = d\log M/d\log\Omega$ for a power law fluid $S \sim n$.

In a shearing disk viscometer, the torque M on the disk may be expressed as
$$M = 2\left[\underbrace{\int_0^r 2\pi r^2 \sigma_{\theta z}\, dr}_{\text{surface}}\right] + \underbrace{2\pi R^2 H (\sigma_{\theta r})_R}_{\text{periphery}} \tag{7-80}$$
where H is the thickness of the disk. Eq. (7-80) may be solved for the torsional shear stress at the outer radius of the disk and an expression analogous to Eq. (7-73) obtained:
$$\sigma_{\theta z}(R) = \frac{S+3}{4}\left(\frac{2M}{\pi R^3}\right)F \tag{7-81}$$
Where, the function F is given by Nakajima and Harrel.

The quality-control shearing disk viscometer has always had only a single rotor speed (2 rpm), though Mooney as early as the 1940s built a multispeed unit. The use of Eq. (7-81) to determine $\sigma_{\theta z}(R)$ and shear viscosity implies a multispeed instrument. There has long been interest in making measurements of viscoelastic properties in this instrument. Mooney described measurements of elastic recoil, and Koopmann and Kramer of Bayer AG developed a Mooney viscometer allowing stress relaxation after flow. Such an instrument was later discussed by Montes et al.. Monsanto and later Alpha Technologies have commercialized a quality-control instrument similar to that of Koopmann and Kramer.

7.6.4 Capillary Rheometer

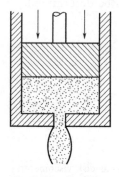

Fig. 7.15 Capillary rheometer.

The capillary rheometer, one the oldest and most widely applied experimental tools for measuring the viscosity of fluids (Fig. 7.15), was used extensively on almost all classes of complex fluids by the founders of modern rheology during the 1920s and 1930s. Its application to elastomer dates to the work of Marzetti and Dillon and coworkers, and it has been widely used through the years.

The basic idea of the instrument is to relate the pressure loss for extrusion through a small diameter tube of diameter D and length L to the shear stress at the capillary wall and the extrusion rate to a wall shear rate. The total pressure drop p_T through a die is the sum of a pressure loss within the die Δp and a second pressure drop Δp_e at the ends (i.e., at the entrance and the exit):

$$p_T = \Delta p_e + \Delta p \tag{7-82}$$

We may relate Δp to the wall shear stress $(\sigma_{12})_w$ by a simple force balance

$$\pi D L (\sigma_{12})_w = (\pi D^2/4) \Delta p \tag{7-83}$$

which allows us to write

$$(\sigma_{12})_w = D \Delta p / 4L \tag{7-84a}$$

$$p_T = \Delta p_e + 4(\sigma_{12})_w L/D \tag{7-84b}$$

As noted by Mooney and Black and subsequently Bagley, Δp_e can be large relative to Δp. By using a series of dies with varying L/D, one may determine $(\sigma_{12})_w$ from the slope of a graph of p_T versus L/D, which is called a Bagley plot. The wall shear rate must be kept constant, but this may be ensured by using dies of constant diameter and maintaining Q.

Implicit in Eq. (7-84) is the idea that the diameter of the reservoir preceding the die is much greater than the die. If this is the case, reservoir pressure losses need to be considered. This correction has been considered by Metzger and Knox.

The capillary wall shear rate may be obtained in a manner devised by Weissenberg and coworkers (see also Mooney). Noting that Eq. (7-83) applies to a telescoping flow at each radius, we see that the shear stress varies linearly with the radius

$$\sigma_{12}(r) = (\sigma_{12})_w (r/R) \tag{7-85}$$

This allows us to rewrite the extrusion rate as

$$Q = \int_0^r 2\pi r v_1 \, dr = \frac{\pi D^3}{8(\sigma_{12})_w} \int_0^{(\sigma_{12})_w} z^2 \left(-\frac{dv_1}{dr}\right) dz \tag{7-86}$$

where we have integrated by parts and presumed the material to adhere to the wall.

Differentiation of $32Q/\pi D^3$ with respect to $(\sigma_{12})_w$ and application of the Leibnitz rule for the differentiation of integrals allow us to solve for the capillary wall shear rate:

$$\dot{\gamma}_w = \left(-\frac{dv_1}{dr}\right)_w = \left(\frac{3n'+1}{4n}\right) \frac{32Q}{\pi D^3} \tag{7-87a}$$

with

$$n' = \frac{d\ln(\sigma_{12})_w}{d\ln(32Q/\pi D^3)} \tag{7-87b}$$

Application of Eq. (7-84) and Eq. (7-87b) allows the evaluation of the viscosity function $\eta(\dot{\gamma})$. Mooney modified Eq. (7-86) to allow for slip and has shown how slip at the capillary wall may be determined. Basically, in place of Eq. (7-86), one has

$$Q = \int_0^r 2\pi r v_1 \, dr = \frac{\pi R^2}{4} v_s + \frac{\pi D^3}{8(\sigma_{12})_w} \int_0^{(\sigma_{12})_w} \left(-\frac{dv_1}{dr}\right) dz \tag{7-88a}$$

$$\frac{8Q}{\pi D^3} = \frac{2v_s}{D} + \frac{1}{(\sigma_{12})_w} \int_0^{(\sigma_{12})_w} z^2 \left(-\frac{dv_1}{dr}\right) dz \tag{7-88b}$$

The integral in Eq. (7-88b) depends only on the die wall shear stress so that

$$v_s = \left[\frac{\partial(4Q/\pi D^3)}{\partial(1/D)}\right]_{(\sigma_{12})_w} \tag{7-89}$$

This requires experiments at constant $\Delta Dp/4L$ with varying diameter D, which means variable pressure. Capillary rheometers have been widely used for quality control in the elastomer industry since the 1920s. As late as the 1970s, they were advocated by Monsanto. The major problem with the instrument is that it gives values at high shear rates where the viscosity is relatively insensitive to molecular mass and its distribution. Furthermore, gum elastomers are often in an unstable flow region in such instruments.

7.6.5 Compression Rheometer

The basic theory of flow of a non-Newtonian fluid in a compressional rheometer is due to Scott, whose analysis is in terms of a power law fluid (Fig. 7.16).

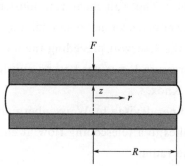

Fig. 7.16 Compression rheometer.

One presumes the flow is basically laminar shearing, with the fluid being driven radially outward from the approaching disks. By continuity at any radius r

$$\pi r^2 \left(-\frac{dH}{dt}\right) = \int_0^H 2\pi r v_r(z) dz \qquad (7\text{-}90)$$

A pressure field exists between the disks with the maximum pressure at the center. The pressure gradient is related to the shear stress through

$$\sigma_{rz}(r) = \left(z - \frac{H}{2}\right)\frac{\partial p}{\partial r} \qquad (7\text{-}91)$$

for z greater than $H/2$. In Scott's model as stated earlier, a power law model is presumed between shear stress and shear rate:

$$K\left(-\frac{dv_r}{dz}\right)^n = \left(z - \frac{H}{2}\right)\frac{\partial p}{\partial r} \qquad (7\text{-}92)$$

The velocity field $v_{r(z)}$ was determined. Using Eq. (7-90), this may be converted to the rate of downward platen movement $(-dH/dt)$. The compression force F is related to $p(r)$ through

$$F = -\int_0^R 2\pi r p(r) dr \qquad (7\text{-}93)$$

This leads to

$$F = 2r\left(\frac{2n+1}{n}\right)^n K \frac{R^{n+3}}{n+3} \frac{1}{H^{2n+1}}\left(-\frac{dH}{dt}\right)^n \qquad (7\text{-}94)$$

The shear rate and shear stress at the outer radius of the disk are

$$\dot{\gamma}(R) = -\frac{dv_r}{dz} = \left[\frac{2n+1}{n}\right]\left(\frac{R}{H}\right)\left(-\frac{1}{H}\frac{dH}{dt}\right) \qquad (7\text{-}95)$$

$$\sigma_{12}(R) = \sigma_{rz}(R) = \frac{n+3}{4}\left(\frac{H}{R}\right)\frac{F}{\pi R^2} \qquad (7\text{-}96)$$

Compressional flow rheometers have been studied since the time of Williams et al. and van Rossem and van der Meijden. In the 1920s, the Firestone Tire and elastomer Company and the B. F. Goodrich Company used the compression plastometer as a quality-control instrument. Subsequently in the 1930s, Baader and Continental Gummiwerke developed a similar instrument called the Defo for quality control which was used by the I. G. Farbenindustrie during World War II to test Buna S. synthetic rubber. After the war interest in the Defo fell off and the Mooney viscometer Section 6.3 of the victorious side came to be generally accepted. Koopmann and his coworkers at Bayer AG have developed a much improved Defo which is manufactured by Haake Messetechnik in Karlsruhe.

7.6.6 Elongational Rheometer

A wide range of instruments have been developed as elongational rheometers. The earliest instrument of this type was by Ballman and involved inserting a vertical specimen and stretching it at an accelerating cross-head speed. The stretch rate is

$$\dot{\gamma}_E = \frac{1}{L}\frac{dL}{dt} \tag{7-97}$$

where L is the sample length. If this is to be constant in time, $L(t)$ must increase as

$$L(t) = L(0) e^{\dot{\gamma}_E t} \tag{7-98}$$

The stress is

$$\sigma = F(t)/\pi R^2(t) \tag{7-99}$$

where R is the sample radius. A second type of experiment was devised in which samples were floated on a silicone bath and clamped at one end and drawn out at the other end by a weight, moving clamp, or rollers. In one version, two sets of rollers are used. In the experiment with the weight, a cam is used so as to lead to constant stress. In the experiments with a fixed clamp and one set of rollers, the stretch rate is

$$\dot{\gamma}_E = \frac{V}{L} \tag{7-100a}$$

and if two sets of rollers are used,

$$\dot{\gamma}_E = \frac{V}{L/2} \tag{7-100b}$$

If an accelerating clamp is used with a bath, then Eq. (7-98) must be reproduced. All of the experiments just described are intended for thermoplastics. Only Cotton and Thiele have developed a uniaxial elongational flow instrument intended for elastomers. In this instrument an extruded elastomer strand is placed over two pulleys. The two ends descend between a pair of knurled pulleys which draws the strand ends downward.

7.7 Processing Technology

7.7.1 Internal Mixers

The first step of elastomer processing is combining the ingredients of a compound in a mixing device. In the early years of the industry, this was done on a two-roll mill open to the environment. From the second decade of the century there was a transition, now long completed, to carrying out this mixing in an internal mixer of a design basically developed by Banbury (see also White and pioneered by the new Farrel Corp.). The modern internal mixer consists of a mixing chamber containing two counter-rotating rotors. At the center of the top of the mixing chamber is a shaft through which compounding ingredients gain access. During most of the mixing cycle, the shaft contains a ram which presses the compounding ingredients into the chamber. At the bottom of the mixing chamber is a door which opens at the end of the cycle and dumps the mixed compound. This machine is shown in Fig. 7.17.

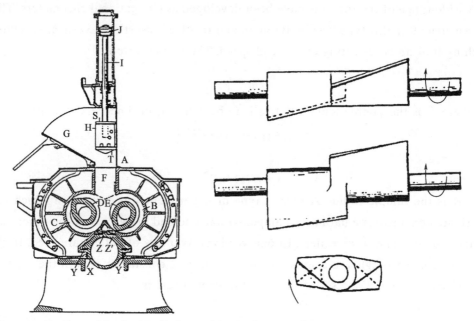

A—Upper case; B—Upper mixing chamber; C—Lower mixing chamber; D—Left rotor; E—Right rotor; F—Pressure chamber; G—Hopper; H—Ram; I—Shaft; J—Piston; S—Piston head; T—Ram end; X—Block bolt; Y—Lock; Z and Z'—Output

Fig. 7.17 F. H. Banbury's internal mixer and rotor design. (From Banbury.)

The rotors of the internal mixer in the original design of Banbury were separated and had two curved flights which tended to pump in opposite direction. In subsequent years, there were many efforts to improve the design of internal mixers. Lasch and Frei, of Werner and Pfleiderer, Tyson and Comper of Goodyear, and Sato et al. of Bridgestone/Kobe Steel have described internal mixers with rotors having four rather than two flights. These give better dispersing character than two-flight rotors. Today most internal mixers with separated rotors contain four flights (Fig. 7.18).

Internal mixers with a much different mixing chamber design were proposed by Cooke of Francis Shaw and Company and by Lasch and Stromer of Werner and Pfleiderer. These internal mixers possess inter-meshing counter-rotating rotors (Fig. 7.19). In these internal mixers, both rotors must move at the same angular velocity. The Francis Shaw mixer, the "Intermix", was marketed first and received considerable attention. In time, the inter-meshing mixers dominated the mechanical elastomer goods industries in Europe and Japan.

Passoni of Pomini has developed an inter-meshing rotor design where the rotors may be moved transverse to their axes to control the inter-rotor clearance. Separated rotor in-

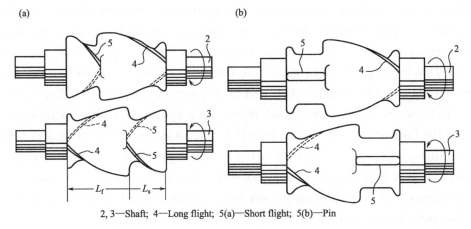

2, 3—Shaft; 4—Long flight; 5(a)—Short flight; 5(b)—Pin

Fig. 7.18 Internal mixer rotors from the designs of Sato et al. of Bridgestone and Kobe Steel.

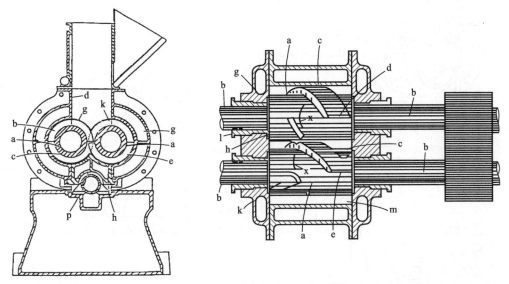

a—Long flight; b—Shaft; c—Rotor; d—Flight end; e—Flight head; g—Short flight; h—Lock; k—Bolt; l—Bearing; p—Discharge port; x—Pin

Fig. 7.19 Inter-meshing internal mixer design of Cooke. (From Coleman and Noll)

ternal mixers are produced by Farrel Corp. in the United States and England. Kobe Steel and its Kobelco-Stewart Bolling subsidiary make these machines in Japan and the United States. Werner and Pfleiderer Gummitechnik (recently a wholly owned Krupp firm) make internal mixers in Germany and Techint-Pomini in Italy.

Mitsubishi Heavy Industries, a licensee of Werner and Pfleiderer Gummitechnik, produce separated rotor internal mixers in Japan. Inter-meshing rotor internal mixers are produced by Francis Shaw and Company in England (now owned by Farrel). Werner and Pfleiderer Gummitechnik (and their licensee Mitsubishi Heavy Industries) produces such internal mixers in Germany and Japan. Pomini manufactures a variable-clearance intermeshing internal mixer in Italy.

There used to be little basic study of the internal mixer before 1979. Freakley and Wan

Idris published the first flow visualization investigation of a non-intermeshing rotor internal mixer using a glass window normal to the rotor axes in 1979. A more extensive study was subsequently reported by Bridgestone and Kobe Steel for the purpose of optimizing rotor design. This investigation used adjustable rotors in a transparent polycarbonate chamber. An aqueous polymer solution was used as a process fluid. Later studies by Min and White, Morikawa et al., and Kim and White using apparatus similar to those. Freakley and Wan Idris focused not only on the material motions inside of the internal mixer but on comparing different elastomers and following the mixing cycle (Fig. 7. 20). Cho et al. have compared two-wing Banbury rotors and four-wing Tyson-Comper rotors in flow visualization experiments. The four-wing rotors mix more rapidly. An investigation of flow visualization was published by Toh et al.

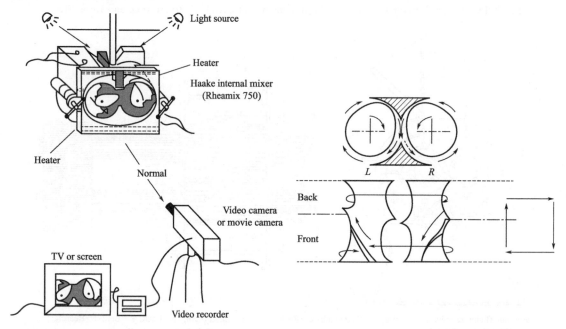

Fig. 7. 20 Flow visualization in internal mixer.   Fig. 7. 21 Circulation of material in internal mixer.

Basically it is found that material in the internal mixer chamber for double-and four-flight rotors circulated around the mixing chamber (Fig. 7. 21). Natural rubber tends to cling to rotors, and most synthetic rubber are torn into chunks. The addition of carbon black and oil to an internal mixer has been studied by flow visualization techniques for a wide range of elastomers. The tearing of elastomers into crumbs disrupts the mixing of carbon black, which proceeds more surely with natural rubber. The rate of oil absorption depends on available surface area of the elastomer and proceeds more rapidly with torn elastomer crumbs and slowly with natural rubber.

P. S. Kim and White compared mixing in R. T. Cooke design intermeshing and separated rotor internal mixers using flow visualization. Bale homogenization, black incorporation, and oil absorption were found to be more rapid in intermeshing rotor mixers. Koolihiran and White compare the mixing in double-flighted Banbury rotors and Cooke intermeshing rotors of silica, talc, and carbon black into SBR. The intermeshing rotors are much better for dispersive mixing.

7.7.2 Screw Extrusion

Screw extruders were first introduced into the elastomer industry in the late 19th century by Francis Shaw, John Royle, and Paul Troester and became the basis of the machinery companies they created (Fig. 7.22). The original elastomer extruders were hot-feed screw extruders, for which strips of elastomer compound preheated on a two-roll mill were added to the extruder. The screws had deep channels and short length/diameter ratios.

In the 1930s, Paul Troester Maschinenfabrik introduced the cold-feed screw extruder to which cold strips of elastomer compound were added. The cold-feed extruder had shallow screw channels and a much longer length/diameter ratio. The original cold-feed extruder produced extrudates which had large temperature distributions and became distorted as they emerged from the die. The issue in cold-feed extrusion of elastomer was to produce a thermally homogenized product in the extruder screw which would yield thermally uniform extrudate that does not distort. From the 1960s, there have been several improvements in cold-feed extruders. First, Maillefer and Geyer (of Uniroyal Inc.) introduced the barrier screw design. This introduces a second flight which isolates the less easily flowing material and prevents it from emerging from the screw until it softens. This was commercialized in the United States by NRM (under Uniroyal license) as the Plastiscrew. Designed for the purpose of controlling the melting of the solid bed in thermoplastics by Maillefer and for isolating scorched elastomer compound by Geyer, this screw, when used in cold-feed elastomer extruders, gave more uniform extrudates but with relatively high temperatures.

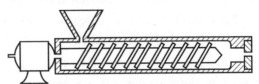

Fig. 7.22　Screw extruder.

Subsequently many new designs of special sections for cold-feed screw extruders, such as the Troester shear section and the various designs of Lehnen and Menges, gave further improvements. In the 1970s, Menges and Harms of the Institut für Kunststoffverarbeitung and of Uniroyal, proposed the use of a pin barrel cold-feed extruder (Fig. 7.23). This was found to give greatly improved extrudate uniformity. The machine was first licensed to Paul Troester Maschinenfabrik and subsequently to other machinery firms including Hermann Berstorff Maschinenbau and Krupp Gummitechnik in Germany, Techint-Pomini SpA in Italy, Nakatazoki in Japan, and Farrel Corp. in the United States.

The earliest studies of the flow of elastomers in screw extruders is found in the work of Vila and Pigott. The observations of Pigott are especially interesting. There were relatively few basic investigations in succeeding years. The characteristics of flow and starvation in devolatilizing screw extruders were considered by Brzoskowski et al. and by Vergnes et al. A more substantial study of starvation cold-feed elastomer screw extruders was given by

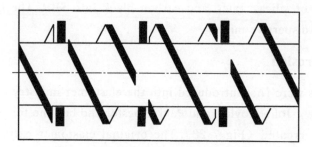

Fig. 7.23 Pin barrel screw extruder.

Kubota et al., who showed that the extent of fill was proportional to the die pressure.

Flow marker studies in screw extruders for elastomer compounds were initiated by Menges and Lehnen to evaluate the mixing and homogenizing characteristics of various screw designs. These studies feed equal-viscosity pigmented and white strips containing curatives into the extruder. After a steady state is achieved, the extruder is shut down, heated up, and cured in place. The screw is then removed, and the elastomer compound strip peeled off and sectioned.

Flow marker studies in a cold-feed elastomer pin barrel extruder were first reported by the inventors Menges and Harms and in more recent years by Yabushita et al. and Shin and White. It was generally found that pin barrel sections do a much superior job of homogenizing (mixing pigmented and white strips) than ordinary screw sections (Fig. 7.24). In the investigations of Yabushita et al. and Shin and White comparisons are also made with screws containing grooves on their flights but possessing no pins. These are found to be in-

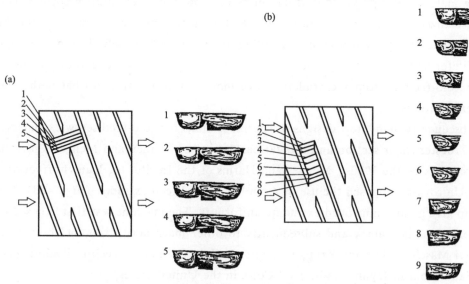

Fig. 7.24 Flow marker experiments for pin barrel extruder of Yabushita et al.
(a) Cross sections of vulcanized elastomer strip removed from the area between the first and second grooves (extruder run without pin application). Operating conditions $T_b = 80°C$, $T_s = 80°C$, $N = 10\,\mathrm{min}^{-1}$. (b) Cross-section of vulcanized elastomer strip removed from the area of the first groove (extruder run without pin application). Operating conditions: $T_b = 80°C$, $T_s = 80°C$, $N = 10\,\mathrm{min}^{-1}$.

termediate in homogenizing character. The latter authors found that introducing slices into screw flights reduces screw pumping characteristics. They also found that introducing pins has little effect on pumping. Shin and White compare a range of elastomer compounds and show that although introduction of a pin always improves homogenization, the level of homogenization varies considerably from compound to compound. It is apparently this homogenizing character, as related to temperature fields in large extruders, which allows pin barrel extruders to produce uniform extrudates to higher outputs.

7.7.3 Die Extrusion

The design of dies has been a problem since the beginning of extrusion technology. The mid-19th-century patent literature gives considerable attention to the development of cylindrical, annular, and wire coating dies for ram and later screw extruders. In this century, much emphasis has been on the development of sheeting dies including the invention of various control systems to maintain thickness uniformity. Such dies have either (1) converging/diverging sections, which convert the cylindrical orifice at the end of the extruder to a slit shape, or (2) a manifold, which distributes melt across the die, providing a uniform feed of polymer melt to the die (Fig. 7.25). They are often called "T" and coat hanger dies.

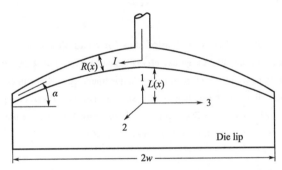

Fig. 7.25 Coat hanger die design to produce slit extrudate.

Die systems which may coextrude tread and sidewall or tread/sidewall/have received considerable attention since the 1930s (Fig. 7.26). Generally today several screw extruders are constructed in a "piggyback" design. These feed the various individual components to the die system.

There have been many advances in the development of profile dies. Techniques have been developed which allow the extrusion of curved sections. The work of Goettler and Miller et al. describe different technologies for extruding curved tubes and profiles. They are often called "T" and coat hanger dies.

Elastomer compounds often contain chopped fibers. It is important to control the orientation of fibers in extrudates emerging through dies. An example is in elastomer tubing/hose where fibers should be circumferential not axial. Various methods have been developed to achieve this. These include various dies with rotating mandrels, and a die with a diverging section developed by Goettler et al. of Monsanto has been devised for the same

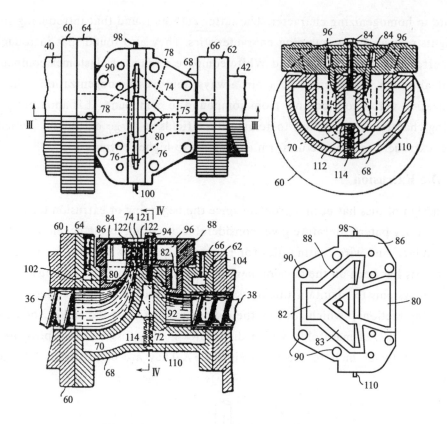

Fig. 7.26 Coextrusion dies for tread/sidewall.

36—Large screw; 38—Small screw; 40—Large barrel; 42—Small barrel; 60—Large flange; 62—Small flange; 64—Large flange; 66—Small flange; 68—Back plate; 70—Major liner; 72—Minor liner; 74—Torpedo; 75—Block; 76—Edge slit; 78—Edge slit; 80—Large ledge; 82—Small ledge; 84—Bolt; 86—Pin; 88—Front plate; 90—Lead hole; 94—Key; 96—Die case; 98—Adjust pin; 100—Adjust pin; 104—Lead; 110—Adjust pin; 112—Liner; 114—Bolt; 121—Center thru; 122—Side thru

purpose.

Experimental studies of flow of elastomers through dies date to the work of Marzetti in the 1920s. Any reasonable review of the present state of knowledge must note the very extensive literature in this area developed for thermoplastics. Investigations for thermoplastics have included both studies of extrudate swell and quality and flow marker/birefringence investigations of fluid motions within dies. The studies of Bagley and Birks, Ballenger and White, and Han and Drexler are typical of the latter activities, which have yielded information on the occurrence of circulating flows inside dies, stress fields in flowing melts, and mechanisms of extrusion instabilities. Low-density polyethylene (LDPF) is the most proment of polymer melts showing large circulating flows in the die entrance. The effect is also found, however, in polystyrene.

Generally, elastomer is processed in the form of highly filled compounds and is opaque. This has tended to suppress activities on elastomer compounds except for observations of extrudate swell and quality. It is, however, important to say what has been done which is relevant.

The first study of pertinence of flow of elastomer compounds within dies is that of Mooney and Black who noted the occurrence of large die entrance pressure losses. Similar observations for thermoplastics were subsequently made by Bagley. White and Crowder reported extensive die end pressure losses for elastomer compounds. Their results suggest the effects may be larger than for thermoplastics. Experimental studies by Ma et al. have shown that addition of small particles including carbon black suppresses the large circulating vortices found in LOPE. Studies by Ma et al. and Song et al. on elastomer compounds have found no evidence of circulating flow in the die entrance.

A most important effect of the addition of carbon black and other fillers to elastomers and thermoplastics is to reduce extrudate swell and shrinkage and improve extrudate surface quality, i. e. , the tendency to eliminate extrudate distortion.

7.7.4 Molding

Molding represents not a single technology but a host of them (Fig. 7.27). These range from compression molding, where chunks or sheets are placed in an open mold which is then closed around it. After the part is formed (and, for elastomer, vulcanized), the mold is opened and the part removed. In injection molding, material is injected under pressure into a closed mold. After the part is formed and vulcanized, the mold is again opened. Transfer molding represents a combination of injection and compression molding in which elastomer is injected from a reservoir into a compression mold. Blow molding involves the preparation of a preform which is inserted into a mold. This is subsequently inflated (by hot air/steam or some other fluid perhaps contained in a bag) to fill out the mold.

These technologies date to the first half of the 19th century and were developed by the Hancock brothers (Thomas and Charles in England) and the Goodyear brothers (Charles and Henry in the United States). Much of this development can be read in the autobiographies and patents of Thomas Hancock and Charles Goodyear published in the mid-19th century. In 1846—1847 patents, Hancock described the process technologies for injection molding, blow molding (including using a flexible bag made of vulcanized elastomer to contain the pressurized hot air), and molding of foams using gutta percha, natural rubber, and their blends.

The 20th century has seen the clear development of two distinct types of elastomer molding technology. One of these involves the molding of tires; the second involves mechanical goods. Tire manufacturing technology involves two processes in its shaping step. The first is building tire preforms on a drum, and the second, the molding of the preform into a final tire with associated vulcanization. In elastomer mechanical goods production one sees a wide range of manufacturing operations including compression molding, transfer molding, and screw injection molding.

Basic studies of flow mechanisms in molding have been lacking for elastomer com-

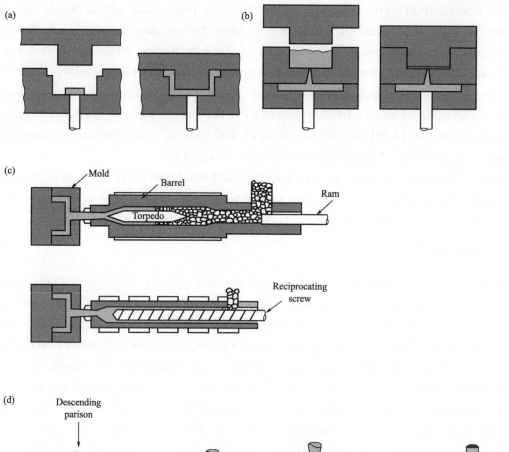

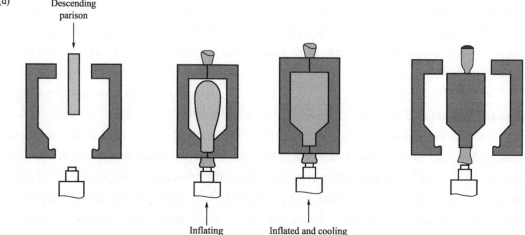

Fig. 7.27 Different types of molding technologies. (a) Compression molding. (b) Transfer molding. (c) Injection molding. (d) Blow molding.

pounds. Almost all published studies have been for thermoplastics. Most of these investigations involve injection mold filling.

It has been long realized that there are two regimes of injection mold filling. These involve a slowly expanding front or jetting into a mold. Oda et al. have shown that jetting into a mold is the result of the polymer or compound having a low extrudate swell. The often complex occurrence of this swell is associated with the melt temperature and its variations in runners. Isayev and his coworkers have extensively studied pressure development and injection mold filling by elastomer compounds when the front is slowly expanding in the mold.

7.8 Engineering Analysis of Processing

There are many ways in which elastomer rheology may be applied to processing operations. This is because processing is in large part simply flow and shaping of compounds. The most obvious use is in determining the sources of problems when difficulties arise in factory production. Frequently the breakdown of regular production, can be traced to some cause such as high viscosity levels or elastic memory of compounds. This is, of course, how the early elastomer rheologists were first employed and, to a good extent, is what they do to this day.

7.8.1 Dimensional Analysis

One of the most useful ideas for the industrial polymer rheologist is dimensional analysis. In this traditional method of engineering analysis, the conservation laws (force-momentum, energy, etc.) are placed in dimensionless form, and the dimensionless groups that arise are interpreted. There were applications of this approach to the development of model basins for ships and to the flow of liquids in pipes and the 19th century. Indeed, the general philosophy of dimensional analysis in engineering was developed by Osborne Reynolds for the latter application. These ideas were extended in the early years of the 20th century by Wilhelm Nusselt. The dimensionless correlations developed by these authors can be used as a guide in the design of pipelines, heat exchangers, and packed columns. Numerous texts describe this subject matter and are an integral part of university engineering education.

Unfortunately, the Reynolds-Nusselt dimensional analysis studies are not directly applicable to polymer melt and elastomer processing for two important reasons. First, they are based on Newtonian fluid behavior, and second, they do not include viscous dissipation heating.

The earliest considerations of the dimensional analysis of isothermal viscoelastic fluids were by Karl Weissenberg during the period from 1928 to 1948. Weissenberg suggested that the recoverable elastic strain represents fluid elasticity in the same way that the Reynolds number represents inertia. A large recoverable strain represents high viscoelastic forces. Reiner later suggested that the basic dimensionless group correlating viscoelastic phenomena was a "Deborah number," which may be represented as a ratio of the characteristic relaxation time τ_{ch} to a residence time t. A large t value of τ_{ch} or a small value of t will, according to Reiner, accentuate viscoelastic response. A more quantitative formulation of the intuitive ideas of Weissenberg and Reiner resulted after the modern theory of nonlinear viscoelasticity had been developed. White, White and Tokita, and Metzner et al. published analyses that have brought these ideas together into a formal scheme. This approach would be expected to be valid for thermoplastics and most applications of elastomer gums but not compounds.

We begin with Cauchy's law of motion, Eq. (7-10), and a general constitutive equation such as Eq. (7-47), which yields

$$\rho \left[\frac{\partial \mathbf{v}}{\partial t}+(\mathbf{v} \cdot \nabla)\mathbf{v}\right]=-\nabla p+\nabla \cdot \int_0^\infty m(z)\,[\mathbf{c}^{-1}+\kappa \mathbf{c}]\,\mathrm{d}z+\rho g \qquad (7\text{-}101)$$

A characteristic velocity U and length L are introduced into this equation and the entire expression is put in dimensionless form. These dimensionless groups are

$$\frac{LU\rho}{G_m \tau_m},\quad \frac{U^2}{gL},\quad \tau_m \frac{U}{L},\quad \frac{\tau_m}{t},\quad \frac{\tau_i}{\tau_m},\quad \frac{G_i}{G_m},\quad a_j \qquad (7\text{-}102)$$

where τ_m is the maximum relaxation time [see Eq. (7-14)] and Gm the corresponding modulus. Here $LU\rho/G_m\tau_m$ is a Reynolds number signifying the ratio of inertial to viscous forces; U^2/gL is a Froude number representing a ratio of inertial to gravitational forces; $\tau_m U/L$ is known as the Weissenberg number and τ_m/t as the Deborah number, both representing the ratio of viscoelastic to viscous forces. The other groups are viscoelastic ratio numbers representing the detailed character of the functional.

Generally, inertial forces are negligible in polymer melts and elastomers, and gravitational forces need not be considered. This leaves the Weissenberg, Deborah, and viscoelastic ratio numbers. If we consider an application to a processing operating in which the material flows through some equipment, it must be recognized that t is a residence time and

$$\tau_m/t = [\tau_m/(L_{11}/U)] = \tau_m U/L_{11} \qquad (7\text{-}103)$$

L_{11} is a characteristic length in the direction of flow. The Deborah number is equivalent to a Weissenberg number with the characteristic length in the direction of flow. Our dimensionless groups thus reduce to

$$\tau_m U/L,\quad \tau_i/\tau_m,\quad G_i/G_m/G_j \qquad (7\text{-}104)$$

The results of the preceding paragraph have some important implications. First, the characteristic velocity and length arise in only one group, and this occurs in the form of a ratio U/L. This means that for any particular viscoelastic fluid there will be dynamic similarity between large and small geometrically scaled isothermal systems only if velocities are scaled with lengths. Furthermore, if any instabilities or changes in regime occur within such a system, they will occur at a critical value of a dimensionless group of the form $\tau U/L$ where the characteristic time τ is given by

$$\tau = \tau_m F\,[\tau_i/\tau_m, G_i/G_m, a\gamma] \qquad (7\text{-}105)$$

Here the function F may be quite complex, so that τ may differ significantly from material to material (with differing viscoelastic ratio numbers) even if they possess the same τ_m.

7.8.2 elastomer Compounds

Now we turn to elastomer compounds and specifically to plastic viscous and plastic viscoelastic fluids. Dimensional analysis of plastic fluids with yield values was first considered by Oldroyd and then subsequently by Prager. Oldroyd pointed out that a new dimensionless group of form

$$\frac{YL}{\eta U} \tag{7-106}$$

arises. This group has become known as the Bingham number.

For a plastic viscoelastic fluid, the dimensionless groups of Eq. (7-102) should be replaced by

$$\frac{YL}{\tau_m G_m U}, \quad \tau_m \frac{U}{L}, \quad \frac{\tau_i}{\tau_m}, \quad \frac{G_i}{G_m}, \quad a_j \tag{7-107}$$

As with viscoelastic fluids in Eq. (7-100) and plastic viscous fluids in Eq. (7-106), dynamic similarity involves systems with the same U/L. Both viscoelastic and plastic flow phenomena are then associated with dimensionless groups involving U/L. Attention needs also to be given to thixotropic effects. The preceding dynamic similarity formulation must involve equivalent periods of deformation.

7.8.3 Nonisothermal Effects

The discussion of the foregoing paragraphs is limited to isothermal systems, with which, however, we are generally not concerned. We must also consider the energy equation, which we may write as

$$\rho c \left[\frac{\partial T}{\partial t} + (\mathbf{v} \cdot \nabla) T\right] = k \nabla^2 T + \sigma : \nabla \mathbf{v} \tag{7-108}$$

Introducing characteristic velocity U, length L, and temperature difference θ, we may show that the dimensionless groups arising in this equation are

$$LU\rho c/k, \quad \eta U^2/k\theta \tag{7-109}$$

as well as $G_m \tau_m/\eta$ and viscoelasticity-related groups of Eq. (7-102) or plastic viscoelastic parameters of Eq. (7-107) as appropriate. Here, $LU\rho c/k$ is known as the Peclet number and signifies the ratio of convective heat flux to heat conduction, whereas $\eta U^2/k\theta$ is the Brinkman number, representing the ratio of viscous heating to heat conduction.

The Brinkman number is of great importance in the interpretation of polymer processing because it represents the ability of the system to conduct away heat produced by viscous dissipation. It is useful to rewrite it in the form

$$\frac{\eta U^2}{k\theta} = L^2 \left(\frac{\eta}{k\theta}\right) \left(\frac{U}{L}\right)^2 \tag{7-110}$$

If we compare small and large systems that have been scaled at constant U/L, as we would expect for dynamic similarity, the Brinkman number increases as L^2. The larger the system, the greater the tendency for viscous heating to override heat conduction and for the temperature to rise. Another way of stating this is that a large system will tend to behave adiabatically. Thus the isothermal dimensional analysis and interpretation are most suitable to systems with small dimensions. For large systems, we have to consider viscous dissipation as well.

The origin of the effect described in the preceding paragraph may be viewed in another way. Viscous dissipation heating is proportional to the volume of the system L^3, whereas heat transfer is proportional to the surface area, which varies as L^2. As one goes to larger systems, surface area per unit volume decreases, making heat transfer more difficult and the system more adiabatic.

A final point of importance, which has already been implied but which should be stated more explicitly, is the inability to scale at both constant Weissenberg number (or Bingham number) and constant Brinkman number, i. e., at constant

$$\frac{Y}{\tau G}\frac{L}{U}, \quad \tau U/L, \quad \eta U^2/k\theta \tag{7-111}$$

The only way one may scale at constant Brinkman number is to maintain the characteristic velocity U constant as L increases, which would decrease the Weissenberg number. Quite often nonisothermal effects are dominant and one must mainly worry about Eq. (7-110).

7.8.4 Simulations of Internal Mixers

Efforts to model flow in an internal mixer may be traced to the 1950s; however, it was not until the late 1980s that there was a substantial effort involving several different independent investigations. Generally, two different approaches were being used, one presuming hydrodynamic lubrication theory to be valid and a second based on the complete equations of motion and applying finite-element analysis.

In the first of these approaches, early work concentrated on simulations in Cartesian coordinates reminiscent of journal bearing theory. More recently these activities have concentrated on cylindrical coordinate formulations essentially based on solving the set of equations

$$0 = -\frac{\partial p}{\partial z} + \eta \frac{\partial}{\partial r}\left(r\frac{\partial v_z}{\partial r}\right) \tag{7-112a}$$

$$0 = -\frac{1}{r}\frac{\partial p}{\partial \theta} + \eta \frac{1}{r}\frac{\partial}{\partial r}\left(r\frac{\partial v_\theta}{\partial r}\right) \tag{7-112b}$$

and integrating the velocity fields to obtain expressions for fluxes q_z and q_θ defined by

$$q_z = \frac{1}{R^*}\int_{R_1}^{R_2} rv_z \, dr \tag{7-113a}$$

$$q_\theta = \int_{R_1}^{R_2} v_\theta \, dr \tag{7-113b}$$

The fluxes q_z and q_θ are balanced according to

$$\frac{\partial q_z}{\partial z} + \frac{1}{4}\frac{\partial}{\partial \theta}(rq_r) = 0 \tag{7-114}$$

which is then solved numerically.

It is necessary to solve the preceding equations both with boundary conditions suitable for flow (1) between the chamber wall and the rotor and (2) in the inter-rotor region, and to match boundary conditions. The solution to Eq. (7-113) yields a pressure field which may be converted to a flux field. Basically, it is found that the two rotors act as screw pumps, pumping in opposite directions and creating a circulational flow in the mixing chamber. Kim et al. and Hu and White have sought to represent flow patterns by defining mean fluxes F_L, F_C, F_{tr}, F_D which represent flow patterns which respectively represent fractional flows that are longitudinal along the roads (F_L) and

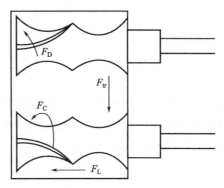

Fig. 7.28 Definition of fractional flow fluxes F_L, F_C, F_{tr}, F_D in an internal mixer.

circumferential around the rotors (F_C), transfer flow between rotors (F_{tr}), and "dispersive" flow going over rotor flights F_D (Fig. 7.28). They have computed these functions for various geometries including many rotors listed in the patent literature. Typical results are summarized in Table 7.3, where Banbury's double-flighted rotors are compared to the four-flighted rotors of Lasch and Frei, Tyson and Comper, and Sato et al. The Sato et al. rotors compared to the old Banbury rotors exhibit a 20% more rapid circulation and a 180% greater flow over flight tips. P. S. Kim and White have modeled the flow in intermeshing rotor internal mixers.

Table 7.3 Flow Characteristics of Double-and Four-Flight Internal Mixer Rotors from the Patent Literature

Inventor	Company	Year	F_L+F_{tr}	F_C	F_D	$|Q_L^-/Q_L^+|$	t_{cirl}/min
Banbury	W & P Birmingham Iron Foundry	1916	0.321	0.674	0.005	0.31	2.7
Lasch and Frei	Werner & Pfleiderer	1943	0.295	0.689	0.016	0.21	3.7
Tyson and Comper	Goodyear Tire & Rubber	1966	0.375	0.615	0.010	0.78	2.3
Sato et al.	Kobe Steel	1981	0.398	0.588	0.014	0.50	2.0

7.8.5 Simulations of Screw Extrusion

Simulations of flow in a single screw extruder date to the 1920s, with early efforts being largely associated with pumps for lubricating oils. In the 1950s, the flow in screw extruders became a subject of interest in flow in many industrial laboratories. The earliest work was largely repetitive of the activities on screw pumps in the 1920s.

The simplest Newtonian fluid model based on hydrodynamic lubrication theory involves the solution of the equations

$$0 = -\frac{\partial p}{\partial x_1} + \eta \frac{\partial^2 v_1}{\partial x_2^2} \tag{7-115a}$$

$$0 = -\frac{\partial p}{\partial x_3} + \eta \frac{\partial^2 v_3}{\partial x_2^2} \tag{7-115b}$$

where direction 1 is along the screw channel, 2 is normal to the screw surface, and 3 is perpendicular to the screw flights. The velocity fields v_1 and v_3 are, if the coordinate system is fixed in the screw,

$$v_1(x_2) = \pi DN \cos\phi \left(\frac{x_2}{H}\right) - \frac{H^2}{2\eta}\left(\frac{\partial p}{\partial x_1}\right)\left[\left(\frac{x_2}{H}\right) - \left(\frac{x_2}{H}\right)^2\right] \tag{7-116a}$$

$$v_3(x_2) = -\pi DN \sin\phi \left[3\left(\frac{x_2}{H}\right) - 2\left(\frac{x_2}{H}\right)^2\right] \tag{7-116b}$$

where H is the depth of the screw channel, N screw speed, D screw diameter, and ϕ screw helix angle. $\partial p/\partial x_3$ has been determined by the boundary condition of no screw fight leakage:

$$\int_0^H v_3(x_2) \, dx_2 = 0 \tag{7-117}$$

The screw throughput is

$$Q = W\int_0^H v_1(x_2) \, dx_2 = \frac{1}{2}\pi DN \cos\phi - \frac{H^3 W}{12\eta}\frac{\partial p}{\partial x_1} \tag{7-118}$$

By the 1960s, there were efforts to simulate flow in screw extruders including both non-Newtonian and nonisothermal character. This is found notably in the work of Griffith and Zamodits and Pearson. They basically formulate in Cartesian coordinates a lubrication-based model with equations of motion of the form

$$0 = -\frac{\partial p}{\partial x_1} + \frac{\partial}{\partial x_2}\left[\eta \frac{\partial v_1}{\partial x_2}\right] \tag{7-119a}$$

$$0 = -\frac{\partial p}{\partial x_3} + \frac{\partial}{\partial x_2}\left[\eta \frac{\partial v_3}{\partial x_2}\right] \tag{7-119b}$$

and the energy equation (with neglect of heat convection):

$$0 = k\frac{\partial^2 T}{\partial x_2^2} + \eta\left[\left(\frac{\partial v_1}{\partial x_2}\right)^2 + \left(\frac{\partial v_3}{\partial x_2}\right)^2\right] \tag{7-120}$$

The non-Newtonian viscosity has the form as would be expected from the Criminale-Erickson-Filbey theory [Eqs. (7-50) and (7-51)] of

$$\eta = F_1(T) f_2\left[\left(\frac{\partial v_1}{\partial x_2}\right)^2 + \left(\frac{\partial v_3}{\partial x_2}\right)^2\right] \tag{7-121a}$$

where approximately for a power law fluid

$$\eta = A e^{-B(T-T_0)}\left[\left(\frac{\partial v_1}{\partial x}\right)^2 + \left(\frac{\partial v_3}{\partial x_2}\right)^2\right]^{\frac{n-1}{2}} \tag{7-121b}$$

Griffith and Zamodits and Pearson simultaneously solve Eq. (7-119), Eq. (7-120), and Eq. (7-121) to obtain the screw pumping characteristics. These show that decreasing power law exponent n and making the material more non-Newtonian reduces pumping characteristics

(Fig. 7.29). Similarly, increasing the viscosity and the temperature dependence of the viscosity by increasing $B(T)$ also reduces pumping characteristics. The above formulation may be improved by adding a term in $\rho c v_1 \, \partial T / \partial x_1$ on the left-hand side of Eq. (7-110).

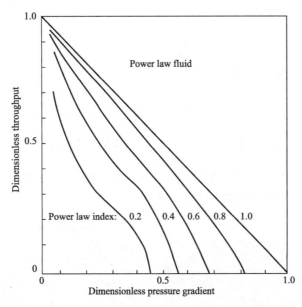

Fig. 7.29 Effect of non-Newtonian viscosity on screw pumping characteristics.

Elastomer compounds have yield values, so the actual behavior in screw extruders must be more complex than indicated by the preceding analysis. If the elastomer compound is a plastic material, then Eq. (7-119a) and Eq. (7-119b) may be replaced by

$$0 = -\frac{\partial p}{\partial x_1} + \frac{\partial \sigma_{12}}{\partial x_{12}} \tag{7-122a}$$

$$0 = -\frac{\partial p}{\partial x_3} + \frac{\partial \sigma_{32}}{\partial x_2} \tag{7-122b}$$

or

$$\sigma_{12} = x_2 \frac{\partial p}{\partial x_1} + C \tag{7-122c}$$

$$\sigma_{32} = x_2 \frac{\partial p}{\partial x_3} + C \tag{7-122d}$$

Consider the case of no die: the pressure gradient $\partial p / \partial x_1$ is zero, and the shear stress σ_{12} is uniform through the screw channel depth. When σ_{12} exceeds Y, a simple shear flow will arise. This will in turn set up a pressure gradient $\partial p / \partial x_3$ induced by the screw flights, and an associated circulational flow will develop. If there is a pressure gradient $\partial p / \partial x_1$, then from Eq. (7-122c) the shear stress will vary across the channel, reaching a maximum at the barrel, and can reach low values at intermediate values of the cross-section. If these stresses decrease below the yield value, a solid plug will develop at an intermediate position in the cross-section. However, as may be seen from experiments of Merges and Lehnen and Brzoskowski et al., only circulating flows and no plugs are seen. This is because of the small magnitudes of the yield value.

The flow in pin barrel extruders has been simulated in recent years by Brzoskowski et al. using hydrodynamic lubrication theory methods for a Newtonian fluid. Basically they embed a coordinate system in the screw channel and set up a flux balance of the form [equivalent to Eq. (7-114)]

$$\frac{\partial q_1}{\partial x_1} + \frac{\partial q_3}{\partial x_3} = 0 \tag{7-123}$$

where the fluxes have the form

$$q_1 = \int_0^H v_1(x_2) \, dx_2 = \frac{1}{2} U_2 H - \frac{H^3}{12\eta} \frac{\partial p}{\partial x_1} \tag{7-124a}$$

$$q_3 = \int_0^H v_3(x_2) \, dx_2 = \frac{1}{2} U_3 H - \frac{H^3}{12\eta} \frac{\partial p}{\partial x_3} \tag{7-124b}$$

Equation (7-121) is solved numerically by finite-difference methods. Fig. 7.30 shows computed flow fields in the neighborhood of pins.

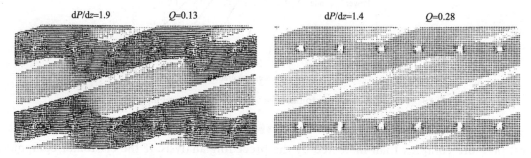

Fig. 7.30 Flow patterns in a pin barrel extruder.

Kumazawa et al. have determined mean temperature fields in screw extrusion by introducing the heat balance

$$\rho c q_1 \frac{d\overline{T}}{dx_1} = -[U_s(\overline{T}-T_s) + U_b(\overline{T}-T_b)] + (\sigma_{12})_b \pi DN \cos\phi + (\sigma_{32})_b \pi DN \sin\phi - q_1 \frac{dp}{dx_1} \tag{7-125}$$

where $\overline{T}$ is a "cup mixing" velocity averaged, temperatures T_s and T_b are screw and barrel temperatures, and U_s and U_b are overall screw and barrel heat transfer coefficients. It is possible to simulate complete velocity and temperature fields through the screw channel cross-section and not simply be limited to mean fluxes. Here one generalizes Eq. (7-119a), Eq. (7-119b), and Eq. (7-120) to

$$0 = -\frac{\partial p}{\partial x_1} + \frac{\partial}{\partial x_2}\left[\eta \frac{\partial v_1}{\partial x_2}\right] + \frac{\partial}{\partial x_3}\left[\eta \frac{\partial v_1}{\partial x_3}\right] \tag{7-126a}$$

$$0 = -\frac{\partial p}{\partial x_3} + \frac{\partial}{\partial x_2}\left[\eta \frac{\partial v_3}{\partial v_2}\right] \tag{7-126b}$$

$$\rho c \left[v_1 \frac{\partial T}{\partial x_1} + v_3 \frac{\partial T}{\partial x_3}\right] = k\left[\frac{\partial^2 T}{\partial x_2^2} + \frac{\partial^2 T}{\partial x_3^2}\right] + \eta\left[\left(\frac{\partial v_1}{\partial x_2}\right)^2 + \left(\frac{\partial v_3}{\partial x_2}\right)^2 + \left(\frac{\partial v_1}{\partial x_3}\right)^2\right] \tag{7-126c}$$

This allows evaluation of the effects of the screw flights and pins, but it is a much more difficult problem to solve.

7.8.6 Simulations of Die Extrusion

Efforts to model flow of polymer melts and elastomer compounds in dies are associated with the development of viscometry. The first extended effort at simulation of flow in dies was by Pearson who used the same lubrication theory formulation described in the two previous sections, i.e., basically Eq. (7-123a), Eq. (7-123b), Eq. (7-124a), and Eq. (7-124b).

Consider the following simple analysis intended for coathanger and T dies. In dies there is only pressure flow. For a Newtonian fluid, Eqs. (7-115a) and (7-115b) would have the simple solution (if the coordinate system is set in the center of the die channel)

$$v_1 = \frac{H^2}{2\eta}\left(-\frac{\partial p}{\partial x_1}\right)\left[1-\left(\frac{x_2}{H}\right)^2\right] \tag{7-127a}$$

$$v_3 = \frac{H^2}{2\eta}\left(-\frac{\partial p}{\partial x_3}\right)\left[1-\left(\frac{x_2}{H}\right)^2\right] \tag{7-127b}$$

This leads to fluxes

$$q_1 = 2\int_0^{H/2} v_1(x_2)\,\mathrm{d}x_2 = \frac{H^3}{12\eta}\left(-\frac{\partial p}{\partial x_1}\right) \tag{7-128a}$$

$$q_3 = 2\int_0^{H/2} v_3(x_2)\,\mathrm{d}x_2 = \frac{H^3}{12\eta}\left(-\frac{\partial p}{\partial x_3}\right) \tag{7-128b}$$

The fluxes may vary with position so that q_1 and q_3 must be balanced as in Eq. (7-123). This leads to

$$\frac{\partial}{\partial x_1}\left(H^3\frac{\partial p}{\partial x_1}\right)+\frac{\partial}{\partial x_1}\left(H^3\frac{\partial p}{\partial x_1}\right)=0 \tag{7-129}$$

governing the pressure field in a die and the flux field being obtained from Eq. (7-128).

Specifying the geometry of a die is equivalent to specifying $H(x_1,x_3)$. This allows computation of $p(x_1,x_3)$ and $q_j(x_1,x_3)$ for different designs. A common problem might be to determine what functions $H(x_1,x_3)$ will make $q_1(L,x_3)$ a constant, i.e., the die exit flux independent of position along the die lips.

The preceding problem needs to be more formulated generally using Eq. (7-119a) and Eq. (7-119b), including non-Newtonian character such as Eq. (7-121a). The fluxes q_1 and q_3 will depend on H and pressure gradients in a different manner for non-Newtonian fluid systems. Eq. (7-123) will remain valid but Eq. (7-124) must be replaced by

$$\frac{\partial}{\partial x_1}\left(S\frac{\partial p}{\partial x_1}\right)+\frac{\partial}{\partial x_3}\left(S\frac{\partial p}{\partial x_3}\right)=0 \tag{7-130}$$

where S is defined by

$$S=\frac{H^3}{12\overline{\eta}} \tag{7-131}$$

$\overline{\eta}$ is an effective non-Newtonian viscosity.

It should be noted that elastomer compounds are plastic materials and that stresses have low values in the center of die cross-sections. When the stresses drop below stress Y, the velocity gradients will go to zero, and there will be a localized solid plug.

It is possible to obtain information on temperature changes as well. To a first approximation, if adiabatic behavior is considered, one may use the logic leading to Eq. (7-125) to surmise that the increase in temperature in flow through a die is

$$\overline{T} - \overline{T}_0 = \frac{\Delta p}{\rho c} \tag{7-132}$$

where $\overline{T}$ is a "cup mixing" temperature.

More generally one needs to solve Eq. (7-130) together with

$$\rho c q_1 \frac{d\overline{T}}{dx_1} = -U(\overline{T} - T_{sn}) + q_1 \frac{dp}{dx_1} \tag{7-133}$$

while recognizing that the viscosity is temperature independent. Shah and Pearson have analyzed this type of problem and shown that instabilities are possible leading to θ_1 becoming a function of x_3. Velocity and temperature profiles may be determined through a die cross-section using Eq. (7-126), but it is again a difficult problem.

7.8.7 Simulations of Molding

The only available literature on simulation of molding of elastomer has involved injection mold filling, i.e., the filling of a fixed mold shape from a gate. The approach to simulation is very close to what we have discussed for dies in the previous section. The filling of a mold for an isothermal Newtonian fluid may be mathematically described by Eq. (7-129), and that for a nonisothermal non-Newtonian fluid, by introducing Eq. (7-130). This type of problem has been simulated by Isayev and his coworkers. The treatment basically resembles that of injection molding thermoplastics except for the necessity of considering vulcanization kinetics during mold filling. Thermal modeling including heat transfer in the thickness direction must he considered. This requires inclusion of Eq. (7-120) with heat convection terms.

References

[1] S. Ahn and J. L. White, Int. Polym. Process. 18, 243 (2003); J. Appl. Polym. Sci. 90, 1555 (2003); Ibid. 91, 651 (2004).

[2] R. D. Andrews, N. Hofman-Bang, and A. V. Tobolsky, J. Polym. Sci. 3, 669 (1948).

[3] R. D. Andrews and A. V. Tobolsky, J. Polym. Sci. 7, 221 (1952).

[4] D. J. Angier, R. J. Ceresa, and W. F. Watson, J. Polym. Sci. 34, 699 (1958).

[5] D. J. Angier, W. T. Chambers, and W. F. Watson, J. Polym. Sci. 25, 129 (1957).

[6] D. J. Angier, E. Farlie, and W. F. Watson, Trans. Inst. elastomer Ind. 34, 8 (1958).

[7] D. J. Angier and W. F. Watson, J. Polym. Sci. 20, 235 (1956).

[8] D. J. Angier and W. F. Watson, J. Polym. Sci. 25, 1 (1957).

[9] D. J. Angier and W. F. Watson, Trans. Inst. elastomer Ind. 33, 22 (1956).

[10] T. Araki and J. L. White, Polym. Eng. Sci. 38, 616 (1998).

[11] G. Ardichvili, Kautschuk 14, 23 (1938).

[12] G. Ardichvili, Kautschuk 19, 41 (1938).

[13] T. Asai, T. Fukui, K. Inoue, and M. Kuriyama, Paper III-4 presented at International elastomer Conference, Paris, June 1982.

[14] G. Ayrey, C. G. Moore, and W. F. Watson, J. Polym. Sci. 19, 1 (1956).

[15] Th. Baader, Kautschuk 14, 223 (1938).

[16] Th. Baader, Kautsch. Gummi 3, 159 (1950).

[17] Th. Baader, Kautsch. Gummi 3, 205 (1950).

[18] Th. Baader, Kautsch. Gummi 3, 245 (1950).

[19] Th. Baader, Kautsch. Gummi 3, 279 (1950).

[20] Th. Baader, Kautsch. Gummi 3, 323 (1950).

[21] Th. Baader, Kautsch. Gummi 3, 361 (1950).

[22] E. B. Bagley, J. Appl. Phys. 28, 624 (1957).

[23] E. B. Bagley and A. M. Birks, J. Appl. Phys. 31, 556 (1960).

[24] T. F. Ballenger and J. L. White, J. Appl. Polym. Sci. 15, 1949 (1971).

[25] R. L. Ballman, Rheol. Acta 4, 137 (1965).

[26] F. H. Banbury, U. S. Patent 1, 200, 070 (1916).

[27] F. H. Banbury, U. S. Patent 1, 227, 522 (1917).

[28] B. Bernstein, E. A. Kearsley, and L. J. Zapas, Trans. Soc. Rheol. 7, 391 (1963).

[29] B. Bernstein, E. A. Kearsley, and L. J. Zapas, J. Res. Natl. Bur. Stand. , Sect. B 68, 103 (1964).

[30] J. P. Berry, R. W. Sambrook, and J. Beasley, Plast. elastomer Process. p. 97 (1977).

[31] H. Bewley, English Patent 10, 825 (1845).

[32] E. C. Bingham, J. Wash. Acad. Sci. 6, 177 (1916).

[33] E. C. Bingham, "Fluidity and Plasticity," McGraw-Hill, New York, 1922.

[34] R. B. Bird and P. J. Carreau, Chem. Eng. Sci. 23, 427 (1968).

[35] R. B. Bird, W. E. Stewart, and E. N. Lightfoot, "Transport Phenomena," Wiley, New York, 1960.

[36] D. C. Bogue, IEC Fundam. 5, 253 (1966).

[37] D. C. Bogue, T. Masuda, Y. Einaga, and S. Onogi, Polym. J. 1, 563 (1970).

[38] D. C. Bogue and J. L. White, "Engineering Analysis of Non-Newtonian Fluids," NATO Agardograph.

[39] W. R. Bolen and R. E. Colwell, SPE ANTEC Tech. Pap. 4, 1008 (1958).

[40] L. Boltzmann, Sitzungsber, Kaiserl. Akad. Wiss. Wien 70, 275 (1874).

[41] C. Booth, Polymer 4, 471 (1963).

[42] H. H. Bowerman, E. A. Collins, and N. Nakajima, elastomer Chem. Technol. 47, 307 (1974).

[43] R. A. Brooman, English Patent 10, 582 (1845).

[44] R. Brzoskowski, S. Montes, J. L. White, and N. Nakajima, J. Non-Newtonian Fluid Mech. 31, 43 (1989).

[45] R. Brzoskowski and J. L. White, Int. Polym. Process. 5, 191 (1990).

[46] R. Brzoskowski and J. L. White, Int. Polym. Process. 5, 238 (1990).

[47] R. Brzoskowski, J. L. White, and B. Kalvani, Kunststoffe 80, 922 (1990).

[48] R. Brzoskowski, J. L. White, W. Szydlowski, N. Nakajima, and K. Min, Int. Polym. Process. 3, 134 (1988).

[49] R. Brzoskowski, J. L. White, W. Szydlowski, F. C. Weissert, N. Nakajima, and K. Min, elastomer Chem. Technol. 60, 945 (1987).

[50] R. Brzoskowski, J. L. White, F. C. Weissert, N. Nakajima, and K. Min, elastomer Chem. Technol. 59, 634 (1986).

[51] W. F. Busse, Ind. Eng. Chem. 24, 140 (1932).

[52] W. F. Busse and R. N. Cunningham, Proc. elastomer Technol. Conf. London, p. 288 (1938).

[53] W. F. Busse and R. Longworth, Trans. Soc. Rheol. 6, 179 (1962).

[54] J. F. Carley, R. S. Mallouk, and J. M. McKelvey, Ind. Eng. Chem. 45, 974 (1953).

[55] P. J. Carreau, Trans. Soc. Rheol. 16, 99 (1972).

[56] B. Caswell and R. I. Tanner, Polym. Eng. Sci. 18, 416 (1978).

[57] R. J. Ceresa, Polymer 1, 477 (1960).

[58] L. C. Cessna, U. S. Patent 3, 651, 187 (1972).

[59] E. M. Chaffee, U. S. Patent 16 (1836).

[60] E. M. Chaffee, U. S. Patent 1939 (1841).

[61] F. M. Chapman and S. L. Lee, Soc. Plast. Eng. J. 26, 37 (1970).

[62] I. J. Chen and D. C. Bogue, Trans. Soc. Rheol. 16, 59 (1972).

[63] J. J. Cheng and I. Manas-Zloczower, Polym. Eng. Sci. 29, 701, 1059 (1989).

[64] J. J. Cheng and I. Manas-Zloczower, J. Appl. Polym. Sci. Appl. Polym. Symp. 44, 53 (1989).

[65] J. W. Cho, J. L. White, and L. Pomini, Kautsch Gummi Tech 50, 496 (1997); J. W. Cho, J. L. White, and L. Pomini, Kautsch Gummi Tech 50, 728 (1997).

[66] F. N. Cogswell, Plast. Polym. April, 109 (1968).

[67] B. D. Coleman, M. E. Gurtin, and I. Herrera, Arch. Rat. Mech. Anal. 19, 1, 239, 266 (1965).

[68] B. D. Coleman and H. Markovitz, J. Appl. Phys. 35, 1 (1964).

[69] B. D. Coleman and W. Noll, Arch. Rat. Mech. Anal. 3, 289 (1959).

[70] R. T. Cooke, British Patent 431, 012 (1935).

[71] F. H. Cotton, Trans. Inst. elastomer Ind. 5, 487 (1931).

[72] G. Cotton and J. L. Thiele, elastomer Chem. Technol. 51, 749 (1978).

[73] W. P. Cox and E. H. Merz, J. Polym. Sci. 28, 619 (1958).

[74] W. O. Criminale, J. L. Erickson, and G. L. Filbey, Arch. Rat. Mech. Anal. 1, 410 (1958).

[75] M. Crochet, elastomer Chem. Technol. 62, 426 (1989).

[76] B. Darrow, U. S. Patent 1, 282, 767 (1918).

[77] B. David, T. Sapir, A. Nir, and Z. Tadmor, Int. Polym. Process. 7, 204 (1992).

[78] B. David, A. Nir, and Z. Tadmor, Int. Polym. Process. 5, 155 (1990).

[79] G. E. Decker and F. L. Roth, India elastomer World 128, 399 (1953).

[80] J. S. Deng and A. I. Isayev, elastomer Chem. Technol. 64, 296 (1991).

[81] T. deWitt, J. Appl. Phys. 26, 889 (1955).

[82] J. H. Dillon, Physics (NY) 7, 73 (1936).

[83] J. H. Dillon and L. V. Cooper, elastomer Age (NY) p. 1306 (1937).

[84] J. H. Dillon and N. Johnston, Physics 4, 225 (1933).

[85] H. J. Donald, U. S. Patent 3, 279, 501 (1966).

[86] R. A. Dunell and A. V. Tobolsky. J. Chem. Phys. 17, 100 (1949).

[87] S. Eccher and A. Valentinotti, Ind. Eng. Chem. 50, 829 (1958).

[88] E. R. G. Eckert and R. M. Drake, "Heat and Mass Transfer," McGraw-Hill, New York, 1959.

[89] J. E. Eilersen, U. S. Patent 3, 099, 859 (1963).

[90] R. Eisenschitz, B. Rabinowitsch, and K. Weissenberg, Mitt. Deutsch. Materialpruf. Son-derh, 9, 9 (1929).

[91] F. Farrel III, "Solid Men of the Place Gave Strong Roots," Newcomen Society, New York, 1956.

[92] P. Fay, U. S. Patent 2, 569, 373 (1951).

[93] J. D. Ferry, "Viscoelastic Properties of Polymers," Wiley, New York, 1969.
[94] V. Folt, elastomer Chem. Technol. 42, 1294 (1969).
[95] T. G. Fox and P. J. Flory, J. Am. Chem. Soc. 70, 2384 (1949).
[96] T. G. Fox, S. Gratch, and E. Loeshaek, in "Rheology," Academic Press, New York, Vol. 1, 1956.
[97] L. C. Frazier, U. S. Patent 3, 485, 692 (1969).
[98] P. K. Freakley and N. Y. Wan Idris, elastomer Chem. Technol. 52, 134 (1979).
[99] H. Freundlich and A. D. Jones, J. Phys. Chem. 40, 1217 (1936).
[100] S. M. Freeman and K. Weissenburg, Nature 161, 324 (1948).
[101] I. Furuta, V. M. Lobe, and J. L. White, J. Non-Newtonian Fluid Mech. 1, 207 (1976).
[102] T. L. Garner, India elastomer J. 78, (1929).
[103] R. E. Gaskell, J. Appl. Mech. 73, 334 (1950).
[104] P. Geyer, U. S. Patent 3, 375, 549 (1968).
[105] L. A. Goettler and A. J. Lambright, U. S. Patent (filed Feb. 2, 1976) 4, 056, 591 (1977); U. S. Patent (filed July 25, 1975) 4, 057, 610 (1977).
[106] L. A. Goettler, A. J. Lambright, R. I. Leib, and P. J. D. Mauro, elastomer Chem. Technol. 54, 277 (1981).
[107] L. A. Goettler, R. I. Leib, and A. J. Lambright, elastomer Chem. Technol. 52, 838 (1979).
[108] H. J. Gohlisch, W. May, F. Ramm, and W. Ruger, Extrusion of Elastomers, in "Plastics Extrusion Technology," F. Hensen (Ed.), Hanser, Munich, 1988.
[109] C. Goldstein, Trans. Soc. Rheol. 18, 357 (1974).
[110] S. Goldstein (Ed.), "Modern Developments in Fluid Dynamics," Clarendon Press, Oxford, 1938.
[111] C. F. Goodeve, Trans. Faraday Soc. 35, 342 (1939).
[112] C. F. Goodeve and G. W. Whitfield, Trans. Faraday Soc. 34, 511 (1938).
[113] C. Goodyear, U. S. Patent 3633 (1844).
[114] C. Goodyear, U. S. Patent 5536 (1848).
[115] C. Goodyear, "Gum Elastic and Its Varieties with a Detailed Account of Its Applications and Uses and the Discovery of Vulcanization," Privately printed, New Haven, CT, 1855.
[116] M. Gray, British Patent 5056 (1879).
[117] A. E. Green and R. S. Rivlin, Arch. Rat. Mech. Anal, 1, 1 (1957).
[118] F. M. Griffith. Ind. Eng. Chem. Fundam. 1, 180 (1962).
[119] R. W. Griffiths, Trans. Inst. elastomer Ind. 1, 308 (1926).
[120] F. F. Grosevnor, U. S. Patent (filed Aug. 17, 1916) 1, 254, 685 (1918).
[121] J. T. Gruver and G. Kraus, J. Polym. Sci. A. 2, 797 (1964).
[122] H. G. Gurhin and W. Spreutels, Kautsch. Gummi Kunstst. 43, 431 (1990).
[123] H. Hagen, Kautschuk 15, 88 (1939).
[124] C. D. Han and L. H. Drexler, J. Appl. Polym. Sci. 17, 2329 (1973).
[125] M. H. Han, J. L. White, N. Nakajima, and R. Brzoskowski, Kautsch. Gummi Kunstst. 43, 1060 (1990).
[126] C. Hancock, English Patent 11, 147 (1846).
[127] C. Hancock, English Patent 11, 208 (1847).
[128] C. Hancock, English Patent 11, 575 (1847).
[129] C. Hancock, English Patent 11, 874 (1847).
[130] T. Hancock, English Patent 4768 (1823).
[131] T. Hancock, English Patent 7344 (1837).
[132] T. Hancock, English Patent 9952 (1843).

[133] T. Hancock, "Personal Narrative of the Origin and Progress of the Caoutchouc or India elastomer Manufacture in England," Longman, Brown, Green, Longmans & Roberts, London, 1857.

[134] E. G. Harms, Eur. elastomer J. 6, 23 (1978); Kunststoffe 74, 33 (1984).

[135] E. G. Harms, G. Menges, and R. Hegele, German Offenlegungsshift 2, 235, 784 (1974); U. S. Patent 4, 178, 104 (1979); U. S. Patent 4, 199, 263 (1980).

[136] E. R. Harrel, J. P, Porter, and N. Nakajima, elastomer Chem. Technol. 64, 254 (1991).

[137] H. Herrmann, "Schneckenmaschinen in der Verfahrenstechnik," Springer, Berlin, 1972.

[138] H. Herrmann, in "Kunststoffe ein Werkstoff macht Karriere," W. Glenz (Ed.), Hanser, Munich, 1985.

[139] W. H. Herschel and R. S. Bulkley, Proc. ASTM 26, 62 (1926).

[140] J. Hoekstra, Physics 4, 295 (1933); Kautschuk 9, 750 (1933).

[141] K. Hohenemser and W. Prager, Z. AMM 12, 216 (1932).

[142] B. Hu and J. L. White, Int. Polym. Process. 8, 18 (1993).

[143] B. Hu and J. L. White, elastomer Chem. Technol. 66, 257 (1993).

[144] B. Hu and J. L. White, Kautsch Gummi Kunstst 49, 285 (1996).

[145] Y. Ide and J. L. White, J. Appl. Polym. Sci. 22, 1061 (1978).

[146] A. I. Isayev, Injection Molding of elastomer Compounds, in "Injection and Compression Molding Fundamentals," A. I. Isayev (Ed.), Dekker, New York, 1987.

[147] A. I. Isayev, Injection Molding of elastomers, in "Comprehensive Polymer Science," G. Allen, J. C. Bevington, and S. L. Aggarwal (Eds.), Pergamon Press, Oxford, Vol. 7, 1989.

[148] A. I. Isayev and Y. H. Huang, Adv. Polym. Technol. 9, 167 (1988).

[149] A. I. Isayev and X. Fan, J. Rheol. 34, 35 (1990).

[150] A. I. Isayev, M. Sobhanie, and J. S. Deng, elastomer Chem. Technol. 61, 906 (1988).

[151] M. W. Johnson and D. Segalman, J. Non-Newtonian Fluid Mech. 2, 255 (1977).

[152] M. R. Kamal and S. Kenig, Polym. Eng. Sci. 12, 294 (1972).

[153] E. Karrer, Ind. Eng. Chem. Anal. Ed. 1, 158 (1929); Ind. Eng. Chem. 21, 770 (1929).

[154] E. Karrer, J. M. Davies, and E. O. Dieterich, Ind. Eng. Chem., Anal. Ed. 2, 96 (1930).

[155] D. H. Killeffer, "Banbury, The Master Mixer," Palmerton, New York, 1962.

[156] J. K. Kim, International Seminar on Elastomers, J. Appl. Polym. Sci., Appl. Polym. Symp. 50, 145 (1992).

[157] J. K. Kim and J. L. White, Nihon Reoroji Gakkaishi 17, 203 (1989).

[158] J. K. Kim and J. L. White, Int. Polym. Process. 6, 103 (1991).

[159] J. K. Kim, J. L. White, K. Min, and W. Szydlowski, Int. Polym. Process. 4, 9 (1989).

[160] K. J. Kim and J. L. White, Polym. Eng. Sci. 39, 2189 (1999).

[161] P. S. Kim and J. L. White, elastomer Chem. Technol. 67, 871 (1994).

[162] P. S. Kim and J. L. White, Kautsch Gummi Kunstst 49, 10 (1996).

[163] P. S. Kim and J. L. White, elastomer Chem. Technol. 69, 686 (1996).

[164] R. G. King, Rheol. Acta 5, 35 (1966).

[165] E. Konrad, Angew. Chem. 62, 491 (1950).

[166] R. Koopmann, Kautsch. Gummi Kunstst. 36, 108 (1983).

[167] R. Koopmann, Polym. Testing 5, 341 (1985).

[168] R. Koopmann, Kautsch. Gummi Kunstst. 38, 281 (1985).

[169] R. Koopmann and H. Kramer, J. Testing Eval. ASTM 12, 407 (1984).

[170] R. Koopmann and J. Schnetger, Kautsch. Gummi Kunstst. 39, 131 (1986).

[171] G. Kraus and J. T. Gruver, J. Polym. Sci. A 3, 105 (1965).
[172] C. Koolihiran and J. L. White, J. Appl. Polym. Sci. 78, 1551 (2000).
[173] K. Kubota, R. Brzoskowski, J. L. White, F. C. Weissert, N. Nakajima, and K. Min, elastomer Chem. Technol. 60, 924 (1987).
[174] T. Kumazawa, R. Brzoskowski, and J. L. White, Kautsch. Gummi Kunstst. 43, 688 (1990).
[175] H. Lamb, "Hydrodynamics," 6th ed., Cambridge University Press, London, 1932.
[176] A. Lasch and K. Frei, German Patent 738, 787 (1943).
[177] A. Lasch and E. Stromer, German Patent 641, 685 (1937).
[178] H. Leaderman, Ind. Eng. Chem. 35, 374 (1943).
[179] H. Leaderman, "Elastic and Creep Properties of Filamentous Materials and Other High Polymers," Textile Found., Washington, D. C., 1943.
[180] L. J. Lee, J. F. Stevenson, and R. M. Griffith, U. S. Patent 4, 425, 289 (1984).
[181] P. W. Lehman, U. S. Patent 2, 096, 362 (1937).
[182] J. P. Lehnen, Kunststofftechnik 9, 3, 90, 114, 198, 352 (1970).
[183] A. I. Leonov, Rheol. Acta 15, 85 (1976).
[184] A. I. Leonov, J. Rheol. 34, 155 (1990).
[185] A. I. Leonov, J. Non-Newtonian Fluid Mech. 42, 343 (1990).
[186] G. H. Lewis, U. S. Patent 1, 252, 821 (1918).
[187] L. L. Li and J. L. White, elastomer Chem. Technol. 69, 628 (1996).
[188] P. W. Litchfield, "Industrial Voyage," Doubleday, Garden City, NY, 1954.
[189] V. M. Lobe and J. L. White, Polym. Eng. Sci. 19, 617 (1979).
[190] A. Lodge, Trans. Faraday Soc. 52, 120 (1956).
[191] C. Y. Ma, J. L. White, F. C. Weissert, A. I. Isayev, N. Nakajima, and K. Min, elastomer Chem. Technol. 58, 815 (1985).
[192] C. Y. Ma, J. L. White, F. C. Weissert, and K. Min, Polym. Composites 6, 215 (1985).
[193] C. Y. Ma, J. L. White, F. C. Weissert, and K. Min, J. Non-Newtonian Fluid Mech. 17, 275 (1985).
[194] C. Macintosh, English Patent 4804 (1823).
[195] H. Mahn, H. Orth, K. H. Roitzsch, and W. Woeckener, in "Kunststoffe ein Werkstoffe macht Karriere," W. Glenz (Ed.), Hanser, Munich, 1985.
[196] C. Maillefer, Swiss Patent 363, 149 (1962).
[197] H. Mark and A. V. Tobolsky, "Physical Chemistry of High Polymeric Systems," 2nd ed., Interscience, New York, 1950.
[198] B. Marzetti, India elastomer World 68, 776 (1923).
[199] B. Marzetti, Chim. Ind. Appl. 5, 342 (1923).
[200] B. Marzetti, Atti Della Reale Acad. Naz. Lincei 32, 399 (1923); elastomer Age 19, 454 (1924).
[201] B. Marzetti, elastomer Age 20, 139 (1925).
[202] T. Masuda, K. Kitagawa, T. Inoue, and S. Onogi, Macromolecules 4, 116 (1970).
[203] T. Masuda, Y. Ohta, and S. Onogi, Macromolecules 4, 763 (1971).
[204] J. C. Maxwell, Philos. Trans. R. Soc. 157, 249 (1866).
[205] J. M. McKelvey, Ind. Eng. Chem. 45, 982 (1953).
[206] F. M. McLaughlin, U. S. Patent 2, 325, 001 (1943).
[207] J. Meissner, Rheol. Acta 8, 78 (1969).
[208] G. Menges and E. G. Harms, Kautsch. Gummi Kunstst. 25, 469 (1972); 27, 187 (1974).
[209] G. Menges and J. P. Lehnen, Plastverarbeiter 20, 31 (1969).

[210] W. Meskat, Kunststoffe 41, 417 (1951).

[211] W. Meskat, Kunststoffe 45, 87 (1955).

[212] A. P. Metzger and J. R. Knox, Trans. Soc. Rheol. 9, 13 (1965).

[213] A. B. Metzner, J. L. White, and M. M. Denn, AIChE J. 12, 863 (1966); Chem. Eng. Prog. 62, 12, (1966).

[214] S. Middleman, Trans. Soc. Rheol. 13, 125 (1969).

[215] W. H. Miller, U. S. Patent 3, 069, 853 (1991).

[216] W. H. Miller, C. C. Lee, and J. F. Stevenson, Int. Polym. Process. 6, 253 (1991).

[217] K. Min and J. L. White, elastomer Chem. Technol. 58, 1024 (1985).

[218] K. Min and J. L. White, elastomer Chem. Technol. 60, 361 (1987).

[219] N. Minagawa and J. L. White, J. Appl. Polym. Sci. 20, 501 (1976).

[220] W. W. Minoshima, J. L. White, and J. E. Spruiell, Polym. Eng. Sci. 20, (1980).

[221] S. R. Moghe, elastomer Chem. Technol. 39, 247 (1976).

[222] W. D. Mohr and R. S Mallouk, Ind. Eng. Chem. 51, 765 (1959).

[223] S. Montes and J. L. White, elastomer Chem. Technol. 55, 1354 (1982).

[224] S. Montes and J. L. White, Kautsch. Gummi Kunstst. 44, 731 (1991).

[225] S. Montes and J. L. White, Kautsch. Gummi Kunstst. 44, 937 (1991).

[226] S. Montes and J. L. White, J. Non-Newtonian Fluid Mech. , 49, 277 (1993).

[227] S. Montes, J. L. White, and N. Nakajima, J. Non-Newtonian Fluid Mech. 28, 183 (1988).

[228] S. Montes, J. L. White, N. Nakajima, F. C. Weissert, and K. Min, Elastomer Chem. Technol. 61, 698 (1988).

[229] M. Mooney, J. Rheol. 2, 210 (1931).

[230] M. Mooney, Ind. Eng. Chem. , Anal. Ed. 6, 147 (1934).

[231] M. Mooney, Physics 7, 413 (1936).

[232] M. Mooney, J. Colloid Sci. 2, 69 (1947).

[233] M. Mooney, personal communication (ca. 1965); unpublished U. S. elastomer Company report (1953).

[234] M. Mooney, in "Rheology," F. R. Eirich (Ed.), Academic Press, New York, Vol. 2, 1958.

[235] M. Mooney, Proc. Int. elastomer Conf. (Washington), p. 368 (1959).

[236] M. Mooney, elastomer Chem. Technol. 35, xxvii (1962).

[237] M. Mooney and S. A. Black, J: Colloid Sci. 7, 204 (1952).

[238] A. Morikawa, K. Min, and J. L. White, Adv. Polym. Technol. 8, 383 (1988).

[239] A. Morikawa, J. L. White, and K. Min, Kautsch. Gummi Kunstst. 41, 1226 (1988).

[240] A. Morikawa, K. Min, J. L. White, Int. Polym. Process 4, 23 (1989).

[241] L. Mullins, J. Phys. Colloid Chem. 54, 239 (1950).

[242] L. Mullins and R. W. Whorlow, Trans. Inst. elastomer Ind. 27, 55 (1951).

[243] N. Nakajima, elastomer Chem. Technol. 54, 266 (1981); 55, 937 (1982).

[244] N. Nakajima, H. H. Bowerman, and E. A. Collins, J. Appl. Polym. Sci. 21, 3063 (1977).

[245] N. Nakajima and E. R. Harrel, elastomer Chem. Technol. 52, 962 (1979).

[246] K. Ninomiya, J. Colloid Sci. 14, 49 (1959).

[247] W. Noll, J. Rat. Mech. Anal. 4, 3 (1955).

[248] W. Noll, Arch. Rat. Mech. Anal. 2, 197 (1958).

[249] R. H. Norman, Plast. elastomer Int. 5, 243 (1985).

[250] K. Oda, J. L. White, and E. S. Clark, Polym. Eng. Sci. 16, 585 (1976).

[251] K. Oda, J. L. White, and E. S. Clark, Polym. Eng. Sci. 18, 25 (1978).

[252] J. G. Oldroyd, Proc. Cambridge Philos. Soc. 43, 100 (1947).
[253] J. G. Oldroyd, Proc. Cambridge Philos. Soc. 43, 382 (1947).
[254] J. G. Oldroyd, Proc. Cambridge Philos. Soc. 45, 595 (1949).
[255] J. G. Oldroyd, Proc. R. Soc. A 200, 523 (1950).
[256] J. G. Oldroyd, Q. J. Mech. Appl. Math 4, 271 (1951).
[257] J. G. Oldroyd, Proc. R. Soc. A 245, 278 (1958).
[258] S. Onogi, T. Masuda, I. Shiga, and F. Costaschuk, U. S.-Japan Seminar on Polymer Processing and Rheology, Appl. Polym. Symp. 20, 37 (1973).
[259] G. Osanaiye, A. I. Leonov, and J. L. White, J. Non-Newtonian Fluid Mech., 49, 87 (1993).
[260] G. Passoni, U. S. Patent 4, 775, 240 (1988).
[261] J. R. A. Pearson, Trans. J. Plast. Inst. 30, 230 (1962).
[262] J. R. A. Pearson, Trans. J. Plast. Inst. 31, 125 (1963).
[263] J. R. A. Pearson, Trans. J. Plast. Inst. 32, 239 (1964).
[264] J. R. A. Pearson, "Mechanical Principles of Polymer Melt Processing," Pergamon Press, New York, 1966.
[265] J. R. A. Pearson and C. J. S. Petrie, Plast. Polym. 38, 85 (1970); J. Fluid Mech. 40, 1 (1970); J Fluid Mech. 42, 609 (1970).
[266] N. Phan Tien, J. Non-Newtonian Fluid Mech. 22, 259 (1978).
[267] N. Phan Tien and R. I. Tanner, J. Non-Newtonian Fluid Mech. 2, 353 (1977).
[268] W. R. Phillips, U. S. Patent 3, 195, 183 (1965).
[269] W. T. Pigott, Trans. ASME 73, 947 (1951).
[270] M. Pike and W. F. Watson, J. Polym. Sci. 9, 229 (1952).
[271] G. H. Piper and J. R. Scott, J. Sci. Inst. 22, 206 (1945).
[272] R. S. Porter and J. F. Johnson, Proc. 4th Int. Rheol. Cong. 2, 467 (1965).
[273] R. S. Porter and J. F. Johnson, Chem. Rev. 66, 1 (1966).
[274] W. Prager, "Mechanics of Continua," Ginn, Boston, 1961.
[275] W. Prager, in Appendix to the article by J. T. Bergen in "Processing of Thermoplastic Materials," E. C. Bernhardt (Ed.), Reinhold, New York, 1959.
[276] W. Prager and P. Hodge, "Theory of Perfectly Plastic Solids," Wiley, New York, 1951.
[277] V. C. Ratliff, U. S. Patent 2, 720, 679 (1955).
[278] M. Reiner, "Deformation and Flow," Lewis, London, 1948, and later editions.
[279] M. Reiner, "Lectures on Theoretical Rheology," North Holland, Amsterdam, three editions (1943), (1949), (1960).
[280] M. Reiner, Phys. Today 17, 62 (1964).
[281] O. Reynolds, Philos. Trans. R. Soc. London, Ser. A 174, 935 (1883).
[282] O. Reynolds, Philos. Trans. R. Soc. London, Ser. A 177, 157 (1886).
[283] J. G. Richardson, U. S. Patent 3, 761, 553 (1973).
[284] S. Richardson, J. Fluid Mech. 56, 609 (1972).
[285] R. S. Rivlin, J. Rat. Mech. Anal. 5, 179 (1956).
[286] R. S. Rivlin and J. L. Ericksen, J. Rat. Mech. Anal. 4, 323 (1955).
[287] N. Sato, M. Miyaoka, S. Yamasaki, K. Inoue, A. Koriyama, T. Fukui, T. Asai, K. Nakagawa, and T. Masaki, U. S. Patent 4, 284, 358 (1981); U. S. Patent 4, 300, 838 (1981).
[288] J. R. Schaefgen and P. J. Flory, J. Am. Chem. Soc. 70, 2709 (1948).
[289] G. H. Schanz, U. S. Patent 2, 807, 833 (1957).
[290] G. Schenkel, "Kunststoffe Extruder-Technik," Hanser, Munich, 1963.

[291] J. R. Scott, Trans. Inst. elastomer Ind. 7, 169 (1931).

[292] J. R. Scott, Trans. Inst. elastomer Ind. 10, 481 (1935).

[293] G. Schramm, Kaustch. Gummi Kunstst. 40, 756 (1987).

[294] T. Schwedoff, J. Deo Phys. 9, 34 (1890); Phys. Z. 1, 552 (1900).

[295] F. A. Seiberling and C. A. Carlton, U. S. Patent 2, 032, 508 (1936).

[296] Y. T. Shah and J. R. A. Pearson, Chem. Eng. Sci. 29, 1485 (1974).

[297] K. C. Shin and J. L. White, elastomer Chem. Technol. 66, 121 (1993).

[298] K. C. Shin, J. L. White, R. Brzoskowski, and N. Nakajima, Kautsch. Cummi Kunstst. 43, 181 (1990).

[299] A. H. P. Skelland, "Non-Newtonian Fluid and Heat Transfer," Wiley, New York, 1967.

[300] R. W. Snyder and J. I. Haase, U. S. Patent 1, 952, 469 (1934).

[301] M. Sobhanie, J. S. Deng, and A. I. Isayev; J. Appl. Polym. Sci. , Appl. Polym. Symp. 44, 115 (1989).

[302] M. Sobhanie and A. I. Isayev, elastomer Chem. Technol. 62, 939 (1989).

[303] L. E. Soderquist, U. S. Patent 2, 296, 800 (1942).

[304] H. J. Song, J. L. White, K. Min, N. Nakajima, and F. C. Weissert, Adv. Polym. Technol. 8, 43 (1988).

[305] J. F. Stevenson, AIChE J. 18, 540 (1972).

[306] J. F. Stevenson, Extrusion of elastomer and Plastics, in "Comprehensive Polymer Science," G. Allen, J. C. Bevington, and S. L. Aggarwal (Eds.), Pergamon Press, Oxford, Vol. 7, 1989.

[307] J. F. Stevenson and W. H. Miller, U. S. Patent 5, 067, 885 (1991).

[308] G. G. Stokes, Trans. Cambridge Philos. Soc. 8, 287 (1845).

[309] G. G. Stokes, "Report on Recent Researches in Hydrodynamics," Report of the British Association (1846).

[310] G. G. Stokes, Trans. Cambridge Philos. Soc. 9, (1851).

[311] Y. Suetsugu and J. L. White, J. Appl. Polym. Sci. 28, 1481 (1983).

[312] Y. Suetsugu and J. L. White, J. Non-Newtonian Fluid Mech. 14, 121 (1984).

[313] H. Tanaka and J. L. White, Polym. Eng. Sci. 20, 949 (1980).

[314] R. I. Tanner, U. S.-Japan Seminar on Polymer Processing and Rheology, Appl. Polym. Symp. 20, 201 (1973).

[315] R. H. Taylor, India elastomer World 112, 582 (1945).

[316] R. H. Taylor, J. H. Fielding, and M. Mooney, "Symposium on elastomer Testing," ASTM, New York, 1947, p. 36.

[317] A. V. Tobolsky, "Properties and Structure of Polymers," Wiley, New York, 1960.

[318] M. Toh, T. Gondoh, T. Mon, S. Haren, and Y. Murakami, International Seminar on Elastomers, J. Appl. Polym. Sci. , Appl. Polym. Symp. 50, 133 (1992).

[319] S. Toki and J. L White, J. Appl. Polym. Sci. 27, 3171 (1982).

[320] N. Tokita, personal communication (1975).

[321] N. Tokita and J. L. White, J. Appl. Polym. Sci. 10, 1011 (1966).

[322] C. Truesdell and R. A. Toupin, The Classical Field Theories, in "Handbuch der Physik," Vol. III/1, Springer, Berlin, 1960.

[323] L. G. Turk, U. S. Patent 3, 782, 871 (1974).

[324] D. M. Turner and M. D. Moore, Plastic elastomer Proc. , (Sept. /Dec.), 81 (1980).

[325] D. Tyson and L. Comper, U. S. Patent 3, 230, 581 (1966).

[326] A. van Rossem and H. van der Meijden, Kautschuk 3, 369 (1927); elastomer Age (NY) 23, 443 (1928).

[327] B. Vergnes, N. Bennani, and C. Guichard, Int. Polym. Process. 1, 19 (1986).

[328]   G. R. Vila, Ind. Eng. Chem. 36, 1113 (1944).
[329]   G. V. Vinogradov, Pure Appl. Chem. 39, 115 (1974).
[330]   G. V. Vinogradov, V. D. Fikham, R. D. Radushkevich, and A. Ya Malkin, J. Polym. Sci. A-28, 657 (1970).
[331]   G. V. Vinogradov and A. Y. Malkin, J. Polym. Sci. A-2 4, 135 (1966).
[332]   G. V. Vinogradov and A. Y. Malkin, "Rheology of Polymers," Mir, Moscow, 1980.
[333]   G. V. Vinogradov, A. Y. Malkin, E. P. Plotnikova, O. Y. Sabasi, and N. E. Nikolayava, Int. J. Polym. Mater. 2, 1 (1972).
[334]   M. H. Wagner, Rheol. Acta 15, 136 (1976).
[335]   W. F. Watson, Trans. Inst. elastomer Ind. 34, 237 (1958).
[336]   K. Weissenberg, Nature 159, 310 (1947).
[337]   G. S. Whitby (Ed.), "Synthetic rubber," Wiley, New York, 1954.
[338]   J. L. White, J. Appl. Polym. Sci. 8, 1129 (1964).
[339]   J. L. White, J. Appl. Polym. Sci. 8, 2339 (1964).
[340]   J. L. White, J. Inst. elastomer Ind. (elastomer Ind.) 8, 148 (1974).
[341]   J. L. White, Polym. Eng. Sci. 15, 44 (1975).
[342]   J. L. White, J Non-Newtonian Fluid Mech. 5, 177 (1979).
[343]   J. L. White, International Seminar on Elastomers, J. Appl. Polym. Sci., Appl. Polym. Symp. 50, 109 (1992).
[344]   J. L. White, Int. Polym. Process. 7, 2 (1992).
[345]   J. L. White, Int. Polym. Process. 7, 110 (1992).
[346]   J. L. White, elastomer Chem. Technol. 65, 527 (1992).
[347]   J. L. White, Int. Polym. Process. 7, 290 (1992).
[348]   J. L. White, Int. Polym. Process. 8, 2 (1993).
[349]   J. L. White, "elastomer Processing: Technology, Materials, and Principles," Epic Press, Hanser, Munich, 1995.
[350]   J. L. White and J. W. Crowder, J. Appl. Polym. Sci. 18, 1013 (1974).
[351]   J. L. White, M. H. Han, N. Nakajima, and R. Brzoskowski, J. Rheol. 35, 167 (1991).
[352]   J. L. White and J. K. Kim, J. Appl. Polym. Sci., Appl. Polym. Symp. 49, 59 (1989).
[353]   J. L. White and A. Kondo, J. Non-Newtonian Fluid Mech. 3, 41 (1977).
[354]   J. L. White and Y. Lobe, Rheol. Acta 21, 167 (1982).
[355]   J. L. White and A. B. Metzner, J. Appl. Polym. Sci. 7, 1867 (1963).
[356]   J. L. White and H. Tanaka, J. Non-Newtonian Fluid Mech. 8, 1 (1981).
[357]   J. L. White and N. Tokita, J. Appl. Polym. Sci. 9, 1921 (1965).
[358]   J. L. White and N. Tokita, J. Appl. Polym. Sci. 11, 321 (1967).
[359]   J. L. White and N. Tokita, J. Appl. Soc. Japan 22, 719 (1967); 24, 436 (1968).
[360]   J. L. White and N. Tokita, J. Appl. Polym. Sci. 12, 1589 (1968).
[361]   J. I. White, Y. Wang, A. I. Isayev, N. Nakajima, F. C. Weissert, and K. Min, elastomer Chem. Technol. 60, 337 (1987).
[362]   I. Williams, Ind. Eng. Chem. 10, 324 (1923).
[363]   M. L. Williams, R. F. Landel, and J. D. Ferry, J. Am. Chem. Soc. 77, 3701 (1955).
[364]   R. F. Wolf, India elastomer World, Aug. 1, p. 39 (1936).
[365]   Y. Yabushita, R. Brzoskowski, J. L. White, and N. Nakajima, Int. Polym. Process. 6, 219 (1989).
[366]   M. Yamamoto, J. Phys. Soc. Japan 11, 413 (1956).
[367]   H. Yamane and J. L. White, Polym. Eng. Rev. 2, 167 (1982).
[368]   H. H. Yang and I. Manas-Zloczower, Int. Polym. Process. 7, 195 (1992).

[369] N. V. Zakharenko, F. S. Tolstukhina, and G. M. Bartenev, elastomer Chem. Technol. 35, 326 (1962).
[370] H. Zamodits and J. R. A. Pearson, Trans. Soc. Rheol. 13, 357 (1969).
[371] L. J. Zapas, J. Res. Natl. Bur. Stand., Sect. A 70, 525 (1966).
[372] L. J. Zapas and T. Craft, J. Res. Natl. Bur. Stand., Sect. A 69, 541 (1965).
[373] S. Zaremaha, Bull Acad. Sci. Cracow, June, p. 594 (1903).

Chapter 8
Thermoplastic Elastomers

Thermoplastic elastomers (TPE), sometimes referred to as thermoplastic rubbers, are a class of copolymers or a physical mix of polymers (usually a plastic and a rubber) that consist of materials with both thermoplastic and elastomeric properties. While most elastomers are thermosets, thermoplastics are in contrast relatively easy to use in manufacturing, for example, by injection molding. Thermoplastic elastomers show advantages typical of both rubbery materials and plastic materials. The benefit of using thermoplastic elastomers is the ability to stretch to moderate elongations and return to its near original shape creating a longer life and better physical range than other materials. The principal difference between thermoset elastomers and thermoplastic elastomers is the type of cross-linking bond in their structures. In fact, crosslinking is a critical structural factor which imparts high elastic properties.

TPE materials have the potential to be recyclable since they can be molded, extruded and reused like plastics, but they have typical elastic properties of rubbers which are not recyclable owing to their thermosetting characteristics. They can also be ground up and turned into 3D printing filament with a recyclebot. TPE also require little or no compounding, with no need to add reinforcing agents, stabilizers or cure systems. Hence, batch-to-batch variations in weighting and metering components are absent, leading to improved consistency in both raw materials and fabricated articles. Depending on the environment, TPEs have outstanding thermal properties and material stability when exposed to a broad range of temperatures and non-polar materials. TPEs consume less energy to produce, can be colored easily by most dyes, and allow economical quality control.

8.1 Introduction

Thermoplastic elastomers (TPEs) are an extremely fast growing segment of polymer manufacturing. A rate of 5% growth per year is expected until 2020, at which time the total domestic demand for these materials will reach 1.0 million tons at a total annual sales of approximately 3 billion dollars per year. The majority of this growth comes in the form of replacements for other types of materials, and the growth of so-called "soft-touch" surfaces. In the approximately 10 years since the second edition of this book appeared, there has been an important technological advancement in this area: the vastly increased production of thermoplastic polyolefin elastomers as a result of the worldwide adoption of metallocene catalysts.

The primary advantage of TPE over conventional rubber is the ease (and therefore low cost) of processing, the wide variety of properties available, and the possibility of recycling and reuse. Besides the obvious environmental benefits of a recyclable raw material, TPE scrap material can be reprocessed. The disadvantage of these materials relative to thermosets is the relatively high cost of raw materials, the general inability to load TPEs with low cost fillers such as carbon black, and poor chemical and temperature resistance. This last property prevents TPEs from being used in automobile tires.

In order to qualify as a thermoplastic elastomer, a material must have three essential characteristics:

(1) The ability to be stretched to moderate elongations and, upon the removal of stress, return to something close to its original shape.

(2) Processable as a melt at elevated temperature.

(3) Absence of significant creep.

In nearly all cases, thermoplastic elastomers will be a copolymer, i. e., there will be at least two monomers in the polymer chain. A thermoplastic elastomer will generally have the modulus versus temperature curve shown in part (c) of Fig. 8.1. The plateau region must include the service temperature of the material. Typically through changes in comonomer composition or identity, the plateau can be shifted upward or downward, giving the manufacturer a great deal of flexibility.

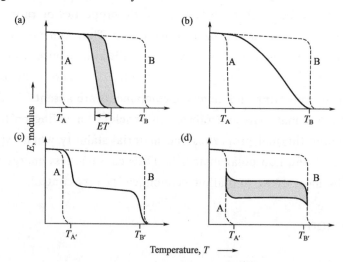

Fig. 8.1 Diagram of the dependence of modulus on temperature for copolymers. In all the sketches, the dashed lines refer to the behavior for the pure materials. The hatched area shows that the range of the behavior can vary depending on the relative amounts of A and B. (a) Random A-B copolymer. (b) Block copolymer of A and B with extremely short blocks. (c) Segmented block copolymer with imperfect phase separation. (d) Segmented block copolymer with perfect phase separation.

Most TPEs have certain similar structural characteristics. The comonomers typically have long runs, making the material a block copolymer. The comonomers are almost always dissimilar, leading to microphase separation on a nanometer length scale, which means these materials are properly termed nanomaterials. The driving force for phase separation is always enthalpic and usually one to two orders of magnitude weaker than primary valence bonds. Crystallinity, hydrogen bonding, ionic, and van der Waals driving forces all have been shown to cause microphase separation in these systems.

The two phases in these systems have different properties. One phase, the soft phase, contains a component that is above its glass transition temperature (T_g) and melting temperature (T_m) so that chains have a high amount of mobility. The other phase, the hard phase, contains chains that are rigidly locked in place, because the service temperature is below either T_m or T_g. The relative amount of the two phases controls the physical properties of the TPE by deter-mining which phase is isolated or continuous. The ability to easily vary these parameters through stoichiometry allows TPEs to be used in the wide variety of applications alluded to earlier. A large number of structures fall into the category of thermoplastic elastomers. The structures of some common thermoplastic elastomers are shown in Fig. 8.2.

Shell Development Company developed the first commercially available thermoplastic elastomer in the early 1960s, which became the KRATON family of materials. These materials are either poly(styrene-butadiene-styrene) (SBS), poly(styrene-isoprene-styrene) (SIS), or poly(styrene-ethylenebutylene-styrene) (SEBS) triblock copolymers. Phase separation occurs because of the incompatibility between the hard and soft segment. The styrene-rich domains serve as the hard phase since T_g for polystyrene is approximately 100℃. The molecular mass polydispersity is low because these triblocks are typically anionically polymerized. The terminal styrene anchors the polymer, which gives this material the necessary toughness while the flexible soft segment imparts elasticity. Approximately 50% of all thermoplastic elastomers produced are SBS, SIS, or SEBS triblock copolymers.

Another major category of thermoplastic elastomers, accounting for approximately 30% of the thermoplastic elastomer market, are based polyolefins. The three most important materials that comprise this category are copolymers of ethylene and propylene (EP), copolymers of propylene and higher α-olefins such as 1-butene and 1-octene, and copolymers of ethylene and α-olefins. In the latter two cases, the propylene or ethylene is the major component. Two important differences between the ethylene-rich and propylene-rich materials are flexibility and softening point; the ethylene-rich materials are more flexible but also soften at a temperature roughly 50℃ below that of the propylene-rich materials. These types of materials are important enough to be given a very common abbreviation, TPO, which stands for thermoplastic elastomer-olefinic. A specific example of these types of TPEs are the Engage® family of materials, which are copolymers of ethylene and α-olefins. Metallocene catalysts, mentioned earlier, have allowed better control over the run lengths of the normal EP copolymers. Generally EP copolymers and copolymer blends are slightly higher cost and higher performance than the triblock copolymers.

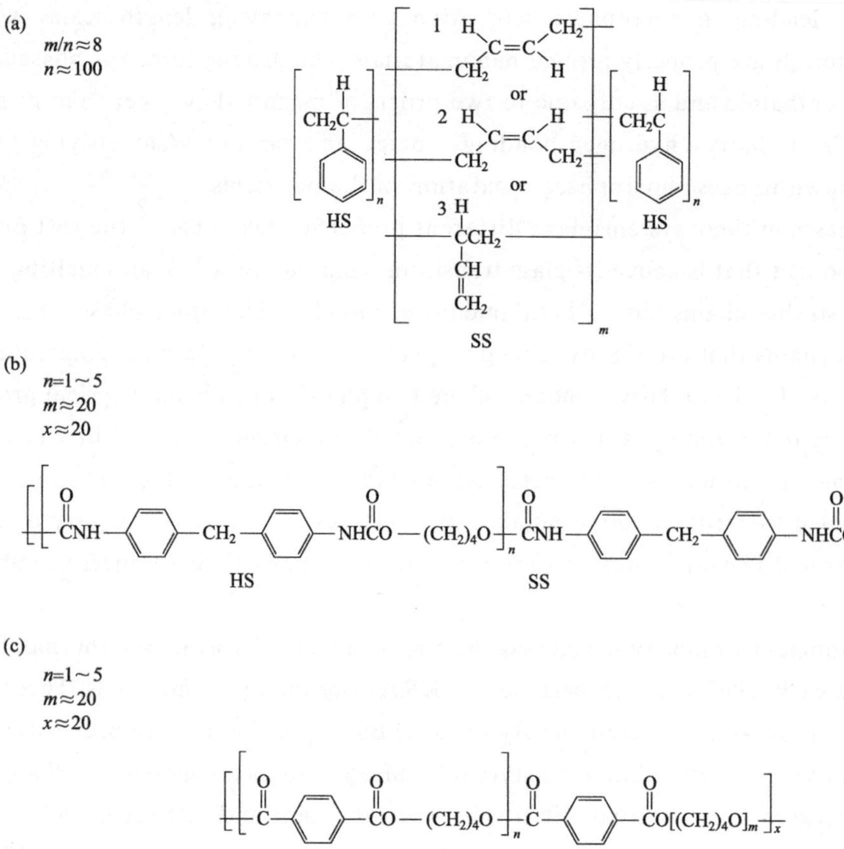

Fig. 8.2 Structures of commercially important thermoplastic elastomers; HS = hard segment, SS = soft segment. (a) SBS; (b) MDI-BD-PTMO polyurethane; (c) PTMT and PTMO copoly-ester; (d) Nylon 66 and PTMO copolyamide; (e) random copolymer ionomer E-MAA neutralized with sodium.

The EP thermoplastic elastomers are distinguished from the crosslinked analogues, which are not thermoplastics since reforming is impossible. A very important thermoplastic elastomer is comprised of a blend of an EP copolymer with an ethylene-propylene-diene (EPDM) terpolymer. This latter mate rial is, of course, a crosslinkable thermoset; however, these materials can be processed as thermoplastics if the crosslinkable component is present at low enough concentration to be present as an isolated phase. Melt-processing causes the formation of chemical bonds within the isolated rubber phase, a process called dynamic vulcanization. A commercial example of this type of material is Santoprene® manufactured by Advanced Elastomer Systems. Other blends of noncrosslinkable TPEs with crosslinkable materials are used commercially. These materials are classified as elastomer blends.

Polyurethane elastomers are copolymers with a hard segment that contains aromatic rings and a polyether or polyester soft segment. Polyurethane elastomers are part of a family of materials termed segmented block copolymers, which is defined as a material with alternating hard and soft segments that repeat multiple times in a single polymer chain. Microphase separation occurs because of incompatibility between the aromatic rings and the soft segment. In some cases, the hard segment may crystallize as well. Segmented block copolymers have the general formula $(AB)_x$ where a triblock copolymer has the general formula ABA. Polyurethanes are generally manufactured from an aromatic diisocyanate, an oligomeric diol, and a low molecular mass diol. The low molecular mass diol is typically called the chain extender because it links AB segments together. A typical polyurethane based on diphenylmethylene-4,4'-diisocyanate (MDI), poly(tetramethylene oxide) (PTMO), and butanediol (BD) is shown in Fig. 8.2. A commercial example of a polyurethane is the MDI-BD family of materials manufactured by the Dow Chemical Company under the commercial name Pellethane®. Polyurethanes are generally expensive and have found uses in high performance structural applications as well as foams. Approximately 15% of the thermoplastic elastomer market is claimed by polyurethanes.

Another class of segmented block copolymers is segmented block copolymers containing an aromatic polyester hard segment and a polyether or polyester soft segment. Hard segment crystallization provides the driving force for phase separation in this system. A copolyester made from poly(tetramethylene terephtalate) and poly(tetramethylene oxide) (4GT-PTMO), which is a member of the Hytrel® high performance thermoplastic elastomers manufactured by DuPont, is also shown in Fig. 8.2. These materials are oil resistant and stable to higher temperatures than other thermoplastic elastomers, which makes these materials more suitable for applications such as automobile engine parts.

Two other thermoplastic elastomers are shown in Fig. 8.2. The copolyamide thermoplastic elastomers are comparable to the copolyesters in structure. Crystallization provides the driving force for phase separation in these materials as well. These materials have especially low chemical permeability and can offer good properties at low temperatures. A commercial example of a co-polyamide is PEBAX® marketed by Atofina. Co-polyamides compete with polyurethanes and copolyesters for market share. Ionomers are the final material that will be discussed. Ionomers are materials where a small mole fraction of monomers, usually less than 10%, contain an ionic functionality.

These materials are not segmented like most of the other materials discussed in this chapter, rather the ionic groups are distributed randomly along the polymer backbone. Incompatibility between the ionic groups and the nonpolar polymer backbone leads to the formation of ionic-rich domains. A commercial example of an ionomer is Surlyn® manufactured by DuPont, shown in Fig. 8.2.

The general synthetic concepts will be augmented by a discussion that is concerned only with synthetic techniques important in TPE synthesis. The bulk of this chapter will be concerned with the TPE morphology, since the morphology determines their physical and mechanical properties. This discussion will be followed by specific examples illustrating the effect of structure on the physical properties. The thermodynamics of phase separation, which includes detailed discussions on morphology as well as thermal behavior, will follow. The rheology and processing of these materials will be discussed next. Finally, some applications for thermoplastic elastomers will be highlighted. Emphasis will be given to those topics that are common to all thermoplastic elastomers; however, some discussion specific to commercially important materials will also be included. Also, the discussions are not meant to be exhaustive; for further information, the interested reader can consult some of the references provided after this chapter.

8.2 Synthesis of Thermoplastic Elastomers

8.2.1 Step-Growth Polymerization: Polyurethanes, Polyether-esters, Polyamides

Polyurethanes, co-polyesters, and co-polyamides are all produced via step-growth polymerization. In step-growth polymerizations relevant to the production of thermoplastic elastomers, a molecule containing two reactive functional groups of one type (e.g., a diisocyanate) reacts with another molecule containing two reactive functional groups of another type (e.g., a diol) to form a polymer. Step-growth polymerizations require extremely high conversions (>99%) to produce high molecular mass product. Generally, TPE properties are only weakly dependent on overall molecular mass, so the breadth of the distribution is usually not very important, although it is important to achieve high molecular masses.

Controlling the ratio of functional reactive groups is critical to achieving high molecular masses as the following formula shows in the case of a stoichiometric imbalance between the two reacting functional groups:

$$\bar{n}_n = \frac{1+r}{1+r-2rp} \tag{8-1}$$

where r is the ratio of the initial imbalance of the functional groups and is defined to be always less than 1, p is the extent of reaction, and n is the number average degree of polymerization. Even to achieve moderate molecular masses, stoichiometric imbalances of more than a few percent cannot be tolerated.

Before embarking on a short description of the synthesis of polyurethanes, the reader should be aware that polyurethanes are generally divided into three classes: foams, coatings, and TPEs. This chapter concerns only the latter, and the morphological difference between a TPE urethane and others is the fact that the chain is not crosslinked, and the segment lengths are longer. Synthetically, crosslinked materials tend to use water, whereas water must be excluded from a reaction that wishes to produce a TPE polyurethane. Books on the subject generally cover all three types of polyurethanes, sometimes without a clear distinction between the different uses.

Polyurethanes can be synthesized in solution or in bulk. Solution polymerized polyurethanes generally have more uniform hard and soft segment distributions. Bulk polymerized polyurethanes generally have higher molecular masses, partially caused by side reactions that cause crosslinking. The majority of industrially produced polyurethanes are made in bulk. Bulk synthesized polyurethanes are reacted at temperatures between 80℃ and 120℃. The isocyanate-alcohol reaction is highly exothermic, which means that heat must be removed from the reaction mixture so that the temperature will be kept below the degradation temperature of 140℃. Generally, higher temperatures mean more side reactions and crosslinking. To produce totally linear polyurethanes, temperatures under 50℃ should be used. Two methods are used in the bulk, the "one-shot" method and the prepolymer method. In the "one-shot" method, all the ingredients are mixed together; in the prepolymer method, the diisocyanate and oligomeric diol are allowed to react before the chain extender is added. Generally, the oligomeric diol and the diisocyanate are not miscible, so the reaction occurs at the interface between the two components, which can lead to large compositional variations over the course of the reaction. A common solvent for the diols, the diisocyanate, and the polymer is needed for solution polymerization. A relatively polar organic solvent such as N,N-dimethyl acetamide or dimethyl sulfoxide will generally suffice. Usually the analogue of the prepolymer method, described earlier for bulk polymerization, is used for solution polymerized polyurethanes. Unlike the bulk reaction, an organotin catalyst is employed in solution polymerization. Organotin compounds are generally not utilized in the bulk method since incomplete catalyst removal will lead to poor hydrolytic stability of the final product.

Another common thermoplastic elastomer produced from a step-growth polymerization are the copolyesters. The general reaction scheme is very similar to the polyurethanes except the hard segment reactants are usually diesters rather than diisocyanates, which means that a small molecule by-product must be removed in order to achieve high conversions. Generally, co-polyesters are produced in a melt phase trans-esterification polymerization. In the first step, the oligomeric diol, chain extender, and diester are mixed and allowed to react at elevated temperature (~200℃) in the presence of a titanate catalyst in order to produce prepolymer. The high temperature serves to drive the reaction toward completion as well as remove the low boiling byproduct through fractional distillation. Polymers are produced by raising the temperature to 250℃ and lowering the pressure to less than 133Pa, which causes the second trans-esterification. The chain extender is removed as a by-product from this second stage. The temperature must be kept below 260℃ in order to prevent substantial degradation. Reaction completion is monitored by the viscosity of the reaction mixture. Molecular masses of approximately 25000g/mol generally result from this procedure.

Different pairs of functional groups can be reacted to form amide linkages, and all of them have been used to produce copolyamides. These include reactions between carboxylic acids and diamines, acid chlorides and diamines, and carboxylic acids and isocyanates. The latter is especially useful for producing copolyamides with aromatic hard segments. The copolyamides are most commonly produced with either ester or amide linkages between the amide hard segment and soft segment. Again, high temperatures are sometimes needed to produce high molecular mass material.

8.2.2 Anionic Polymerization: Styrene-Diene Copolymers

Anionic polymerization is used to produce the styrenic block copolymers and produces a polymer with an extremely narrow block and overall molecular mass distributions. The narrow molecular mass distributions are extremely useful in fundamental studies of polymers, and have led to a great deal of study of anionic polymerization, much more than justified by its commercial importance. In fact, the styrenic-block copolymers are the only polymers produced in large quantities via anionic polymerization. The extremely narrow polydispersity is evident in the following expression for the polydispersity index:

$$\frac{n_w}{n_x} = 1 + \frac{n_x}{(n_x + 1)^2} \tag{8-2}$$

where n_w is the weight-average degree of polymerization and n_x is the number-average degree of polymerization. This analysis assumes no chain transfer and no chain termination reactions, which are easily achieved using low temperatures and pure ingredients. An extremely fast initiation rate relative to propagation is also assumed.

The triblock copolymers SBS or SIS are produced using a high vacuum process that serves to eliminate both oxygen and water from the reaction mixture. Besides the general features that apply to any anionic synthesis, it is important to produce a polydiene block with a high 1,4-structure so that the glass transition temperature of the soft phase will be sufficiently low. This is generally accomplished by using a relatively non-polar solvent, e.g., cyclohexane or toluene, which favors 1,4-formation. SEBS is formed through the hydrogenation of SBS.

Generally three anionic polymerization methods can be used to produce the styrene-diene triblock copolymers: a three-stage process with monofunc tional initiator, a two-stage process with monofunctional initiator and a difunc tional linking agent, and a two-stage process with a difunctional initiator. In all cases, the initiator of choice is an organolithium compound. In the first process, the styrene monomer is anionically polymerized, followed by the butadiene monomer, and then finally more styrene monomer is added to produce a triblock polymer. The butadiene rapidly initiates when added to the styrene blocks; however, a small amount of polar solvent is needed to initiate the final styrene anionic polymerization. In the second process, the styrene and diene blocks are polymerized separately then combined using the difunctional linking agent, which produces symmetric triblocks. The key in this method is to control the stoichiometry exactly so that no diblocks are formed. A small amount of diblock material has an extremely adverse effect on mechanical properties. Finally, a difunctional initiator can be used to initiate the diene block followed by the styrene endblocks. This method suffers from the difficulty in finding an appropriate difunctional initiator.

8.2.3 Catalytic Polymerization

Ethylene-propylene copolymer thermoplastic elastomers and other ethylene-α-olefin copolymers are produced from either Ziegler-Natta or metallocene processes. Ziegler-Natta polymerizations are reactions catalyzed by a mixture of alkyl metal halides, e.g., $Al(C_2H_5)_2Cl$ and transition metal salts, e.g., $TiCl_4$. These polymerizations typically produce crystallizable stereospecific products. For example, in propylene-α-olefin copolymers, polypropylene crystallites provide the hard phase for the TPE. Long blocks of one component, which are necessary for phase separation, result from the inherent reactivity ratios of the components. In spite of the industrial importance of Ziegler-Natta polymerizations, their kinetics are not well understood. This is a result of the often heterogeneous nature of this reaction as well as the possibility of multiple mechanisms. Blends of EP copolymers with isotactic polypropylene are often made in a two-step reaction process. The first reactor contains only propylene monomer, while the second reactor contains both propylene and ethylene. Other blends are made by mixing the pure components together in the melt. Blending conditions have a large effect on the resulting properties. Compatibilizers, which are often triblock copolymers, can adjust the blend's characteristics.

Metallocene polymerizations have many of the same properties as Ziegler-Natta polymerizations, but as the name implies, the catalyst is very different. Metallocene catalysts have two general properties that lead to stereospecific products: first, the catalyst is rigid, and second, the catalyst is chiral. These properties are obtained by having a metal cation sandwiched between two negatively charged cyclopentadienyl anions. Fig. 8.3 shows a representative metallocene catalyst, 1,1'-ethylenedi-η^5-indenylzirconium dichloride.

A modification to this approach is to replace one of the cyclopentadienyl anions with a ring containing a nitrogen atom, and then constraining the ring by bonding the nitrogen via a bridging atom, typically silicon. Just as in the Ziegler-Natta polymerization, a compound with labile alkane groups is required; one typical compound used is methyl alumoxane. Unfortunately for the economics of the process, the amount of aluminum required is

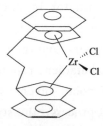

Fig. 8.3 Structure of 1,1'-ethylenedi-η^5-indenylzirconium dichloride, a common metallocene catalyst. Only one enantiomer is shown; both enantiomers produce isotactic product for this particular catalyst.

much higher in the metallocene materials than in the Ziegler-Natta materials. In spite of the higher cost of these materials, metallocene polymerizations are gaining in market share because of the ability to better control the number of defects, which in turn can be used to tune copolymer properties. Just as with Ziegler-Natta polymerizations, metallocene polymerizations are used to produce non-stereospecific polymers because of changes in molecular weight distribution and reactivity ratios relative to other types of polymerizations.

8.2.4 Free Radical Polymerization

Free radical copolymerization is used to produce ionomers that are used commercially as thermoplastics elastomers. There are two types of TPE ionomers, copolymers of ethylene and methacrylic acid, and copolymers of ethylene and acrylic acid. The mole fraction of the acid monomer is typically 5% or less. The property difference between the two types of copolymers are small; copolymers of ethylene and methacrylic acid are slightly more resistant to the formation of anhydrides that can crosslink the polymer. The usual method uses high pressure (~1500atm) and high temperatures (~130℃) similar to the method used for low density polyethylene. In contrast with Ziegler-Natta polymerizations, the kinetics are well known. The neutralizing cation after synthesis is hydrogen; this material does not phase separate into the nanometer-size aggregates required for toughness improvements. Therefore neutralization with an appropriate metal salt occurs industrially by mixing the polymer and the salt together in an extruder or roll mill. In commercial materials, not all the hydrogen atoms are replaced by a metal cation, or else a material with too high a viscosity is produced; typical neutralization levels in most ionomers is around 50%.

8.2.5 Molecular Weight and Chain Structure

Determining the molecular weight and the chain microstructure is very difficult for some thermoplastic elastomers. Nearly all molecular weight characterization techniques rely on the ability to dissolve the polymer in a solvent in such a manner that the polymer chains behave individually. Since TPEs are usually composed of two dissimilar components, a good solvent for one of the components may be a poor solvent for the other component, which leads to aggregation of similar components from different chains. Hence the molecular weight is an aggregate molecular weight, rather than a true molecular mass. Even if this problem is overcome, methods that measure molecular weight relative to a standard, such as gel permeation chromatography (GPC), cannot be converted to actual molecular weight because the proportionality constant depends on the copolymer composition.

For polyurethanes and polyesters, end group analysis, either by titration or infrared analysis, can be used to monitor the extent of reaction and the final molecular mass if the reaction stoichiometry is properly controlled. This method only works well if the molecular mass is below about 20000g/mol. Membrane osmometry and light scattering can give reliable molecular masses if the appropriate solvent is used; however, a long time is required to make these measurements. Different solvents must be tested so that data can be collected without significant curvature that indicates aggregation. Ultra centrifugation is almost never used because, among other issues, by definition a theta solvent does not exist for thermoplastic elastomers. The molecular weight calculated from viscosity measurements will also depend on the copolymer composition since the radius of gyration (R_g) of the polymer will generally be a function of composition. Fractionation methods also typically fail because the fractionation efficiency depends not only on molecular mass distribution but also composition distribution. Most often, GPC is used and the molecular weight is usually reported relative to a standard and for a certain solvent.

Molecular mass characterization of polyurethanes using GPC was studied in detail. Three different molecular mass standardization methods were tested, and it was found that a multidetector method using an refractive index and an UV spectrometer in series provided the most accurate results. The UV spectrometer was used to calculate the variation of the derivative of refractive index with respect to concentration at each point in the chromatogram, which accounts for the effect of changing copolymer composition with molecular mass. The authors found that the normal polystyrene standards used to calibrate the GPC give an upper limit on the actual molecular mass. Nevertheless, the authors were hesitant to call these values absolute. A related issue is the molecular mass of the linear prepolymers if a staged reaction scheme is used; a procedure for determining the molecular mass and soft segment-hard segment distribution of end-capped diisocyanates using GPC and a double detection method was given recently.

Accurate molecular masses can be easily measured for the triblock copolymers. Because these materials are produced via anionic polymerization, the theoretical molecular mass is usually very close to the actual molecular mass. Depending on the polymerization method, the pure styrene or pure diene block can be removed and the molecular mass measured before making the triblock. Finally, molecular mass standards of polystyrene, polybutadiene, and polyisoprene are commercially available, which means measuring absolute molecular mass of these blocks is easily done using GPC.

The chain microstructure has a very important influence on the properties of TPEs. As mentioned earlier, production of SBS or SIS with a high 1,4-content is necessary. TPO properties also depend quite heavily on any deviations of the microstructure from the ideal head-to-tail, pure isotactic, or syndiotactic microstructure. Properties such as tacticity, *cis-trans* isomerization, and copolymerization content are usually characterized using NMR. Peak positions and peak intensities are used to quantitatively ascertain microstructure to a high degree of accuracy. Copolymer composition can also be determined using NMR. Infrared spectroscopy can also be employed to determine microstructural characteristics in some polymers.

In the segmented block copolymers, the average molecular mass of the hard and soft segments is very important, and the number of studies for the various types of TPE that have investigated this variable are too numerous to list here, although the anionic block copolymers deserve special mention in this regard because of their importance in elucidating fundamental thermodynamic information. In general, the higher the molecular mass of the blocks, the more complete the phase separation. However, it should be noted that complete phase separation is not always desired, since a decrease in toughness will eventually occur as phase separation becomes more complete.

A related parameter, which is particularly important in both segmented block copolymers as well as the TPOs, is the distribution (as opposed to the average) of hard and soft segment lengths. In the latter, the hard segments are typically crystalline domains of either propylene or ethylene, and studies of crystallite size distribution are an important topic not just to TPEs. For the step-growth block copolymers, the distribution of soft segments tends to be fairly trivial to characterize, since the fully formed soft segment is typically one of the ingredients fed to the reaction. The hard segment, on the other hand, is typically formed as part of the polymerization, and more sophisticated approaches are required. Two approaches have been applied to study the effect of the distribution of hard segment lengths: the first is to synthesize hard segments with a known length distribution and the second is to attempt to measure the hard segment lengths. Of course, the latter is necessary for all commercial materials. A number of experimental methods have been used to determine hard segment length distributions. One method is to chemically cleave the structure at the point where the hard segment and the soft segment join and analyze the residual fractions using GPC or HPLC. ^{13}C NMR and mass spectroscopy have been used to determine hard segment length in MDI-PTMO polyurethane chains extended with ethylene diamine. Monte Carlo simulations of Markov processes have also been used to derive hard segment molecular mass distributions under ideal and non-ideal conditions.

8.3 Morphology of Thermoplastic Elastomers

8.3.1 General Characteristics

In spite of the wide variety in structure of two-phase thermoplastic elastomers, the number of underlying morphologies in commercially important materials is surprisingly small. If we assume that A is initially the minor component and B the major component, then the seven equilibrium morphologies given in increasing A content are:

(1) Isolated spheres of A in a continuous matrix of B;
(2) Hexagonally packed isolated cylinders of A in a continuous matrix of B;

(3) Alternating lamellae of A and B Hexagonally packed isolated cylinders of B in a matrix of A;

(4) Isolated spheres of B in a matrix of A.

Other, more complicated morphologies can be generated with anionically synthesized block copolymers; a complete review of this subject is beyond the scope of this chapter, and the interested reader is referred to reviews and monographs on the subject. Transmission electron micrographs are shown in Fig. 8.4~Fig. 8.6 for each of the phases for a SBS block polymer.

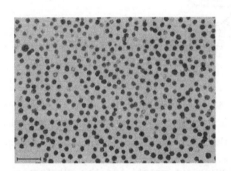

Fig. 8.4 TEM of spherical butadiene domains in SBS (80% styrene) film cast from toluene. The same patterns were observed in both normal and parallel sections, confirming the periodicity of butadiene domains.

Fig. 8.5 TEM of cylindrical microdomains in extruded and annealed SBS sample where the micrograph was taken perpendicular to the extrusion direction. An electron micrograph taken parallel to the extrusion direction had a striated structure.

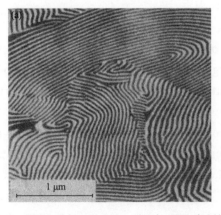 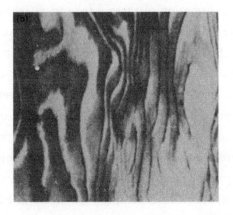

Fig. 8.6 TEM of lamellar domains in SBS (40% styrene) film cast from cyclohexane. (a) Normal section; (b) parallel section with lamellar layers orienting their surface parallel to film surface.

In general, the morphology in commercial materials are not nearly as well developed as the micrographs imply. Once the general morphology (spheres, cylinders, etc.) has been specified, specific questions remain concerning domain spacing, radii, and arrangement in space.

The underlying morphology will have a large effect on the physical properties. The soft phase is usually the continuous phase to maintain the elastomeric behavior of the material. In materials that have crystalline hard segments, such as copolyesters and polyurethanes, both phases are essentially continuous. Thermoplastic elastomers almost never have the soft segment as the isolated phase and the hard segment as the continuous phase. The most common morphology for TPEs is illustrated in Fig. 8.7 for SBS triblock copolymers.

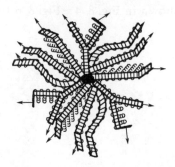

Fig. 8.7 Illustration of phase separation in SBS triblock copolymers normally used as a thermoplastic elastomer. The isolated spherical styrene domains form the hard phase, which acts as both intermolecular tie points and filler. The continuous butadiene imparts the elastomeric characteristics to this polymer.

The isolated hard phase acts as intermolecular tie points for the elastomeric soft phase. Normally the isolated hard domains are between 1 nm and 20 nm. The ability to crystallize changes the morphology dramatically; the crystalline regions form rectangular thin sheets of material termed lamellae and hence will not be constrained into isolated domains and form more of a continuous structure. As with all polymers, lamellae can organize into spherulites and, a bit surprisingly, the segmented nature of the TPE does not prevent the formation of these large superstructures.

The interface between the hard and soft segment is often considered as a separate phase. The influence of the interface on the properties of TPEs is considered to be very important, although not well understood. The higher cost of TPEs relative to commodity polymers is justified in different ways for different materials, but in general a property worth a high premium is toughness, i.e., a great deal of energy is required for failure. Toughness is achieved by creating stress-relieving processes in a material, and in TPEs much of these processes occur at the hard and soft domain interface resulting in the deflection and bifurcation of cracks. The adhesion between the interface and the polymer matrix is especially important in developing toughness in EP copolymer blends since failure often occurs in blends because of poor inter facial adhesion.

As implied earlier, kinetic considerations play an extremely important role in the morphology of TPEs. Predicting the precise morphology of any TPE without some knowledge of sample history is impossible. To produce a material that is near the equilibrium morphology, the TPE can be annealed above the dissolution transition of the hard segment, but below the order-disorder temperature. Polydispersity in block lengths may also give morphologies that deviate substantially from the underlying morphologies. Finally, many industrially produced parts have oriented microdomains remaining from processing.

The remainder of this section contains an introduction to analytical methods used to characterize the morphology of TPEs. A short description of each method is given, with emphasis on those characteristics relevant to TPEs. Important examples of each method are also discussed, so the reader can understand how each method is used to characterize the morphology of TPEs.

8.3.2 Studies of Morphology

8.3.2.1 Transmission Electron Microscopy (TEM)

In transmission electron microscopy (TEM), a sample is bombarded with electrons, and the number of electrons that travel through the sample, which is proportional to the sample thickness and the electron density, is measured as a function of position. If the electron densities are not identical for each phase in a TPE, the number of electrons passing through the two phases may be different and hence contrast will be created. TEM can be used on samples with extremely well ordered morphologies where the length scale of electron density variations is greater than the resolution of the method. If the morphology is not well ordered, then the average number of domains of one type that a single electron passes through does not vary with position. The thickness of the film can be reduced, but a practical limit to thickness reductions exists. Since the electron density difference between atoms that comprise polymers is generally very small, a low molecular mass compound with a high electron density that will preferentially bind or migrate to one of the two phases is almost always added. Typical compounds used for such staining are OsO_4 and RuO_4.

Styrene-diene block copolymers represent the optimum material for TEM. If prepared properly, these materials can be almost 100% phase separated with the staining agent residing almost exclusively in the diene phase. The morphology is generally well ordered, and individual domain sizes tend to be large. Examples of electron micrographs of SBS block copolymers were given in Fig. 8.4~Fig. 8.6. TEM is probably the most important characterization method used for the determination of bulk morphology, so most papers in the field published TEM micrographs of their systems. The review papers and books referenced earlier contain a thorough description of TEM studies of block copolymers.

Copolyesters and polyurethanes have also been imaged. These two materials have not been the focus of electron microscopy studies for many years because of the difficulties in excluding artifacts. In fact, one must cast a very questioning eye on published results of these systems for this very reason. A number of papers have appeared describing results of TEM studies of ionomers. The heavy metal atoms are used to provide the necessary contrast, and high-resolution TEM is used to image these systems. A much wider variety of ion aggregate morphologies have been found than previously thought, including spheres and vesicles.

8.3.2.2 Infrared and Raman Spectroscopy

In infrared spectroscopy, a beam of infrared light is passed through a sample, and light is absorbed if the frequency of the light is the same as the frequency of a normal mode of vibration. This method is sensitive to molecular bonding between atoms. Beer's law relates the absorbance (A) of a vibration to its absolute concentration:

$$A = \varepsilon t c \tag{8-3}$$

where ε is the absorption coefficient, t is the path length, and c is the concentration. As indicated previously, infrared spectroscopy can be used to explore chain microstructure. Beyond chemical characterization, infrared spectroscopy can aid in describing interchain interactions such as hydrogen bonding or crystallization.

Hydrogen bonding is the secondary bonding of hydrogen atoms to an atom containing unpaired electrons. Hydrogen bonds generally have strengths on the order of 12.6kJ/mol, which is a half order of magnitude less than covalent bond strengths but an order of magnitude greater than simple van der Waals interactions. Hydrogen bonding is extremely important in polyurethanes and co-polyamides and occurs between the urethane or amide hydrogen and the carbonyl oxygen. Since hydrogen bonding occurs primarily between hard segments, it provides a strong driving force for phase separation and hard segment crystallization.

Hydrogen bonding causes a shift toward lower wave number (lower energy) in the vibration of the bonds involving the hydrogen donating group and the hydrogen accepting group, which indicates that both primary bond strengths have been diminished because of the secondary hydrogen bond. In a typical thermoplastic elastomer that can hydrogen bond, vibrations of functional groups participating in hydrogen bonding are split into bands termed *bonded* and *free*. Table 8.1 gives wave numbers that have been assigned to the bonded and free bands for the C=O and N—H vibrations for a variety of polymers. There is a small but noticeable effect of the polymer type on the exact wave number for the vibration. The two vibrations listed for some C=O stretches are due to the ordered (crystalline) and disordered bonded states.

Table 8.1 Band assignments of hydrogen bonded functional groups in TPEs

Material	Free	Bonded
Polyurethane: hexamethylene diisocyanate/butanediol. C=O	1720 cm^{-1}	1685 cm^{-1}
		1700 cm^{-1}
N—H	3440 cm^{-1}	3320 cm^{-1}
Polyurethane: 2,6-toluene diisocyanate/butanediol/PTMO C=O	1740 cm^{-1}	1700 cm^{-1}
N—H	3460 cm^{-1}	3300 cm^{-1}
Polyurethane: MDI/butanediol/PTMO C=O	1733 cm^{-1}	1703 cm^{-1}
N—H	3420 cm^{-1}	3320 cm^{-1}
Polyamide: Nylon 11 C=O	1680 cm^{-1}	1636 cm^{-1}
	1645 cm^{-1}	
N—H	3450 cm^{-1}	3300 cm^{-1}

Since hydrogen bonding is extremely important and infrared studies are relatively inexpensive and simple, a large number of papers have been published in this field. Many studies have tried to relate the amount of hydrogen bonding to the degree of phase separation using Beer's law. Harthcock explored the carbonyl stretching region in model polyurethanes with monodisperse hard segment lengths. A detailed morphological model that relates the hard segment order to the infrared vibrational frequency was given and is reproduced in Fig. 8.8. A similar study was undertaken for polyurethane ureas. Changes in hydrogen bonding due to structural changes such as hard segment and soft segment type have also been studied. FTIR has aided in the discovery of structure-mechanical property relationships for segmented polyurethanes. The thermal behavior of hydrogen bonding has been investigated in some detail. A series of papers published on polyamides and polyurethanes concluded that many of the studies involving thermal behavior were based on incorrect reasoning concerning the nature of hydrogen bonds. These authors showed that quantitative analysis of the N—H region was impossible because of the large difference in extinction coefficient between the bonded and free bands along with the large change in the bonded N—H extinction coefficient as a function of temperature. The simple analysis normally employed to describe the dependence of the extinction coefficient with temperature could not be used because the vibrational frequency of the N—H bond shifted, which indicated that the hydrogen bond strength changed. Therefore the reader should be very careful interpreting papers that quantitatively analyze the N—H stretching region.

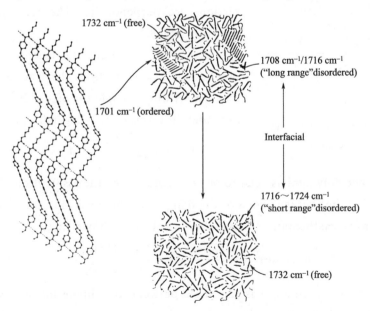

Fig. 8.8 Morphological model developed by Harthcock that shows the wave number of the C=O stretching vibration absorbed intensity to the local environment.

Raman spectroscopy in general gives complementary information to infrared spectroscopy and, because of the historical difficulty in performing these experiments, Raman studies of TPEs have been much less frequent than infrared spectroscopy. Raman experiments have become much simpler in the last 20 years with the development of powerful lasers and CCD detection systems, although the cost of a Raman spectrometer is still much more than an FTIR spectrometer. Raman spectroscopy has two distinct advantages to IR spectroscopy: first, the signal can be measured remotely using fiber optics, which makes it possible to use Raman for process control, and second, laser light can be focused to an approximately 1 micron spot, allowing researchers to image very small cross-sectional areas. The spatial-resolving capabilities of micro-Raman has been used to probe the composition heterogeneity in polyurethanes.

Another use of infrared spectroscopy uses linearly polarized infrared radiation to determine information about oriented samples, an experiment that has been termed infrared dichroism. The absorbance will be a maximum when the electric field vector and the dipole moment vector are in the same direction, and the absorbance will be zero when the two are perpendicular. Only uniaxial orientation will be considered because this situation is normally found in the literature. R, the dichroic ratio, is defined as follows:

$$R = \frac{A_\parallel}{A_\perp} \tag{8-4}$$

where $A_\parallel$ is the absorbance when the orientation direction and the polarization direction are parallel to each other and $A_\perp$ is the absorbance when the two directions are perpendicular. The relevant geometry is shown in Fig. 8.9.

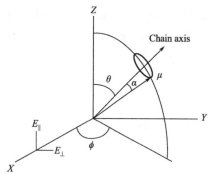

Fig. 8.9 Geometry for uniaxial extended sample.

Let $f(q)$ represent the orientation distribution function, i.e., $f(q)dq$ is the fraction of chain axis that lies within an angle dq of q. This function can be expanded in terms of the Legendre polynomials:

$$f(\theta) = \frac{1}{2\pi} \sum_{n=0}^{\infty} \frac{2n+1}{2} \langle P_n(\cos\theta) \rangle P_n(\cos\theta) \tag{8-5}$$

Only even number polynomials need to be considered since the assumption of uniaxial symmetry implies that the average values of the odd powers of cos q are zero. From Fig. 8.9 after rather extensive mathematical manipulation it can be shown that:

$$\langle P_2(\cos\theta) \rangle = \frac{3\langle \cos^2\theta \rangle - 1}{2} = \frac{(R-1)(R_0+2)}{(R+2)(R_0-1)} \tag{8-6}$$

R_0 is given by $2\cot^2\alpha$, where α is the angle between the dipole moment vector and the chain axis. $<P_2(\cos\theta)>$ is the one parameter measure of orientation normally given in the literature for uniaxially oriented samples. α is a function of the vibration in question and is known for a number of vibrations useful in studying thermoplastic elastomers.

Infrared dichroism can be used to follow the orientation of each phase independently if nonoverlapping bands can be found in each domain where a is known. For example, in polyurethanes, the bonded carbonyl band is typically used to monitor hard segment orientation, while the free carbonyl band or C—H stretching bands are used to follow soft segment orientation. The hard segment orientation is almost always less than the soft segment orientation at the same draw ratio. In semicrystalline and more highly ordered polyurethanes, the hard segments are ordered transverse to the stretch direction initially and later become aligned in the elongation direction. Morphologically, at low elongations, the radial arms of the spherulite become oriented in the stretch direction, which means that the chain axes become oriented perpendicular to the stretch direction. At higher elongations, hard segments are physically removed from the arms and align with the elongation. In relaxation experiments, the soft segments tend to relax very quickly to a nearly unperturbed conformation, while hard segments relax much more slowly, especially at high strains. The transverse orientation of the hard segments is reversible, while physically removed hard domains cannot be restored to their previous environment without heating the sample. Dichroism measurements have been made on 4GT-PTMO copolyesters that showed the same negative orientation at low elongations followed by positive orientation at higher elongations. Dichroism measurements at higher temperatures in MDI-BD polyurethanes showed changes in behavior; the onset of positive orientation occurs at a much lower elongation, and the hard segment orientation becomes much greater at a given draw ratio. These results were interpreted as a weakening of hard segment domain cohesion at higher temperatures. The response to elongation is also altered when polyurethanes are hydrolytically degraded or plasticized.

8.3.2.3 Wide Angle X-ray Scattering (WAXS)

When an electron density difference occurs periodically over a distance that is the same order of magnitude as the wavelength of X-rays, X-rays will be scattered coherently from a sample. A peak or peaks corresponding to this distance will appear in the scattering pattern if the periodicity occurs enough times. The width and number of the peaks will be proportional to the variation of this repeat distance about its average value, as well as the number of times this periodicity occurs before ending. WAXS measures electron density variations with distances on the order of angstroms, which corresponds to interatomic distances. Therefore, WAXS is utilized to study thermoplastic elastomers with crystalline hard or soft segments.

The fundamental relationship that relates the repeat distance of electron density variation and the scattering angle is Bragg's law

$$n\lambda = 2d\sin\theta \tag{8-7}$$

where n is an integer, λ is the wavelength of radiation, 2θ is the angle between the incident and exiting radiation, and d is the repeat distance between crystallographic planes. Further details regarding the crystallographic analysis of polymers are quite complicated and beyond the scope of this chapter.

Detailed crystallographic studies have been performed on copolyesters, copolyamides, and polyurethanes. Regarding the former, hard segment crystallites of 4GT are identical to

4GT homopolymer crystallites in the quiescent state. A different crystalline form is found in the hard segment when the TPE is extended due to the methylene sequences forming an all-*trans* configuration. Using electron microscopy and SAXS, the lamellar thickness has been shown to be smaller than the average hard segment length, which means that chain folding must occur. Although the unit cell is insensitive to soft segment fraction or soft segment composition, the overall amount of hard segment crystallites decreases as the soft segment fraction increases, as shown in Fig. 8.10. However, the fraction of 4GT units that are crystalline increases as the soft segment fraction increases. These materials will also show strain-induced crystallization under stretching.

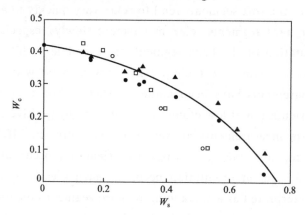

Fig. 8.10 Weight fraction of crystallites, W_c, vs. weight fraction of soft segments, W_s, for 4GT-PTMO copolyester. ▲ from density measurements, ● from WAXS, open symbols from differential scanning calorimetry.

Copolyamides show many of the same features as the co-polyesters in WAXS studies. The crystal structure in the hard segment (nylon) domains is the same as in the homopolymer. Whether chain folded crystals occur depends on the block length of the hard segment, at short block lengths a fringed micelle model was postulated to occur, while at long block lengths chain folding was present. At sufficiently low temperatures (below 0℃), the soft segment will crystallize if the soft segment is PTMO. Finally, the PTMO segments show strain-induced crystallization, which is reversible if the sample is heated slightly above room temperature. The crystallization is great enough so that a permanent set will occur.

Polyurethanes show many of the same features as the copolyesters and copolyamides with respect to both hard and soft segment crystallinity. One substantial difference between polyurethanes and the previous two materials is that not all commercial polyurethane TPEs show hard segment crystallinity; MDI-based polyurethanes show crystallinity while TDI-based urethanes do not. Two crystalline structures have been found for MDI-BD hard segments in the unoriented state with another distinct form found in the oriented state. Other studies have examined the effect of chain extender length on the crystal structure. The differences between polyurethanes with chain-folded and chain-extended hard segment crystallites has been extensively studied.

Copolymers of isotactic propylene (i-PP) with α-olefins also exhibit diffraction peaks due to i-PP crystallites. As one would predict, increasing the α-olefin content decreases the percent crystallinity. When the copolymer is blended with i-PP homopolymer, the copolymer will cocrystallize with the homopolymer, a phenomena that is rare in polymers. Cocrystallization is believed to substantially contribute to the improved mechanical properties found in the blend. Large spherulites are not generally found in these blends as opposed to the homopolymer, and the crystal form is monoclinic rather than smectic.

8.3.2.4 Small-Angle X-ray Scattering (SAXS)

WAXS measures the scattering of X-rays at distances relatively far away from the primary beam, while SAXS measures the scattering very close to the primary beam. Since the scattering distance is inversely related to the distance in real space, SAXS is sensitive to length scales on the order of nanometers rather than angstroms. SAXS is used to probe the two phase morphology in TPEs. In other words, WAXS characterizes intraphase morphology, while SAXS characterizes interphase morphology.

Precise collimation systems are needed to make measurements close to the primary beam. In addition, scattering at small angles is usually very weak. Both pinhole and line collimation has been used; block collimation systems require mathematically transforming the data from line collimation to point collimation. During the last 15 years, the development of synchrotron sources and better area detectors has led to a proliferation of scattering experiments that occur in real time.

The interpretation of SAXS curves is more difficult than WAXS curves. In WAXS, the atoms are so small that they can be considered as point scatterers. In SAXS, however, the individual domains are not insignificant on the length scale of X-rays, and this scattering must be considered. Scattering due to individual domains is called *form factor scattering*, while scattering due to the spatial arrangement of the domains is called *structure factor scattering*. The total scattering can be considered as the product of structure and form factor scattering if and only if the domains are spherically symmetric. For a spherically symmetric two-phase system with uniform electron densities in each phase and one phase discretely immersed in a sea of the other, the scattered intensity can be written as:

$$\frac{I(q)}{I_e(q)V} = nV_0^2(\rho_1 - \rho_0)^2 \phi^2(qR)S(q) \tag{8-8}$$

where $I(q)$ is the scattered intensity at the scattering vector q ($q = 4\pi\sin\theta/\lambda$), $I_e(q)$ is the scattering of one electron if it were the sample, V is the irradiated volume, n is the number density of discrete domains, $(\rho_1 - \rho_0)$ is the electron density difference between the two phases, $\phi(qR)$ is the form factor scattering, and $S(q)$ is the structure factor scattering. $\phi(qR)$ has been calculated for a number of common shapes.

$S(q)$ is generally not a simple function unless the system is very well ordered. If the system is well ordered, then multiple peaks should appear in the SAXS pattern. The relative spacing of these peaks can be used to tentatively identify the domain packing arrangement (Table 8.2). The only TPEs that show multiple peaks are the anionically synthesized materials, and typically only if laboratory processing procedures are used. A great many studies in the literature have used SAXS to study these types of materials; an excellent listing is found in the review papers given earlier.

In commercially important TPEs, a much more featureless pattern is typically found; the most common SAXS pattern from a TPE is a very broad single peak. Bragg's law can be used to calculate an interdomain spacing; of course this calculation is convoluted with form-factor scattering. Two more quantitative approaches are typically applied to the analysis of the data. One approach is to develop a morphological model, calculate the scattering pattern, and change adjustable morphological parameters until the predicted pattern matches the experimental pattern. The second approach is to Fourier transform the data and calculate a radial distribution function for electron density. A description of these approaches is beyond the scope of this review; the interested reader should examine monographs on the subject.

Table 8.2 Relative peak positions in structure factor for common well-ordered morphologies

Arrangement	Relative peak positions
Simple cubic packed spheres	$1, \sqrt{2}, \sqrt{3}, \sqrt{4}, \sqrt{5}, \sqrt{6}, \sqrt{8}, \sqrt{9}, \cdots$
Body-centered cubic packed spheres	$1, \sqrt{2}, \sqrt{3}, \sqrt{4}, \sqrt{5}, \sqrt{6}, \sqrt{7}, \sqrt{8}, \cdots$
Face-centered cubic packed spheres	$1, 1.155, 1.633, 1.915, 2, \cdots$
Diamond packed spheres	$1, 1.633, 1.915, 2.309, 3.416, \cdots$
Hexagonally packed cylinders	$1, \sqrt{3}, \sqrt{4}, \sqrt{7}, \sqrt{9}, \cdots$
Lamellae	$1, 2, 3, 4, 5, \cdots$

SAXS can be used to study the interfacial region between the two phases. SAXS gives a one-parameter measure of the interfacial thickness if some concentration profile is assumed. In a two-phase system with sharp interfaces, the scattering at high angles will be given by (after background subtraction):

$$\lim_{q \to \infty} \left[q^4 \frac{I(q)}{I_e(q)V} \right] = 2\pi(\rho_1 - \rho_0)^2 \frac{S}{V} \quad (8\text{-}9)$$

where S is the total interfacial surface area. The presence of a diffuse interface causes the intensity to fall off more rapidly than a q^{-4} dependence predicted above. Now Eq. (8-9) is modified as shown below:

$$\lim_{q \to \infty} \left[q^4 \frac{I(q)}{I_e(q)V} \right] = [H^2(q)] 2\pi(\rho_1 - \rho_0)^2 \frac{S}{V} \quad (8\text{-}10)$$

$H^2(q)$ has been calculated for sigmoidal and linear concentration gradients. Because of the errors associated with background determination, the use of SAXS to study interface properties should be considered to be relative rather than absolute.

TPE investigations that involve SAXS are numerous. Because the information from SAXS is often ambiguous, the most effective studies are often done in conjunction with a different morphological probe such as electron microscopy or small-angle neutron scattering. Because the number of studies that have used SAXS to study TPEs are far too numerous to list, and in order to give the reader some flavor for SAXS experiments on TPEs, three interesting examples will be highlighted.

Deformed SBS triblock copolymers have been extensively investigated with SAXS. However, patterns were collected after these samples were allowed to relax, which has been shown to substantially affect the morphology. The use of synchrotron radiation along with two-dimensional detectors enables intensity measurements while the sample is being drawn. In one such study, short polystyrene cylinders were imbedded in a continuous polybutadiene phase. The deformation was found to be affine in the meridional direction until an elongation of 3, which corresponds to the inflection point of the stress-strain curve. Above an elongation ratio of 4, it was shown that the cylinders were aligned with their long axis parallel to the stretch direction. The cylinders were not disrupted up to an elongation ratio of 8.

SAXS determinations of interfacial thickness have been used to show that MDI-BD polyether urethanes exhibit narrower interfaces than polyester urethanes. The difficulty of correct background subtraction has been discussed in detail, nevertheless the interfacial thickness assuming a linear gradient profile was approximately twice as large for the polyester versus the polyether soft segments using two different procedures applied in the same way to the patterns. The absolute magnitude of the numbers must be questioned since other authors have found substantially larger values for similar materials.

Small-angle X-ray scattering has been used to follow morphological changes in copolyesters as a function of temperature. In order to slow down the crystallization kinetics, 4GT was replaced by poly(tetramethylene isophthalate) (4GI) as the hard segment. Similar to the 4GT systems, the crystallization rate was found to only weakly influence the morphology of these copolyesters. At a fixed temperature, the Bragg spacing increased with decreasing hard segment concentration, and the long spacing was roughly proportional to the inverse of the undercooling. Annealing at temperatures near the melting point led to morphological reorganization through the melting of imperfect crystallites and recrystallization into more perfect crystallites, which was accompanied by an irreversible increase in the Bragg spacing.

8.3.2.5 Small-Angle Neutron Scattering (SANS)

The only difference between SAXS and SANS is that the contrast for neutron scattering is a variation in scattering density rather then electron density. Scattering density is a function of the nucleus (not the atomic number!) and varies in a complex way. Because the difference in scattering density between hydrogen and deuterium is large, isotopic substitution is used to create the contrast required for SANS. Perhaps the most famous use of SANS in polymer science was the experimental verification that polymer chains in the bulk assume an unperturbed random coil conformation.

SANS can be used to look at the same sorts of things that SAXS is used for, i.e., domain size, distance between adjacent domains, interphase sizes, etc., and a number of studies have used SANS in this manner. However, SANS has a capability unrealizable with SAXS: if a fraction of the chains in the system contain deuterium, with the rest containing hydrogen, or the reverse is true, then it is possible to examine scattering from individual chains. Therefore in two-phase systems with isotopic substitution, scattering will be from two sources: chain scattering, both interchain and single chain, and interphase contrast. By matching the scattering density of the two phases through partial labeling of one or both phases, it is possible to eliminate scattering due to inter-phase contrast (which gives information similar to SAXS) and study only single-chain scattering. Since the amount of phase mixing is unimportant (assuming no volume change upon mixing) and the compositions of the pure phases are well known in most TPEs, contrast matching is relatively easy to perform. Methods have also been developed for noncontrast matched systems to isolate the single chain scattering by subtracting the interdomain scatter-ing, using either SAXS or unlabeled SANS patterns.

For a two-phase system where interphase scattering has been eliminated and only one phase has been partially labeled, the coherent scattering inten-sity can be written as:

$$I(q) = \left[\frac{(\Delta\beta)^2 m_0}{\rho_m N_A}\right] v_1 v_d (1 - v_d) N S(q) \tag{8-11}$$

where $\Delta\beta$ is the coherent neutron scattering length density difference between the fully hydrogenous and fully deuterous materials, v_1 is the volume fraction of the labeled phase, v_d is the volume fraction of deuterous material in the labeled phase, N_A is Avogadro's number, m_0 and N are the monomer molecular mass and number of repeat units respectively, and $S(q)$ is the single-chain scattering function. Further review of SANS theory and experimental studies of polymers is found in monographs on the subject.

The effect of temperature and composition on chain conformation has been investigated in MDI-BD-PPO polyurethanes, 4GT-PTMO copolyesters, and MDI-BD-PTMO polyurethanes. At room temperature, R_g, of the soft segments in the polyurethane, TPEs are approximately 25% larger than in bulk, while for the copolyester the increase is only approximately 10%. The soft segment radius of gyration decreased as the temperature increased above room temperature for the all materials except for a MDI-BD-PTMO material at a 7 : 6 : 1 mole ratio. Evidence of phase mixing was found when the temperature reached a high enough value for all materials except the 7 : 6 : 1 material as evidenced by an increase in R_g of the soft segment with temperature. SANS measurements of the 4GT segments indicate that substantial chain folding occurs in the copolyesters. The hard segment R_g in the copolyester increased dramatically with temperature, which indicated that the amount of chain folding and/or the degree of phase separation was changing. The hard segment R_g decreased as the temperature was raised in the 7 : 6 : 1 polyurethane, which the authors were unable to satisfactorily explain. Measurements of the entire chain dimensions in the copolyesters indicated that the chain initially contracted then expanded as the temperature was raised.

In lamellar styrene-diene diblock copolymers, SANS studies showed that the segment

R_g contracts to 70% of the unperturbed value parallel to the interface and expands to 160% of the unperturbed value perpendicular to the interface. These values were found for both the styrene and diene blocks. A study of stretching SIS block copolymers having spherical styrene phases showed that the deformation in the direction of stretch was greater than affine, while the deformation perpendicular to the stretch was much less.

8.3.2.6 Nuclear Magnetic Resonance (NMR)

Solid-state NMR has the capability of providing information on a wide variety of characteristics in TPEs, encompassing both static and dynamic properties as well as orientation information. As mentioned previously, NMR can be used to determine chain microstructure information such as tacticity and sequence distributions. NMR has been used in TPEs to investigate spatial interactions between atoms as well as the relative mobility of particular segments. Deuterium labeling significantly expands the capabilities of NMR. Using pulse sequences, relaxation times of segments on different time scales can be probed. Given its capabilities, the information gained from this technique, although significant and important, has not been as great as one might expect because of the extreme difficulty in spectral interpretation.

NMR measures change in the spin magnetic moment of nuclei. A strong magnetic field ($\sim 10^5$ Gauss) along one axis, usually taken as the z direction, causes a net population distribution of nuclear spins aligned parallel to the magnetic field. Polarized electromagnetic energy in the radio frequency region (10^2 MHz) with the magnetic field vector perpendicular to the z direction causes transitions to a higher energy spin state. A voltage that is proportional to the relaxation of nuclei from the higher energy state to a lower energy state is measured as a function of electromagnetic radiation frequency in typical NMR experiments. However, chemical shift rather than frequency is reported where chemical shift is defined as

$$\frac{\nu(\text{sample}) - \nu(\text{standard})}{\nu(\text{standard})} \tag{8-12}$$

where ν represents frequency and the standard is a material that contains the same atomic species. Only nuclei with nonzero spin quantum numbers can be studied with this technique. In TPEs, the most common nuclei that fulfill this requirement are ^{1}H, ^{2}D, and ^{13}C.

The chemical shift, which is due to electronic shielding by atoms in the surrounding environment, is generally anisotropic in solids. Three other anisotropic interactions can occur in solid-state NMR: dipole-dipole interactions, which is a through-space interaction with other nuclei; spin-spin coupling, which is a through-bond interaction of two spins; and the quadropole interaction, which occurs when the spin quantum number is greater than and generally obscures the spectrum if present. Due to the fast tumbling of the molecules in solution, the dipole-dipole interactions and the quadropole interaction generally vanish, leaving the isotropic chemical shift and the spin-spin interaction. Much of solid-state NMR involves creating experimental conditions that reduce or eliminate some of these inherent complications.

In order to remove anisotropic interactions, a technique called "magic angle spinning" (MAS) is used. Solid interactions have an approximate angular dependence of $(3\cos^2\theta - 1)$, where θ is the angle between the z-direction and the sample. If a sample is spun around its axis and the axis is at the magic angle with respect to the magnetic field ($\theta \approx 54.7°$), then much of the anisotropy is removed. If the rotation frequency is less than the characteristic frequency of the averaged interaction, then spinning side bands will appear at integral multiples of the spinning speed. Usually, solid-state NMR spectra are obtained at two or more rotation speeds to identify the true features. Other interactions can be removed using radio frequency pulses. High-power proton decoupling removes proton dipole-dipole and spin-spin coupling to different nuclei, such as ^{13}C. Cross polarization enhances magnetization of rare spins from abundant spins. ^{13}C nuclei are often enhanced using ^{1}H.

Radiofrequency pulses are also utilized to measure relaxation times. Three relaxation times have been measured in TPEs, and each is sensitive to different phenomena. T_1, the spin-lattice relaxation time in the laboratory frame, is the relaxation from the nonequilibrium population distribution created by the pulse to the equilibrium Boltzmann distribution. T_1 is sensitive to molecular motions that rate in the range of $10^6 \sim 10^9$ Hz. T_2, the spin-spin relaxation time, is the relaxation caused by the establishment of equilibrium between nuclear spins within the system. Spin-spin relaxation measurements also probe motions with rates in the range of $10^6 \sim 10^9$ Hz; however, low frequency motions ($10^2 \sim 10^3$ Hz) also affect T_2. Generally, T_2 is one to three orders of magnitude smaller than T_1 in solid polymers. T_{1r}, the spin-lattice relaxation time in the rotating frame, probes motions with rates on the order of $10^3 \sim 10^4$ Hz. Cross polarization is usually used in T_{1r} measurements.

$T_2(H)$ measurements have been made with a series of SBS block copolymers with styrene contents between 14% and 30%. Three different T_2 relaxation processes were found: a fast relaxation due to hard segment material, an intermediate relaxation due to interfacial material, and a slow relaxation due to soft segment material. T_2 for the interface was approximately two orders of magnitude greater than for the hard segment, while T_2 for the soft segment was approximately one order of magnitude greater than for the interfacial region. T_2 for the soft butadiene phase was smaller than T_2 for the pure rubber, which indicates that copolymerization restricted the mobility of the butadiene. In all cases, less material was found in the styrene phase than would be predicted; however, the approximate stoichiometric amount of material was found in the butadiene phase. Interfacial calculations showed that the interface consisted predominantly of butadiene segments, therefore a substantial amount of styrene must be dissolved in the butadiene. The thickness of the interface was calculated as 20Å, which agrees roughly with SAXS results on similar systems.

NMR can be used to look at the two phases and the interface of a TPE through deuterium labeling. Spiess *et al.* analyzed a series of model polyurethanes based on a monodisperse piperazine hard segment shown in Fig. 8.11. The synthesis was very carefully controlled so that only specific units in the hard segment were labeled. The area of the narrow component relative to a broad NMR signal was used to quantify the fractional amount of piperazine units that were mobile, which was termed the motionally averaged component (ϕ_{MA}). The motionally averaged component depended on the position of the piperazine unit in the hard segment, as shown in Fig. 8.11. Based on studies with materials containing only single piperazine units, the premature rise of the motionally averaged component of the three curves between 250K and 300K was assigned to piperazine units dissolved in the hard segment. Based on this assumption, the authors calculated that only 15% of the hard segment was dissolved in the soft segment, while the remaining 85% was contained within the hard segment domains.

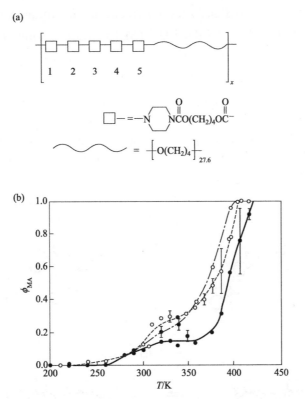

Fig. 8.11 (a) Structure of piperazine-based polyurethanes. Deuterium labeling was only on piperazine units 1 and 5, or only on rings 2 and 4, or only on ring 3. (b) Fraction of motionally averaged material ϕ_{MA} as a function of temperature. •piperazine at the center (third ring), ○ at the second and fourth ring, at the outer (first and fifth ring).

The quadropolar characteristic of deuterium was used to study molecular motions in 4GT-PTMO where the two interior carbons on the PTMO were replaced with deuterium. Previous ^{13}C NMR experiments showed that the 4GT groups could be thought of as molecular anchors because the motions of these groups occurred on a much slower time scale. The shape of the NMR curve after a series of pulses was used to follow three-bond types of motion by the methylene sequences in PTMO. An activation energy of 24.4kJ/mol was calculated for the motion, which is slightly greater than one C—C bond rotation in butane (15.5kJ/mol). Six different trans-gauche conformational transitions were presented and based on NMR data; four of these possibilities were eliminated so that only two remained, as shown in Fig. 8.12. The authors argued that both of the remaining mechanisms may contribute to the observed motion.

A number of relaxation times were measured on MDI-BD-PTMO polyurethanes and a MDI-PTMO commercial polyurethane that is chain extended with ethylene diamine (ED) in order to differentiate the interfacial, hard segment, and soft segment phases. It was shown that $T_1(C)$ and $T_{1r}(C)$ measurements distinguish between hard segments dissolved in the soft phase and hard segments in the hard phase for the BD chain extended material, while $T_{1r}(C)$ measurements for the ED chain extended material can distinguish between carbonyls located in all three regions. Eleven percent of the urethane groups were dissolved in the soft segment in the 7 : 6 : 1 MDI-BD-PTMO material, while only 20% of the MDI units were crystalline.

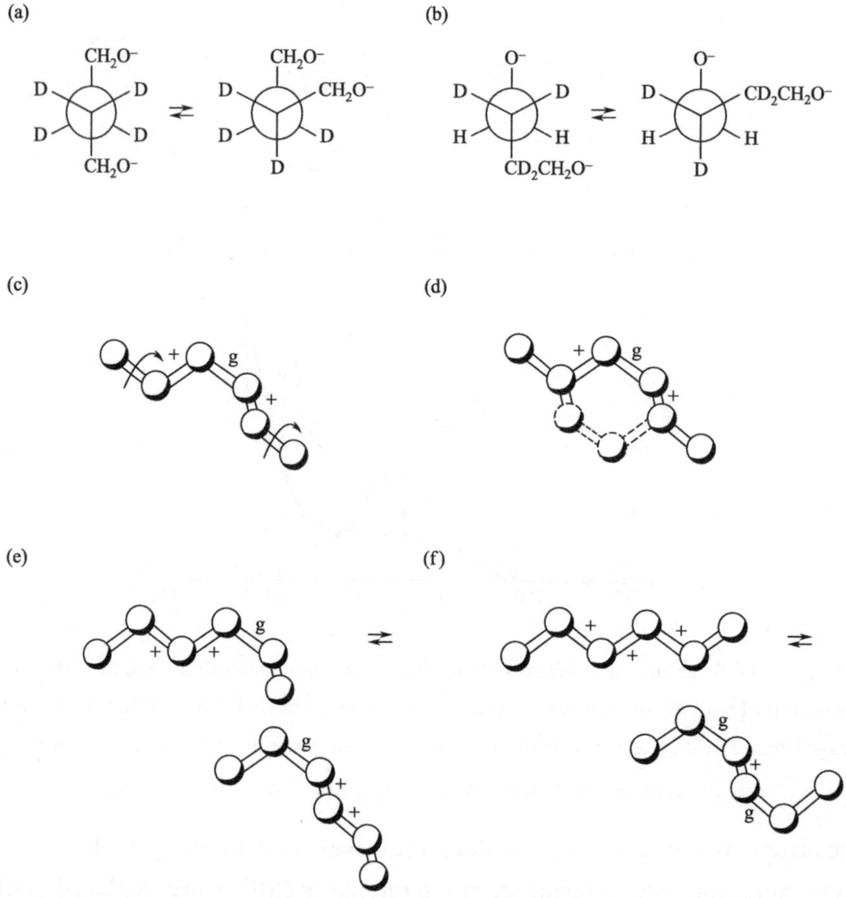

Fig. 8.12 Possible mechanisms for PTMO trans-gauche conformational changes in 4GT-PTMO. Only (e) and (f) were found to be consistent with NMR results.

In the ED material, it was found that 50% of the hard segments were able to move as free rotors, which suggests that the degree of phase separation is poorer in this material than in the MDI/BD material. Since poly (urethane ureas) generally show better phase separation than polyurethanes, this result was probably due to effects such as sample preparation method and starting materials. Variable temperatures were studied using a polymethylene adipate soft segment instead of PTMO, and it was shown that the mobile phase component of the hard segment increases with increasing temperature. The amount of increase depended on the composition of the polyurethane, which the authors attributed to the imperfection of the hard domains.

8.4 Properties and Effect of Structure

8.4.1 General Characteristics

Table 8.3 lists the properties of some typical thermoplastic elastomers and other common rubbery polymer materials. TPEs generally extend to high elongations without failure and have a high tensile stress at break, i.e., they are extremely tough. As mentioned earlier a thermoplastic elastomer should return to its initial shape after the removal of the stress. The range of extensibilities where a TPE will recover its original shape after stress is removed will generally not be as large as for conventional crosslinked rubbers. At high elongations upon the removal of stress, TPEs will often maintain some residual elongation, termed *permanent set*. As discussed in the section on infrared spectroscopy, the domains can flow under high stress. The toughness of these materials (area under stress-strain curve) is usually many times that of conventional crosslinked rubbers.

Table 8.3 Properties of typical TPEs relative to other rubbery polymers

Material	Relative Cost	Tensile Strength at Break/MPa	Tensile Strain at Break/%	Service Temperature/℃	Hardness (Shore A)
Styrene-butadiene rubber	1	15(reinforced)	500(reinforced)	high	35~100
Natural rubber	1	30(reinforced)	500(reinforced)	high	30~100
Silicone	1.2	5(reinforced)	150	high	40~100
Polyethylene	1	10	high	−10~50	100
Fluorinated elastomer	1.5	10	200	high	50~90
SBS	2	25	800	−20~80	50~90
Polypropylene-EPM blend	1.5	20	500	0~110	70
Polyurethane	6	50	600	−20~80	50~100
Copolyester	7	40	600	−40~150	>100
Ionomer	5	15	500	−20~100	50~90

The Mullins effect, also called stress softening, occurs in most TPEs. If a TPE sample is strained and then released, then less stress will be required to strain the sample a second time. This effect is generally quantified by the hysteresis energy, which is the difference in the areas under the stress-strain curve for a loading cycle followed by an unloading cycle. The Mullins effect can lead to heat buildup in a material, which is undesired in most applications. However, the ability of a TPE to dissipate energy is related to its strength and toughness.

Fracture of a polymer involves initiation, slow crack growth, and catastrophic crack propagation. The extreme toughness in thermoplastic elastomers is due to the inhibition of catastrophic failure from slow crack growth rather than the prevention of initiation or prevention of slow crack growth. Table 8.4 lists the mechanisms that can strengthen two phase materials. The hard domains in TPEs act as both filler and inter molecular tie points. Hard domains are effective fillers if the volume fraction exceeds 0.2, their size is less than 100 nm, and the softening temperature is well above the test temperature. The filler effect is one key for the extreme toughness inherent in most TPEs. Another key is the interface between the hard and soft segments; in polymer blends, it has been hypothesized that the strongest materials result from two polymers on the edge of miscibility. These results suggest that broad interfacial zones lead to improved properties in two-phase block copolymers.

Table 8.4 Strengthening processes in polymers

Matrix	Dissipation of energy near crack tip Strain-induced crystallization Orientation processes
Filler particles	Increased dissipation of energy Deflection and bifurcation of cracks Cavitation Deformation of domains

Due to the dual filler and crosslinking nature of the hard domains in TPEs, the molecular deformation process is entirely different than the Gaussian network theories used in the description of conventional rubbers. Chain entanglements, which serve as effective crosslinks, play an important role in governing TPE behavior. The stress-strain results of most TPEs have been described by the empirical Mooney-Rivlin equation:

$$\sigma = \left(\frac{\rho RT}{M_c} + \frac{2C_2}{\lambda}\right)\left(\lambda - \frac{1}{\lambda^2}\right) \quad (8\text{-}13)$$

where σ is the stress (force/unit area), R is the universal gas constant, T is the absolute temperature, λ is the extension ratio, and M_c is the average molecular mass between crosslinks. C_2 is an empirical constant that depends on the material. Stress-strain and swelling measurements have shown that M_c is closer to the molecular mass between entanglements than the soft segment length in SBS. The filler effect is quantified by the Guth-Smallwood equation:

$$\frac{E_F}{E} = (1 + 2.5\phi + 14.1\phi^2) \quad (8\text{-}14)$$

where E_F/E is the ratio of the moduli for the filled and unfilled elastomer and ϕ is the volume fraction of filler. Combination of these two equations gives reasonable values for M_c in SBS copolymers; however, objections have been raised to using this analysis. More complicated theories have been proposed to explain the stress-strain behavior of thermoplastic elastomers.

The general tensile behavior of TPEs as the temperature changes is shown in Fig. 8.13. TPEs become more rigid as the temperature nears the soft segment T_g, and a discontinuous change in brittleness will occur at this temperature. Normally, the soft segment T_g is never reached in service, i.e., this temperature is far below room temperature. As the temperature rises in a typical TPE, the modulus and strength decrease slightly due to softening of the hard domains. At the hard segment dissolution temperature, the modulus will decrease dramatically, and the material can no longer be used as a thermoplastic elastomer. The two-phase structure may persist in the melt however. The precise temperature of the dissolution depends on the nature of the hard block. Plasticizers or other additives may be added to reduce the softening temperature; however, these materials will also tend to disrupt the domain structure.

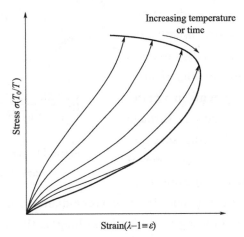

Fig. 8.13 Effect of temperature and draw rate on the stress-strain curve of thermoplastic elastomers. σ is stress, T_o is some reference temperature, T is the test temperature, λ is the draw ratio, and ε is the strain.

The chemical resistance of many TPEs is poor compared to that of conventional rubbers. Polyurethanes, copolyesters, and copolyamides are very susceptible to oxidation, especially at elevated temperatures. Antioxidants and other additives are added to commercial products to improve the chemical resistance of these materials. Carbon black can be added to improve stability to UV light if the color of the material is unimportant. Hydrolytic stability is poor for the polyester-based polyurethanes and the copolyamides because the ester linkage can be attacked by water. For all TPEs, certain organic solvents can degrade these materials if one or both of the blocks will dissolve in the particular solvent. The resistance to many common oils and greases is high for the more polar TPEs.

8.4.2 Mechanical Properties

Because the structures of TPEs are diverse, the influence of structure on mechanical properties may not be universal for all materials. Nevertheless, some general characteristics do hold for most TPEs.

The relative amount of hard and soft segment influences the mechanical behavior of TPEs. As the hard segment content increases, the material will change from a flexible rubber to a tough, rigid plastic. This is illustrated in Fig. 8.14, which shows tensile curves for SBS as a function of styrene content.

Similar data has been presented for 4GT-PTMO copolyesters and polyurethanes. TPEs also show a qualitative change in the shape of tensile curves as the hard phase changes from discrete to continuous. The yield point at approximately 40% styrene content has been interpreted as evidence of a continuous polystyrene phase; usually the existence of a yield point in a TPE implies a continuous hard phase. Altering the relative amounts of hard and soft segments in a material can be accomplished either through changing the hard segment length, soft segment length, or both. The effect of segment length has been investigated at constant relative amounts of hard and soft segment.

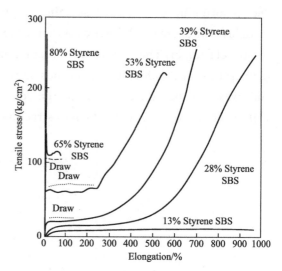

Fig. 8.14 Tensile properties of SBS as a function of styrene content.

In the styrene-diene triblock copolymers, the tensile properties do not depend on segment lengths as long as the polystyrene is of sufficient molecular mass to form strong hard domains (>8000). However, the molecular mass of the blocks will influence the kinetics of phase separation in styrene-diene systems, which can in turn affect the mechanical properties.

In polyurethanes, copolyesters, and copolyamides, the molecular mass of each block is a key factor in determining the mechanical properties of the material. Soft segment block lengths between 1000 and 5000 have been found to produce materials with optimum properties. Below this value, the materials are unable to adequately phase separate. Lower molecular masses also inhibit the ability of the soft segment to strain crystallize, which provides an important strengthening mechanism. Finally, lower molecular mass soft segments at a constant overall soft segment content requires short hard domains, which can severely reduce hard domain crystallization. Higher soft-segment molecular masses introduce practical problems. Since these materials are synthesized industrially in the bulk (without solvent), high molecular mass soft segments result in extremely high viscosities during polymerization, which leads to mixing and pumping difficulties as well as problems in achieving high conversions. Also, higher molecular mass blocks promote phase separation, which will reduce fractional conversion. Since all step-growth polymerizations require high fractional conversions in order to achieve high molecular masses, high molecular mass segments are used with difficulty. The effect of hard segment block length is also important. Longer hard segment blocks generally lead to better phase separation and better properties. However, at the same relative hard segment fraction, longer hard segment blocks contribute to difficulties in synthesis. Some of these difficulties can be overcome with solution polymerization methods, which may or may not lend themselves to commercial scale synthesis.

The overall molecular mass can be altered in TPEs. The overall molecular mass, as long as it is greater than some threshold value, has little effect on the mechanical properties. Two important cases can occur where the overall molecular mass is important. If soft segments in multiblock copolymers are used and have low functionality, then the resulting molecular mass distribution will be quite broad, and the properties will be reduced substantially. Most industrially important soft segments have a functionality near 2, so this effect is usually not important. However, of more importance, especially in polyurethanes, is the possibility of side reactions that can lead to network formation. A small amount of crosslinking can actually improve the properties, which is presumed to occur through strengthening of the hard domain. Large amounts of crosslinking are undesired since crosslinking can inhibit phase separation and increase the brittleness of the TPE.

The effect of hard and soft segment polydispersity has been investigated in different polyurethanes. Soft segment polydispersity does not significantly alter the mechanical properties, while materials with monodisperse hard segments have a higher modulus and increased tensile strength. An important underlying assumption of these studies is that the soft-segment molecular mass distribution is unaffected by synthesis, which was proven to be the case using a novel analytical technique. The improved mechanical properties can be attributed to improved phase separation as shown in synchrotron SAXS studies; better packing of the hard segments may also be playing a role.

A parameter that can be easily changed, which can have a dramatic effect on the properties, involves the constituents of the TPE. In SBS triblock polymers, this includes changing the center block to isoprene or the end blocks to α-methyl styrene. In the first case, the mechanical properties are unaffected except that in SBS the tensile strength depends on the styrene content, while in SIS the tensile strength does not vary with styrene content. Substitution of a-methyl styrene leads to a tougher polymer, which is at least partially due to the higher glass transition (160°C vs. 100°C) of the new end block. Even though T_g for isoprene and butadiene differ substantially, no corresponding effect due to the different glass transition temperature is found in the mechanical properties of the TPEs. This difference emphasizes the importance of the hard phase in determining the mechanical properties of the material.

For the segmented block copolymers such as the polyurethanes or copolyesters, the composition of the hard segment or the soft segment can also easily be changed. Changing hard segment type will change the crystallinity of the hard phase. For example, copolyesters can be produced using a mixture of tetramethylene terephthalate and another diacid, such as tetramethylene isophthalate, rather than just pure tetramethylene terephthalate. The ultimate properties of the mixed hard segment material are better if the total hard segment content is low, with little effect on hardness or stiffness. The disadvantage of using mixed acids is that crystallization is limited and proceeds much more slowly. In polyurethanes, symmetric diisocyanates produce stronger TPEs. The presence of substituents on the aromatic ring tends to reduce the tensile properties of polyurethanes. In BD-polyester polyurethanes, the following diisocyanates give tensile strengths according to the following: MDI > hexane diisocyanate > isophorone diisocyanate > toluene diisocyanate (TDI).

Soft segment type also plays an important role in the physical properties of multiblock copolymers. Soft segments that strain crystallize produce tougher materials with higher tensile strengths and tear resistance. Since incorporation into a block copolymer reduces crystallization kinetics and slightly lowers the melting temperature, the unstrained elastomer may not contain any soft segment crystallites. Upon deformation, crystallization may occur, which will cause a large permanent set in the material. An upturn in the tensile curve at high elongations is often taken as evidence of crystallization, but WAXS provides the most direct and conclusive evidence. Soft segments that strain crystallize generally have a melting point slightly above the service temperature, and higher soft segment molecular masses favor strain crystallization. PTMO and polycaprolactone are two common soft segments that can strain crystallize.

Soft segment type also influences the driving force for phase separation. However, improved phase separation does not necessarily lead to improved properties, since both polyether and polybutadiene soft segments generally show more complete phase separation than polyesters, yet the polyester-based materials have better mechanical properties. The most likely explanation for this result is poor interfacial adhesion in well-phase-separated systems.

8.4.3 Thermal and Chemical Properties

The response to changes in thermal or chemical environment is largely the result of the underlying chemical structure of the material. As discussed in the introduction, lack of resistance to chemical or thermal stimuli limit the use of TPEs, especially as compared to conventional rubbers. Unfortunately it is difficult, if not impossible, to modify thermal and chemical characteristics substantially. The use of small amounts of antioxidants or fillers can improve these properties somewhat, but the overuse of these materials can result in a large change in mechanical behavior.

The sensitivity to hydrolysis is a key issue in many applications. The ester bond in 4GT-PTMO copolymers is sensitive to hydrolysis; however, it is fairly protected since most of the ester is contained in a crystalline structure. The addition of a small amount (1%~2%)

of a hindered aromatic polycarbodiimide substantially increases the lifetime of this material in the presence of hot water or steam. Polyurethanes are susceptible to hydrolytic attack, especially those with polyester soft segments. However, polyester soft segment polyurethanes are generally more resistant to oils, organic solvents, and thermal degradation. Ionomers will swell when exposed to water; in fact, a commercial hydrated perfluorosulfonic ionomer (Nafion) is used as a membrane separator in chlor-alkali cells. Styrene-diene copolymers and polyolefin TPEs are insensitive to water.

The ability of a TPE to withstand variations in temperature depends almost entirely on the chemical structure. The maximum service temperature is usually about 40℃ below the hard segment glass transition or melting temperature. Because of hysteresis, excessive heat buildup can occur during use so that the local temperature of the material can be much higher than the nominal temperature. Changing the maximum service temperature involves changing the structure of the hard block. Using α-methyl styrene in styrene-diene triblock copolymers or ethylene diamine chain extenders in polyurethanes can extend the service temperature in these TPEs substantially. The minimum service temperature is usually about 10℃ above the soft segment T_g. Below this temperature, the material will become brittle. If the soft segment can crystallize, then low temperatures can cause crystallization and a corresponding increase in stiffness and brittleness. Using a mixed or copolymer soft segment will eliminate this problem, but at a cost of reducing strain-induced crystallization at higher temperatures.

8.5 Thermodynamics of Phase Separation

This section will present the theoretical framework and understanding about the thermodynamics of phase separation in block copolymers. Most theories consider four factors that influence the phase separation of block copolymers:

(1) the Flory-Huggins interaction parameter χ,

(2) the overall degree of polymerization N,

(3) architectural constraints such as the number of blocks and linear vs. starblock polymers,

(4) the weight fraction f of one component.

The thermodynamic theories are conveniently divided into three cases: the strong segregation limit (SSL), the weak segregation limit (WSL), and the intermediate segregation region (ISR). In the SSL, the equilibrium state of the material consists of relatively pure phases of A separated from relatively pure phases of B. In the WSL, the two phases are intimately mixed. The ISR is essentially a region that arises because of finite molecular mass; in the case of infinite molecular mass no ISR exists. Whether the ISR should be considered as part of thermodynamic phase space is questionable.

The transition between the WSL and the SSL is termed the order-disorder transition (ODT), which is also called the microphase separation transition (MST). The analogous transition in small molecules is the solid-to-liquid transition. The reader should be aware of the order-order transition (OOT), which is a shift of morphology from one type to another (e.g., spheres to cylinders), that can occur with changes in temperature in anionically synthesized block copolymers at very specific block lengths. If the ODT temperature is below the hard segment T_g or T_m, then the material is one phase above the hard segment dissolution temperature. If the ODT temperature is above the hard segment T_g or T_m, then the material will exist as a two-phase melt. The phase state of a TPE above the dissolution temperature has a substantial effect on the rheological properties since one-phase mixtures have a viscosity much smaller than two-phase mixtures.

A great deal of effort has been spent on developing theories for diblock or triblock copolymers, much of which has been driven by the measurement of phase diagrams, which in turn provide good tests of theories. It should be noted that the rigorous application of these theories to TPEs used in commercial applications is limited, since in nearly all systems some arrest due to slow kinetics occurs. The most rigorous theories were developed for monodisperse block lengths, e.g., anionically polymerized TPEs. In some cases these theories have been extended to multiblock copolymers that have a distribution of block sizes. Further, amorphous systems have had a significantly larger focus on them than crystalline systems.

In the WSL, the chain configuration is unperturbed; e.g., R_g scales as $N^{1/2}$. However, the probability of finding an A or a B segment at a distance r from a particular point does not only depend on f, it also depends on whether the original point sits on an A or B segment. If the original point lies on an A segment, then at short distances the probability is much higher of finding another A segment and at distances comparable to the chain's radius of gyration, the probability is higher of finding a B segment. This concept is called the correlation hole effect. At extremely large distances, the probability of finding an A segment reduces to the volume fraction of A segments in the WSL. The chains generally do not assume their unperturbed conformation in the SSL. The probability of finding a segment at a distance r depends strongly on the type of morphology and can also have a directional dependence, whereas in the WSL the probability is isotropic.

The theoretical development of Helfand and Wasserman contains all the necessary ingredients for a complete description of phase separation in the strong segregation limit; in fact, the theory has been modified slightly so that the results are considered to be quantitatively reliable.

Three energetic contributions are included in this theory:
(1) confinement entropy loss due to a concentration of AB joints to the interface,
(2) conformational entropy loss due to extended chains, and
(3) enthalpy due to mixing of A and B segments.

An expression is written for a function that is proportional to the probability density that a chain with N segments has one end at r_0 and another at r. The resultant equation is identical to the form for the time-dependent diffusion equation where the differential with respect to time is replaced by a differential with respect to segment. The remainder of the development involves solving this equation with the appropriate boundary conditions. Complete analytical solutions have not been derived, but numerical solutions to the equations have been calculated. This theory was originally developed for diblock copolymers and has been extended to triblock systems.

The interfacial thickness (t) was predicted to be approximately equal to the following:

$$t = 0.816 a \chi^{1/2} \tag{8-15}$$

where a is the statistical segment length. This equation predicts that the inter-facial thickness is independent of molecular mass. In the limit as $N \to \infty$, the domain spacing D was found through numerical analysis to scale as:

$$D \sim a N^{9/14} \chi^{1/7} \tag{8-16}$$

In this limit, the confinement entropy of a junction is insignificant compared to the stretching entropy of the chain. Numerical predictions of the predicted periodicity have been made for diblock and triblock copolymers for the cylindrical, spherical, and lamellar morphologies. In general, the agreement between the theoretical predictions and the experimentally observed results have been good for cylindrical and lamellar systems, while not as good for spherical systems. Because spheres are isolated, changes in domain size can only occur by transport of segments through an incompatible matrix, which provides a substantial diffusional resistance to changes in morphology. Hence, the morphology tends to not change once it is formed, and the agreement between theory and experiment is poor. Numerical procedures were given for predicting the phase diagram including calculation of the underlying morphology (spheres, cylinders, etc.) as a function of composition. The underlying morphology type was predicted to be almost temperature independent.

Numerous studies were completed on block copolymers in the SSL long before a theory of the detail inherent in the Helfand and Wasserman approach existed. In the WSL, experiments have been driven by the mean-field theories originally outlined by Leibler for diblock copolymers with monodisperse blocks. This method uses the random phase approximation (RPA) to calculate the free energy in terms of an order parameter $\Psi(r)$ which describes the average deviation from the uniform distribution at any point r. The thermodynamic state of the material was found to depend only on χN and f, just as in the Helfand and Wasserman approach. Only three two-phase morphologies are predicted from this model: body-centered-cubic packed (bcc) spheres, hexagonally-closest-packed (hcp) cylinders, and lamellae. Since c depends on temperature, the morphology can be changed from spheres to cylinders or lamellae simply by changing temperature, in contrast with the Helfand and Wasserman theory, which predicts the underlying morphology is independent of temperature. A critical point (second-order transition) is predicted at $\chi N = 10.495$ for the symmetric diblock copolymer, which also corresponds to the minimum χN that divides the ordered and disordered state. Since the point of demixing for two homopolymers of identical molecular mass is $\chi N = 2$, the joining of two individual chains at the ends means that the individual chains need to be over 2.5 times as long in the diblock copolymer than in the homopolymers to phase separate.

Leibler's theory outlined an experimental method for testing its conclusions. The structure factor for scattering of radiation by the disordered phase was given by:

$$\frac{1}{S(q)} = \frac{F[(qR_g)^2, f]}{N} - 2\chi \qquad (8\text{-}17)$$

where F is a dimensionless function that is related to the Debye correlation function of a Gaussian chain. A Lorentzian peak is predicted for the disordered phase. The position of the peak maximum at $q*$ does not change with temperature. The periodicity of the concentration fluctuations is given by $2p/q*$, which is approximately the radius of gyration of the chain. The height of the peak is a function of temperature because of changes in c. Generally, χ has been found to depend on temperature according to the following:

$$\chi = \frac{A}{T} + B \qquad (8\text{-}18)$$

where A and B are empirically determined constants. According to the above analysis, a plot of 1/peak intensity in SAXS or SANS experiments vs. reciprocal temperature should be linear when the polymer is in the disordered state. Deviation from linearity will mark the order-disorder transition temperature, while an extrapolation to zero intensity will allow the calculation of the spinodal decomposition temperature. The ODT according to this theory can also be determined from the temperature where $q*$ changes as a function of temperature since $q*$ is predicted to be independent of temperature in the disordered state.

The WSL theory developed by Leibler has been shown to be incorrect because of deviations from the fundamental underlying mean-field assumption. Fig. 8.15 shows experimental results for a poly (ethylene-propylene/ethyl ethylene) (PEP-PEE) diblock copolymer that has been fit to the predictions of the Leibler theory without any adjustable parameters, since the ODT and c were calculated from rheological measurements. This mean-field theory does not qualitatively describe the behavior of this material. Other experiments have indicated that the RPA approximation and the Gaussian coil assumption are inaccurate near the ODT.

In the original analysis, it was understood that this theory should not be applied near the critical point. Mean-field theories ignore concentration fluctuations at distances

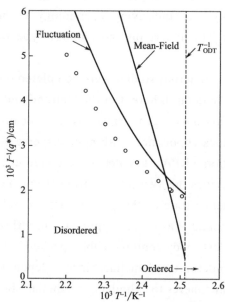

Fig. 8.15 Comparison of 1/SANS peak intensity vs. inverse temperature for PEP-PEE diblock copolymer and comparison to theory.

other than $q = q*$. Near the critical point, concentration fluctuations on very large length scales become increasingly important. A modification to this theory that includes concentration fluctuations has been developed. A critical point is not predicted by the fluctuation theory, rather a first-order phase transition is predicted for all compositions. A molecular weight dependence is found for χN, which delineates the ordered from the disordered phase shown below for the symmetric diblock copolymer:

$$\chi N = 10.495 + 41.022 N^{-1/3} \tag{8-19}$$

The minimum χN value corresponding to the disordered phase still occurs at $f = \frac{1}{2}$. The fluctuation theory predicts an ODT at a slightly lower temperature than the Leibler theory. The lamellar and hcp cylinder phases are directly accessible from the disordered state, which seems to be confirmed by experiment, rather than having to pass through the bcc sphere phase as in the Leibler theory. The rather simple structure factor presented in Eq. (8-17) is retained; however, c is replaced with a χ_{eff} that depends on temperature, composition, and molecular mass in a complicated manner. At temperatures slightly above the ODT, a partially ordered morphology is predicted. Figure 8.15 also compares the results of experiment with the fluctuation theory without any adjustable parameters. The qualitative shape of the experimental data is described better by the fluctuation theory; however, quantitative agreement is still not found.

The previously discussed theories were developed for monodisperse diblock copolymers, which are not TPEs. However, Leibler's mean-field theory has been extended to include polydispersity and to include triblock, star, and graft copolymers. In the former case, polydispersity corrections tend to lower χN corresponding to the ODT. As would be expected from the analogy between blends and diblocks, triblocks will phase separate at higher χN values than the corresponding diblocks. This theory predicts a monotonic increase in the critical value of χN as the symmetry of the triblock increases, to a maximum of about 18 for the symmetric triblock. Surprisingly, the minimum χN value that separates the order and disordered regions in triblocks does not necessarily correspond to the critical point.

The development of the mean-field theory for triblocks is very similar to the approach followed by Leibler. A second parameter, t, defines the asymmetry of the triblock. If the block copolymer is labeled ABA, then starting at the center of the B block, t is the fraction of A units going in one direction along the chain divided by the overall number of A units in the copolymer. By convention, t is always less than 0.5, so $t = 0$ defines a diblock copolymer, while $t = 0.5$ is a symmetric triblock. f is defined as the overall fraction of A units in the triblock copolymer. For all t values, a critical point is predicted at a certain composition and χN. The critical point does not occur at $f = 0.5$ like in the case of symmetric materials, rather the copolymer composition at the critical point is a function of t. The ordered phase following the ODT is bcc spheres followed by hcp cylinders and finally lamellae except at the critical point. The same experimental methods and analysis that are used for diblock copolymers can also be used for triblocks. $q*$ seems to be weakly temperature dependent, and the qualitative shape of the scattering curve is different because of an upturn at low q. The difference in the ODT temperature between the diblock PEE-PEP and the triblock PEP-PEE-PEP was 72℃, which is near 61℃ predicted from the mean-field theories.

8.6 Thermoplastic Elastomers at Surfaces

8.6.1 General Characteristics

The two phases of TPEs also will affect applications that are sensitive to surface or interfacial properties. Generally, the fraction of a component at an interface can be substantially different than the overall bulk fraction. The presence of an interface introduces another thermodynamic consideration that can also alter the morphology that exists in the bulk of the sample. The compositional and morphological variation depends on the processing conditions that were used to generate the surface, as well as the nature of the other surface. The energetic driving force for these variations is a desire to minimize interfacial energy. The primary parameter that characterizes the interfacial characteristics of a material is the surface tension or surface energy.

Although the two terms are often used interchangeably, strictly speaking the two are not identical. The surface tension is defined as:

$$\gamma = \left(\frac{\partial A}{\partial \Omega}\right)_{V, T, N_i} \tag{8-20}$$

where A is the Helmholtz free energy of the entire system, Ω is the interfacial energy, V is the system volume, T is temperature, and N_i is the number of moles of the i_{th} component. This derivative is defined such that the surrounding medium is a vacuum. A higher surface tension means a stronger opposition to the formation of a larger surface area. The expression for surface energy is identical except the change in the Helmholtz free energy of the surface replaces the Helmholtz free energy of the system. For relatively deformable polymers, the difference between the two expressions is small and is usually ignored. Surface tension is experimentally measured either from polymer melts or from contact angle experiments. The former requires extrapolation to the solid, while the latter cannot be measured directly and must be calculated from semiempirical equations. However, the values calculated from the two methods usually agree well. The Polymer Handbook lists surface energies for many segments commonly found in TPEs. Most polymers have surface energies in the range 30～45mN/cm at room temperature. Exceptions to this simple generalization are fluorine-and silicon-containing polymers, which have significantly lower surface energies.

This section describes analytical methods that are used to characterize the atomic composition or morphology at TPE surfaces. Measurement of surface composition in TPEs can suffer from one major drawback. Since the soft phase is mobile at room temperature, the surface composition can change depending on the medium in which the measurement is being made. Many analytical methods are performed in ultra high vacuum (UHV), which is not a normal atmosphere for TPE applications. Typically the difference in a TPE surface exposed to UHV or exposed to air is small; however, a significant difference will exist if the surface is exposed to a liquid such as water. Different strategies have been devised to help overcome this problem, and some of these will be discussed in this section. Analytical methods for surface composition also require a very clean surface to give quantitative results. Small amounts of additives normally present in commercial materials may also significantly alter the surface composition. Finally, some methods can alter the surface during measurement.

Another extremely important parameter of any surface technique is the depth of penetration (d_p), which is a characteristic distance from the surface that the measurement will probe. Table 8.5 gives approximate depths of penetration for the methods discussed in this section for measurement of TPE surfaces. Since Table 8.5 shows that the depth of penetration varies widely, the change in composition can be monitored from the surface almost continuously to thousands of angstroms. Alternatively, some methods discussed in this section measure the morphology near the surface. Studies of TPE surfaces are less numerous than investigations of bulk properties, which in general speaks to the relative lack of importance of the surface properties of TPEs for most applications. However, exceptions to the latter generalization definitely exist, for example in the medical device area.

Table 8.5 Depths of penetration for surface-sensitive analytical methods

Method	Depth of Penetration
Attenuated total internal reflection	1000~10000 Å
X-ray photoelectron spectroscopy	10~100 Å
Static secondary ion mass spectrometry	1~10 Å
Dynamic secondary ion mass spectrometry	Varies
Neutron reflectivity	Varies
Atomic force microscopy	Surface

8.6.2 Studies of Surfaces

8.6.2.1 Scanning Electron Microscopy (SEM)

In SEM, electrons are reflected from the surface, and detection is done on the same side of the surface as the source; in TEM, the electrons pass through the sample and are detected on the other side. SEM is done under UHV conditions, generally $10^{-5} \sim 10^{-7}$ Pa. Probably the most common use of SEM in polymer science is to examine fracture surfaces. The morphology of a fracture surface gives details about the fracture mechanism. In TPEs, SEM is not appropriate for determining whether interphase or intraphase fracture occurs, for example; the domains are too small and the large amount of phase mixing may also play a role as well.

In well-phase-separated systems, i.e., anionically synthesized block copolymer cast from solution, information about composition at the surface can be obtained by microtoming perpendicular to the surface and then directing electrons parallel at the just-cut surface. The top (or bottom) of the image contains information about the composition at the free surface. An excellent study on a lamellar SI diblock copolymer with 52% styrene that was cast from toluene showed that generally the lamellae oriented parallel to the air-polymer interface in agreement with earlier studies and studies of other copolymer systems. A micrograph of SI diblocks with lamellae parallel to the interface is shown in Fig. 8.16. The lower surface energy component polyisoprene (dark component in Fig. 8.16) was always located at the interface between the polymer and the vacuum. Some micrographs showed perpendicular orientation of the lamellae relative to the surface as also shown in Fig. 8.16. However, a thin polyisoprene layer is still at the air-polymer interface in this micrograph, i.e., the lower surface energy component was located at the interface. A very similar system, except with a crystallizable block, did not show the same behavior, i.e., the surface composition depended on the orientation of the crystallites. This type of interaction at a surface can be used in anionically synthesized block copolymers to give very unique morphologies on a surface when very thin films are cast from solution.

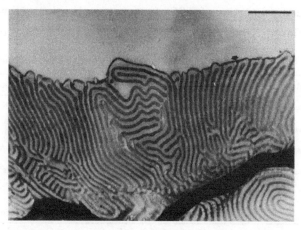

Fig. 8.16 Electron micrographs of SI diblock copolymer
(M_w = 524000g/mol, 52% styrene) at a free surface.

8.6.2.2 Attenuated Total Internal Reflection Infrared Spectroscopy (ATR)

In ATR, a beam of infrared light is totally reflected inside a specially cut infrared transparent material that has a high index of refraction. Typical materials used for ATR prisms are Ge, Si, and ZnSe. Because the index of refraction differs between the polymer and the prism, an evanescent wave penetrates the polymer if it intimately contacts the prism. The infrared radiation will interact with molecular vibrations in the same manner as in conventional infrared spectroscopy. The amplitude of the evanescent wave decays exponentially from the surface, so the depth of penetration is arbitrarily taken as the point where the amplitude decays to $1/e$ (37%) of its initial value. The depth of penetration depends on the ratio of the refractive indexes between the polymer and the prism, the angle of incidence, and frequency of radiation in the following manner:

$$d_p = \frac{1}{2\pi\nu \left[\sin^2\theta - \left(\frac{n_2}{n_1}\right)^2\right]^{1/2}} \tag{8-21}$$

where ν is the frequency, n_2 and n_1 are the refractive index of the polymer and prism respectively, and θ is the angle of incidence of the infrared wave. The dependence of the depth of penetration is not as simple as the above relationship suggests because the refractive indices also have a frequency dependence. For nonisotropic materials, d_p also depends on the direction of the electric field vector of the radiation relative to the surface of the prism. ATR spectroscopy is a very versatile technique and can theoretically be used in almost any medium. The latter capability has made ATR a very important method for the characterization of TPEs used in the medical device area.

Although ATR has been used to quantify the variation in composition at the surface in TPEs, a related utility is its ability to monitor *in situ* processes such as reaction injection molding (RIM) and protein adsorption on to a polyurethane substrate. In the latter, the effect of shear rate on the kinetics of protein adsorption and desorption from phosphate buffered saline (PBS) was studied in a specially designed flow cell. A very thin film of the commercial MDI-ED-PTMO polyurethane Biomer was cast from solution onto a Ge ATR prism. The thickness of the film was less than the penetration depth so the protein concentration could be monitored after the infrared absorption of the polymer and PBS was subtracted. The study found that increasing the wall shear rate does not affect the rate of protein adsorption, but the desorption rate is slowed. In later studies using essentially the same apparatus, spectral changes corresponding to protein conformational changes were followed as the protein absorbed to a polyurethane substrate.

8.6.2.3 X-ray Photoelectron Spectroscopy (XPS)

XPS (also termed electron spectroscopy for chemical analysis) takes place in a vacuum and is able to quantify the relative amount of atomic constituents on the surface of a material. With the exception of hydrogen and helium, any chemical element can be identified, and XPS is also sensitive to chemical bonding effects. In XPS, a surface is irradiated with low-energy X-rays (usually aluminum or magnesium K_a), which results in the ejection of core level electrons. Qualitatively, the depth of penetration is limited by inelastic scattering of the ejected photoelectron as it travels toward the surface. Quantitatively, d_p is an extremely complicated function that depends on the energy of X-rays and the structure of the material; typical depths of penetration in a polymer are on the order of a few Ångstroms. The number of ejected photoelectrons that reach the detector corresponding to a particular distance from the surface decays exponentially as the distance increases. d_p is defined where the number of ejected electrons reaches 37% of its initial value and is proportional to $\cos\theta$, where θ is the angle of emission with respect to the sample normal.

The most common use of XPS on TPEs has been to quantify the fractions of each phase at the surface. By varying the emission angle, the composition can be probed at different depths of penetration. XPS studies have consistently shown that the low surface energy component will predominate at a free surface. Even a small difference in surface energy provides a strong driving force for surface enrichment of the lower energy component. The effect of evaporation rate and overall molecular mass was studied in symmetric poly (methyl methacrylate-styrene) (PMMA-S) cast from toluene. The surface energy difference between these two components is less than 1mN/m. The amount of polystyrene at the surface was greater for the slowly evaporated film. The fraction of polystyrene at the surface increased as the overall molecular mass increased and reached a constant value of approximately 90% polystyrene at the surface when the overall degree of polymerization was approximately 1000. A later study on annealed films of these same materials showed that the polystyrene layer completely covered the surface when the degree of polymerization was greater than approximately 2500. In both studies, the surface excess of polystyrene F_{PS} was found to obey the following relationship:

$$\Phi_{PS} = a - bN^{-1/2} \qquad (8-22)$$

where a and b are empirical constants. This functional form agrees with the mean-field theory of block copolymers at surfaces developed by Fredrickson and the constants were given a physical interpretation regarding the surface energy differences and the ability of the surface to modify the energetic interactions between the two polymers. The surface excess of polystyrene was also found to decrease with increasing temperature, which agrees with qualitative expectations.

Ultra-high vacuum is required to perform XPS, which is a major disadvantage of this technique. As mentioned earlier, samples can show substantially different surface compositions in UHV than under normal conditions, especially if the material is immersed in a liquid, e. g., blood-contacting applications. Recent techniques that utilize freeze-drying and measurements at temperatures lower than the glass transition of the soft segment have been devised in order to quantify the surface composition at liquid interfaces. In a number of different polyurethanes, the low surface energy component was in excess when spectra were collected under normal conditions while the high energy component was in excess when the freeze-drying method was used. This generalization held whether the low energy surface component was the soft or hard segment. This study illustrates surface rearrangement that can occur in any study of polymer surfaces.

8.6.2.4 Secondary Ion Mass Spectroscopy (SIMS)

When a surface is bombarded with a beam of high energy primary ions, atomic collisions between the beam and the solid cause the ejection of secondary ions, which can be characterized using a mass spectrometer. These secondary ions provide information on the atomic and molecular species present at the surface. The ions O_2^+, O^-, and Ar^+ are generally used for SIMS analysis, with the first ion being the most common. SIMS experiments are characterized as either dynamic or static. In static SIMS, the ion beam currents are low (less than $5nÅ/cm^2$) so that the surface etches very slowly. The depth of penetration for static SIMS is approximately 10Å. Beam currents are much higher for dynamic SIMS and successive layers are etched away rapidly during the test. This method eventually produces a pit in the sample due to this etching process. Atomic species can be quantitatively characterized as a function of distance from the surface if the relationship between the etching rate can be measured. In SIMS experiments, the sample surface can become charged because most polymers are non-conductive, and this charge must be removed during the experiment. UHV conditions are required for SIMS experiments.

SIMS is extremely sensitive, even to the part per billion range, and is used widely in the semiconductor industry. SIMS and XPS are complementary since they provide very similar information, but the depth resolution of static SIMS is substantially higher than XPS. However, the quantitative use of SIMS in polymer science is currently not as well developed as XPS. SIMS has been used to study the surface of polyurethanes and PMMA-S lamellar diblocks. In the latter study, dynamic SIMS was used to show that the lamellae aligned parallel to the surface after annealing. The thickness of the layer nearest the free surface as well as near the polymer-substrate surface was shown to be approximately—the thickness of the interior layers. The experimental data were fit with a model that included the lamellar period of the block copolymer, and the results were compared with SAXS results from the same material.

8.6.2.5 Atomic Force Microscopy

The concept of atomic force microscopy is extremely simple. Essentially a very small probe (50~100nm in size) is rastered over a surface such that the force between the surface and the probe is maintained constant; for studies of TPEs normally tapping mode is used, which means that a vertical oscillation is superimposed on the rastering. Essentially, the deflection of the probe is measured as a function of position. The interaction between the probe and the surface is not well understood, but the AFM pattern does not seem to be very sensitive to probe characteristics. The resolution of AFM is in the 1nm range. A primary advantage of AFM is the scanning range, which allows for the investigation of variations in surface roughness and free surface characteristics over a large lateral area.

Different TPEs imaged using AFM include polyurethanes, polypropylene-based thermoplastic elastomers, copolyamides, and anionically synthesized block copolymers. Not only can one visualize topographical differences, the AFM can also distinguish between hard and soft segments, and hence can determine the shape of these regions at the interface. The contrast is provided by the difference in modulus of the hard and soft segments; this difference allows AFM to distinguish between the two components. Alternatively, the sample can be heated and imaged simultaneously to explore changes in morphology as the system softens. In systems where crystallization is the main driving force toward phase separation, the resulting AFM is qualitatively identical to AFMs from normal semicrystalline polymers. In systems where this is not the case, e. g., polyurethanes, the morphology can be very different. For example, on a system with a monodisperse distribution of hard domain lengths, such as the spherulitic hard-domain superstructure, the orientation of the chain axis is the same as in normal spherulitic structures, i. e., in a direction tangential to the radius, but no chain folding occurs, as shown in Fig. 8.17.

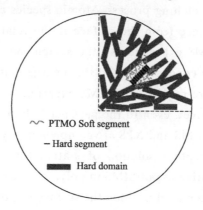

Fig. 8.17 Schematic model for a polyurethane with monodisperse hard segment lengths showing spherulitic structure of hard domains and the orientation of chain segments within the arms of the spherulite.

8.7 Rheology and Processing

The theoretical development and characterization techniques apply equally well to TPEs. However, there are fundamental differences in the response to stress between a conventional rubber and a TPE. Of course the biggest dissimilarity is that a TPE will flow at higher temperatures while a crosslinked rubber will not. Some of the similarities and differences between the two types of materials will be discussed later. Also, the processing methods used for TPEs will also be introduced; the processing methods for these materials are no different than for any other thermoplastic. .

At service temperature and frequency, the solid-state rheological properties of a TPE and a conventional rubber are similar, although at large strains the former will generally have some permanent set, while the latter will not. Only by changing the temperature and/or the frequency do substantial differences between the two emerge. If the temperature is lowered or the frequency raised, the qualitative response of the two materials is still very similar. Both materials become more resistant to stress until a temperature is reached that corresponds to the T_g of the soft segment in a TPE or the rubbery matrix in a conventional rubber. Below this temperature, both a conventional rubber and a TPE will behave as a one-phase brittle glassy polymer. As the temperature is raised, the two materials show qualitatively different behavior. In a conventional rubber, the material will continue to behave as an elastomer until the degradation temperature is reached. In a TPE, the hard domains will eventually weaken and flow, which allows the material to be processed as a conventional thermoplastic. The processing characteristics will be intimately related to the

ODT temperature since the viscosity is substantially lower above the ODT.

More specifically, the dynamic mechanical properties are significantly different in TPEs vs. conventional elastomers. The WLF equation does not generally apply to TPEs since two phases make the material rheologically complex. Dynamic mechanical experiments are used as a macroscopic characterization of phase separation since a more highly phase separated material will have a larger temperature difference between the soft and hard segment transitions and a flatter plateau modulus. Fig. 8.18 shows E' and tanδ for a series of polyurethane ureas as a function of increasing hard segment content that demonstrates the fore mentioned behavior in the more highly phase separated higher hard segment materials.

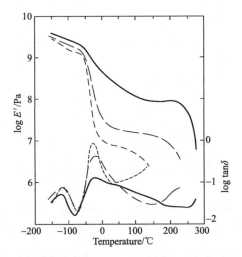

Fig. 8.18 Storage modulus (E') and tan delta (E'/E'') vs. temperature for MDI-ED-PTMO (M_w = 1000) polyurethane ureas. The molar ratio of the components MDI-ED-PCL is (—) 3 : 2 : 1; (—) 2 : 1 : 1; (----) 1.3 : 0.3 : 1.

Fig. 8.18 also exhibits the general features common to all dynamic mechanical measurements of TPEs, a soft segment glass transition temperature below room temperature and a hard segment dissolution temperature above room temperature. The normal operating region of a TPE is the flat plateau region between the two transition temperatures. As suggested by the diagram, one can shift the upper transition temperature by increasing phase mixing, which is usually accomplished by reducing the length of the hard segments. Rheological measurements can also supply a measure of the ODT temperature independent of scattering measurements. In this method, the storage and loss moduli are measured as a function of frequency at different temperatures above and below the MST.

One strain level should be used for these experiments since it has been shown that below the ODT, the dynamical mechanical properties are highly strain dependent. The WLF equation is then used to collapse all of the data onto one curve in a modulus vs. reduced frequency plot, if the frequency is high enough. The frequency at which the curves begin to diverge marks the critical frequency (ω_c) of the material. The critical frequencies are not necessarily identical for the storage and loss moduli, but generally differ by less than an order of magnitude. After finding ω_c, a storage or loss modulus is measured as a function of temperature at constant frequency, which is significantly below the critical frequency. The temperature at which a discontinuity occurs in the modulus marks the ODT temperature, as shown in Fig. 8.19.

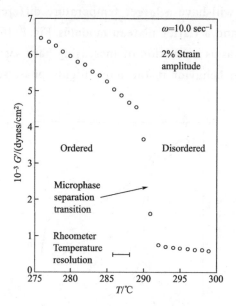

Fig. 8.19 Temperature of G', the shear storage modulus, at a frequency of 1.6Hz for the diblock copolymer PEP-PEE. The ODT temperature was calculated as $(291 \pm 1)\,\text{℃}$.

The authors recommend using the storage modulus because the discontinuity is sharper. This method relies on the different behavior of the storage and loss moduli at low frequencies for the disordered and ordered systems. For the disordered system, the behavior is identical to normal terminal behavior in one-phase materials, i.e., $G' \sim w^2$ and $G'' \sim w$, while in the ordered system, the dependency depends on the morphology and orientation of the material; for a macroscopically unoriented lamellar system, $G' \sim G'' \sim w^{0.5}$.

The viscosity of a TPE depends heavily on whether the system is above or below the ODT temperature. For example, in styrene-diene triblock systems, commercially important materials have ODT temperatures above a reasonable processing temperature; therefore, melt viscosities are usually orders of magnitude higher in the block copolymer than in the homopolymers of the same molecular mass. Typical values at high shear rates are 10^4 poise, which can increase orders of magnitude as the shear rate is decreased.

Generally, TPEs are processed identically to any other thermoplastic, i.e., injection molding, extrusion, calendaring, etc. Some TPEs, for example ionomers, tend to adhere to surfaces, so special equipment and techniques may be necessary to remove an injection-molded part. The processing temperature must be higher than the hard segment dissolution temperature, but even two-phase melts can generally be handled with conventional equipment. The viscosities of TPEs above the ODT are rather low in comparison to other thermoplastics, because of the low overall molecular masses. In fact, in some one-phase materials with low overall molecular masses such as the copolyesters, blow molding cannot be utilized because of low viscosities. The resultant morphology and properties of the material depend heavily on the processing method.

Reaction injection molding (RIM) is a specialized technique very popular in polyurethane manufacture. Two streams, one containing the diisocyanate and the other containing the diols, are mixed by impingement mixing at high pressures. Normally high pressure piston pumps are needed to provide the high pressures as well as the strict metering required. The diisocyanate may have already been reacted with some polyol prior to impingement mixing, which would correspond to a prepolymer method discussed earlier. The mixed material is quickly allowed to flow into the closed mold, but the flow is always regulated to be laminar. The material is cured for a short time, resulting in a finished part.

8.8 Applications

As stated previously, styrene-diene triblock copolymers are the most important category of thermoplastic elastomer. Unlike most other TPEs, they can be blended with large quantities of additives without a drastic effect on properties. In almost all applications, the actual triblock copolymer content is less than 50%. Oils are used as a processing aid and do not result in a significant loss of properties if the polystyrene domains are not plasticized. For this reason, napthalenic oils are preferred. The use of inert fillers such as clays or chalks reduce the cost of the final material. Unlike conventional rubbers, inert fillers do not have a substantial effect on the mechanical properties of TPEs. Thermoplastics such as polyethylene or polypropylene are also used to improve the solvent resistance and can increase the upper service temperature. Polystyrene homopolymer is used as a processing aid, which also increases the hard phase weight fraction and causes the material to stiffen.

KRATON Polymers, the world's largest single producer of styrene-diene block copolymers, produces four types of compounds. SBS compounds are found in shoe soles and as property modifiers for asphalt. In this latter application, the block copolymer is a minor component ($\sim 10\%$) and improves the flexibility at low temperature and increases the softening temperature of the asphalt. SIS materials are employed almost exclusively as adhesives while SEBS triblock copolymers are used as structural materials. The polystyrene content tends to be low in pressure-sensitive adhesives in order to produce a material with more tack; although with hot melt adhesives tack may be undesirable and a harder product is appropriate. The absence of a double bond in SEBS copolymers substantially improves its resistance to UV light and high temperatures, allowing SEBS to be substituted for SIS or SBS in applications where this resistance is necessary. Finally, through postsynthesis processing with maleic anhydride, acid groups are introduced to the SEBS, which improves adhesion to a variety of substrates. These maleated materials are often used in composites, i.e., a polymer containing an inorganic filler.

Applications for TPOs, polyurethanes, copolyesters, and copolyamides are substantially different than for the triblock copolymers. Only small amounts of additives are used in these materials since large amounts of any additive tend to substantially reduce the mechanical properties of these materials. The applications for these multiblock copolymers take advantage of their abrasion resistance, tear strength, and toughness. The TPOs are marginally more expensive then the commodity thermoplastics; however, the segmented block copolymers are substantially more expensive and hence the application must require the improved mechanical properties to justify the increased cost. In the automobile industry, TPEs are used in both interior and exterior parts. For example, TPEs are used in boots and bearings for joints and some tubing as well as exterior bumpers and some paneling. Other industrial uses include hoses, gears, and cables.

Although not a large volume market, the medical device area has certainly been a very important user of TPEs. One very visible application is the use of polyurethanes in angioplasty devices. Polyurethanes (as well as polyamides and polyesters) comprise the balloon material that is expanded to increase the increase the size of opening for blood flow within an artery. Further application of polyurethane TPEs in medical devices can be found in a monograph on this subject by one of the authors of this review.

A very visible consumer product group for TPEs has been in sporting goods. Footwear, including ski boots and soccer shoes, often contain a substantial fraction of TPE. Athletic shoe soles are an especially large application for TPEs. Athletic equipment such as skis and golf ball covers are an area where TPEs are used. Polyurethanes are used for track and field surfaces. Finally, TPEs are beginning to be used in outdoor equipment such as tents because of their low permeability to water coupled with their inherent toughness.

References

[1] N. R. Holden, R. P. Quirk Legge, and H. E. Schroeder, Thermoplastic Elastomers, 2nd Edition, Hanser Publishers, New York, 1996.
[2] C. Hepburn, Polyurethane Elastomers, Applied Science Publishers, New York, 1982, Chap. 1.
[3] M. Morton, Anionic Polymerization: Principles and Practice, Academic Press, New York, 1983.
[4] A. Noshay and J. E. McGrath, Block Copolymers, Academic Press, New York, 1977.
[5] C. W. Lantman, MacKnight W. J., and Lundberg R. D., Annual Review of Material Science 19, 295 (1989).
[6] N. Aust and G. Gobec, Macromolecular Materials and Engineering 286, 119 (2001).
[7] I. W. Hamley, Developments Block Copolymer Science and Technology, Wiley, West Sussex, 2004.
[8] I. W. Hamley, The Physics of Block Copolymers, Oxford University Press, Oxford, 1998.
[9] M. M. Coleman, J. F. Graf, and P. C. Painter, Specific Interactions and the Miscibility of Polymer Blends, Technomic Publishing Company, Lancaster, PA, 1991.

Chapter 9
Elastomer Blends

Blending or mixing of elastomers is undertaken for three main reasons: improvement of the technical properties of the original elastomer, achievement of better processing behavior, and lowering of compound cost. All elastomers have deficiencies in one or more properties and blending is a way of obtaining optimum all-round performance. Elastomer blends are one class of composite materials used in polymer-based industrial products. They are macro scale mixes of high molecular weight elastomer phases. On a microstructural scale, the elastomer could contain blocky or crystalline chain segments which act as reinforcing agents or stiffeners. More than 93% of polymer blends are combinations of thermoplastic and elastomers. However, automotive tires (5000kt/yr in 1995), a blend of elastomers, are the largest application for any immiscible polymer blend.

9.1 Introduction

Blends of elastomers are of technological and commercial importance since they allow the user to access properties of the final blended and vulcanized elastomer that are not accessible from a single, commercially available elastomer alone. These potentially improved properties include chemical, physical, and processing benefits. In reality, all blends show compositionally correlated changes in all of these properties compared to the blend components. The technology of elastomer blends is largely focused on the choice of individual elastomers and the creation of the blends to achieve a set of final properties. This chapter shows some of the instances of the uses of elastomer blends. Empirical guidelines for the creation of novel blends of elastomers is a comparatively more difficult proposition.

Blends provide an acceptable technological process for accessing properties not available in a single elastomer. In elastomers composed of a single monomer in a single insertion mode (e.g., 1,2-polybutadiene), there are no other procedures available except blends. In the case of elastomers that are copolymers (e.g., styrene-butadiene rubber, SBR), changes in intramolecular composition, such as formation of a block polymer instead of random copolymer at the same composition, are effective. However, intramolecular changes are limited by available synthesis processes. Intermolecular changes, in either composition or distribution of monomers, such as in blends, are not limited by such systemic or synthetic limitations.

Theoretically, blends of elastomers can attain a wide variation in properties. Combinations of elastomers can lead to changes in properties due to either intrinsic differences in the constituents or differences in the reinforcement and vulcanization of the constituents. Miscible blends of elastomers that consist of a single elastomeric phase with microscopically uniform crosslinking and distribution of reinforcing agents reflect a compositionally weighted average of the intrinsic properties of the constituents. Miscible blends are commonly used though they have been very rarely recognized. Analysis of such blends, particularly after vulcanization, is difficult. The current analytical techniques are only slightly more capable than the classical techniques of selective precipitation of the components of an unvulcanized elastomer blend from solution.

Common examples of miscible blends are ethylene-propylene copolymers of different composition that result in an elastomer comprising a semi-crystalline, higher ethylene content and an amorphous, lower ethylene content components. These blends combine the higher tensile strength of the semicrystalline polymers and the favorable low temperature properties of amorphous polymers. Chemical differences in miscible blends of ethylene-propylene and styrene-butadiene copolymers can also arise from differences in the distribution and the type of vulcanization site on the elastomer. The uneven distribution of diene, which is the site for vulcanization in blends of ethylene-propylene-diene elastomers, can lead to the formation of two distinct, intermingled vulcanization networks.

Immiscible blends show additional, more complex changes due to a microscopically inhomogeneous phase structure of the two component elastomers. The two separate phases typically have differences in the retention of the fillers and plasticizers as well as vulcanization in the presence of the curative. Changing the properties of elastomers by uneven distribution of fillers and vulcanization is, however, the more common use of blends of immiscible elastomers. The engineering properties of elastomers (i. e. , tensile strength, hysteresis) in vulcanized compounds depend not only on the elastomer itself but also on the amount and identity of the fillers and plasticizers, as well as on the extent of cure. In an immiscible blend, the amount of these additives in any phase can be modulated by changes in the viscosity and chemical identity of the elastomer, the surface chemistry of the filler, the chemical nature of the plasticizer, and the sequence of addition of the components as well as the details of the mixing procedure. A large body of experimental procedures (*vide infra*) has been developed to attain a thermodynamically metastable, inter-phase distribution of additives in blends. On vulcanization, this distribution is rendered immobile and leads to desirable engineering properties of the blend.

Two notable reviews of elastomer blends exist. The first, by Hess *et al.*, reviews the applications, analysis, and the properties of the immiscible elastomer blends. The second, by Roland, has in addition a discussion of the physics of mixing immiscible polymer blends and a more recent account of the analytical methods. Other reviews by Corish and McDonel *et al.* deal with specific aspects of elastomer blends. These reviews are focused on immiscible blends of elastomers. In this section we will complement this with information on miscible blends.

9.2 Miscible Elastomer Blends

9.2.1 Thermodynamics

The extension of thermodynamics to a blend of elastomers has been discussed by Roland. Miscible blends are most commonly formed from elastomers with similar three-dimensional solubility parameters. An example of this is blends from copolymer elastomers (e.g., ethylene-propylene or styrene-butadiene copolymers) from component polymers of different composition, microstructure, and molecular weights. When the forces between the components of the polymer blend are mostly entirely dispersive, miscibility is only achieved in neat polymers with a very close match in Hansen's three-dimensional solubility parameter.

Miscible blends of elastomers differ from corresponding blends of thermoplastics in two important areas. First, the need for elastic properties require elastomers to be high molecular weight polymers with a limited polydispersity. This reduces the miscibility of dissimilar elastomers by interdiffusion of the low molecular components of the blends. Second, elastomers are plasticized in the conventional compounding with process oils. The presence of plasticizers leads to a higher free volume for the blend components and stabilizes, to a small extent, blends of dissimilar elastomers.

9.2.2 Kinetics

The formation of miscible rubber blends slows the rate of crystallization when one of the components is crystallizable. This phenomenon accounts for data that shows lower heats of fusion that correlate to the extent of phase homogeneity in elastomer blends. Additionally, the melting behavior of a polymer can be changed in a miscible blend. The stability of the liquid state by formation of a miscible blend reduces the relative thermal stability of the crystalline state and lowers the equilibrium melting point. This depression in melting point is small for a miscible blend with only dispersive interactions between the components.

9.2.3 Analysis

9.2.3.1 Glass Transition

The principal effect of miscibility of elastomer blends of dissimilar elastomers is alteration of the glass-transition temperature. Since miscible blends should have negligible changes in the conformation of the polymer chains, the entanglement density of miscible blends should be a compositionally weighted average of entanglement density of the pure components.

9.2.3.2 Magnetic Resonance Imaging

Nuclear magnetic resonance (NMR) has been applied to the study of homogeneity in miscible polymer blends and has been reviewed by Cheng and Roland. When the components of a blend have different T_g's, proton NMR can be used to assess the phase structure of the blend by taking advantage of the rapid decrease of proton-proton coupling with nuclear separation. For blends containing elastomers of almost identical T_g, proton MAS NMR is applied to blends where one of the components is almost completely deuterated. Another technique is cross polarization MAS ^{13}C NMR. The transfer of spin polarization from protons to the ^{13}C atoms of the deuteriated component can occur if these carbons are in proximity (nanometers) to the protons.

9.2.3.3 Crystallinity

Changes in polymer crystallinity have also been employed to study the homogeneity of elastomer blends. Morris studied the rate of crystallization of cis-1,4-BR in blends with SBR. At any given blend composition, the BR crystallization rate diminishes with greater blend homogeneity. Sircar and Lamond also studied the changes in BR crystallinity in blends with NR, IR, EPM, CIIR, NBR, and CR (Fig. 9.1). The nature of the blend component had the greater effect since the more compatible blends (smaller domains) had the greater the loss in BR crystallinity.

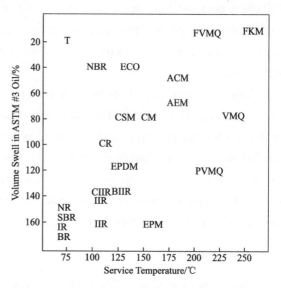

Fig. 9.1 Elastomer distribution by resistance to aromatic solvents (ordinate) and temperature.

9.2.3.4 Interdiffusion

Interdiffusion between a pair of polymers is a demonstration of their thermodynamic miscibility. The adhesion between contacted rubber sheets parallels the extent of any interdiffusion of the polymer chains. If the contacted sheets are comprised of immiscible rubbers, no inter-diffusion occurs.

Natural rubber (NR) and 1,2-polybutadiene (1,2-BR) are miscible even at high molecular weights. When NR is brought into contact with 1,2-BR, they interdiffuse spontaneously. When some form of scattering contrast exists between the materials, interdiffusion will enhance the scattering intensity (either X-ray or neutron) measured from the plied sheets. A variety of spectroscopic methods have been used to detect the interdiffusing species.

9.2.3.5 Mechanical Properties

Miscible blends should have greater mechanical integrity than a comparable multiphase structure. Miscible rubber blends that react chemically have a densification and a higher cohesive energy density. This may provide improved mechanical properties but has been observed only below the T_g.

9.2.4 Compositional Gradient Copolymers

A significant development in the last decade is the use of miscible blends of compositionally different EPDM. The blends are designed to balance viscoelastic properties such as the rate of extrusion or adhesion with physical properties such as tensile strength by changes in the relaxation characteristics. This is achieved with the components having different average molecular weight or differences in the crystallinity (due to extended ethylene sequences) or both. An example of the components of these blends is shown in Table 9.1.

Table 9.1 Ethylene-propylene blend components differing in molecular mass and crystallinity

Sample	Composition C2 Wt%	Viscosity ML(1+4)125℃
A	60	41
B1	74	72
B2	76	247
B3	78	1900
B4	68	189
B5	84	291

Note: Viscosity determined according to ASTM D1646.

The blends of component A with one of the B polymers are miscible and are made by mixing hexane solutions of the elastomers. Fig. 9.2 and Fig. 9.3 show the effect on tensile strength of the compounded but unvulcanized blends from the components in Table 9.1.

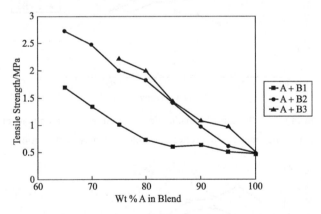

Fig. 9.2 Variation in tensile strength of unvulcanized, compounded blends of ethylene-propylene copolymer due to differences in molecular mass distribution.

The blends in Fig. 9.2 are different in the molecular mass distribution while in Fig. 9.3 they are different in composition and crystallinity. This dispersion in molecular weight and crystallinity is apparent in other viscoelastic properties such as peel adhesion. Fig. 9.4 and Fig. 9.5 show self-adhesion measured by the force needed for the failure of a spliced portion in blends identical to those in Fig. 9.2 and Fig. 9.3, respectively. Adhesion increases with increasing molecular weight and compositional dispersity of the blend. This is shown

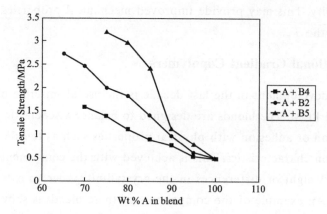

Fig. 9.3 Variation in tensile strength of unvulcanized, compounded blends of ethylene-propylene copolymer due to differences in composition and crystallinity distribution.

by the comparative data for B1 and B2 in Fig. 9.5, and B4 and B2 in Fig. 9.4. Small increases in the cystallinity and molecular weight dispersion by blending promote adhesion by slowing down the latter without substantial effect on the former. Further increases in either severely retard the intermingling of chains necessary for the self-adhesion.

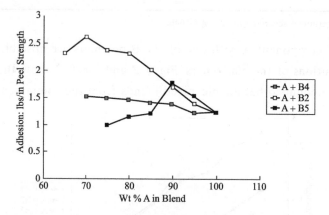

Fig. 9.4 Variation in peel adhesion of unvulcanized, compounded blends of ethylene-propylene copolymer due to differences in molecular mass distribution.

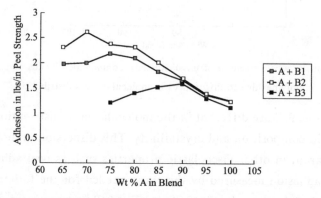

Fig. 9.5 Variation in peel adhesion of unvulcanized, compounded blends of ethylene-propylene copolymer due to differences in composition and crystallinity distribution.

Table 9.2 Ethylene-propylene blend components differing in crosslink density

Sample	Composition C2 wt%	Composition ENB wt%	Viscosity ML(1+4)125°C
A1	57.0	3.2	20
A2	60.2	2.9	32
A3	60.3	2.8	41
A4	59.4	2.6	51
A5	60.5	3.2	67
B	64	0.9	2100

Note: Viscosity determined according to ASTM D1646.

Nonuniform vulcanization networks in miscible blends of elastomers have a strong effect on tensile strength and elongation. These networks have an intermolecular distribution in crosslink density and are composed of different concentrations of crosslinkable sites in the components of the blend. Differences in the level of the enchained diene (5-ethylidene-2-norbornene) for EPDM copolymers or differences in the level of the vulcanizable chain end unsaturation for siloxane polymer define these blends. These networks lead to an increase in both the elongation as well as the tensile strength at high elongation compared to vulcanizates of similar viscosity having a uniform network. In particular, nonuniform networks display a non-linear increase in the tensile modulus at high elongation. This non-linearity is due to a nonaffine deformation of the network at the high elongation, which continually reallocates the stress during elongation to the lightly crosslinked component of the blend that is most able to accommodate the strain. Blends of one A and various amounts of polymer B in Table 9.2 were blended in hexane solution, compounded, and vulcanized. Both of the polymers are amorphous, and the A polymers differ in the molecular weight and contain approximately 3% of vulcanizable diene (ENB). The B polymer is much lower in diene and has 0.7% ENB. The tensile strength of the blends derived from all of the A with varying amounts of B are shown in Fig. 9.6. In all cases where small amounts (<25%) of B are included in the blends, the tensile strength is higher for the blends than for the parent A, even though the extent of vulcanization is lower in the blends than in the blend components A due to reduction in the total amount of diene.

9.2.5 Distinct Polymers

9.2.5.1 IR-BR Blends

Blends of 1,4-IR with predominantly 1,2-BR are unique since they are miscible, chemically distinct, high-molecular weight homopolymers, even though there is no dipole or specific interaction between the components. As the concentration of 1,4-units in the BR increases, there is a decrease in miscibility with IR. The interaction parameter measured for blends of IR with BR of varying 1,2-and 1,4-geometry indicates that the exchange enthalpy between IR and BR becomes more endothermic as the concentration of 1,4-units increases. Differences in structure are apparent from the thermal expansion coefficients of the components. The difference in thermal expansion coefficients of IR and BR is greatly diminished as the 1,2-content of the latter increases.

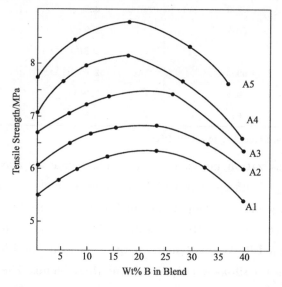

Fig. 9.6　Tensile strength of blends in Table 9.2 differing in crosslink density.

9.2.5.2　Epoxidized PI-CPE Blends

The intramolecular epoxidation of 25 mole % of PI leads to miscible blends with CPE containing 25wt % chlorine. The origin of the miscibility is the specific interactions involving the oxirane of the epoxy ring with the chlorine.

9.2.6　Reactive Elastomers

The formation of miscible blends of elastomers by mutual chemical reaction of the blend components has been explored by Coleman *et al*. Chemical reaction in a miscible blend provides the negative free energy needed to compensate for the unlike *i-j* contacts between dissimilar elastomers. Miscible blends were formed by the use of stabilizing weak hydrogen bonding between a 2,3-dimethyl butadiene-4-vinyl phenol copolymer and carbonyl group of polymers such as EVA and poly alkyl methacrylates. The extent of the H-bonding stabilization was modulated by the concentration of the phenol residues in the unsaturated copolymer. The choice of the stabilizing chemical bond between the elastomers is important. A stable, localized chemical bond would lead to a loss of viscoelastic properties characteristic of an unvulcanized elastomer. Weak hydrogen bonding is fluxional and allows flow in response to stress in the elastomer blend.

9.3　Immiscible Elastomer Blends

9.3.1　Formation

Immiscible blends can be made from any two dissimilar elastomers. They can be made by *in situ* copolymerization, blending polymer solutions or suspensions, and mechanical mixing. These procedures have been reviewed by Corish. In contrast to blends of other polymers, elastomers can be either "preblended" or "phase mixed" with each other. "Preblended" systems are those where the compounding ingredients (e. g. , fillers, plasticizers, and curatives) are added simultaneously to a mixture of the immiscible polymers. In "phase mixed" blends, ingredients are added separately to each of the individual polymers in separate compounding operations. The compounded elastomers are then blended together. Phase mixed blends provide a greater certainty of the initial interphase location of fillers, plasticizers, and curatives than the preblends.

9.3.2 Kinetics of Blend Morphology

The kinetics of the formation of the phase morphology is dominated by two competitive factors. First, the viscoelastic flow of elastomers, the presence of plasticizers, the long resting times (days) between mixing and crosslinking as well as low shear processing steps (e. g. , calendering) are ideal for the development of the equilibrium morphology. However, the sluggish diffusion of high molecular weight elastomers, the presence of particulate fillers, and the ultimate crosslinking generate persistent nonequilibrium morphologies. Tokita and Avgeropoulos *et al.* have studied the mixing of EPDM with the diene rubbers —NR or BR— as a function of relative viscosity.

The detailed morphology of elastomer blends depends on

(1) the mixing procedure,

(2) the rheology of the blend components, and

(3) the interfacial energy. As with other polymer blends, the elastomer of lower viscosity tends to be the continuous phase.

Cocontinuous blend morphology is observed only for elastomers with similar viscosities. The viscoelastic forces developed during the formation of the compounded blend from two rheologically dissimilar elastomers are principally responsible for the morphology. Interfacial tension due to chemical differences between the elastomers is less important.

9.3.3 Analysis

9.3.3.1 Microscopy

Phase contrast light microscopy has been applied extensively to the analyses of unfilled binary elastomer combinations. This method is based on differences in the refractive indices of the polymers and has been reviewed by Kruse. Callan *et al.* have shown the versatility of the method for a wide range of binary blends containing NR, SBR, BR, CR, NBR, EPDM, IIR, and CIIR. The results of these experiments are shown in Table 9.3, which lists the measured areas of the disperse phase in more than 50 combinations of Banbury-mixed 75 : 25 binary blends containing eight different elastomers. Blends of IIR-CIIR and SBR-BR are excluded since the contrast was low. It can be seen that NBR produced the greatest heterogeneity in all blends except those with CR.

Transmission electron microscopy (TEM) is applicable to both filled and unfilled elastomer blends. However, for most elastomer combinations, there is no contrast between the polymer phases in a TEM. If the polymers differ significantly in unsaturation, osmium tetroxide (OsO_4) or ruthenium tetroxide (RuO_4) staining is the best method for obtaining contrast. The metal oxides selectively oxidize the unsaturation and the location of the metal atoms in the phase with the greater unsaturation provides the electron density contrast for TEM. Contrast for TEM analysis of elastomer blends of diene rubbers is achieved through the sulfur hardening method developed by Roninger and utilized by Smith and Andries. Small rubber specimens are immersed in a molten mixture of sulfur, an accelerator, and zinc stearate. The selective absorption of the zinc salts into the SBR phase renders it darker than the BR in a TEM.

Table 9.3 Average areas (in mm^2) of the dispersed phase in 75 : 25 pure elastomer blends

Disperse phase 25%	Matrix(75%)							
	NR	CR	BR	SBR	NBR	EPDM	IIR	CIIR
NR	—	45	1.5	1.2	300	1.5	2.0	3.2
CR	35	—	4.0	2.5	1.5	25	20	15
BR	0.7	4.5	—	—	15	2.2	2.1	2.5
SBR	0.5	2.7	—	—	20	2.1	12	10
NBR	400	1.3	17	30	—	250	100	225
EPDM	3.5	75	2.8	2.6	225	—	2.0	1.5
IIR	3.0	15	3.0	4.2	75	1.0		
CIIR	2.2	25	2.3	2.5	85	1.2		

Scanning electron microscopy involves simpler specimen preparations than TEM. Both OsO_4 and RuO_4 staining techniques work with SEM and can be applied to both bulk specimens and films. Atomic force microscopy is an extension of scanning tunneling microscope (STM) and has the potential for atomic resolution. In its simplest form, the AFM acts as a "miniature surface profilometer" and provides topographical images. The potential advantages of these techniques include higher resolution, simplicity of specimen preparation, and greater versatility in varying the mechanisms for achieving image contrast. Digital image analysis has expedited particle size analysis from micrographs at resolution limits of a few Angstroms.

9.3.3.2 Glass Transition Temperature

The most common method of estimating the degree of homogeneity in elastomer blends is by measurement of the temperature of transitions from rubber to glass. For elastomers this is usually at subambient temperatures. Glass transition measurements do not provide any information on blend morphology. The observation of distinct transitions corresponding to the respective components of the blend indicates the existence of a multiphase structure. A potential source of error is that vulcanization tends to raise the T_g due to restricted motion of the chains; this might be interpreted as miscibility.

9.3.3.3 Magnetic Resonance Imaging

NMR imaging of solids has been used to characterize the phase sizes with a spatial resolution of less than 50mm in immiscible mixtures containing polybutadiene. Butera *et al.* have reported improvement over the ^{13}C NMR method by using magic angle spinning (MAS) in polymer blends containing a deuterated and a protonated polymer. More recently, Koenig *et al.* have used NMR for the determination of vulcanization efficiencies in elastomer blends.

9.3.3.4 Light, X-ray, and Neutron Scattering

Roland and Bohm used both SAXS and SANS in conjunction with TEM to assess the size of butadiene domains (5% by weight) in a CR matrix. Roland *et al.* used different master batches containing deuterated and protonated butadiene, respectively. These were each blended on a two-roll mill to study the coalescence of the polybutadiene phase. Coalescence (measured by SANS) was minimized by low temperature mixing, or by significantly increasing the molecular weight of the disperse or continuous phase to create a greater disparity in viscosity. Some preliminary SANS work on IR blended with deuterated 1,2-BR has been reported. All scattering techniques are affected by the presence of fillers, which lead to more extensive scattering than the phase morphology. This is a severe limitation to the wide applicability of these techniques to most elastomers.

9.3.4 Interphase Distribution of Filler, Curatives, and Plasticizers

At ambient and processing temperatures, elastomers are viscous fluids with persistent transport phenomenon. In immiscible blends, these lead to change in the size and shape of the elastomer phases and migration of the fillers, plasticizers, and curatives from one phase to another. These changes are accelerated by processing and plasticization but retarded by the ultimate vulcanization. Retention of the favorable properties of a metastable blend, which is often attained only at a select interphase morphology and filler/plasticizer distribution, thus requires careful control of both the processing and the vulcanization procedures.

9.3.4.1 Curative and Plasticizer Migration in Elastomer Blends

During mixing, curatives are initially located within the continuous phase. Since the curatives dissolve in the elastomer, curative migration across phase boundaries can occur. Owing to the higher solubility of sulfur in elastomers containing diene or styrene groups and the greater affinity of many accelerators for polar rubbers, large differences in crosslink density of the different phases result on vulcanization. Further, increased rate of vulcanization in the diene or styrene containing elastomers can cause depletion of the curatives in this component, leading to even greater curative migration (Le Chatelier's principle). For most elastomer blends, these effects are in concert, and the result is a large cure imbalance between the phases. In addition, the boundary layers of the two elastomers, adjacent to the interface, can be different than the bulk since the greatest migration of curatives and plasticizers occurs in the proximity to this region. The transfer of curatives to one phase at the expense of the other creates a morphology where the boundary layers separates a curative-depleted, slightly crosslinked elastomer near a curativeenriched, tightly crosslinked elastomer.

Amerongen has an early review on curative migration in heterophasic elastomer blends based on optical and radiochemical analyses. A later, more detailed work by Gardiner used optical analysis to study curative diffusion across the boundaries of elastomer blends consisting of binary combination of polymers of CIIR, IIR, EPDM, CR, SBR, BR, and NR. Gardiner measured a diffusion gradient for the concentration change as a function of distance and time. His measurements for the diffusion of accelerator and sulfur from IIR to other elastomers are listed in Table 9.4.

Table 9.4 Curative diffusion coefficients in elastomer blends

Curative	From	To	$D \times 10^7 / (cm^2/s)$
Accelerator(TDDC)	IIR	BR	12.66
		EPDM	1.09
		CR	1.08
		SBR	0.58
		NR	0.70
Sulfur	IIR	SBR	4.73
		SBR & 50 PHR N700 CB	17.2
		NR	2.82

9.3.4.2 Cure Compatibility

Adequate properties in a vulcanized rubber blend depend on the covulcanization across the phases. Covulcanization is the formation of a single network structure including crosslinked macromolecules of both polymers. The degree of vulcanization is at similar levels in both phases with crosslinking across the phase interfaces. Shershnev and earlier Rehner and Wei have summarized the experimental requirements for covulcanization of the components of elastomer blends. These are:

(1) Phase blending followed by very short, high-temperature cure cycles;
(2) Vulcanization agents chemically bound to the elastomer;
(3) Accelerators similar in solubility parameter to less reactive elastomers;
(4) Insoluble vulcanizing systems that cannot migrate;
(5) Nonpolar vulcanizing agents that distribute uniformly and have similar reactivities toward different elastomers (e.g., peroxides and reactive resins).

Gardiner and Woods and Davidson have improved the covulcanization of EPDM-NR and IIR-NR blends by using an insoluble lead-based curative for the less unsaturated elastomer. The blends with soluble curatives from the same elastomers were much inferior. Coran covulcanized EPDM-NR blends by modifying the EPDM with maleic anhydride. The modified EPDM can be crosslinked with zinc oxide to form an ionic crosslink network. The migration of curatives was investigated by Bhowmick and De in binary blends of NR, BR, and SBR. Cure rates decreased in that order, and the migration of curatives was the greatest for NR-SBR.

The state of cure of the phases in a blend can be determined from changes in the magnitude of the damping peaks and from freezing-point-depression measurements on swollen networks. Honiball and McGill have employed the freezing point depression of solvent-swelled NR-BR blends to determine the crosslink densities of the individual polymer phases.

The freezing point depression of a solvent in a swollen vulcanizate is dependent on the volume fraction of solvent in the elastomer phase.

9.3.4.3 Interphase Filler Transfer

The distribution of fillers (e.g., carbon black, silica) and various processing aids in heterogeneous elastomer blends can be nonuniform. Interphase transfer of fillers has been observed in blends of both diene and saturated elastomers. This migration is due to the greater solvation between a filler and one of the polymers in the melt-mixed blends. Callan *et al.* showed significant transfer of carbon black from NR to BR and silica from BR to NR in solution and latex blends. Extensive carbon black transfer from a low unsaturation elastomer (IIR and EPDM) to other high unsaturation (e.g., diene) elastomers occurs regardless of the master batch mixing procedure. This is seen in blends of EPDM or IIR with NR or SBR. During mechanical mixing, chemisorption of the unsaturated polymer onto the carbon black occurs, thus preventing any subsequent transfer of the carbon black. In all instances, the carbon black was located almost entirely in the NR phase when added to the preblended elastomers. Transfer of carbon black from EPDM to NBR could be reversed if the EPDM was functionalized with a primary amine.

Amines are tenaciously chemisorbed to the surface of carbon black, and the functionalization allows the EPDM phase to retain significant amounts of filler. The carbon black distribution is shown in Fig. 9.7 (a) for the blend of EPDM with NBR and is compared to Fig. 9.7 (b) for the blend of the amine functionalized EPDM with NBR.

Callan *et al.* used the differential swelling technique (*vide infra*) for a ranking of the relative affinity of carbon black to various elastomers. The data was obtained for 50 : 50 binary blends, and the carbon black was added to about 25 binary elastomer combinations. The carbon black retention in each phase was estimated from TEM micrographs of the blend. The ranking of carbon black affinity for different elastomers was in the order SBR > BR, CR, NBR > NR > EPDM > CIIR, IIR. Cotton and Murphy, in a complementary experiment, have determined the carbon black distribution for seven different carbon blacks in preblended SBR-BR and SBR-NR blends. The data is shown in Table 9.5. The carbon blacks differ in the surface structure and size; they ranged in surface area

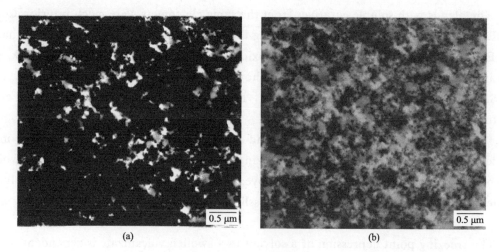

Fig. 9.7 Micrographs of a N234 carbon black distribution in 70 : 30 blend of NBR and EPR.
(a) The EPR is a copolymer of 42 mole % ethylene.
(b) The EPR has 43 mole% ethylene and 0.9 mole % primary amine functionality

(CTAB) from 43 to 136m²/g. For all cases of SBR-NR blends, the carbon black was preferentially located in the SBR phase.

Table 9.5 Effect of carbon black on the distribution in 50 : 50 natural-rubber/SBR blend containing 45phr black

Carbon black type	Surface area(CTAB)m²/g	Bound rubber/%	Carbon-black loading/phr	
			NR	SBR
N560	43	17.0	41.2	48.8
N347	87	28.0	33.9	56.1
N339	95	30.2	31.1	58.9
N234	120	30.4	30.0	60.0
Vulcan(10H)	136	33.1	27.2	62.8
N326	84	22.5	31.9	68.1
N330	81	25.0	34.0	66.0

9.3.5 Analysis of Interphase Transfer

9.3.5.1 Microscopy

Electron microscopy is a common technique for determining the filler distribution in a heterogeneous elastomer blends. Dias *et al.* have used time of flight secondary mass spectrometry (TOF-SIMS) to simultaneously deter-mine the morphology as well as curative diffusion in BIMS-diene elastomer blends.

9.3.5.2 Differential Swelling

The method of differential swelling of thin section of blends of EPDM and IIR was first utilized by Callan *et al.* In blends with EPDM, the IIR phase absorbs more solvent; therefore, the IIR domains are thinner and appear lighter in the TEM. Hess *et al.* applied the same differential swelling method to the analysis of carbon black distribution at low filler loadings. Wang *et al.* have improved the technique by swelling separately with two solvents—one for each of the phases.

9.3.5.3 Staining

Staining with volatile reactive metal oxides, OsO_4 and RuO_4, is the preferred method for achieving inter-phase contrast for TEM analyses. It is applicable to blends of elastomers with different degrees of unsaturation such as NR-EPDM blends.

9.3.5.4 Differential Pyrolysis

This method is applicable to blends containing polymers with significantly different thermal degradation temperatures. It has been used for analysis of carbon black distribution in NR-SBR and NR-BR blends.

9.3.5.5 GC Analysis of Bound Rubber

Bound rubber is the elastomer insoluble in solvents due to chemisorption onto the carbon black during mixing. It is extracted by swelling the unvulcanized polymer in a solvent for an extended period. Any soluble lower molecular weight polymer that is not bound to the carbon black is removed. This method was first used by Callan *et al.* for a number of elastomer blends.

9.3.5.6 Mechanical Damping

The value of $\tan\delta$ (at T_g) is lower for a filled elastomer than the pure elastomer. This is due to the increase in dynamic elastic modulus of the filled compound for the higher temperature side of the T_g peak. The effect is governed by filler concentration and loading. Maiti *et al.* have used this lowering of $\tan d$ at T_g to estimate the distribution of filler in an immiscible elastomer blend.

9.3.6 Compatibilization

Compatibilization of highly incompatible elastomers has only been used to a limited extent. Compatibilization is the addition of a minor amount of an interfacial agent and serves only to stabilize the extended surface of the dispersed phase in a finely dispersed morphology. The amount and composition of the interfacial agents are not designed to affect the bulk properties of either of the phases. Properties of elastomer blends are determined by intensive properties such as cohesive energy, crosslink densities, and chemisorption of fillers of the components that are unaffected by the addition of compatibilizers. However, in binary blends of elastomers with large differences in solubility parameters, as in polyolefin elastomers with polar elastomers, properties are dominated by the large domain size and the lack of interfacial adhesion. These elastomer blends are significantly improved by the addition of a compatibilizer.

Setua and White used CM (chlorinated polyethylene) as a compatibilizer to improve the homogeneity of binary and ternary blends of CR, NBR, and EPR. NBR-EPM and CR-

EPDM blends homogenize more rapidly when small amounts of CM are added. The presence of the compatibilizer leads to reductions in both the time needed for mixing, observed by flow visualization, and the domain size of the dispersed phase, observed by SEM. Arjunan *et al.* have used an ethylene acrylic acid copolymer and an EPR-*g*-acrylate as a compatibilizer for blends of EPDM-CR. The addition of the compatibilizer leads to the reduction in the phase size of the dispersed EPDM phase as well as increase in the tensile tear strength of the blend.

Intensive properties of the blend components that dominate vulcanizate properties of the elastomer are improved if the compatibilizer is the predominant fraction of the elastomer blend. Davison *et al.* describe the formation of compatibilized blends of poly(alkyl acrylates) and preformed EPDM-*g*-acrylate that on vulcanization are resistant to solvents. The acrylate-grafted EPDM are copolymers of methyl-, ethyl-, or *n*-butyl acrylate. These vulcanized blends have excellent tensile strength and modulus, similar to a single elastomer. Cotton and Murphy were able to generate the graft polymer *in situ* during compounding. A primary amine copolymerized EPDM blended with NBR to form a graft polymer by the amidine reaction of the amine with the nitrile. The grafting reaction was catalyzed by the presence of the Lewis acidic phophite plasticizer for the NBR phase. The compatibilizer promotes the formation of very small dispersed phase domains. Fig. 9.8 (a) and Fig. 9.8 (b) are micrographs of the dispersion of the EPDM and amine functionalized EPDM, respectively, in the NBR matrix. In this case, the previous micrographs (see Fig. 9.7) also show the retention of carbon black filler in the bulk of the EPDM phase.

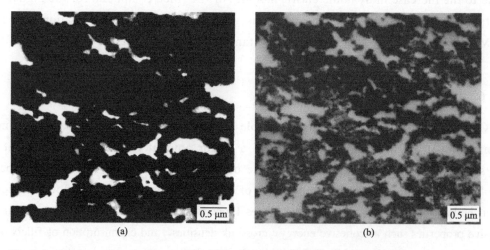

Fig. 9.8　Micrographs of a dispersion of 70 : 30 blend of NBR and EP elastomer. (a) The EP elastomer is a copolymer of 42 mole % ethylene. (b) The EP elastomer has 43 mole % ethylene and 0.9 mole % primary amine functionality.

Vulcanization of the amine functional EPDM and NBR blends with nonpolar peroxides that are expected to distribute equally in both phases lead to blends with excellent solvent and temperature resistance.

9.3.7 Properties of Immiscible Blends

While true miscibility may not be required for elastomer properties, adhesion between the immiscible phases is required. Immiscible polymer blends that fulfill this criteria provide a significant opportunity to change the rheological, tensile, and wear properties of elastomer blends compared to miscible blends.

9.3.7.1 Processing

Blends are often used to improve the processability of rubbers. This improvement may consist of either lowering the viscosity or producing a material less prone to melt fracture during flow. Secondary elastic effects such as die swell can also be affected by blending. Avgeropoulos *et al.* showed that the low viscosity phase in a binary blend tends to become continuous. The effect is accelerated under shear as the morphology of the blend responds to the applied stresses. In the vicinity of a wall, the shear rates tend to be highest, and the lower viscosity component will accumulate at the surface. Incorporation of only a few percent of EPDM to a fluoroelastomer or of PDMS into an SBR was found to reduce steady state viscosities. This is due to the lower viscosity component residing at the interface.

Much of the knowledge in this area is either derived from practical experience or anecdotal. Theoretical predictions of the viscosity of elastomer blends are of limited use since (1) the inhomogeneous phase morphology of an elastomer blend changes easily in response to applied stress, and (2) the nonuniform distribution of fillers and plasticizers in the phases responds to flow. These structural changes in elastomer blends under shear lead to anomalous rheological properties quite different from the expected average of pure components.

9.3.7.2 Modulus

In a heterogeneous blend, the details of the morphology do not generally exert much influence on the stress-strain tensile response. Contrary to expectation that the continuous phase would have more influence, the stress-strain response of unfilled EPDM-BR blends was found to be unaffected by a change in the BR domains from continuous to disperse.

Carbon black distribution has a profound effect on modulus. Meinecke and Taftaf have shown that an increase in the nonuniformity of this filler distribution results in a lower blend modulus. The transfer of a portion of the carbon black from one phase lowers its modulus more than the increase in modulus of the phase with the higher carbon black concentration. This effect is related to the nonlinear dependence of rubber modulus on carbon-black loading.

9.3.7.3 Tack and Adhesion

Adhesion is a surface phenomenon due to entangling of chains at the surface. Adhesion of different elastomers before (tack) and after vulcanization (co-cure) can often be obtained only through the blending of the dissimilar elastomers with a single component. However, the composition of the surface and thus the adhesion characteristics can be altered without using a high concentration of a particular elastomer in the blend.

Blends with components that differ in viscosity tend to have the lower viscosity rubber concentrated at the surface during processing. The most common procedure for increasing the tack and co-cure of elastomer blends is to have a single elastomer as the predominant phase in each of the blends. Morrisey showed that tack of dissimilar elastomers blended with NR improves monotonically with the amount of NR in the blend. Increasing the proportion of a single elastomer in both blends enables the ability of the plied surfaces to fuse together.

9.3.7.4 Hysteresis

A principal use of elastomer blends is in sidewalls of automotive tires. The reduction of hysteresis losses ("rolling resistance") is a principal design target. Lower hysteresis in a single elastomer requires reduced carbon-black loading or increased crosslink density. These changes adversely affect other aspects of performance. Blending of elastomers provides a lower hysteresis with few of these adverse effects. The hysteresis of a filled elastomer containing zones of different carbon black concentrations is lower than a uniformly filled elastomer. Blends of dissimilar elastomers, with differing ability to retain carbon black, provide an easy experimental process to obtain this nonuniform carbon-black distribution.

9.3.7.5 Failure

Improved failure properties can result from the blending of elastomers, including performance that exceeds either pure component. An important aspect of the structure of a rubber blend is the nature of the interphase bonding. The mechanical integrity of an interphase crosslinked morphology will usually lead to superior performance. In blends of SBR and chlorobutyl rubber, for example, an increase in fatigue life was obtained by the introduction of interphase crosslinking. Similarly, providing for inter-facial coupling improves the tensile strength of EPDM/silicone-rubber blends.

There is improved abrasion resistance associated with a preferential carbon-black BR phase distribution in blends of NR-BR and SBR-BR. The first abrasion studies on the effects of carbon black phase distribution in NR-BR blends were reported by Krakowski and Tinker. Tread wear resistance was found to increase progressively with increasing carbon black in the BR phase, which was determined from TEM analyses. Tse *et al.* have shown in blends of dispersed BIMS in BR matrix failure due to fatigue can be retarded if the mean distance between the crosslinks of the BIMS is less than 60nm.

The incidence of cracking due to ozone attack has been investigated for NR-EPDM blends. Andrews showed that small zones of EPR in an EPR-NR blend provide a barrier that inhibits ozone crack growth. Ambelang *et al.* found the importance of small EPM domain size in EPDM-SBR blends. Matthew has shown that carbon black improves the ozone resistance of NR-EPDM. An improvement was obtained in the blends with a balanced carbon-black phase obtained by phase inversion. This is because (1) there is better reinforcement of the EPDM phase and (2) carbon black in the EPDM expands the volume of that phase. In BIMS-BR blends, the ozone failure can be retarded by reducing the size of the

BIMS dispersion.

The transport properties of polymer blends are of interest both for the practical application of blends and for providing insight into the morphology of the blend. Measurement of the effect of blend composition on permeability in various rubbers has been described. The permeability of elastomer blends depends on the concentration of the continuous phase and the morphology of the dispersed phase. Extended disperse phase structures, particularly lying in a stacked or lamellar configuration, can lead to a reduced permeability because of the more tortuous path that must be taken by penetrants.

9.3.8 Applications

9.3.8.1 Unsaturated Elastomer Blends

The most common blends of unsaturated elastomers are those used in various sections of automotive tires. Table 9.6 lists the important component of tires and the typical blends used for them. Much of the literature of elastomer blends reflects this important application. It is outside the scope of this chapter to discuss each of the applications. We have outlined most of the important principles used in the generation of the blends.

Table 9.6 Elastomer blends in automotive tires

Component	Passenger tires	Commercial vehicle tires
Tread	SBR-BR	NR-BR or SBR-BR
Belt	NR	NR
Carcass	NR-SBR-BR	NR-BR
Sidewall	NR-BR or NR-SBR	NR-BR
Liner	NR-SBR-IIR	NR-IIR

9.3.8.2 Saturated and Unsaturated Elastomer Blends

The use of blends of polyolefin elastomers, such as IIR and EPDM, as substantial components in blends of unsaturated elastomers is a rapidly developing area. Mouri *et al.* have compared the properties of EPDM-NR and BIMS-NR blends as sidewall components. In many of the applications, the saturated elastomer is considered a polymeric antioxidant for the diene rubber. It is believed that the higher molecular weight polyolefins are better in these applications due to limited interdiffusion and a more stable morphology. Some of the benefits in tensile properties and abrasion resistance of the blends may be due to the interdiffusion of high molecular chains of dissimilar elastomers across the phase interface. Significant advances have been made in modifying the structure of polyolefin elastomers to increase the compatibility to unsaturated elastomers. Tse *et al.* have shown that uncompatibilized blends of saturated elastomers and unsaturated elastomers are possible if the former contains substantial amounts ($>12\%$) styrene residues. This is expected to be an important area of development in the future with the advent of new synthesis procedures for polyolefins.

9.4 Conclusion

The formulation and use of elastomer blends is technologically demanding. Miscible blends are widely used but usually not recognized since analytical separation of the vulcanized elastomer is experimentally impossible. Immiscible blends require excellent phase dispersion and interfacial adhesion typical of all polymer blends. In addition, they require control of filler distribution and crosslink density in each component. This is due to the need for mechanical integrity in vulcanized elastomers. The current design criteria of North American automotive tires require treads to last for 80000 miles with less than 0.4 inch of wear. Vulcanized roofing membranes require 35 years of outdoor exposure with minimal change in elongation and tensile strength. The technical complexity of analysis and use of elastomer blends has lead to secrecy for many of the formulations and uses. In spite of the difficulties of analysis and gaps in understanding, the use of blends containing elastomers continues to be an active and increasing area of research. Part of the impetus is the availability of directed synthesis of many of the older elastomers. These new synthetic tools include new catalysts ("single sited") and process designs for the manufacture of ethylene-based polyolefin elastomers, group transfer polymerization for acrylates, and anionic polymerization for diene elastomers. The availability of elastomers with a narrow compositional and molecular mass distribution in these syntheses makes the utility of blending more apparent and useful.

References

[1] R. German, R. Hank, and G. Vaughan, *Rubber Chem. Technol.* 40, 569 (1967).

[2] S. Datta and D. J. Lohse, The sampling of polymer blends in "Polymeric Compatibilizers" Hanser-Gardner, 1996.

[3] W. M. Hess, C. R. Herd, and P. C. Vergari, *Rubber Chem Technol.* 66, 329 (1993).

[4] C. M. Roland, *Rubber Chem. Technol.* 62, 456 (1989).

[5] P. J. Corish, in "Science and Technology of Rubber," F. R. Eirich (Ed.), Academic Press, New York, 1978, Chap. 12.

[6] E. T. McDonel, K. C. Baranwal, and J. C. Andries, in "Polymer Blends," D. R. Paul and S. Newman (Eds.), Academic Press, New York, Vol. 11, 1978, Chap. 19.

[7] (a) C. M. Hansen, "The Three Dimensional Solubility Parameter and Solvent Diffusion Coef-ficients," Danish Technical Press, Copenhagen, 1967. (b) C. M. Hansen and A. Beerbower, Solubility Parameters, in "Encyclopedia of Chemical Technology, Supplement Volume," 2nd ed., Wiley, New York, 1971.

[8] (a) J. P. Runt and L. M. Martynowicz, *Adv. Chem. Ser.* 211, 111 (1985). (b) H. D. Keith and F. J. Padden, *J. Appl. Phys.* 36, 1286 (1964).

[9] A. Ghijsels, *Rubber Chem. Technol.* 60, 278 (1977).

[10] (a) T. Nishi and T. T. Wang, *Macromolecules* 8, 909 (1975). (b) B. Rim and J. P. Runt, *Macromolecules* 17, 1520 (1984).

[11] (a) H. N. Cheng, in "Polymer Analysis and Characterization IV," H. G. Barth and J. Janca (Eds.), John Wiley and Sons, New York, 1992, p. 21. (b) C. M. Roland, *Rubber Chem. Technol.* 62, 456 (1989). (c) B. Albert, R. Jerome, P. Teyssic, G. Smyth, N. G. Boyle, and V. J. McBrierty, *Macromolecules* 18, 388 (1985). (d) C. M. Roland and C. A. Trask, *Polym. Mater., Sci. Eng.* 60, 832 (1989).

[12] D. L. Vander Hart, W. F. Manders, R. S. Stein, and W. Herman, *Macromolecules* 20, 1726 (1987).

[13] (a) J. Schaefer, M. D. Sefcik, E. O. Stejskal, and R. A. McKay, *Macromolecules* 14, 188 (1981). (b) T. R. Steger, J. Schaefer, E. O. Stejskal, R. A. McKay, and M. D. Sefcik, *Annals NY Acad. Sci.* 371, (1981).

[14] (a) M. C. Morris, *Rubber Chem. Technol.* 40, 341 (1967). (b) A. K. Sircar and T. G. Lamond, *Rubber Chem. Technol.* 46, 178 (1973).

[15] (a) C. M. Roland and G. G. A. Bohm, *Macromolecules* 18, 1310 (1985). (b) C. M. Roland, *Rubber Chem. Technol.* 61, 866 (1988). (c) C. M. Roland, *Macromolecules* 20, 2557 (1987). (d) Klein, *Prilos. Mag.* 43, 771 (1981). (e) J. Klein, D. Fletcher, and L. J. Fetters, *Nature (London)* 304, 526 (1983). (f) Y. Kumugai, K. Watanabe, T. Miyasaki, and J. Hata, *J. Chem. Eng. Jpn.* 12, 1 (1979). (g) T. Gilmore, R. Falabella, and R. L. Lawrence, *Macromolecules* 13, 880 (1980). (h) T. Hashimoto, Y. Tsukahara, and H. Kawai, *Macromolecules* 14, 708 (1981). (i) J. E. Anderson and J.-H. Jou, *Macromolecules* 20, 1544 (1987). (j) G. C. Summerfield and R. Ullman, *Macromolecules* 20, 401 (1987).

[16] L. W. Kleiner, F. E. Karasz, and W. J. MacKnight, *Polym. Eng. Sci.* 19, 519 (1979).

[17] (a) S. Datta and E. N. Kresge, U. S. Patent 4, 722, 971 (1988). (b) S. Datta, L. Kaufman, and P. Ravishankar, U. S. Patent 5, 571, 868 (1996).

[18] F. Morrar, L. L. Ban, W. G. Funk, E. N. Kresge, H. C. Wang, S. Datta, and R. C. Keller, U. S. Patent 5, 428, 099 (1996).

[19] (a) Z. M. Zhang and J. E. Mark, *Polym. Sci. Polym. Phys. Ed.* 20, 473 (1982). (b) J. E. Mark and A. L. Andrady, *Rubber Chem. Tech.* 54, 366 (1981).

[20] (a) D. I. Livingston and R. L. Rongone, *Proc. 5th, Int. Rubboer Conf.*, Brighton, England, Paper No. 22 (1967). (b) G. M. Bartenev and G. S. Kongarov, *Rubber Chem. Technol.* 36, 668 (1983). (c) P. A. Marsh, A. Voet, L. D. Price, T. J. Mullens, and L. Kongarov, *Rubber Chem. Technol.* 41, 344 (1968). (d) R. Couchman, *Macromolecules* 20, 1712 (1987). (e) T. K. Kwei, E. M. Pearce, J. R. Pennacchia, and M. Charton, *Macromolecules* 20, 1174 (1987). (f) M. -J. Brekner, H. A. Schneider, and H.-J. Cantow, *Polymer* 29, 78 (1988). (g) M. Aubin and R. E. Prud'homme, *Macromolecules* 21, 2945 (1988).

[21] (a) C. M. Roland, *J. Polym. Sci., Polym. Phys. Ed.* 26, 839 (1988). (b) C. A. Trask and C. M. Roland, *Polym. Comm.* 29, 332 (1988). (c) C. M. Roland, *Macromolecules* 20, 2557 (1987). (d) C. M. Roland and C. A. Trask, *Rubber Chem. Technol.* 62, 456 (1989). (e) C. A. Trask and C. M. Roland, *Macromolecules* 22, 256 (1988).

[22] (a) I. R. Gelling, *NR Technol.* 18, 21 (1987). (b) A. G. Margaritis, J. K. Kallitsis, and N. K. Kalfoglou, *Polymer* 28, 2122 (1987).

[23] (a) M. M. Coleman, G. J. Pehlert, and P. C. Painter, *Macromolecules* 29, 6820 (1996). (b) G. J. Pehlert, P. C. Painter, B. Veytsman, and M. M. Coleman, *Macromolecules* 30, 3671 (1997). (c) M. M. Coleman, G. J. Pehlert, X. Yang, J. Stallman, and P. C. Painter, *Polymer* 37, 4753 (1996). (d) G. J. Pehlert, X. Yang, P. C. Painter, and M. M. Coleman, *Polymer* 37, 4763 (1996).

[24] P. J. Corish and B. D. W. Powell, *Rubber Chem. Technol.* 47, 481 (1974).

[25] (a) N. Tokita, *Rubber Chem. Technol.* 60, 292 (1977). (b) G. N. Avgeropoulos, R. C. Weissert,

P. H. Biddison, and G. G. A. Bohm, *Rubber Chem. Technol.* 49, 93 (1976).

[26] (a) M. Takenaka, T. Izumitani, and T. Hashimoto, *Macromolecules* 20, 2257 (1987). (b) H. Yang, M. Shibayama, and R. S. Stein, *Macromolecules* 19, 1667 (1986). (c) J. Kumaki and T. Hashimoto, *Macromolecules* 19, 763 (1986).

[27] (a) J. E. Kruse, *Rubber Chem. Technol.* 46, 653 (1973). (b) J. E. Callan, W. M. Hess, and C. E. Scott, *Rubber Chem. Technol.* 44, 814 (1971). (c) C. E. Scott, J. E. Callan, and W. M. Hess, *J. Rubber Res. Inst. Malays.* 22, 242 (1969). (d) J. E. Callan, B. Topcik, and F. P. Ford, *Rubber World* 161, 60 (1965).

[28] (a) F. H. Roninger, Jr., *Ind. Eng. Chem., Anal. Ed.* 6, 251 (1933). (b) R. W. Smith and J. C. Andries, *Rubber Chem Technol.* 47, 64 (1986).

[29] (a) E. W. Stroup, A. Pungor, V. Hilady, and J. D. Andrade, *Polym. Prepr. ACS., Div. Polym. Chem.* 34, 86 (1993). (b) W. Stocker, B. Bickmarm, S. N. Magonov, and H. J. Cantow, *Ultrami-croscopy* 42, 1141 (1992). (c) Y. H. Tsao, S. X. Yang, and D. F. Evans, *Langmuir* 8, 1188 (1992). (d) E. Hamada and R. Kaneko, *Ultramicroscopy* 42, 184 (1992). (e) J. E. Sax and J. M. Ottino, *Polymer* 26, 1073 (1985). (f) T. Nishi, T. Hayashi, and H. Tanaka, *Makromol. Chem., Macro-mol. Symp.* 16, 91 (1988). (g) J. Kruse, *Rubber Chem. Technol.* 46, 653 (1973). (h) D. Vesely and D. S. Finch, *Makromol. Chem., Macromol. Symp.* 16, 329 (1988).

[30] (a) R. Buchdahl and L. E. Nielsen, *J. Polym. Sci.* 15, 1 (1955). (b) M. C. Morris, *Rubber Chem. Technol.* 40, 341 (1967). (c) H. K. de Decker and D. J. Sabatine, *Rubber Age* 99, 73 (1967). (d) L. Bohn, *Rubber Chem. Technol.* 41, 495 (1968). (e) W. Scheele, *Kautsch. Gummi Kunstst.* 24, 387 (1971). (f) A. R. Ramos and R. E. Cohen, *Polym. Eng. Sci.* 17, 639 (1977). (g) K. A. Mazich, M. A. Samus, P. C. Kilgoar, Jr., and H. K. Plummer, Jr., *Rubber Chem. Technol.* 59, 623 (1986).

[31] (a) R. J. Butera, J. B. Lando, and B. Simic-Glavaski, *Macromolecules* 20, 1724 (1987). (b) M. R. Krejsa and J. L. Koenig, *Rubber Chem. Technol.* 64, 635 (1991). (c) M. R. Krejsa and J. L. Koenig, *Rubber Chem. Technol.* 65, 956 (1992). (d) S. R. Smith and J. L. Koenig, *Macromolecules* 24, 3496 (1991). (e) S. N. Sarkar and R. A. Komoroski, *Macromolecules* 25, 1420 (1992).

[32] (a) C. M. Roland and G. G. A. Bohm, *J. Polym. Sci., Polym. Phys. Ed.* 22, 79 (1984). (b) C. A. Trask and C. M. Roland, *Macromolecules* 22, 256 (1989). (c) C. M. Roland and G. G. A. Bohm, *J. Polym. Sci., Polym. Phys. Ed.* 22, 79 (1984).

[33] (a) J. L. Leblanc, *Plast. RubberProcess. Appl.* 2, 361 (1982). (b) G. J. Van Amerongen, *Rubber Chem. Technol.* 37, 1065 (1964). (c) V. A. Shershnev, *Rubber Chem. Technol.* 66, 537 (1982). (d) M. G. Huson, W. J. McGill, and R. D. Wiggett, *Plast. Rubber Process. Appl.* 6, 319 (1985). (e) R. F. Bauer and A. H. Crossland, *Rubber Chem. Technol.* 61, 585 (1988). (f) A. K. Bhowmick and S. K. De, *Rubber Chem. Technol.* 63, 960 (1980). (g) R. L. Zapp, *Rubber Chem. Technol.* 46, 251 (1973).

[34] (a) G. J. Amerongen, *Rubber Chem. Technol.* 37, 1065 (1964). (b) J. R. Gardiner, *Rubber Chem. Technol.* 41, 1312 (1968). (c) J. R. Gardiner, *Rubber Chem. Technol.* 42, 1058 (1969). (d) J. R. Gardiner, *Rubber Chem. Technol.* 43, 370 (1970).

[35] (a) V. A. Shershnev, *Rubber Chem. Technol.* 66, 537 (1982). (b) J. Rehner, Jr. and P. E. Wei, *Rubber Chem. Technol.* 42, 985 (1969). (c) A. K. Bhowmick and S. K. De, *Rubber Chem. Technol.* 63, 960 (1980). (d) K. C. Baranwal and P. N. Son, *Rubber Chem. Technol.* 47, 88 (1974). (e) K. Hashimoto, M. Miura, S. Takagi, and H. Okamoto, *Int. Polym. Sci. Technol.* 3, 84 (1976). (f) A. Y. Coran, *Rubber Chem. Technol.* 61, 281 (1988). (g) R. W. Amidon and R. A. Gencar-elli,

U. S. Patent 3, 674, 824 (1972). (h) R. P. Mastromatteo, J. M. Mitchell, and T. J. Brett, *Rubber Chem. Technol.* 44, 1065 (1971). (i) Sumitomo Chemical Company, British Patent 1, 325, 064 (1973). (j) M. E. Woods and T. R. Mass, U. S. Patent 3, 830, 881 (1974).

[36] M. E. Woods and J. A. Davidson, *Rubber Chem. Technol.* 49, 112 (1976).

[37] (a) M. G. Huson, W. J. McGill, and P. J. Swart, *J. Polym. Sci., Polym. Lett. Ed.* 22, 143 (1984). (b) D. Honiball and W. J. McGill, *J. Polym. Sci., Polym. Phys.* 26, 1529 (1988). (c) L. D. Loan, *Rubber Chem. Technol.* 40, 149 (1967).

[38] (a) J. E. Callan, W. M. Hess, and C. E. Scott, *Rubber Chem. Technol.* 44, 814 (1971). (b) J. E. Callan, W. M. Hess, and C. E. Scott, *Rev. Gen. Caoutch. Plast.* 48, 155 (1971). (c) A. K. Sircar and T. J. Lammond, *Rubber Chem. Technol.* 46, 178 (1973). (d) G. Cotton and L. J. Murphy, *Rubber Chem. Technol.* 61, 609 (1988).

[39] A. J. Dias and A. Galuska, *Rubber Chem. Technol.* 69, 615 (1996).

[40] (a) J. E. Callan, B. Topcik, and E. P. Ford, *Rubber World* 151, 60 (1951). (b) W. M. Hess, P. A. Marsh, and F. J. Eckert, presented at a meeting of the Rubber Division, American Chemical Society, Miami Beach, FL, 1965. (c) Y. F. Wang and H. C. Wang, *Rubber Chem. Technol.* 70, 63 (1997).

[41] (a) W. M. Hess, R. A. Swor, and P. C. Vegvari, *Kautsch. Gummi Kunstst.* 38, 1114 (1985). (b) W. M. Hess and V. E. Chirico, *Rubber Chem. Technol.* 50, 301 (1977).

[42] (a) A. Medalia, *Rubber Chem. Technol.* 51, 437 (1978). (b) W. P. Fletcher and A. N. Gent, *Br. J. Appl. Phys.* 8, 1984 (1957). (c) S. Maiti, S. K. De, and A. K. Bhowmick, *Rubber Chem. Technol.* 65, 293 (1992).

[43] (a) M. E. Woods and T. R. Mass, *Adv. Chem. Ser.* 142, 386 (1975). (b) E. T. McDonel, K. C. Baranwal, and J. C. Andries, in "Polymer Blends," D. R. Paul and S. Newman (Eds.), Acade-mic Press, New York, Vol. 2, 1978, p. 263. (c) L. Leibler, *Makromol Chem. Macromol Symp.* 16, 1, (1982). (d) J. Noolandi and K. M. Hong, *Macromolecules* 16, 482 (1982).

[44] (a) D. K. Setua and J. L. White, *Kautsch. Gummi Kunstst.* 44, 821 (1991). (b) D. K. Setua and J. L. White, *Poly. Eng. & Sci.* 31, 1742 (1991). (c) P. A. Arjunan, R. B. Kusznir, and A. Dekmezian, *Rubber World* 21 (1997).

[45] J. A. Davison, W. Nudenberg, and Y. Rim, U. S. Patent 4, 316, 971 (1982).

[46] (a) G. N. Avgeropoulos, R. C. Weissert, P. H. Biddison, and G. G. A. Bohm, *Rubber Chem. Technol.* 49, 93 (1976). (b) C. M. Roland and M. Nguyen, *J. Appl. Polym. Sci.* 36, 141 (1988). (c) A. C. Pipkin and R. I. Tanner, *Ann. Rev. Fluid Mech.* 9, 13 (1977). (d) W. M. Hess, P. C. Vegvari, and R. A. Swor, *Rubber Chem. Technol.* 68, 350 (1985). (e) W. M. Hess, R. A. Swor, and P. C. Vegvari, *Kautsch. Gummi Kunstst.* 38, 1114 (1985). (f) R. F. Heitmiller, R. Z. Maar, and H. H. Zabusky, *J. Appl. Polym. Sci.* 8, 873 (1964). (g) S. Uemura and M. Takayanagi, *J. Appl. Polym. Sci.* 10, 113 (1966).

[47] (a) H. Yang, M. Shibayama, and R. S. Stein, *Macromolecules* 19, 1667 (1986). (b) E. A. Meinecke and M. I. Taftaf, *Rubber Chem. Technol.* 37, 1190 (1988). (c) A. I. Medalia, *Rubber Chem. Technol.* 61, 437 (1978). (d) E. A. Meinecke and M. I. Taftaf, *Rubber Chem. Technol.* 61, 534 (1988).

[48] R. T. Morrisey, *Rubber Chem. Technol.* 44, 1029 (1971).

[49] (a) J. M. Mitchell, *Rubber Plast. News*, June 3, 1985, p. 18. (b) N. M. Matthew, *J. Polym. Sci. Polym. Lett. Ed.* 22, 135 (1984). (c) W. von Hellens, Rubber Division, American Chemical Society, Detroit, MI, Fall 1989. (d) F. C. Cesare, *Rubber World* 201, 14 (1989). (e) W. Hong, *Rubber Plast. News* 19, 14 (1989). (f) W. von Hellens, D. C. Edwards, and Z. J. Lobos, *Rubber Plast. News* 20, 61 (1990). (g) E. H. Andrews, *J. Polym. Sci.* 10, 47 (1966). (h) B. Ambelang, F. H.

Wilson, Jr., L. E. Porter, and D. L. Turk, *Rubber Chem. Technol*. 42, 1186 (1969). (i) J. Barrier, *Rubber Chem. Technol*. 28, 814 (1956). (j) J. Kinning, E. L. Thomas, and J. M. Ottino, *Macromolecules* 20, 1129 (1987).

[50] (a) F. J. Krakowski and A. J. Tinker, *Elastomerics* 122, 34; 122, 24 (1990). (b) M. F. Tse, K. O. McElrath, H.-C. Wang, Y. F. Wang, and A. L. Tisler, Paper 24, 153rd Rubber Division Meeting, ACS, Indianapolis, IN, May 5-8, 1998.

[51] H. Mouri and Y. Tonosaki, Paper 65, 152nd Rubber Division Meeting, ACS, Cleveland, OH, October 21-24, 1997.

Chapter 10
Chemical Modification of Elastomers

10.1 Introduction

The terms elastomer embrace those polymers which have useful elastomer-like, highly elastic properties at ambient temperatures; however, many polymers which are not elastomers by themselves can be chemically modified to a relatively small extent to give products with very useful viscoelastic properties. For example, the introduction of a few chlorine atoms and sulfoxide groups into polyethylene changes the macromolecules so that they no longer tend to form crystalline regions, but become elastomers over a wide temperature range. These groups also allow vulcanization so that reversibly deforming, solvent-resistant products can be formulated. On the other hand, starting with a conventional elastomer, it is possible, by means of chemical reactions, to convert the macromolecular chains so that they lose their viscoelasticity and become thermoplastic at ambient temperatures (e.g., the cyclization of polyisoprenes). Quite apart from the question of the temperature at which observations are carried out, the borderline between the viscoelastic and the plastic states is a relatively ill-defined one. The common factor of elastomers and plastics is, of course, their macromolecular nature. Now that we have a better understanding of their structure at a molecular level (not forgetting elastomer-modified plastics and "plasticized" elastomers), we are able to say that, with some exceptions, a modification that can be carried out on one polymer species may, under suitable conditions, be carried out on a polymer of a different species.

Whether the same type of chemical modification will give us the properties we are seeking without the loss of properties we would wish to retain is a matter, first, for conjecture and, subsequently, for experimental verification. In the early days of polymer chemistry and technology, a new chemical modification successfully applied to one polymer was quickly evaluated with a whole range of chemically similar reagents (e.g., the esterification of cellulose), and the same type of reaction was attempted with the available range of similar material. Today, the number of homopolymers available runs into the hundreds, and the number of random copolymers runs into the thousands; therefore, the number of possible chemical modifications, including block and graft copolymers, runs into astronomical figures. Only a fraction of these potential systems has been evaluated (and then frequently only in a superficial way to obtain patent coverage) so that the field is wide open for research and development. Because of its breadth and depth, the field of the chemical modification of polymers can be treated only in outline in a single chapter, and only the more important reactions can be described. To serve the interest of the reader, however, this broad survey is punctuated by discussions in greater depth of areas that are of interest to elastomer chemists and that, in the opinion of the writer, have considerable potential for further development.

The emphasis is on the principles underlying the chemical modification of polymers; specific details of reactions conditions are deliberately omitted. A number of aspects of the chemical modification of elastomers have been covered in detail elsewhere in this book, and the reader is referred to the appropriate chapters on chemistry, vulcanization, characterization, block copolymers, and so on.

10.2 Chemical Modification of Polymers within Backbone and Chain Ends

Polymer properties are dependent on many factors, including chain end interactions with substrates such as carbon black or silica fillers, as well as clay and calcium carbonate. At this point, there is a large volume of work that has been done on chain end modification, particularly those made by anionic polymerization with group I or group II metals, as seen below:

$$\underset{R''}{\overset{R'}{N}}-\underset{}{\bigcirc}-\underset{O}{\overset{}{C}}-R \qquad \underset{R''}{\overset{R'}{N}}-\underset{}{\bigcirc}-\underset{O}{\overset{}{C}}-\underset{}{\bigcirc}-\underset{R''}{\overset{R'}{N}}$$

This approach was made possible via the reaction of metallic chain ends with active metal halide as shown on the following page, such as tin tetrachloride and silica terachlorides. These types of reactions led to increased molecular mass as well as improvements in polymer filler interactions that result in improved properties of the polymers.

Modification of polymers through the introduction of polar moieties such as amine siloxy groups made by anionic catalysts or so-called "living polymerizations" are made by either functional initiators or by chain functional monomers, as has been reported by several groups. The functional initiators are cyclic amines attached to alkyl groups and the functional monomers are based on styrene and disoproenyl benzene.

A U. S. Patent issued to the Goodyear Tire and elastomer Company claimed that polar functional monomers could be copolymerized with conjugated dienes and vinyl aromatics to chemically modify the polymer chain. Functional monomers, such as 3-(2-pyrrolidinoethyl) styrene and 3-(pyrrolidino-2-methylethyl) α-methylstyrene, have been copolymerized with solution and emulsion styrene butadiene copolymers in ratios from 1% to 10% and have imparted major improvements in polymer properties resulting in lower hysterisis and better wet grip of tires.

10.3 Esterification, Etherification, and Hydrolysis of Polymers

The chemical modifications discussed in this section are historically and scientifically so closely linked to one polymer, cellulose, that although the latter occurs primarily as a fiber and not an elastomer, a discussion of this group of cellulose modifications seems appropriate. Apart from the fact that some cellulose derivatives, like ethyl cellulose, when plasticized, can be quite elastomeric, the effects of modification of a basic polymer are particularly well demonstrable on a substance as stiff and highly crystalline as cellulose. Moreover, in view of the expected hydrocarbon shortage, cellulose may soon gain a new role as a polymeric starting commodity.

Cellulose, identified chemically as β-1,4-glucan, is the most widely found natural polymer, constituting the permanent structure of plant cell walls. For the general properties and chemistry of cellulose itself the reader is referred to standard textbooks and recent reviews.

Much of the early history of the chemical modification of cellulose is related to the attempts to find a solvent for it, as its macromolecular structure was not understood at that time. In 1844, Mercer discovered and commercialized the interaction of alkali with cellulose fibers, a process still in use under the name mercerization. The initial product of the reaction, alkali cellulose, is not a chemical modification but a physical form in which water and sodium ions penetrate the macromolecular structure and reduce the hydrogen bonding, with consequent swelling of the fibers. The initial product, cellulose I, is converted to cellulose II, a complex physicochemical modification, in the final washing stage. The degree and rate of swelling in this process are dependent on the source of the cellulose, and if the fibers are stretched prior to and during the reaction, optimum interaction is achieved. Many other inorganic salt solutions swell cellulose, and of these, zinc chloride has found the widest application. Aqueous solutions of thiourea, resorcinol, chloral hydrate, and benzene sulfonates also lead to limited swelling of cellulose. In all cases, the reduction in physical crosslinking can be followed by a study of the X-ray diffraction diagrams of the crystalline content.

The complete solubility of cellulose in cuprammonium solutions, discovered in 1857 by Schweizer, led to the development of the rayon industry, but, as in the case of alkali cellulose, the regenerated polymer is chemically the same as the precursor. Regeneration via cellulose zanthate solutions, invented by Cross and Bevan in 1893, is another process still in use; it forms the basis for the manufacture of Cellophane.

The first "chemical" modification of cellulose was achieved by Braconnot in 1833 with the production of cellulose nitrate from a wide range of cellulosic materials. The products were highly inflammable powders which could be dissolved in concentrated acetic acid to give clear tough varnishes (Note the conversion of a fiber to a film by chemical modification.). In 1847, highly nitrated cellulose, gun cotton, was discovered by Bottger and Schonbein, and in 1870, Schutzenberger produced acetylated cellulose using hot acetic anhydride as the reaction medium. These reactions have a common mechanism, namely, the esterification of the hydroxyl groups in the basic cellulose moiety:

$$\text{(cellulose structure)} \longrightarrow \sim\sim\text{Cellulose}\sim\sim\text{-OR} \tag{10-1}$$

Since this early work, a very large range of organic acids have been used to prepare cellulose esters, mixed esters, and ether esters. A typical example of considerable commercial importance is the acetylation of cellulose. As in all esterifications of macromolecular materials, the accessibility of the hydroxyl groups to the esterifying acid is of prime importance. Reaction (10-1) represents complete esterification, a process which is probably never fully achieved. The identification of the esterified products is, therefore, dependent not only on the content of acetyl groups but also on the location of these groups on the macromolecular backbone. Both factors are affected by the method of preparation and the esterification conditions.

Although many esterification reactions are based on inorganic acids, for insoluble hydroxyl compounds like cellulose, xanthation is more important. Sodium hydroxide is normally used to produce the swollen alkali cellulose, which (after aging) is reacted with carbon disulfide to form the sodium salt of cellulose xanthate:

$$\sim\sim\text{Cellulose}\sim\sim\text{-O}' + CS_2 \longrightarrow \sim\sim\text{Cellulose}\sim\sim\text{-O}'\text{-C}(=S)S \tag{10-2}$$

The cellulose can be regenerated by spinning (or extruding as a film) into an acid bath containing salts such as sodium and zinc sulfate. During spinning or extrusion, the macromolecules are oriented in the direction of flow to give high strength to the viscose fiber or the Cellophane film. The occurrence of macromolecular orientation during spinning is very important, and it is used in the chemical modification of many polymers.

The cellulose ethers constitute another important group of cellulose derivatives prepared from alkali cellulose by standard etherification reactions between the hydroxyl groups and an alkyl halide. The properties of the ethers depend on the extent of the reaction, i.e., the degree of etherification. In general, the ethyl celluloses are water-insoluble thermoplastic materials, whereas methyl ether, ethyl hydroxyethyl cellulose, and carboxymethyl cellulose are soluble in cold water and are used as viscoelastic thickeners and adhesives.

For the preparation of synthetic hydroxy polymers, hydroxyl groups can be introduced by copolymerization of the base monomer with a hydroxy monomer. These groups can then be used for esterification or etherification, but the relatively high cost of hydroxy monomers detracts from the wide spread use of direct copolymerization. Instead, one introduces the groups required by the complete or partial hydrolysis of the ester groups in an appropriately hydrolyzable polymer, such as poly(vinyl acetate). Complete hydrolysis yields poly(vinyl alcohol) (PVA), a water-soluble polymer with considerable utility as a stabilizer and viscosity modifier for aqueous systems:

$$\sim\!\!CH_2\!-\!CH\!-\!CH_2\!-\!CH\!\sim \atop OHOH \qquad (10\text{-}3)$$

PVA has a unique use as a strengthening fiber in conjunction with weaker materials such as merino wools in the weaving of delicate fabrics, from which it can afterward be removed by water washing. A major portion of the polymer produced is reacted with aldehydes to form the corresponding poly(vinyl formal), poly(vinyl acetal), and poly(vinyl butyral):

$$\underset{\sim\!\!\!\!\!}{\overset{OH}{|}} + RCHO \longrightarrow \underset{\sim\!\!\!\!\!}{\overset{RO}{\underset{C}{\diagdown\!\!\diagup}}} + H_2O \qquad (10\text{-}4)$$

There are different grades of each of these materials according to the overall molecular mass and the degree of substitution. These polymers are used as components of systems with unique adhesive properties, e.g., in the manufacture of safety glass laminates [poly(vinyl butyral) and mixed derivatives] and of metal-to-metal adhesive [poly(vinyl formal) cured with phenolics and other resins]. Reactions of poly(vinyl alcohol) with acids or anhydrides occur as normal esterifications, a route used to synthesize polymers and copolymers which cannot be readily formed by conventional polymerization (e.g., when the reactivity ratios of the monomers are not suitable).

Natural rubber and synthetic elastomers in general do not have hydroxyl groups in sufficient numbers for them to be used for esterification reactions. Terminal hydroxyl groups may be introduced into synthetic elastomers as terminal catalyst or initiator fragments and used for coupling or extension reactions.

10.4 The Hydrogenation of Polymers

Any polymer with unsaturated hydrocarbon groups, either in the main chain or as side groups, can be hydrogenated. Early research on the hydrogenation of elastomers focused on destructive hydrogenation with consequent loss of the macromolecular structure. This is beyond the scope of this chapter so the reader is referred to several references on the subject. The most recent work in hydrogenation has produced excellent products, such as linear polyethylene from the hydrogenation of poly(1,4-butadiene) and poly(ethylene-co-propylene) elastomer from the hydrogenation of polyisoprene.

$$+CH_2-C=C-CH_2\rightarrow \xrightarrow{H_2} +CH_2-CH_2CH_2-CH_2\rightarrow_n \quad (10-5)$$

$$+CH_2-\underset{CH_3}{C}=C-CH_2\rightarrow \xrightarrow{H_2} +CH_2-\underset{CH_3}{C}-CH_2-CH_2\rightarrow_n \quad (10-6)$$

These reactions were carried out using a heterogenous catalyst. Homogeneous soluble transition metal catalysts for hydrogenation have been used to create novel polymers. Homogeneous hydrogenation catalysts are usually generated from Ziegler-type catalysts based on nickel or cobalt organic salts reduced in the presence of organoaluminum or organolithium compounds. These catalysts are used to form saturated elastomers by hydrogenating unsaturated elastomers. The resulting polymers have vastly different viscoelastic properties than their unsaturated parent polymers. For example, the hydrogenation of a 99% poly(1,2-butadiene) has resulted in the formation of polybutene which has a lower glass transition temperature than its parent elastomer. It is interesting to note that hydrogenation does not affect the polymer molecular mass or backbone architecture.

The ease of hydrogenation and the resultant degree of saturation achieved reflect the microstructure of the polymer. Hydrogenation of unsaturated elastomers usually proceeds in a blocky way. This is due to the different reactivities of the various double bonds. In general, double bonds of 1,2-structure are four times more reactive than double bonds of 1,4-structure. *cis*-1,4-units are more reactive than *trans*-1,4units. Chamberline et al. have shown that hydrogenation of 1,2-units is statistically random, whereas the hydrogenation of 1,4-units is not.

The complete hydrogenation of poly(1,4-butadiene), *cis* or *trans* structure, forms a polyethylene with a low melting point of about 115℃. It is believed that this linear polyethylene is a low-density material. Unpublished data obtained on the partial hydrogenation of a 99% *cis* poly(1,4-butadiene) showed that the hydrogenation proceeded in a blocky fashion. The 40% to 50% hydrogenation of poly(1,4-butadiene)(*cis* content 98%) made by nickel catalyst gives a polymer with a melt point of +98℃ and a crystallization temperature of -13℃ as measured by differential scanning calorimetry (DSC). This confirms the fact that hydrogenation of *cis*-poly-1,4-butadiene proceeds in a blocky fashion to produce a block of polyethylene and a block of poly(*cis*-1,4-butadiene).

Polybutadienes made by anionic catalysts in the presence of polar modifiers contain a mixed microstructure of *cis*-1,4, *trans*-1,4- and 1,2-units. Hydrogenation of these polymers leads to interesting products. As mentioned previously, hydrogenation favors the 1,2-units over the 1,4-units by a 3(or 4)-to-1 ratio. Because of this mismatch in reactivity, hydrogenation of a polybutadiene containing 40% to 50%, 1,2-units produces a polymer containing a polyethylene portion with a T_m of 85°C to 95°C and a elastomery portion with a T_g of −62°C.

Block copolymers can also be hydrogenated to produce unique products. Hydrogenated triblock copolymers of poly(styrene-*co*-butadiene-*co*-styrene)(SBS) are commercially available from the Shell Company under the trade name Kraton G. The middle block is usually a mixed microstructure of poly(1,2-butadiene) and poly(1,4-butadiene) units. The resulting product is a hydrogenated unsaturated polymer which exhibits greater thermal and oxidative properties than the parent SBS triblock. Similar procedures have been used by several workers to hydrogenate poly(1,4-butadiene-*co*-1,2-butadiene) diblocks and poly(1,4-butadiene-*co*-1,4-isoprene-*co*-1,4-butadiene) triblocks. Hydrogenation of these diblock and triblock copolymers forms thermoplastic elastomers with crystalline and amorphous segments. All these materials exhibit crystallinity, glass transition, solubility, and dynamic mechanical loss spectra different from those of their unsaturated counterparts.

Another method of preparing saturated elastomers was developed using the diimide reduction. This method can be used to produce a high degree of saturation dependent on the type of reagent used; however, side reactions can occur in this method. Generation of the diimide from *p*-toluenesulfonyl hydrides leaves an acidic fragment which may cause cyclization in some unsaturated elastomers.

10.5 Dehalogenation, Elimination, and Halogenation Reactions in Polymers

10.5.1 Dehydrochlorination of Poly(vinyl chloride)

The dehydrochlorination of poly(vinyl chloride) has been the subject of much investigation, particularly with the view of developing greater stability in PVC polymers and copolymers. Like many polymeric reactions, dehydrochlorination is a complex process. The vinylene groups, created by the elimination of HCl from adjacent carbon atoms in the chain,

$$\sim\sim CH_2-CHCl-CH_2-CHCl\sim\sim \longrightarrow HCl + \sim\sim CH=CH-CH_2CHCl\sim\sim \qquad (10-7)$$

may be the result of free radical ionic, or ion-radical steps. The presence of a small proportion of head-to-head, tail-to-tail, and other configurational irregularities in the backbone structure of poly(vinyl chloride) leads to more complex elimination steps by thermal degradation (alone or in the presence of catalysts such as aluminum chloride). The introduction of ring structures, a major process during dehydrochlorination, is likewise affected by the distribution of the chlorine atoms along the polymeric backbone. Hydrogen bromide can be effectively eliminated thermally from poly(vinyl bromide), as dehydrohalogenation is a universal thermal reaction, the complexities of which increase from chloride to iodide.

10.5.2 Thermal Elimination

The thermal elimination process can be applied to most "substituted" groups in vinyl polymers by controlled pyrolysis at 600℃ to 700℃, producing polyvinylene compounds, e.g., by the splitting off of acetic acid from poly(vinyl acetate). By careful temperature control, one can achieve bifunctional reactions and/or intramolecular cyclizations. This has been developed commercially at relatively high temperatures, in the case of the polymerization of methacrylamide above 65℃, to yield a polymer with a substantial proportion of imide groups:

$$\text{(structure)} + NH_3 \tag{10-8}$$

Polymers and copolymers of multifunctional vinyl monomers (using the term to cover the presence of a halogen or other reactive group in addition to the vinyl group, rather than in the sense of more than one polymerizable group), such as α-chloroacrylic acid, often undergo partial lactonization and hydrolysis during polymerization. Heating in alcohol solution or electrolyzing alcoholic solutions, one obtains, e.g., the introduction of double lactam rings during the acid hydrolysis of poly(α-acetamineoacrylic acid).

$$\text{(structure)} \longrightarrow \text{(structure)} + AcOH \longrightarrow \text{(structure)} + AcOH \tag{10-9}$$

The discoloration of polyacrylonitrile is due to a similar type of elimination reaction, which in this case occurs intra- as well as intermolecularly to give crosslinked insoluble ring products:

$$\text{(structure)} \longrightarrow \text{(structure)} \tag{10-10}$$

The controlled heating of polyacrylonitrile fibers under tension also causes an elimination of nitrogenous products to leave a "carbon fiber" of high tensile strength that can be considered as the end product of the line of chemical elimination reactions. Carbon fibers from cellulosic materials, lignin, and various interpolymers and blends have been developed. The structures of these products consist largely of three-dimensional carbon networks, partially crystalline and partially graphitic or amorphous.

10.5.3 Halogenation of Polymers

Halogenation and hydrohalogenation of elastomers have been reported extensively in literatures. The main problems with these reactions are the cyclization and chain scission that occur parallel to the halogenation reaction. These introduce difficult problems in the characterization of the resulting products. Despite these problems, several products have been prepared and commercialized. Chlorination of poly(1,4-butadiene) to prepare a product similar to poly(vinyl chloride) has been reported by several workers. This process had extensive side reactions and chain degradation. The chlorination of butyl elastomer and conjugated diene-butyl elastomers gives end products that are used in the tire industry as inner liners for air retention.

Ethylene-propylene copolymers (EPDM) are, by their random copolymerization, amorphous in structure and therefore easily halogenated. EPDM has been chlorinated to improve its properties and cocurability with other elastomers. The chlorination was directed toward the termonomer dicyclopentadiene to form the allylic chloride. In this manner, EPDM was chlorinated, and the resulting products had improved properties.

EPDM elastomers are modified by 1,2-addition of N-chlorothiosulfonamides to their olefinic sties. Such additions may be carried out in solution or without solvent in an internal mixer or extruder. The solventless reactions are facilitated by added carboxylic acids or by certain metal salts of weak acids. The modified EPDMs are of interest because of their ability to covulcanize in ozone-resistant blends with polydiene elastomers. Although less fully explored, N-chlorothiocarboxamides and imides also react with EPDM to produce modified products which covulcanize in blends with polydienes.

In the absence of oxygen, the chlorination of polyethylene, with or without a catalyst, can be controlled to provide products with varying chlorine content. The chlorination process is statistically random so that chlorination of polyethylene to the same chlorine content as poly(vinyl chloride) (60%) gives a product which is chemically different from PVC yet fully compatible with it. This random chlorination of polyethylene destroys its crystallinity. At a degree of chlorination corresponding to the loss of all its crystallinity, the chlorinated product becomes soluble at room temperature. The p-bromination of polyethylene follows a similar course to yield a elastomerlike polymer at 55% bromine content.

Both chlorination and bromination of polypropylene and isobutylene lead to degradation of the main chain, with the loss of many useful properties. Degradation during chlorination can, however, be avoided at low temperatures by limiting the reaction to a maximum of about 2%. This procedure forms a useful commercial product. The addition of hydrogen chloride to unsaturated elastomers has also received considerable attention. Extensive work has been done on the hydrochlorination of Hevea [poly(*cis*-1,4-isoprene)] and Balata [poly(*trans*-1,4-isoprene)] elastomers since 1940. Both *cis*-1,4-and *trans*-1,4-polyisoprenes readily add hydrogen chloride following Markovnikov's rules with only a small amount of cyclization.

10.5.4 Cyclization of Polymers

Cationic cyclization of unsaturated elastomers such as poly(*cis*-1,4-isoprene), poly(3,4-isoprene), poly(1,2-butadiene), and poly(1,4-butadiene) usually leads to the formation of cyclized resinous products of no commercial value. Cyclization of unsaturated elastomers, such as polyisoprene, can be carried out in the solid state, in solution, or even in the latex. The process involves the transformation of linear macromolecular chains into much shorter ones consisting of mono-, di-, tri-, and tetrapolycyclic groups distributed randomly along the chain. Cyclization of polyisoprene increases its glass transition temperature by 20℃ to 30℃. The mechanism of cyclization of elastomers depends on the catalyst employed in the process.

10.6 Other Addition Reactions to Double Bonds

10.6.1 Ethylene Derivatives

Besides the addition of halogens and hydrohalogens "across the double bond" just covered, there are many other reagents which will react similarly with unsaturated polymers by free radical, ionic, or radical-ion mechanisms. Of prime importance is the addition of ethylene derivatives to polydienes. One of the earliest reactions of natural rubber to be studied in detail was the combination with maleic anhydride. Depending on the reaction conditions and the presence or absence of free radical initiators, one or more of four basic reactions may take place, with the products shown (the arrows indicate where the addition has taken place and the new bonds formed).

(1) Intramolecular addition to the double bond within polyisoprene chains:

$$\sim\sim\underset{H}{\overset{CH_3}{\overset{|}{C}}}-CH\underset{CH_2-CH_2}{\overset{O\diagdown\overset{O}{\diagup}\diagdown O}{\diagdown\diagup}}\overset{CH_3}{\underset{|}{C}}-CH_2\sim\sim \qquad (10\text{-}11)$$

(2) Intermolecular addition to double bonds in different polymer chains. In this group should be included the statistically possible reaction between widely separated double bonds within the same molecule:

$$\text{(structure 10-12)} \tag{10-12}$$

(3) Addition to α-methylenic carbon atoms of a polyisoprene chain:

$$\text{(structure 10-13)} \tag{10-13}$$

(4) Intermolecular addition to α-methylenic carbon atoms in adjacent chains (or widely spaced α-methylenic carbon atoms in the same macromolecule):

$$\text{(structure 10-14)} \tag{10-14}$$

In general, the overall reaction rates increase with rising temperature and in the presence of oxygen or free radical initiators, but these same conditions promote intermolecular reactions leading to gel formation. Similar reactions take place with gutta-percha, synthetic poly(cis-1,4-isoprene), and poly(cis-1,4-butadiene). Many workers used two-roll mills and other mastication techniques as convenient ways of blending the maleic anhydride with elastomers at elevated temperatures, but where these techniques have been used, mechanochemical reactions have complicated the overall process. Reaction products of natural rubber containing 5% to 10% combined maleic anhydride can be vulcanized by conventional sulfur cures; of greater interest is the possibility of creating crosslinking by the use of oxides of calcium, magnesium, and zinc.

Other compounds reacting similarly via activated double bonds (excluding here block or graft copolymerization) include maleic acid, N-methyl-maleimide, chloromaleic anhydride, fumaric acid, γ-crotonolactone, p-benzoquinone, and acrylonitrile. Other polymers with unsaturated backbones, such as polybutadiene, copolymers of butadiene with styrene and with acrylonitrile, and butyl elastomer, react in similar ways, but the recorded reaction with poly(vinyl chloride) is largely mechanochemical in nature (discussed later).

10.6.2 The Prins Reaction

Another addition to polymers with main-chain unsaturation is the Prins reaction between ethylenic hydrocarbons and compounds containing aldehydic carbonyl groups. Kirchof, in 1923, described the reaction of natural rubber in benzene solution with aqueous formaldehyde in the presence of concentrated sulfuric acid. The general reaction of an aldehyde, RCHO, with a polyisoprene in the presence of an inorganic or organic acid or an anhydrous metal salt is represented by

$$\sim\sim\underset{\underset{CH_3}{|}}{C}=\underset{\underset{H}{|}}{C}\sim\sim \; + \; \underset{R}{\overset{H}{\diagdown}}C\overset{O}{\diagup} \; \xrightarrow{\frac{H_c}{X_c}} \; \sim\sim\underset{\underset{CH_3}{|}}{\overset{\overset{X}{|}}{C}}-\underset{\underset{H}{|}}{\overset{\overset{R}{\diagdown}CHOH}{C}}\sim\sim \qquad (10\text{-}15)$$

In the absence of such catalysts, the reaction leads to a shift in the double bond rather than its elimination:

$$\sim\sim\underset{\underset{CH_3}{|}}{C}=\underset{\underset{H}{|}}{C}\sim\sim \; + \; R-C\underset{H}{\overset{\diagup O}{\diagdown}} \; \longrightarrow \; \underset{\underset{CH_2}{\|}}{\overset{\overset{\sim\sim R}{|}}{C}}-\underset{\underset{H}{|}}{\overset{\overset{\diagdown CHOH}{}}{C}}\sim\sim \qquad (10\text{-}16)$$

These reactions can be carried out in solution or in dispersion or by reaction in the solid phase; in the last case it is again difficult to differentiate the Prins reaction from mechanochemical reactions initiated by chain rupture during mastication. Other aldehydic compounds, such as glyoxal and chloral, also react in a similar way with polyisoprenes and unsaturated elastomers [e. g., poly(*cis*-1,4-isoprenes), poly(*cis*-1,4-butadiene) and copolymers of isobutylene and isoprene]. The use of strong acids, or Lewis acids, causes complications, as the acids themselves, under suitable conditions, catalyze cyclization and *cis-trans* isomerization, and these reactions may occur simultaneously with the addition reactions.

10.7 Oxidation Reactions of Polymers

Uncontrolled oxidation of elastomer is detrimental to its physical properties. Oxidation reactions take place readily at unsaturated groups in polymers and are often referred to collectively as epoxidation; however, oxidation under controlled conditions can lead to useful products such as the epoxidized natural rubber introduced by the Malaysian elastomer Producers Association. Natural rubber in the latex form is treated with hydrogen peroxide dissolved in acetic acid. This gives 50% epoxidized natural rubber. This elastomer shows very interesting physical properties and excellent carbon black dispersion. Similarly, nonaqueous epoxidizations of synthetic polyisoprene can be achieved using either hydrogen peroxide or hypochloride in *t*-butanol. The controlled degree of epoxidation usually leads to some interesting products. For example, the 25% epoxidized synthetic polyisoprene is an elastomer with viscoelastic properties similar to those of the unepoxidized material, but has better carbon black dispersion. This gives high modulus and tensile strength, however, higher degrees of epoxidation (60% to 75%) produce a resinous material which is not elastomery.

10.8 Functionalization of Polymers

Polymers with stable backbones such as polystyrene, polyethylene, and polypropylene can be functionalized. Functionalization of polystyrene has received considerable attention, because it is a unique polymer with aromatic rings which are capable of undergoing many nucleophilic as well as electrophilic reactions. A resin recently introduced on the market is based on sulfonated polystyrene. Applications for this resin include ion-exchange material and catalyst binding materials.

Electrophilic substitution on polystyrene through a chlorometallation reaction yields chlorine functionality. This has opened up the possibilities of making many derivatives of polystyrene. Starting with chlorometallated polystyrene, derivatives such as quaternary, ammonium, or phosphonium salts have been made. Similarly, ethers, esters, sulfonamides, silanes, and ketone derivatives have been made by replacing the chlorine atom on chlorometallated polystyrene. In the case of polystyrene, however, it was discovered that chain end functionalization can be realized if the chain ends were terminated by group I metals such as lithium and potassium.

Both the Japanese Synthetic elastomer Company and Nippon Zeon have reported that anionically prepared elastomers that are functionally terminated by active lithium can be chain terminated with Michler ketone, benzophenone, and a variety of enamide groups. Moreover, these chains can be terminated with silicone or tin metals. Chain end functionalization did not change the viscoelasticity of the polymer chains but rather dramatically improved the elastomer-filler interaction and, therefore, reduced its hysteretic properties.

10.9 Miscellaneous Chemical Reactions of Polymers

Direct replacement of the hydrogen atoms of aromatic rings such as styrene or the allylic hydrogen in poly(1,4-butadiene) or poly(1,4-isoprene) can be carried out via metallation with organometallic compounds of group I such as lithium, sodium, and potassium. Usually, the yield tends to be low and the product is insoluble; however, the use of chelating diamines with organolithium compounds has increased the yield, and the products are soluble in cyclohexane. For example, polystyrene has been metallated in high yields to give polylithiated polystyrene in which several functional groups have been successfully introduced. Similarly, polyisoprene and polybutadiene have been successfully metallated with either s-butyllithium or t-butyllithium in the presence of tetramethylethylenediamine (TMEDA) at 50℃. In the case of the polyisoprene, chain scission and reduction in molecular mass resulted at longer metallation temperatures and times. In many cases, these lithiated polymers have been used to prepare graft and block copolymers. These are discussed in more detail in next section.

10.10 Block and Graft Copolymerization

10.10.1 Effects on Structure and Properties of Polymers

Some of the most significant changes in structure and properties of polymers can be brought about by either block or graft copolymerization. The term block copolymer is applied to macromolecules made up of sequences with different chemical (or physical, i.e., tactic) structures, usually represented by A, B, and C. The sequences are of a molecular mass that would give them polymeric features even if separated. The manner in which these sequences are arranged defines the type of block copolymer prepared. A diblock copolymer is represented by AB, indicating that a segment with chemical composition A is connected to a segment with composition B.

Other possible types of block copolymers include triblocks ABA and ABC, for example. They may be linear or branched; the linear structures are called block copolymers,

$$AAAAAA \cdot BBBB \cdots BBBBB \cdot AAAAA \qquad (10\text{-}17)$$

and the branched structures graft copolymers (10-18),

$$
\begin{array}{cc}
\sim\!\!\sim\!\!AAA \cdots AAA\!\!\sim\!\!\sim & \sim\!\!\sim\!\!BBBBB \cdots BBB\!\!\sim\!\!\sim \\
\begin{array}{cc} B & B \\ B & B \\ B \ (a) & B \\ B & B \\ B & B \end{array} \text{ or }
\begin{array}{cc} A & A \\ A & A \\ A \ (b) & A \\ A & A \\ \vdots & \vdots \\ A & A \end{array}
\end{array} \qquad (10\text{-}18)
$$

Thus the same polymeric sequences may be put together as block or as graft copolymers, with differing properties, though in the author's experience, the major differences between the properties of block and graft copolymers of the same constituent polymers are pronounced only in solution or in the melt. For example, natural rubber may be block copolymerized with poly(methyl methacrylate), or methyl methacrylate monomer may be grafted onto natural rubber. In an attempt to distinguish by nomenclature one structure from the other, insertion of the letters b and g, for block and graft, respectively, between the names of the specific sequences was introduced, e.g., natural elastomer—b-poly(methyl methacrylate) and natural elastomer—g-poly(methyl methacrylate) in the case of the examples cited. The structures of these two macromolecules would be represented by (10-17) and (10-18), where the ~AAAA··· sequences represent natural rubber and the ~BBBBB···sequences represent poly(metyl methacrylate).

The number and order of sequences may be more complicated. Block copolymers are usually made by free radical or living polymerizations. These processes can produce polymers that consist of a pure A block connected to a pure B block, with no interphase zone of mixed A and B structure. The preparation of block copolymers is not limited to monomers A and B, but can also encompass segments of random copolymers. For example, a block of a random copolymer AB can be connected to a block of polymer A or B. Moreover, the point of attachment of the blocks can be either at the end or the middle of the polymer chain. Several examples of the various types of block copolymers possible follow:

$$\sim AAAAAA\ BBBB\ AAAAA \sim \quad \text{or} \quad \sim AAAAAAAAAAA\sim$$
$$\begin{array}{cc} B & B \\ B & B \\ B & B \\ B & B \end{array} \quad (10\text{-}19)$$

$$\sim ABABAB\ AAAAAA\sim \quad \text{or} \quad \sim ABABBABB\ AAAAAAAA\sim$$
$$\begin{array}{cc} A & B \\ A & B \\ A & B \\ A & B \end{array} \quad (10\text{-}20)$$
$$\sim ABABA\ BBBBBBB\sim$$

When the sequences making up the segments are random copolymers, the prefix co may be introduced, with the major component monomer preceding the minor constituent. A backbone polymer of butadiene-styrene elastomer grafted with styrene containing a small percentage of acrylic acid would be described as poly[(butadiene-co-styrene)-(styrene-co-acrylic acid)] and could be schematically represented as

$$\sim\sim AABABAAABBABAAAABBA\sim\sim$$
$$\begin{array}{cc} B & B \\ B & B \\ B & C \\ C & B \\ B & B \\ B & B \\ B & C \\ C & B \\ B & B \\ & B \end{array} \quad (10\text{-}21)$$

where A represents butadiene, B represents styrene, and C represents acrylic acid.

10.10.2 Block Copolymer Synthesis

Several methods can be used to synthesize block copolymers. Using living polymerization, monomer A is homopolymerized to form a block of A; then monomer B is added and reacts with the active chain end of segment A to form a block of B. With careful control of the reaction conditions, this technique can produce a variety of well-defined block copolymers. This ionic technique is discussed in more detail in a later section. Mechanicochemical degradation provides a very useful and simple way to produce polymeric free radicals. When a elastomer is mechanically sheared, as during mastication, a reduction in molecular mass occurs as a result of the physical pulling apart of macromolecules. This chain rupture forms radicals of A and B which then recombine to form a block copolymer. This is not a preferred method because it usually leads to a mixture of poorly defined block copolymers.

Using living polymerizations, the Shell Company was able to commercialize several poly(styrene-*co*-butadiene) and poly(styrene-*co*-isoprene) block copolymers known in the industry as Kraton 1101 and Kraton G. These block copolymers have found many uses in the shoe sole and adhesive industries. The physical properties were dependent on the macrostructure and microstructure of these block copolymers.

10.10.3 Examples

As major examples, let us consider the three monomers butadiene, styrene, and acrylonitrile, and see how they can be block copolymerized together by mechanochemical means. From the large number of theoretical possibilities, 11 have been selected for discussion; these may be prepared by mastication of the following:

① A butadiene-styrene copolymer elastomer with acrylonitrile monomer;
② Polyacrylonitrile (plasticized) with a mixture of butadiene and styrene monomers;
③ A butadiene-acrylonitrile copolymer elastomer with styrene monomer;
④ Polystyrene with a mixture of butadiene and acrylonitrile monomers;
⑤ A styrene-acrylonitrile resin with a mixture of styrene and butadiene monomers;
⑥ Polybutadiene with a mixture of styrene and acrylonitrile monomers;
⑦ A butadiene-styrene elastomer with polyacrylonitrile (best plasticized);
⑧ A styrene-acrylonitrile resin with polybutadiene;
⑨ A butadiene-acrylonitrile elastomer with polystyrene;
⑩ A high styrene-butadiene resin with acrylonitrile monomer;
⑪ A high styrene-butadiene resin with polyacrylonitrile (plasticized).

All of the foregoing reactions except ② and ⑪ have been reported in the patent literature or are known to have been commercially evaluated. In each example, the products would be chemically and physically different in terms of the makeup of the structural sequences, and all properties would also depend on the relative proportions of the initial components. In all mechanochemical reactions, some of the starting polymer or copolymer remains unchanged, mainly the low-molecular-mass fraction, which is not effectively sheared, and some homopolymer may be formed from the polymerizing monomers by chain transfer reactions. Varying the mastication conditions greatly influences the yield and rate of reaction; the chemical nature of the products is less affected, except that the presence of butadiene as one of the constituents (either polymer or monomer) will cause increasing gel contents with continued mastication. Processes ③, ⑤, ⑥, ⑧, and ⑨ are known to give products in which a elastomer phase is dispersed in a resinous matrix; i.e., they are alternative methods for producing an A-B-S-type copolymer. It has been found in practice that a number of monomers that normally do not polymerize by free radical processes in the temperature range 10°C to 50°C can be block copolymerized by cold mastication techniques, indicating ionic initiation via heterolytic scission. The presence of a proportion of block or graft copolymer in the system assists in stabilizing the dispersion of the elastomer phase in the resin matrix by acting at the phase boundary as a "soap," i. e., a compatibilizing agent at the phase boundary.

10.10.4 Other Methods of Effecting Mechanicochemical Reactions

Mechanicochemical degradation is the term used in describing chain scission of polymer backbones through the application of shear during a processing operation. It was previously believed that this type of process led to carbon-carbon chain scission, which usually causes a dramatic change in rheological properties. In the early 1950s, Watson and coworkers showed that, in the absence of oxygen, radicals produced by mechanical shear can be used to initiate the polymerization of vinyl monomers to form block copolymers. For example, vibromilling of natural rubber below its glass transition temperature has enabled block copolymerization of natural elastomer with methyl methacrylate to be carried out on a small scale, with conversions as high as 86%. Similar results were achieved with styrene and with acrylonitrile.

This type of approach has also been used to attach antioxidants to unsaturated polymers. The novel approach of Scott in the 1970s using the technique employed by Watson enabled the attack of substituted allyl mercaptans and disulfides to olefinic double bonds employing the Kharasch reactions. Mechanicochemical reactions can occur during processing, when the polymer is converted to a finished product. The chain scission can occur in both saturated and unsaturated polymer backbones. For example, during processing in a screw extruder, backbone scission in polypropylene produces long tail free radicals which can form macroalkyl radical peroxides. These peroxides are responsible for the observed decrease in melt viscosity. In the absence of oxygen, these monoallyl macroradicals can be used to graft new monomers or such polymers as polypropylene and polyethylene to the backbone. In this manner, maleic adducts of polypropylene have been prepared, giving improved dyeability, hydrophilicity, and adhesion.

10.10.5 Ionic Mechanisms

Ionic mechanisms for the preparation of block copolymers are a very important tool of the synthetic polymer chemist. A feature of many homogeneous anionic polymerizations in solution is that termination can be avoided by careful control of experimental conditions. In fact, an infinite life of the active chain end is theoretically possible, and this has led to the term living polymers. Polymer carbanions can resume growth after the further addition of monomer. By changing the monomer composition, block copolymerization is readily initiated, and this process can be repeated. A major advantage of this type of synthesis over most free radical processes is the ability to control the chain length of the sequences by adjusting the concentrations of initiating sites and of monomer at each stage of the block copolymerization.

Anionic block copolymerization employs organolithium initiators, which have wider use because of their extended range of solubility, which includes hydrocarbons. Organolithium compounds act as initiators by direct attack of the organic anion on the monomer species, again a fast reaction and, in the absence of compound with active hydrogen atoms, without transfer or termination steps. If carefully executed, the reaction permits one to have precise control over molecular mass and (the narrow) molecular mass distribution.

The convenience of this technique has led to the development of many commercial products, including thermoplastic elastomers based on triblocks of styrene, butadiene, and isoprene. The initiator used in these systems is based on hydrocarbon-soluble organolithium initiators. In some cases, a hydrocarbon-soluble dilithio initiator has been employed in the preparation of multi-block copolymers. Several techniques are used to prepare thermoplastic elastomers of the ABA type. A short summary of these techniques is given here.

(1) Three-Stage Process with Monofunctional Initiators

In this technique, used, e. g., for the synthesis of block copolymers of poly(styrene-b-butadiene-b-styrene) (SBS), a polystyrene block is formed by employing n-butyllithium as the initiator in an aromatic solvent. Butadiene monomer is then added to react with the polystyrene-lithium chain end to form the poly(butadiene) block. If the reaction was terminated at this stage, a poly(styrene-b-butadiene) copolymer would result, which has no thermoplastic properties. Therefore, styrene monomer is added to produce the triblock SBS. The process for the preparation of SBS is very carefully controlled to avoid the formation of a diblock, as the presence of any appreciable amount of SB dramatically reduces the thermoplastic properties of SBS.

(2) Two-Stage Process with Difunctional Initiators

Several commercial processes using difunctional initiators based on soluble organolithium compounds have been developed. These compounds can polymerize at both ends. Difunctional initiators are useful in the cases of ABA block copolymers where B can initiate A but A cannot initiate B. These difunctional initiators are useful in the preparation of SBS. The elastomeric butadiene block is polymerized with hexane as a solvent. The added styrene monomer is also soluble in hexane. This method is also useful in preparing triblocks with hydrocarbon middle blocks and polar end blocks such as poly(methacrylonitrile-bisoprene-b-methacrylonitrile).

(3) Monofunctional Initiation and Coupling

In this two-stage process, B is sequentially polymerized onto A, and then the two chains are coupled to yield an ABBA block copolymer. Triblocks of SBS have been prepared using this method, with methylene dichloride as the coupling agent. The disadvantage is the formation of radical anions which can lead to contamination of the triblock with multiblock species.

(4) Tapered Block Copolymers

This method is used to form a block copolymer which consists of two segments of essentially homopolymeric structure separated by a block of a "tapered" segment of random copolymer composition. These are usually prepared by taking advantage of the differences in reaction rates of the component monomers. When polymerized individually in hexane, butadiene reacts six times more slowly than styrene; however, when styrene and butadiene are copolymerized in a hydrocarbon solvent such as hexane, the reaction rates reverse, and the butadiene becomes six times faster than the styrene. This leads to a tapering of the styrene in a copolymerization reaction.

10.10.6 Graft Copolymer Synthesis

The synthesis of graft copolymers is much more diverse but can nevertheless be divided into groups of related processes: ①polymer transfer, ②copolymerization via unsaturated groups, ③redox polymerization, ④high-energy radiation techniques, ⑤photochemical synthesis, and, most importantly, ⑥metallation using activated organolithium with chelating diamines.

(1) Polymer Transfer

In a free radical polymerization, chain transfer is an important reaction. Chain transfer to a monomer, solvent, mercaptan, or other growing chain can take place. When a chain transfer reaction to another chain takes place, it creates a radical which acts as a site for further chain growth and grafting:

$$P \cdot + P \longrightarrow PH + P \cdot$$
$$R \cdot + P \longrightarrow RH + P \cdot \qquad (10\text{-}22)$$

The reaction proceeds by the transfer of a hydrogen or halogen (in the case of halogenated polymers) atom from a macromolecule P to the growing chains P· (or to an excess initiator free radical R·, thereby "terminating" them). The reactivity is now located on the transfer molecule, which in turn initiates copolymerization, i.e., the growth of a grafted side chain of a newly introduced second monomer. A measure of grafting occurs with most monomer-polymer systems, especially those initiated by benzoyl peroxide, if the concentrations of polymer and initiator are high.

The simplest technique is to dissolve the polymer in the appropriate solvent; add the peroxide initiator, which abstracts a hydrogen radical and generates a radical on the polymer chain; and then add fresh monomer for grafting onto this site. This technique has been employed in grafting methylacrylate onto natural rubber and synthetic polyisoprene. In this manner, several commercially useful products such as ABS resins have been prepared; however, tire elastomers are not made in this manner because of the generation of micro and macro gel particles, which are detrimental to physical properties. In many cases when latex grafting has been used, the product has usually been targeted toward thermoplastic applications rather than elastomer applications.

(2) Copolymerization via Unsaturated Groups

Other methods (e. g., most car body repairs) are based on the polyester-styrene copolymerization process (reinforced with various types of inert mesh or glass fabric), a graft copolymerization of styrene onto backbone unsaturated polymer of relatively low molecular mass. In general, for high grafting yields, a reasonably high concentration of pendant vinyl groups is required on the backbone polymer. For glass-reinforced plastics, the polyester resins are selected with this in view. In natural rubber, a few such groups per molecule are always present and these undoubtedly participate during normal grafting. About 0.4% of the unsaturated groups are pendant vinyl groups in an average sample of acetone-extracted pale crepe elastomer. The content of pendant vinyl groups can be increased by mastication of unsaturated elastomers under nitrogen, because the resonance structures recombine as

$$
\begin{array}{c}
\sim\!\!\!\sim\!\!\overset{CH_3}{\underset{\|}{C}}\!\!=\!\!CH\!-\!\dot{C}H_2 \\
\sim\!\!\!\sim\!\!\underset{\dot{C}H_3}{\underset{|}{C}}\!-\!CH\!=\!CH_2
\end{array}
\quad + \quad
\begin{array}{c}
\cdot CH_2\!-\!\overset{CH_3}{\underset{\|}{C}}\!=\!CH\!\sim\!\!\!\sim \\
CH_2\!=\!\underset{CH_3}{\underset{|}{C}}\!-\!\dot{C}H\!\sim\!\!\!\sim
\end{array}
\tag{10-23}
$$

$$
\begin{array}{c}
\sim\!\!\!\sim\!\!\overset{CH_3}{\underset{|}{C}}\!-\!CH_2\!-\!\overset{CH_3}{\underset{|}{C}}\!=\!CH\!\sim\!\!\!\sim \\
\underset{\|}{CH} \\
CH_2
\end{array}
\text{ and }
\begin{array}{c}
\sim\!\!\!\sim\!\!\overset{CH_3}{\underset{|}{C}}\!=\!CH\!-\!CH_2\!-\!CH\!\sim\!\!\!\sim \\
\underset{\|}{C}\!-\!CH_3 \text{ etc.} \\
CH_2
\end{array}
$$

The direct introduction of peroxide groups into the backbone of polymers, such as poly(methyl methacrylate), has been used to produce macromolecular initiators for the synthesis of block copolymers, e. g., poly(methyl methacrylate-b-acrylonitrile) and poly(methyl methacrylate-b-styrene). Ozonization can also be used, with careful control of the degree of ozonolysis, to introduce epoxy ring structures into natural rubber:

$$
\sim\!\!\!\sim\!CH_2\!-\!\overset{CH_3}{\underset{|}{C}}\!=\!CH\!-\!CH_2\!\sim\!\!\!\sim \xrightarrow{O_3}
\begin{array}{c}
H_3C \diagdown \quad \diagup O \diagdown \\
\sim\!\!\!\sim\!CH_2 \diagup C \diagdown \quad CH\!-\!CH_2\!\sim\!\!\!\sim \\
O\!-\!O
\end{array}
\tag{10-24}
$$

By carrying out the reaction to about 4% of the available double bonds in a solvent such as toluene at a low temperature followed by a nitrogen purge, grafting can be effected by addition under nitrogen of methyl methacrylate (MMA) monomer (reacting at 80℃ in sealed ampules) and formation of two MMA chains attached to the oxygens of the opened —O—O— bridge. This technique should be applicable to isoprene and butadiene copolymers.

(3) Redox Polymerization

Redox polymerizations are among the most popular techniques for grafting reactions, and of the possible initiator systems, ferrous ion oxidation and those based on ceric ion reduction are widely used. In a redox polymerization, a hydroperoxide or similar group is reduced to a free radical plus an anion, while the metal ion is oxidized to a higher valency state, and at the same time a monomer is added. When the reducible group is attached to a polymeric chain, the free radical grafting sites thus formed on the macro-molecular backbone act as initiators for graft copolymerization.

$$\sim\!\!CH_2\!-\!\underset{\underset{O-OH}{|}}{CH}\!\sim\!\! + Fe^{2+} \xrightarrow[\text{aldehyde}]{\text{amine of}} \sim\!\!CH_2\!-\!\underset{\underset{O\cdot}{|}}{CH}\!\sim\!\! + Fe^{3+} + OH^- \quad (10\text{-}25)$$

This method has been used to graft methyl methacrylate to natural rubber latex. (Actually, fresh latex contains a few hydroperoxide groups per macromolecule, which can take part in grafting reactions.) Recentrifuged latex concentrate is mixed with methyl methacrylate and a solution of tetraethylene pentamine is added, followed by a small quantity of ferrous sulfate solution. The homogenized blend is allowed to stand, often overnight. The graft copolymer is isolated by coagulation. As practically all free radical sites are formed on the elastomer backbone, there is very little free poly(methyl methacrylate) in the grafted system; on the other hand, some elastomer chains are without grafts, as not all chains have hydroperoxy groups. Higher yields of graft copolymer are obtained by allowing the monomer to dissolve in, and equilibrate with, the latex particles before adding the amine and ferrous ion initiator. It has been claimed that passing oxygen (air) through the latex for several hours reduces the free elastomer content of the polymerization product, but nitrogen purging is then necessary to prevent dissolved oxygen from acting as a polymerization inhibitor.

Hydroxy polymers can be grafted by redox polymerization by using a water-soluble peroxide, such as hydrogen peroxide in conjunction with ferrous ions. The OH radicals thus produced abstract H atoms from the hydroxy groups in the polymer, giving free radical grafting sites on the backbone. This method has been used with starch and cellulose derivatives, but considerable quantities of homopolymer are formed from the initial hydroxyl radicals in parallel with the H abstraction. By introducing a few hydroxyl groups into a copolymeric synthetic elastomer, grafting can be effected, provided the presence of homopolymer can be tolerated. Mixtures of ferrous ammonium sulfate and ascorbic acid are suitable redox initiation systems. Many patents claim the preferred use of ceric ions, which easily oxidize hydroxyl groups by a radical-ion reaction:

$$R\!-\!OH + Ce^{3+} \longrightarrow R\!-\!\overset{+}{\underset{\cdot}{OH}} + Ce^{2+}$$
$$R\!-\!\overset{+}{\underset{\cdot}{OH}} + OH^- \longrightarrow R\!-\!O\cdot + H_2O \quad (10\text{-}26)$$

The advantage of this reaction lies in the fact that only hydroxyls on the polymer are converted into R—O free radicals, so that no homopolymer can be produced and pure graft is obtained.

(4) High-Energy Radiation Techniques

During high-energy irradiation in vacuo, e.g., from a ^{60}Co source, some main-chain degradation of natural rubber and other polyisoprenes occurs:

$$\sim\sim CH_2-\underset{\underset{CH_3}{|}}{C}=CH-CH_2-CH_2-\underset{\underset{CH_3}{|}}{C}=CH-CH_2\sim\sim \rightarrow \quad (10\text{-}27)$$

Much of the irradiation energy is also adsorbed by the removal of hydrogen atoms:

$$\sim\sim CH_2-\underset{\underset{CH_3}{|}}{C}=CH-CH_2\cdot + \cdot CH_2-\underset{\underset{CH_3}{|}}{C}=CH-CH_2\sim\sim \quad (10\text{-}28)$$

$$\sim\sim CH_2-\underset{\underset{CH_3}{|}}{C}=CH-CH_2\sim\sim \xrightarrow{\text{irradiation}} \sim\sim CH_2-\underset{\underset{CH_3}{|}}{C}=CH-\overset{\cdot}{CH}\sim\sim$$

The irradiation of natural rubber in the presence of a vinyl monomer thus leads primarily to a synthesis of graft copolymers, but some block copolymer is certainly always present. Irradiation syntheses may be carried out in solution, either in contact with liquid monomer (with or without a diluent) or in contact with monomer in the vapor phase, or in emulsion or suspension. The elastomer may be preirradiated in the absence of air to produce free radicals for later monomer addition, but the life of these radicals is short as a result of mobility within the elastomer matrix. Irradiation at very low temperatures makes it possible to use the trapped radicals technique for a variety of natural and synthetic elastomers. Plastics and polymers with a crystalline phase are more readily preirradiated to initiate later grafting by trapped radicals. Irradiation may also be carried out in air to introduce peroxide groupings:

$$CH_2-\underset{\underset{CH_3}{|}}{C}=CH-\overset{\cdot}{CH}_2 \xrightarrow{O_2} CH_2-\underset{\underset{CH_3}{|}}{C}=CH-CH_2-O-O\cdot \quad (10\text{-}29)$$

(with "or" branch yielding $CH_2-\underset{\underset{CH_3}{|}}{C}=CH-CH_2-OOH + R\cdot$ and $CH_2-\underset{\underset{CH_3}{|}}{C}=CH-CH_2-O-O-CH_2-\underset{\underset{CH_3}{|}}{C}=CH$)

$$-CH_2-\underset{\underset{CH_3}{|}}{C}=CH-\overset{\cdot}{CH} \rightarrow CH_2-\underset{\underset{CH_3}{|}}{C}=CH-\underset{\underset{O-O\cdot}{|}}{CH}$$

with "or" giving $CH_2-\underset{\underset{CH_3}{|}}{C}=CH-\underset{\underset{OOH}{|}}{CH}+RH$ and $CH_2-\underset{\underset{CH_3}{|}}{C}=CH-\underset{\underset{O}{|}}{CH}\ \ \underset{\underset{CH_2-\underset{\underset{CH_3}{|}}{C}=CH-CH}{|}}{O}$ \quad (10-30)

These groups can then be used to initiate grafting by any of the methods already discussed. Latex phase grafting is generally favored for its simplicity; natural rubber grafts with methyl methacrylate styrene, acrylonitrile, and vinyl chloride have been made in this way.

The irradiation of mixed lattices for subsequent combination of the ruptured chains is another approach; it has been carried out with natural rubber and poly(vinyl chloride) lattices to prepare graft (and block) copolymers in fairly high yields without the problem of monomer recovery. The same method has been used to graft polychloroprene onto synthetic polyisoprene dispersions and onto polybutadiene lattices of various compositions.

(5) Photochemical Synthesis

Macromolecules containing photosensitive groups which absorb energy from ultraviolet frequencies often degrade by free radical processes. The degradative process as a rule is fairly slow, but by the addition of photosensitizers, such as xanthone, benzyl, benzoin, and 1-chloroanthraquinone, the rate can be speeded up to enable graft copolymerization to take place in the presence of methyl methacrylate or other monomers. This can be done in the case of natural rubber in the latex phase with reasonably high yields of graft copolymer. Natural rubber-y-polystyrene and poly(butadiene-y-styrene) have both been prepared by ultraviolet irradiation of sensitized latex-monomer dispersions. A combination of photochemical synthesis and redox-type initiation can also be carried out—a process known as one-electron oxidation—to achieve grafting with minimal homopolymer formation.

Bromine atoms on the backbone of a polymer can be liberated readily by ultraviolet irradiation to give free radical sites for grafting reactions. The bromination can be photochemically induced

$$\sim\sim CH_2-CH-CH_2-CH\sim\sim \xrightarrow[h\nu]{Br_2} \sim\sim CH_2-\underset{\underset{\text{(cyclohexyl)}}{|}}{\overset{Br}{C}}-CH_2-CH\sim\sim + HBR \qquad (10\text{-}31)$$

or a chain transfer agent such as carbon tetrabromide may be used in the polymerization step to introduce the labile groups,

$$\sim\sim CH_2-\overset{\bullet}{CH} + CBr_4 \xrightarrow[h\nu]{Br_2} \sim\sim CH_2-CHBr + \bullet CBr_3 \text{ etc.} \qquad (10\text{-}32)$$

With the aid of suitable sensitizers, polymers such as brominated butyl elastomer, valuable because of their flame retardancy, may act as backbone polymers for a variety of grafting reactions.

An early synthesis of block copolymers was based on the ultraviolet irradiation of poly(methyl vinyl ketone) in the presence of acrylonitrile. The initial degradative step is

$$-\underset{H}{\overset{H}{\underset{|}{C}}}-\underset{\underset{CH_3}{|}}{\overset{H}{\underset{|}{C}}}=O}-\underset{H}{\overset{H}{\underset{|}{C}}}- \longrightarrow \overset{\bullet}{CH_2} + \bullet \underset{\underset{CH_3}{|}}{\overset{H}{\underset{|}{C}}}=O}-CH_2- \qquad (10\text{-}33)$$

This degradation reaction, supplemented by various subsequent oxidation steps, has found renewed interest in the form of the introduction of photodegradable plastics as part of the campaign to reduce plastic litter from throwaway packaging. Although as yet there has been no demand for photodegradable elastomers, the incorporation of a small percentage of a vinyl ketone into a elastomer copolymer or homopolymer would open the way to a useful synthesis of block copolymers.

Many other syntheses of block and graft copolymers have been reported, but enough has been said to indicate the scope of these reactions and to indicate a potential that has still to be thoroughly explored. Many grafting and block copolymerization systems have only been evaluated for plastic materials but are capable of extension to elastomers.

(6) Metallation Using Activated Organolithium with Chelating Diamines

Unsaturated elastomers can be readily metallated with activated organo-lithium compounds in the presence of chelating diamines or alkoxides of potassium or sodium. For example, polyisoprene, polybutadiene, styrene-butadiene copolymers, and styrene-isoprene copolymers can be metallated with n-butyllithium. TMEDA complexes (1 : 1 or 1 : 2 ratio) to form allylic or benzylic anions. The resulting allylic anion can be employed as an initiator site to grow certain branched or comb polymer species. These polymers can include polystyrene, which would form hard domains, or polybutadiene, which forms soft domains.

Research in this area has resulted in the preparation of several comb polymers. The metallation technique is a useful and versatile method as it can be used with any polymeric material which contains a few double bonds. For example, ethylene-propylene was successfully grafted with norbornene. Similar reactions were performed on polymeric materials which contain aromatic rings, such as polystyrene, poly-α-methyl styrene, and polyphenylene oxide (PPO).

In general, polymeric materials that contain either side groups or mainchain allylic or acidic hydrogens can be metallated with organolithium compounds in the presence of chelating diamines. They can also be grafted with ionically polymerizable monomers to produce comb-like materials.

10.10.7 Base Polymer Properties

The properties of natural rubber grafted with poly(methyl methacrylate) cannot be evaluated unless the copolymer is isolated from either homopolymer species. The methods used are based on fractional precipitation, selective solution, or a combination of these basic techniques. In many cases, though, technologists are oncerned with the materials as manufactured, so we consider in this context also the properties of the block and graft copolymers without homopolymer removed.

The presence of two chemically different polymeric sequences in the same chain causes that macromolecule to act as a soap; i.e., it helps to compatibilize two species of homopolymer in a blend by accumulating at their interface, assisting a more gradual transition from one phase to the other, and thus reducing the interfacial energy. Microphase separation, of course, still occurs, the predominant case in practice, but macrophase separation is thereby usually prevented. In high-impact polystyrene and in A-B-S copolymers prepared by grafting reactions, the dispersed elastomer phase in the glassy matrix and the dispersed glassy phase within the elastomery particles are both prevented from forming separate phases by the graft copolymer chains, which on a molecular scale have their elastomery segments associated with the elastomer particles and their plastic segments with the glass phase. In this respect, there is little difference in properties between a graft and a linear block copolymer—the essential feature is the presence of the two types of sequences in the same macromolecule.

The block copolymeric thermoelastic polymers owe their properties to this very structure, whereby the polystyrene end blocks (along with any homopolymeric polystyrene) form the microphase, which is dispersed within a continuous phase of polybutadiene formed from the polybutadiene segments of the central sequences in S-B-S-type block copolymers. For this to happen, the total volume of the polybutadiene segments must exceed the total volume of the polystyrene segments. When the reverse is the case, the product exhibits the properties of a high-impact polystyrene. Although the polystyrene "structures" act as physical crosslinks at low temperature, at processing temperatures above the softening temperature of polystyrene both segment types exhibit viscoelasticity, allowing the material to be extruded, injection molded, etc. On cooling, the polystyrene domains become rigid again and assert their infliuence on the material properties.

When block and graft copolymers are dispersed in solvents, the solutions have properties which depend on whether or not the copolymer is eventually fully solvated. If the solvent is a "good" solvent for both sequences, e. g. , chloroform in the case of natural rubber graft copolymerized with poly(methyl methacrylate), then both segment types are expanded and films cast from dilute solutions will usually be intermediate in properties to the two homopolymers (in this example the properties of a reinforced elastomer film). If the solvent is a good solvent for the elastomer but a poor solvent or nonsolvent for poly (methyl methacrylate), e. g., petroleum ether, then the solutions show the typical turbidity of a block or graft copolymer and the cast film is highly elastic. When the solvent is acetone, a good solvent for poly(methyl methacrylate) but a nonsolvent for elastomer, the cast films are plastic with high tear strengths.

When grafting is carried out on a polymer under conditions such that the physical form of the substrate polymer is maintained, then the original properties of the substrate usually predominate, and supplementary properties accrue as a result of the grafting. This is invariably the case when the substrate is in fibrous form, e. g. , cellulose, nylons, and terylene grafted with various monomer systems. The nature of the grafting reaction to these fibers is usually such as to form a surface coating over the substrate polymer; the surface characteristics, such as dyeing, are therefore usually those of the grafting system.

It is very doubtful that any blends of two polymers, or of chemically different copolymers, can from a thermodynamic point of view ever be fully compatible. Even most block or graft copolymers systems therefore show microphase separation which will be typical for the properties of a given system. Chemical modification, as discussed previously, will in general lead to the formation of polymeric single phases, provided the reaction has been carried out homogeneously. The choice need not be restricted, however, to just these two approaches, as chemical modification can be carried out after block or graft copolymerization or vice versa. Very little has been published on such a consecutive use of these two physically different ways of modifying polymers chemically, so there is considerable scope for developing new modifications of long-established elastomers, as well as generally for changing old into new polymers.

References

[1] D. M. Bielinsk, L. A. Sulsarki, H. M. Stanley, and R. A. J. Pethrick, Appl. Plym. Sci. 56, 853 (1995).
[2] M. Yamato and M. Oahu, Progress in Organic Coating 27, 277 (1996).
[3] D. N. Schulz, A. F. Halasa, and A. E. Oberster, J. Polym. Sci. Chem. Ed. 12, 153 (1974).
[4] C. A. Urneck and J. N. Short, J. Appl. Polymer Sci. 14, 1421 (1970).
[5] A. F. Halasa, J. Polym. Sci. Polym. Ed. 19, 1357 (1981).
[6] A. Yoshioka, A. Ueda, H. Watanabe, and N. Nagata, Nippon Kagaka Kaisha 341.
[7] A. Uda Brain, Proceedings of the International Institute of Synthetic elastomer Producers, 33rd Annual Meeting, 1992, pp. 65-83.
[8] G. Bohm and D. Graves, U. S. Patent 4, 788, 229 (1988).
[9] A. F. Halasa and Wen-ling Hsu (to The Goodyear Tire and elastomer Co.), U. S. Patents 6, 693, 160 and 6, 753, 447 (2004).
[10] H. L. Hsieh and R. P. Quirk, "Anionic Polymerization: Principles and Practical Applications," Marcel Dekker, New York, 1996.
[11] J. E. Hall, T. A. Antikowiak, European Patent 0693 505 (1955).
[12] V. C. Haskell, in "Encyclopedia of Polymer Science and Technology," H. S. Mark et al. (Eds.), Wiley-Interscience, New York, Vol. 3, 1965, p. 60.
[13] G. N. Bruxelles and N. R. Grassie, in "Encyclopedia of Polymer and Technology," H. S. Mark et al. (Eds.), Wiley-Interscience, New York, Vol. 3, 1965, p. 307.
[14] B. P. Rouse, in "Encyclopedia of Polymer and Technology," H. S. Mark et al. (Eds.), Wiley-Interscience, New York, Vol. 3, 1965, p. 325.
[15] W. Jarowenko, in "Encyclopedia of Polymer Science and Technology," H. S. Mark et al. (Eds.), Wiley-Interscience, New York, Vol. 3, 1965, p. 787.
[16] J. Wichlatz, in "Chemical Reaction of Polymers," M. Fetters (Ed.), Interscience, New York, 1964, Chap. 2.
[17] A. Yakubchik, B. Tikhominov, and V. Sumilov, elastomer Chem. Technol. 35, 1063 (1962).
[18] H. Rachapudy, G. Smith, V. Raju, and W. Graessley, J. Polym. Sci. Phys. Ed. 17, 1211 (1979).
[19] A. F. Halasa, elastomer Chem. Technol. 54, 627 (1981).
[20] A. F. Halasa and J. M. Massie, "Kirk-Othemer Encyclopedia of Chemical Technology" 4th ed., Vol. 8, John Wiley & Sons, Inc., New York, 1993.
[21] Y. Chamberline, J. Pascoult, H. Razzouk, and H. Cheradem, Makromol. Chem. Rapid Commun 2, 322 (1981).
[22] A. F. Halasa, L. E. Vescelius, S. Futamura, and J. Hall, "IUPAC Symposium on Macromolecules,"

Vol. 28, Amherst, MA, 1982.
[23] A. F. Halasa (to The Goodyear Tire & elastomer Co.), U. S. Patent 3, 872, 072 (1985) and U. S. Patent 4, 237, 245.
[24] H. Harwood, D. Russell, J. Verthe, and J. Zymons, Macromol. Chem. 163, 1 (1973).
[25] M. Poutsma, in "Methods of Free Radical Chemistry," S. Huyser (Ed.), Dekker, New York, Vol. 1, 1969, Chap. 3.
[26] J. Bevington and L. Ratt, Polymer 16, 66 (1975).
[27] L. Schoen, W. Raajien, and W. Van'twout, Br. Polym. J. 7, 165 (1975).
[28] R. J. Hopper (to The Goodyear Tire & elastomer Co.), U. S. Patent 3, 915, 907 (1975).
[29] R. J. Hooper, elastomer. Chem. Technol. 49, 341 (1976).
[30] R. J. Hooper, R. D. Mcquateand, T. G. Hutchin, Preprints, International Conference on Advances in the Stablization and Controlled Degradation of Polymers, Lucerne, Switzerland, May 23-25, 1984.
[31] R. J. Hooper (to The Goodyear Tire & elastomer Co.), U. S. Patent 4, 820, 780 (1989).
[32] R. J. Hooper (to The Goodyear Tire & elastomer Co.), U. S. Patent 4, 910, 266 (1990).
[33] D. A. White, R. S. Auda, W. M. Davis, and D. T. Ferrughlli (to Exxon Chemical Co.), U. S. Patent 4, 956, 420 (1990).
[34] R. J. Hooper (to The Goodyear Tire & elastomer Co.), U. S. Patent 4, 017, 468 (1977).
[35] A. Staudinger, elastomer Chem. Technol. 17, 15 (1944).
[36] M. Gordon and J. Tyler, J. Polym. Chem. 3, 537 (1953).
[37] D. N. Schults, S. Turner, and M. Golub, elastomer Chem. Technol. 85, 809, (1983).
[38] J. I. Cunneen and M. Porter, in "Encyclopedia of Polymer Science and Technology," H. S. Mark et al. (Eds.), Wiley-Interscience, New York, Vol. 2, 1965, p. 502.
[39] R. J. Ceresa, in "Encyclopedia of Polymer Science and Technology," H. S. Mark et al. (Eds.), Wiley-Interscience, New York, Vol. 2, 1965, p. 502.
[40] C. Avery and W. Watson, J. Appl. Polym. Sci. 19, 1 (1956).
[41] R. J. Ceresa and W. Watson, J. Appl. Polym. Sci., 101, 1 (1959).
[42] G. Scott, Polym. Eng. Sci. 24, 1007 (1984).
[43] W. H. Janes, in "Block Copolymers," D. C. Allport and W. H. Janes (Eds.), Applied Science Publishers, London, 1973, p. 62.
[44] J. E. Morris and B. C. Sekher, Proceedings of the International elastomer Conterence, Washington, D. C., 1959, p. 277.
[45] E. G. Cockbain, T. D. Pendle, and D. T. Turner, J. Polym. Sci. 39, 419 (1959).
[46] A. F. Halasa, Adv. Chem. Ser. 130, 77 (1974).
[47] J. Folk, R. Sclott, D. Hoey, and J. Pendeltor, elastomer Chem. Technol. 46, 1044 (1973).
[48] A. F. Halasa, G. Mitchell, M. Stayer, D. P. Tate, and R. Koch, J. Polym. Sci. Chem. Ed. 14, 297 (1976).
[49] F. M. Merrett, Trans. Faraday Soc. 50, 759 (1964).

Chapter 11
Strength of Elastomers

11.1 Introduction

Fracture is a highly selective process: only a small number of those molecules making up a test piece or a component actually undergo rupture; the great majority are not affected. For example, of the 10^{26} chain molecules per cubic meter in a typical elastomer, only those crossing the fracture plane, about $10^{18}/m^2$, will definitely be broken. Moreover, these will not all break simultaneously but successively as the fracture propagates across the specimen at a finite speed.

Thus, the first questions posed in studying the strength of elastomers (and other materials as well) are: where and under what conditions does fracture begin? Also, what laws govern the growth of a crack once it has been initiated? This chapter seeks to answer such questions, first in a general way and then with particular reference to important modes of failure of elastomers in service. It does not deal with the rather complex problem of the strength of composite structures, such as a pneumatic tire, which involves failure of adhesive bonds at interfaces between the components as well as fracture of the components themselves.

We consider first the initiation of fracture from flaws or points of weakness, where the applied stresses are magnified greatly. The rate of development of cracks after initiation is treated next. Naturally, this depends on the local stress levels but also on the way in which these stresses vary with time. For example, rapid crack growth may take place if stresses are applied and removed frequently, whereas the crack may grow quite slowly, if at all, when the same stresses are held constant and never removed. This phenomenon of accelerated growth under dynamic stressing is termed mechanical fatigue or dynamic crack growth. Because rubber is viscoelastic, or more generally anelastic, to varying extents and because the mechanical properties depend on rate of deformation and temperature, it is not surprising to find that the strength is also dependent on the rate at which stresses are applied and on the temperature of measurement. Other effects of the environment, notably the destructive action of ozone. Finally, a brief survey is given of abrasive wear.

11.2 Initiation of Fracture

11.2.1 Flaws and Stress Raisers

Every solid body contains flaws or points of weakness resulting from heterogeneities of composition or structure. In addition, because of the presence of sharp corners, nicks, cuts, scratches, and embedded dirt particles or other sharp inclusions, applied stresses are magnified (concentrated) in certain regions of the body so that they greatly exceed the mean applied stress. Fracture will begin at such a site where the local stress exceeds a critical level and the small flaw starts to grow as a crack.

The stress concentration factor, i.e., the ratio of the stress at the tip of a sharp flaw σ_t to the applied tensile stress σ, is given by Inglis's relation for elastic solids that obey a direct proportionality between stress and strain:

$$\sigma_t/\sigma = 1 + 2\sqrt{l/r} \qquad (11\text{-}1)$$

where l is the depth of an edge flaw and r is the radius of the tip in the unstressed state. If the flaw is totally enclosed, it is roughly equivalent to an edge flaw of depth $l/2$ (Fig. 11.1). Thus, edge flaws are more serious stress raisers than enclosed flaws of the same size, and they are more usual sources of fracture than inclusions, but not exclusively so. Heterogeneities of composition have been shown to nucleate fatigue cracks internally. Also, some types of cracks cannot form near a free surface.

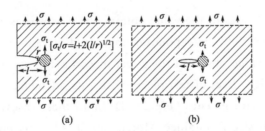

Fig. 11.1 Stresses near a crack of depth l and tip radius r.

When the tip radius is much smaller than the depth of the flaw, as seems probable for the severe stress raisers responsible for fracture, Eq. (11-1) can be approximated by

$$\sigma = \frac{1}{2}\sigma_t \sqrt{r/l} \qquad (11\text{-}2)$$

Thus, the breaking stress, denoted by σ_b, is predicted to vary inversely with the depth of the flaw l, in proportion to $1/l^{1/2}$. This prediction has been tested for brittle polymers, i.e., in the glassy state, and for rubbery materials cracked by ozone (Fig. 11-2). In both cases the breaking stress σ_b was found to vary in accordance with Eq. (11-2) with the depth of a crack or razor cut made in one edge of the test piece. For elastomers broken by mechanical stress alone, however, elongations at break are generally much too large for the assumption of linear stress-strain behavior to be valid, and Eq. (11-2) becomes a relatively poor approximation. Even so, by extrapolating measured values of the breaking stress for different depths of edge cut to the breaking stress for a test piece having no cuts introduced at all, the depth of flaw characteristic of the material may be inferred. (Actually, the value obtained is the depth of a cut equivalent in stress-raising power to natural flaws, which may be smaller and sharper, or larger and blunter, than knife cuts of equivalent stress-concentration power.)

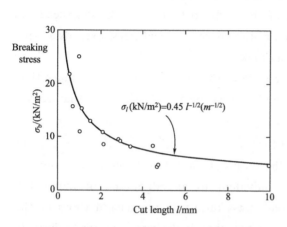
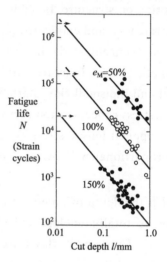

Fig. 11.2　Fracture stresses for test pieces having cuts of depth l, exposed to ozone.

Fig. 11.3　Fatigue lives N for test pieces having initial cuts of depth l, subjected to repeated extensions to the indicated strains.

For both rubbers and glasses the value obtained in this way is about (40 ± 20) mm. The same value is also obtained by extrapolating measured stresses for ozone cracking (see Fig. 11.2) back to that value observed for a test piece having no initial cut in it, and also by extrapolating the fatigue lives of test pieces with cuts introduced in them back to the fatigue life for test pieces with no cuts (Fig. 11.3). In all these cases, substantially the same value of the natural flaw size is obtained. Moreover, it is largely independent of the particular elastomer or mix formulation used, even though these factors greatly alter the way in which the breaking stress or fatigue life changes with cut size, as discussed later. Thus, a variety of fracture processes appear to begin from a natural flaw equivalent to a sharp edge cut 40mm deep.

The exact nature of these failure initiation sites is still not known. They may consist of accidental nicks in molded or cut surfaces, but even if great care is taken in preparing test pieces (e.g., by molding against polished glass), the breaking stress is not greatly increased. Dust or dirt particles or other heterogeneities nearly as effective as mold flaws seem to be present in a sufficient amount to initiate fracture. Only when the test piece size is reduced to about $10^{-8} m^3$ or less is a significant increase in strength observed, suggesting that powerful stress raisers are present only in concentrations of $10^8/m^3$ or less. Of course, if a way could be found to eliminate them, or at least reduce the effective sharpness of these natural flaws, substantial increases in strength, and even more striking increases in fatigue life, might be achieved, as discussed later. At present, however, they appear to be an inevitable consequence of the processes used in making elastomeric materials and components.

11.2.2 Stress and Energy Criteria for Rupture

Equations (11-1) and (11-2) raise several other questions. What is the radius r of a natural flaw? What is the magnitude of the breaking stress at the tip σ_t when the flaw starts to grow as a crack? From a comparison of experimental relations such as that shown in Fig. 11.2 with the predictions of Eq. (11-2), only the product $\sigma_t r^{1/2}$ can be determined and not the two quantities separately. The value of r is, however, unlikely to exceed 1mm for a sharp cut, and hence the tip stress st may be inferred to be greater than 200MPa, taking a value for the product str 1/2 of 0.2MN · m$^{-3/2}$ as representative of fracture under mechanical stress. For ozone cracking this product takes the value 900N · m$^{-3/2}$ (see Fig. 11.2), and hence the tip stress in this case is presumably 1MPa or greater.

We must recognize, however, that a tear that begins to propagate from an initial cut or flaw will soon develop a characteristic tip radius r of its own, independent of the sharpness of the initiating stress raiser. It is therefore more appropriate to treat the product $\sigma_t r^{1/2}$ as a characteristic fracture property of the material. Indeed, Irwin proposed that fracture occurs for different shapes of test piece and under varied loading conditions at a characteristic value of a "stress intensity factor" K_c, defined as

$$K_c = \frac{\sqrt{\pi}}{2}\sigma_t r^{1/2} = \sqrt{\pi}\sigma_b l^{1/2} \qquad (11\text{-}3)$$

when expressed in terms of the applied stress σ_b by means of Eq. (11-2).

An alternative but equivalent view of the critical stress criterion for fracture was proposed by Griffith for elastic solids, and applied by Irwin and Orowan to solids that are globally elastic, even when they exhibit plastic yielding around the crack tip. Griffith suggested that a flaw would propagate in a stressed material only when, by doing so, it brought about a reduction in elastically stored energy W more than sufficient to meet the free energy requirements of the newly formed fracture surfaces. Irwin and Orowan recognized that in practice the energy expended in local plastic deformation during crack growth generally far exceeds the true surface energy; however, provided that the total energy expended is proportional to the amount of surface created by fracture, Griffith's relations may still be employed.

Griffith's fracture criterion takes the form

$$-\frac{\partial W}{\partial A} \geqslant G_c/2 \qquad (11\text{-}4)$$

where A is the surface area of the specimen, which increases as the crack grows, and G_c is the amount of energy required to tear through a unit area of the material. The factor 2 arises on changing from the area torn through to the area of the two newly formed surfaces. The derivative is evaluated at constant length of the sample, so that the applied forces do no work as the crack advances. (An example where this is not appropriate, because the crack will advance only when the applied forces do the necessary work, is given in Section 11.2.4.)

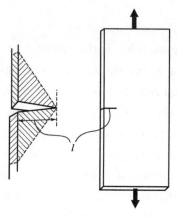

Fig. 11.4 Tensile test piece.

In Griffith's original treatment, the surface free energy per unit area of fracture plane was employed in place of the generalized fracture energy $G_c/2$. His results therefore carried the implication of thermodynamic reversibility. In contrast, G_c merely represents energy dissipated during fracture. Nevertheless, provided that it is dissipated in the immediate vicinity of the crack tip and is independent of the overall shape of the test piece and the way in which forces are applied to its edges, the magnitude of G_c can be employed as a characteristic fracture property of the material, independent of the test method.

This expectation has been borne out by critical experiments on a variety of materials, including elastomers, using test pieces for which the relation between the breaking stress σ_b and the rate of release of strain energy on fracture G_c, defined by Eq. (11-4), can be either calculated or measured experimentally. Two important cases are considered here.

11.2.3 Tensile Test Piece

As shown in Fig. 11.4, a thin strip of thickness t with a cut in one edge of depth l is placed in tension until it breaks. The effect of the cut in diminishing the total stored elastic energy at a given extension may be calculated approximately by considering a small triangular region around the cut (shown shaded in Fig. 11.4) to be unstrained and the remainder of the test piece to be unaffected by the presence of the cut, with stored strain energy U per unit volume. The reduction in strain energy due to the presence of the cut is thus kl^2tU, where k is a numerical constant whose value depends on the applied strain [k is given approximately by $\pi/(1+e)^{1/2}$, where e is the tensile strain]. Thus, as the tensile strain increases, k decreases from a value of π at small strains, to about l at large strains.]

For a tensile test piece, therefore, $-(\partial W/\partial A) = kltU$ and $(\partial A/\partial l) = 2t$. Eq. (11-4) becomes

$$2lkU \geqslant G_c \qquad (11\text{-}5)$$

We note that the breaking stress σ_b does not appear explicitly in this fracture criterion. σ_b is the stress at which the strain energy density U satisfies Eq. (11-5).

It therefore depends on the elastic properties of the material and the length of the initial cut, as well as on the fracture energy G_c. For a material obeying a linear relation between tensile stress σ and extension e, the stored energy U is given by $Ee^2/2$ or $\sigma^2/2E$, where E is Young's modulus. The stress and extension at break are therefore given by

$$\sigma_b = \sqrt{\frac{G_c E}{\pi l}} \tag{11-6}$$

$$e_b = \sqrt{\frac{G_c}{\pi l E}} \tag{11-7}$$

where k has been given the value π appropriate to linearly elastic materials. Equations (11-6) and (11-7) were obtained by Griffith.

On comparing Eq. (11-3) and Eq. (11-6), we see that the critical stress intensity factor K_c and the fracture energy, or critical strain energy release rate G_c, are related to each other and to the breaking stress at the crack tip, as follows:

$$K_c^2 = E G_c = \frac{\pi}{4} \sigma_t^2 r = \frac{\pi}{2} U_t r \tag{11-8}$$

where U_t is the strain energy density at the crack tip. Hence,

$$\sigma_c = \frac{\pi}{2} U_t r \tag{11-9}$$

The fracture energy G_c is thus a product of the energy required to break a unit volume of material at the crack tip, i.e., in the absence of nicks or external flaws, and the effective diameter of the tip, as pointed out by Thomas. These two factors can be regarded as independent components of the fracture energy: an "intrinsic" strength U_t and a characteristic roughness or bluntness of a developing crack, represented by r. Because K_c also involves the elastic modulus E, it is not considered as suitable a measure of the fracture strength as G_c for materials, like elastomers, of widely different moduli.

Eq. (11-5) is more generally applicable than Eq. (11-6) because it is not restricted to linearly elastic materials. It constitutes a criterion for tensile rupture of a highly elastic material having a cut in one edge of length l, in terms of the fracture energy G_c. Two important examples of test pieces of this type are ① the ASTM "tear" test piece for vulcanized rubber (ASTM D624-54) and ② a typical tensile test piece that has accidental small nicks caused, for example, by imperfections in the surface of the mold or die used to prepare it.

Several features of Eq. (11-6) and Eq. (11-7) are noteworthy. For a given value of fracture energy G_c, stiffer materials with higher values of Young's modulus E will have higher breaking stresses and lower extensions at break than softer materials. These correlations are well known in the rubber industry. Less well known is the effect of the size of an initial cut or flaw on both the breaking stress and elongation at break. Finally, if the fracture criterion, Eq. (11-5), is met for an initial flaw of depth l, it will be greatly exceeded as fracture proceeds. As a consequence, a crack will accelerate across the specimen catastrophically.

11.2.4 Tear Test Piece

This test piece, shown in Fig. 11.5, has regions I in the arms that are in simple extension and a region II that is virtually undeformed. If the arms are sufficiently wide, or if

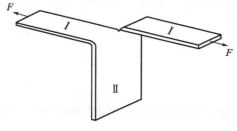

Fig. 11.5　Tear test piece.

they are reinforced with inextensible tapes, their extension under the tear force F will be negligibly small. The work of fracture $G_c \Delta A$ is then provided directly by the applied force F acting through a distance $2\Delta l$, where Δl is the distance torn through. The corresponding area torn through is $t\Delta l$, where t is the thickness of the sheet. Actually, the tear tends to run at 45° to the thickness direction, i.e., at right angles to the principal tensile stress, and thus the tear path has a width of about $\sqrt{2}\,t$ instead of t. On equating the work supplied to that required for tearing, the fracture criterion becomes

$$F \geqslant G_c t/2 \tag{11-10}$$

Because the tear force in this case is a direct measure of the fracture energy G_c and is independent of the elastic properties of the material and of the length of the tear, this test piece is particularly suitable for studying the effects of composition and test conditions on G_c.

It is important to recognize that the fracture energy G_c is not a constant value for a particular material; it depends strongly on the temperature and rate of tear, i.e., the rate at which material is deformed to rupture at the tear tip. Nevertheless, several critical values may be distinguished. The smallest possible value is, of course, twice the surface free energy, about $50 mJ/m^2$ for hydrocarbon liquids and polymers. Values of this order of magnitude are indeed observed for fracture induced by ozone, when the function of the applied forces is merely to separate molecules already broken by chemical reaction.

Another critical value is that necessary to break all of the molecules crossing a plane, in the absence of any other energy-absorbing processes. This minimum energy requirement for mechanical rupture is found to be about $50 J/m^2$; it is treated in the following section. Finally, there are the considerably larger values found in normal fracture experiments, ranging from 100 to $100000 J/m^2$.

11.3　Threshold Strengths and Extensibilities

A threshold value for the fracture energy of elastomers was first pointed out by Lake and Lindley from studies of fatigue crack growth. By extrapolation, they found that a minimum amount of mechanical energy, about $50 J/m^2$ of torn surface, was necessary for a crack to propagate at all. Mueller and Knauss measured extremely low tearing energies directly, by employing low rates of tear, high temperatures, and a urethane elastomer composition swollen highly with a mobile fluid. Under these near-equilibrium conditions, they obtained a lower limit of about $50 J/m^2$ for the tear energy, similar to Lake and Lindley's extrapolated value. More recently, threshold tear strengths have been measured for several elastomers, crosslinked to varying degrees. Again, the values are about $20 \sim 100 J/m^2$, much smaller than tear energies obtained in conventional tearing experiments, which range from about 10^3 to about 10^5 J/m^2, depending on tear rate, test temperature, and elastomer composition. Indeed, they amount to only about 1 lb of force to tear through a sheet several inches thick. Nevertheless, they are much larger than would be expected on the basis of C—C bond strengths alone. For example, about 2×10^{18} molecules cross a randomly chosen fracture plane having an area of $1 m^2$, and the dissociation energy of the C—C bond is about 5×10^{-19} J. Thus, a fracture energy of only about $1 J/m^2$ would be expected on this basis, instead of the observed value of about $50 J/m^2$.

This large discrepancy has been attributed by Lake and Thomas to the polymeric character of elastomers: many bonds in a molecular chain must be stressed equally to break one of them. Thus, the greater the molecular length between points of crosslinking, the greater the energy needed to break a molecular chain. On the other hand, when the chains are long, a smaller number of them cross a randomly chosen fracture plane. These two factors do not cancel out; the net effect is a predicted dependence of the threshold fracture energy G_0 on the average molecular mass M_c of chains between points of crosslinking, of the form

$$G_0 = \alpha \sqrt{M_c} \tag{11-11}$$

Two other features of molecular networks can be taken into account, at least in an approximate way: the presence of physical entanglements between chains, at a characteristic spacing along each chain of molecular mass M_e, and the presence of molecular ends that do not form part of the load-bearing network. Equation (11-11) then becomes

$$G_0 = \alpha [(1/M_c) + (1/M_e)]^{-1/2} [1 - 2(M_c/M)] \tag{11-12}$$

where M is the molecular mass of the polymer before crosslinking. The constant α involves the density of the polymer, the mass, length, and effective flexibility of a monomer unit, and the dissociation energy of a C—C bond, assumed to be the weakest link in the molecular chain. If reasonable values are taken for these quantities, α is found to be about $0.3 J/m^2$ $(mole/g)^{-1/2}$. Thus, for a representative molecular network, taking $M_c = M_e = 15000$ and $M = 300000$, the threshold fracture energy obtained is about $25 J/m^2$, in reasonable agreement with experiment in view of uncertainties and approximations in the theory. Moreover, the predicted increase in fracture energy with molecular mass M_c between crosslinks appears to be correct; increased density of crosslinking (shorter network chains) leads to lower threshold fracture energies. Because, however, the tensile strength σ_b also involves the elastic modulus E [Eq. (11-6)], and E is increased by crosslinking, the threshold tensile strength shows a net increase with increased crosslinking.

Threshold values of tensile strength and extensibility may be calculated by means of Eq. (11-6), using an average threshold fracture energy of $50 J/m^2$, a "natural" flaw size of 40mm (assumed to be independent of composition), and a typical value for Young's modulus E for rubber of 2MPa (corresponding to a Shore A hardness of about 48°). The results are $\sigma_{b,0} = 0.9 MPa$ and $e_{b,0} = 0.45$.

These values are indeed close to experimental "fatigue limits," i.e., stresses and strains below which the fatigue life is effectively infinite in the absence of chemical attack.

A surprisingly large effect has been found on the tensile strength and, by inference, on the tear strength of rubber as a result of the specific distribution of molecular masses M_c in the network. When a small proportion of short chains, about 5 mole%, is combined in a network of long chains, the tensile strength is considerably higher than for other mixtures. It seems likely that the threshold strength is also higher. This remarkable enhancement of strength may be the result of strain redistribution within the network, i.e., the ability to undergo non-affine deformation, so that internal stress concentrations are minimized. Whatever the cause, the phenomenon is clearly of both scientific and practical interest.

Low values of fracture energy, only about one order of magnitude greater than the threshold level, have been obtained by measuring the resistance of rubber to cutting with a sharp knife, a razor blade. Frictional effects were minimized by stretching the sample as it was being cut. By adding the energy supplied by the stretching force to that supplied by the cutter, the total fracture energy was found to be rather constant, and low, of the order of $300 J/m^2$. The value was affected by the sharpness of the blade used, being lower for sharper blades, as would be expected from Eq. (11-9).

11.4 Fracture Under Multiaxial Stresses

Although relatively few studies have been made of the fracture of elastomers under complex stress conditions, some general conclusions can be drawn regarding fracture under specific combined-stress states, as follows.

11.4.1 Compression and Shear

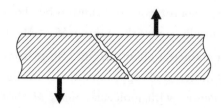

Fig. 11.6 Tearing under shear stresses (schematic).

Elastomers do not appear to fail along shear planes. Instead, fractures develop at 45° to the direction of shear (Fig. 11.6), i.e., at right angles to the corresponding principal tensile stress, at a shear stress theoretically equal to the tensile strength. Indeed, the general condition for rupture appears to be the attainment of a specific tensile stress st at the tip of an existing flaw, and this circumstance can arise even when both applied stresses are compressive, provided that they are unequal. When all the compressive stresses are equal, i.e., under a uniform triaxial compression, the elastomer will merely decrease in volume. No case of fracture under such a loading condition is known.

Under a uniaxial compressive stress, the theory of brittle fracture predicts a breaking stress eight times as large as in tension [Eq. (11-6)] by growth of a crack in an oblique direction. A uniform compressive stress is not, however, readily achieved. Instead, friction at the loaded surfaces of a thin compressed block generally prevents the elastomer from expanding freely in a lateral direction, and a complex stress condition is set up. The outwardly bulging surfaces may split open when the local tensile stress is sufficiently high, but this local fracture does not propagate inward very far because the interior is largely under triaxial compression. Instead, the tear curves around and eventually causes a ring of rubber to break away from the outside of the block, leaving the remainder of the block intact but with a narrower central cross section. Thus, a rubber block in compression is remarkably resistant to fracture, but its stiffness may be seriously reduced after many load cycles by loss of rubber from the outer regions. Failure in simple shear is still more complex. An approximate treatment for an interfacial crack, starting at one edge, yields a relation analogous to Eq. (11-5):

$$G = kUt \qquad (11\text{-}13)$$

Here the constant k is 0.4 initially and then varies between 0.2 and 1.0 as the crack length increases.

11.4.2 Equibiaxial Tension

Quite surprisingly, the breaking stress in equibiaxial tension has been found to be significantly greater than in uniaxial tension, by about 20% to 30%. The breaking elongation is lower but the stored elastic energy at fracture is greater. It should be noted that test sheets put into a state of biaxial extension do not have a cut edge at the desired point of failure, in the central region of the sheet, whereas specimens for uniaxial tests are usually cut from sheets in the form of thin strips. Stress raisers caused by cutting will therefore be present only in uniaxial tests.

Experiments with rather brittle rubber sheets that contained deliberately introduced initial cracks of the same size and type in both uniaxial and biaxial specimens have shown that the breaking stress is still about 20% to 30% higher in equibiaxial tension. The results were, however, consistent with a single value for the fracture energy G_c (about $150 J/m^2$). The difference between the two tests is that when a crack grows in a sheet stretched equibiaxially, only about one-half of the strain energy stored in that area is released, whereas for a crack growing in a uniaxially stretched specimen, all of the energy is released. As a consequence, the strain energy needs to be considerably larger, about twice as large, to cause fracture in equibiaxial stretching.

11.4.3 Triaxial Tension

A small spherical cavity within a block of rubber will expand elastically from its original radius λr_0 to a new radius λr_0 under the action of an inflating pressure P. In the same way, it will expand to an equal degree when the faces of the block are subjected to a uniform triaxial tension of $-P$, i.e., to a negative hydrostatic pressure (Fig. 11.7), provided that the rubber is itself undilatable.

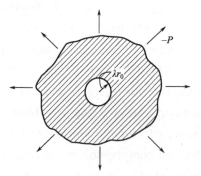

Fig. 11.7 Expansion of a cavity under a triaxial tension.

When the expansion is small, it is proportional to P and given by $\lambda = 1 + \frac{3}{4} P/E$, where E is Young's modulus of the rubber. When the expansion is large, it increases more rapidly than in direct proportion. Indeed, for rubber obeying the neo-Hookean constitutive relation for elasticity, the expansion becomes indefinitely large at a finite value of the applied tension, given by

$$-P = 5/6E \qquad (11\text{-}14)$$

Of course, the original cavity will burst when the expansion of its wall reaches the breaking extension of the rubber, and a large tear will form, governed by an energy requirement for growth. For large precursor voids, the tensile stress for bursting open is much smaller than Eq. (11-14) predicts. And for small precursor voids, there are two further complexities: the actual surface energy of the void needs to be taken into account, and the bursting stresses become so large that the rubber around the cavity will cease to follow elasticity relations valid only for low and moderate strains. However, over a surprisingly wide range of initial radius r_0, from about 0.5mm to about 1mm, Eq. (11-14) is found to be a close approximation to the predicted fracture stress.

Rubber samples are almost invariably found to undergo internal cavitation at the triaxial tensions given by Eq. (11-14). This phenomenon must therefore be regarded as the consequence of an elastic instability, namely, the unbounded elastic expansion of preexisting cavities, too small to be readily detected, in accordance with the theory of large elastic deformations. It does not generally involve the fracture energy, because it is principally a transformation of potential energy (from the loading device) into strain energy. Apparently, rubber contains many precursor voids lying in the critical range, 0.5mm to 1mm (not larger because they would break open at lower stresses).

The critical stress predicted by Eq. (11-14) depends only on the elastic modulus and not at all on the strength of the elastomer. In agreement with this, cavitation stresses in bonded rubber blocks under tension (Fig. 11.8 and Fig. 11.9), and near rigid inclusions, at points where a triaxial tension is set up (Figs. 11.10 and 11.11), are found to be accurately proportional to E and independent of the tear strength of the elastomer, in accordance with the dominant role of an elastic rather than a rupture criterion for failure.

Cavitation near small rigid inclusions is more difficult to induce, probably because the volume of rubber subjected to a critical triaxial tension is too small to contain relatively large precursor voids. And larger stresses are necessary to expand small voids less than about 0.5mm in diameter.

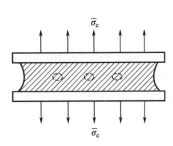

Fig. 11.8 Schematic cavitation in a bonded block.

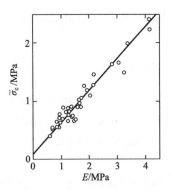

Fig. 11.9 Critical applied stress σ_c for cavitation in bonded blocks versus Young's modulus E of the elastomer.

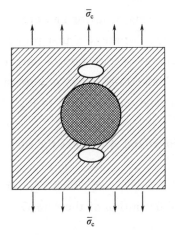

Fig. 11.10 Schematic cavitation near a rigid inclusion.

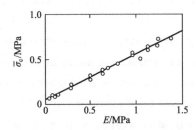

Fig. 11.11 Critical applied stress σ_c for cavitation near rigid inclusions versus Young's modulus E of the elastomer.

If elastomers could be prepared without any microcavities greater than, say, 10nm in radius, they would be much more resistant to cavitation. This seems an unlikely development, however, so Eq. 11-14 remains an important general fracture criterion for elastomers. It predicts a surprisingly low critical triaxial tension, of the order of only a few atmospheres, for soft, low-modulus elastomers. Conditions of triaxial tension should probably be avoided altogether in these cases.

Cavitation is an important practical issue when elastomers are used for containing high-pressure gases. If the gas dissolves in the rubber and migrates to fill the precursor voids, it will be at the (high) external pressure. Then, when the outside pressure is released suddenly, the voids break open in accordance with Eq. (11-14).

11.5 Crack Propagation

11.5.1 Overview

Whereas the initiation of fracture appears to be a similar process for all elastomers, the propagation of a crack is widely different. Three basic patterns of crack propagation, or tearing, can be distinguished corresponding to three characteristic types of elastomeric compound:

① Amorphous elastomers like SBR;

② Elastomers, like natural rubber and Neoprene, that crystallize on stretching, even if only at the crack tip where local stresses are particularly high;

③ Reinforced elastomers containing large quantities, about 30% by volume, of a finely divided reinforcing particulate filler such as carbon black.

Elastomers in the first category show the simplest tearing behavior and are therefore described first. For these materials, once fracture has been initiated, a tear propagates at a rate dependent on two principal factors: the strain energy release rate G and the temperature T. The former quantity represents the rate at which strain energy is converted into fracture energy as the crack advances. It is defined by a relation analogous to Eq. (11-4):

$$G = -2(\Delta W/\Delta A) \tag{11-15}$$

Here W denotes the total strain energy of the specimen and A denotes the surface area (which, of course, increases as the crack advances).

Even if a crack is stationary, because the critical value G_c at which fracture takes place has not been attained, Eq. (11-15) is still a useful definition of the rate G at which energy would be available from the strained specimen.

For a tensile strip with an edge cut, it yields

$$G = 2\pi l U \tag{11-16}$$

by analogy with Eq. (11-5), and for a tear test piece, from Eq. (11-10), we get

$$G = 2F/t \tag{11-17}$$

11.5.2 Viscoelastic Elastomers

Experimental relations between the fracture energy G, the rate of tearing, and the temperature of test are shown as a three-dimensional diagram in Fig. 11.12 for an SBR material. The fracture energy is seen to be high at high rates of tearing and at low temperatures, and vice versa, in a manner reminiscent of the dependence of energy dissipation in a viscous material on rate of deformation and temperature. Indeed, when the rates of tear are divided by the corresponding molecular segmental mobility ϕ_T at the temperature of test, the relations at different temperatures superpose to form a single master curve, as shown in Fig. 11.13. In this figure, the rates have been multiplied by the factor

$$a_T = \phi_{T_g}/\phi_T \tag{11-18}$$

to convert them into equivalent rates of tearing at the glass transition temperature T_g of the polymer, $-57°C$ for the SBR material of Fig. 11.12.

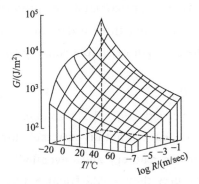

Fig. 11.12 Fracture energy G for an unfilled SBR material as a function of temperature T and rate of tearing R.

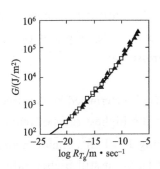

Fig. 11.13 Fracture energy G versus rate of tearing R reduced to T_g for six unfilled amorphous elastomers.

Furthermore, values of fracture energy G for five other amorphous elastomers, two butadiene-styrene copolymers of lower styrene content ($T_g = -72°C$ and $-78°C$) and three butadiene-acrylonitrile copolymers having T_g values of $-30°C$, $-38°C$, and $-56°C$, all fall on a single curve in this representation, increasing with rate of tearing in a similar way to the dissipation of energy internally by a viscous process. We conclude that the fracture energy G is approximately the same for all unfilled amorphous lightly crosslinked elastomers under conditions of equal segmental mobility, and that the dependence of tear strength on temperature arises solely from corresponding changes in segmental mobility.

Thus, internal energy dissipation determines the tear resistance of such elastomers: the greater the dissipation, the greater the tear strength. This point will also emerge in connection with variations in tensile strength of elastomers. It is demonstrated strikingly in the present case by a proportionality between fracture energy G and a direct measure of energy dissipation, namely, the shear loss modulus G'', for the same six elastomers (Fig. 11.14).

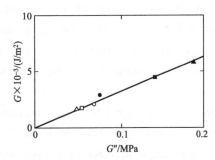

Fig. 11.14 Fracture energy G versus shear loss modulus G'' for six unfilled amorphous elastomers.

For simple C—C crosslinked elastomers, the reduction factors a_T used to transform tear energy results at different temperatures as in Fig. 11.12 to yield a master curve as in Fig. 11.13 are found to correspond closely to the universal form of the WLF rate-temperature equivalence relation:

$$\log a_T = -17.5(T - T_g)(52 + T - T_g) \qquad (11\text{-}19)$$

At low rates of tearing, about $10\sim 20$m/s at T_g, the contribution from viscous losses becomes vanishingly small and the tear strength is reduced to the small threshold value, G_0. As the tear rate increases, the tear strength rises, approximately in proportion to $(R_{a_T})^{0.24}$, until it becomes about $1000\times G_0$, i.e., by about three orders of magnitude. Thus, over wide ranges of rate and temperature the strength of simple rubbery solids can be expressed as:

$$G/G_0 = 1 + 2.5\times 10^4 (R_{a_T})^{0.24} \qquad (11\text{-}20)$$

where G_0 is determined by the structure of the molecular network [Eq. (11-11) and Eq. (11-12)].

At the highest rates of tear, about 1m/s at T_g the tear strength is extremely high, approaching 10^6 J/m². Simultaneously, elastomers become leathery and, eventually, glasslike. Indeed, at still higher rates and lower temperatures they would fracture like typical polymeric glasses, by a failure process in which a narrow craze band forms and propagates, and is followed by a running crack. The fracture energy for this process is relatively small, about $500\sim 1000$J/m², in comparison with that for highly viscous but highly deformable elastomers, so that the curve shown in Fig. 11.13 turns sharply down at higher rates to level off at this value.

Why is the tear strength so strongly dependent on tear rate and temperature, i.e., on the viscoelastic response of the polymer? This striking feature has been explained by Knauss as a consequence of retarded elasticity. His treatment assumes that the stress intensity factor K_c, given by Eq. (11-8), is largely unchanged by changes in rate and temperature, so that

$$K_c^2 = E(t)G_0 = E(t=\infty)G \qquad (11\text{-}21)$$

where $E(t)$ denotes the tensile modulus E at a time t after straining. $E(t=\infty)$ denotes the equilibrium modulus. Energy G obtained from a quasi-equilibrium solid far from the crack tip is expended in breaking material with a time-dependent modulus $E(t)$ at the crack tip. Thus, the fracture energy G is expected to increase with rate of tearing R approximately in proportion to the increase in elastic modulus E with rate of deformation,

$$G/G_0 = E(t=\delta/R)/E(t=\infty) \qquad (11\text{-}22)$$

where δ is a characteristic distance ahead of the crack tip over which the high stress concentration at the crack tip is built up. This simple picture gives a good representation of the observed increase in fracture energy with rate and temperature, but the value of δ obtained by a direct comparison of G and E is too small, about 1Å, to be physically reasonable. The discrepancy may arise because tearing is discontinuous on a microscopic scale, i.e., it may take place in a stick-slip fashion. When rubber is cut at a controlled rate with a sharp knife, the variation of cutting resistance with rate of cutting and temperature is found to be accounted for correctly by Eq. (11-22) using a value for d about equal to the blade tip diameter.

The tear strength of sulfur-crosslinked elastomers is higher and its dependence on temperature is greater than would be expected from Eq. (11-19).

This has been attributed to fracture of weak polysulfide crosslinks rather than the stronger C—C bonds in network chains. As a result, a second temperature-dependent process is introduced, because the strength of S—S bonds falls as the temperature is raised. The strength of practical rubber compounds thus reflects not only internal dissipation of energy from viscous processes but also in detachment from filler particles, from changes in tear tip radius, and from fracture of weak crosslinks.

11.5.3 Strain-Crystallizing Elastomers

As shown in Fig. 11.15, the tear strength of strain-crystallizing elastomers is greatly enhanced over the range of tear rates and temperatures at which crystallization occurs on stretching, at the tear tip. At high temperatures, however, crystallization becomes thermodynamically prohibited because even the high melting temperatures of crystallites in highly stretched elastomers have been exceeded. Conversely, at low temperatures and high rates of tear, molecular reorganization into crystallites cannot take place in the short times of stretching as the crack tip advances. Thus, the strengthening effect of strain-induced crystallization

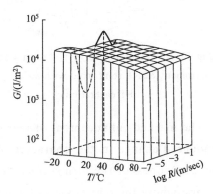

Fig. 11.15 Fracture energy G for a strain-crystallizing elastomer, natural rubber, as a function of temperature T and rate of tearing R.

is limited to a particular range of tear rates and temperatures, as seen in Fig. 11.15. Outside this range, the material has only the strength associated with its viscous characteristics, dependent on $T - T_g$.

The high strength of strain-crystallizing materials has been attributed to pronounced energy dissipation on stretching and retraction, associated with the formation and melting of crystallites under nonequilibrium conditions. Reinforcing particulate fillers have a similar strengthening action, as discussed later, and they also cause a marked increase in energy dissipation.

Whether this is the sole reason for the strengthening effect of crystallites and fillers, and other strengthening inclusions such as hard regions in block copolymers, hydrogen-bonded segments, etc., is not clear, however.

11.5.4 Reinforcement with Fillers

A remarkable reinforcing effect is achieved by adding fine particle fillers such as carbon black or silica to a rubber compound. They cause an increase in tear strength and tensile strength by as much as 10-fold when, for example, 40% by weight of carbon black is included in the mix formulation. But this strengthening action is restricted to a specific range of tear rates and test temperatures—ranges that depend on both the type of filler and the elastomer (Fig. 11.16). Outside this range of effectiveness, the filler does not enhance the observed strength to nearly the same degree.

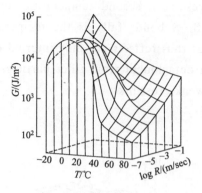

Fig. 11.16 Fracture energy G for an amorphous elastomer (SBR) reinforced with 30% by weight FT carbon black.

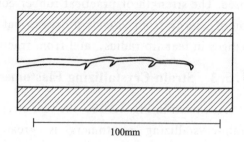

Fig. 11.17 "Knotty" tear in a carbon-black-reinforced elastomer.

The marked enhancement of tear strength in certain circumstances is associated with a pronounced change in the character of the tear process, from relatively smooth tearing with a roughness of the torn surface of the order of 0.1~0.5mm to discontinuous stick-slip tearing, where the tear deviates from a straight path and even turns into a direction running parallel to the applied stress, until a new tear breaks through. This form of tearing has been termed knotty tearing; an example is shown in Fig. 11.17. A typical tear force relation is shown in Fig. 11.18(a); it may be compared with the corresponding relation for an unfilled material in Fig. 11.18(c). The peak tearing force at the "stick" position reaches high values, but the force during catastrophic "slip" tearing drops to a much lower level, only about twice as large as that for continuous tearing of the unfilled elastomer. Indeed, when the tear is prevented from deviating from a linear path by closely spaced metal guides, or is made to propagate in a straight line by stretching or prestretching the sample in the tearing direction, then the tear force is much smaller, only two to three times that for the corresponding unfilled material, see Fig. 11.18(b).

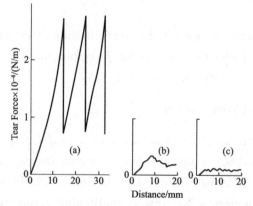

Fig. 11.18 Tear force relations for (a) a filled elastomer without constraints; (b) the same material with the tear confined to a linear path; and (c) the corresponding unfilled elastomer, with and without constraints.

Thus, reinforcement of tear strength by fillers is of two kinds: a small (no more than two-to threefold) increase in intrinsic strength, and a major deviation of the tear path on a scale of several millimeters under special conditions of rate of tearing, temperature, and molecular orientation. The first effect may be attributed to enhanced energy dissipation in filled materials, as discussed in the previous section. The second is attributed to a lowering of the tear resistance sideways, parallel to the stretching direction. If the tear resistance is reduced sufficiently in the sideways direction, then the tear will be deflected sideways and rendered relatively harmless. Paradoxically, rubber is reinforced if its strength is lowered in a certain direction—one that does not result in catastrophic fracture. This mechanism of reinforcement is supported by two observations: measurements of tear strength in the stretching direction show a pronounced decrease as the stretch is increased, and calculations reveal that the energy available to turn a crack into this direction is surprisingly large (40% or more of that for continuing in the straight-ahead direction when the sample is highly stretched). Thus, when the tear strength in the stretching direction falls to 40% or less of that in the straight-ahead direction, the crack is expected to turn sideways.

11.5.5 Repeated Stressing: Dynamic Crack Propagation

Although amorphous elastomers are found to tear steadily, at rates controlled by the available energy for fracture G (as shown in Figs. 11.13 and 11.14), strain-crystallizing elastomers do not tear continuously under small values of G, of less than about 10^4 J/m^2 for natural rubber for example (see Fig. 11.15). Nevertheless, when small stresses are applied repeatedly, a crack will grow in a stepwise manner by an amount Δl per stress application, even though the corresponding value of G is much below the critical level. Experimentally, four distinct growth laws have been observed (Fig. 11.19) corresponding to four levels of stressing:

① $G < G_0$: no crack growth occurs by tearing, but only by chemical (ozone) attack.

② $G_0 < G < G_1$: the growth step Δl is proportional to $G - G_0$.

③ $G_1 < G < G_c$: the growth step Δl is proportional to G^α.

④ $G \sim G_c$: catastrophic tearing.

The transitional value of G between one crack growth law and another, denoted G_1 above, is found to be about 400J/m^2. No explanation has yet been advanced either for the form of these experimental growth laws or for the transition between them. They must therefore be regarded for the present as empirical relations for the growth step Δl per stress application.

In practice it is customary to approximate crack growth over a wide range of G values (but greater than the threshold value G_0) by a "Paris Law" relation that can be put in the form:

$$\Delta l = B'(G/G_0)^\alpha \tag{11-23}$$

where the constant B' is found to be about 1Å per stress application for many rubber

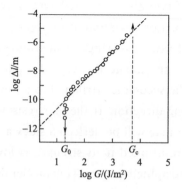

Fig. 11.19 Crack growth step Δl per stress application versus energy G available for fracture, for a natural rubber vulcanizate.

compounds and the exponent α takes different values for different elastomers, ranging from 2 for natural rubber compounds, represented by the broken line in Fig. 11.19, to values of 4 to 6 for noncrystallizing elastomers such as SBR and cis-/trans-polybutadiene.

Crack growth in natural rubber compounds is brought about only by imposing the deformation; if the deformation is maintained, the crack does not grow further under forces insufficient to cause catastrophic tearing. The reason for this is that a crystalline region develops in the highly stressed material at the crack tip and effectively precludes further tearing. This explains a striking feature of crack growth in strain-crystallizing elastomers: the growth steps under repeated stressing become extremely small if the test piece is not relaxed completely between each stress application. In these circumstances, the crystalline region does not melt; it remains intact to prevent further crack growth when the stresses are reimposed. As a result the mechanical fatigue life becomes remarkably prolonged if the component is never relaxed to the zero-stress state. Indeed, failure in these circumstances is a consequence of chemical attack, usually by atmospheric ozone, rather than mechanical rupture.

Amorphous elastomers show more crack growth under intermittent stressing than under a steady stress, and the additional growth step per stress cycle is found to depend on the available energy for fracture G in substantially the same way as for natural rubber. The principal difference is that over region 3, the exponent α in Eq. (11-23) is about 4 for SBR in place of 2 for NR.

Andrews has put forward a general explanation for the slowing down of a crack in an amorphous elastomer (and the complete cessation of tearing in a strain-crystallizing elastomer), after the stresses have been applied, in terms of time-dependent stress changes at the tip of the crack. As a result, the stress concentration at the growing tip is smaller for a viscoelastic material, or for one which is energy-dissipating, than would be expected from purely elastic considerations. Crack growth is correspondingly slowed. This concept has features in common with that of Knauss.

Oxygen in the surrounding atmosphere is found to increase crack growth, presumably by an oxidative chain scission reaction catalyzed by mechanical rupture. The minimum energy G_0 is found to be somewhat larger for experiments carried out in vacuo. When antioxidants are included in the elastomer formulation, then the results in an oxygen-containing atmosphere approach those obtained in vacuo.

11.5.6 Thermoplastic Elastomers

Thermoplastic elastomers derive their physical characteristics from the fundamental immiscibility of different polymers. They consist of triblock molecules having the general structure A—B—A, where A denotes a glassy polymeric strand, e.g., of polystyrene, and B denotes a flexible polymeric strand, e.g., of polybutadiene. The end sequences A are immiscible in polymer B and hence they tend to cluster together to form small domains of a glassy polymer isolated within an elastomeric matrix. Moreover, because the sequences A at each end of one triblock molecule generally become part of different glassy domains, a network of elastomeric strands is formed, linked together by small hard domains, 10~30nm in diameter. Materials of this kind behave in a characteristically rubberlike way, at least at temperatures below the glass temperature of polymer A and above that of polymer B.

The tear strength of a representative thermoplastic elastomer, Kraton 1101, is quite comparable to that of a well-reinforced amorphous or strain-crystallizing elastomer, about 20 kJ/m^2 at room temperature. This remarkably high value is attributed to plastic flow of the hard domains under high local stresses, approaching the breaking point. Indeed, such a deformation process seems essential for these materials to have the capacity to dissipate strain energy, as any tough material must do.

11.6 Tensile Rupture

11.6.1 Effects of Rate and Temperature

In Fig. 11.20 several relations are shown for the breaking stress σ_b of an unfilled vulcanizate of SBR as a function of the rate of elongation at different temperatures. A small correction factor (T_g/T) has been applied to the measured values to allow for changes in the elastic modulus with temperature. The corrected values are denoted σ_b'.

The experimental relations appear to form parallel curves, super imposable by horizontal displacements. The strength at a given temperature is thus equal to that at another temperature

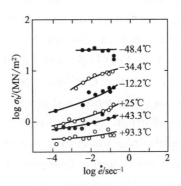

Fig. 11.20 Breaking stress σ_b for an SBR vulcanizate versus rate of elongation e

provided that the rate is adjusted appropriately, by a factor depending on the temperature difference. (Using a logarithmic scale for rate of elongation, a constant multiplying factor is equivalent to a constant horizontal displacement.) As in the case of fracture energy G, this factor is found to be the ratio ϕ_{T_g}/ϕ_T of segmental mobilities at the two temperatures. It is readily calculated from the WLF relation [Eq. (11-19)]. A master curve may thus be constructed for a reference temperature T_s, chosen here for convenience as T_g, by applying the appropriate shift factors to relations determined at other temperatures. The master curve for tensile strength, obtained from the relations shown in Fig. 11.20, is given in Fig. 11.21.

The variation of tensile strength with temperature, like the variation in fracture energy, is thus due primarily to a change in segmental mobility.

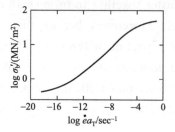
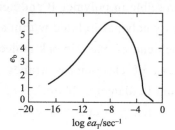

Fig. 11.21 Master relations for breaking stress σ_b as a function of rate of elongation e, reduced to T_g ($-60\,^\circ\mathrm{C}$) by means of the WLF relation, Eq. (11-18) and Eq. (11-19).

Fig. 11.22 Master relations for breaking elongation e_b as a function of rate of elongation e, reduced to T_g ($-60\,^\circ\mathrm{C}$).

Moreover, the master curve has the form expected of a viscosity-controlled quantity: it rises sharply with increased rate of elongation to a maximum value at high rates when the segments do not move and the material breaks as a brittle glass. The breaking elongation at first rises with increasing rate of elongation, reflecting the enhanced strength, and then falls at higher rates as the segments become unable to respond sufficiently rapidly (Fig. 11.22).

Rupture of a tensile test piece may be regarded as catastrophic tearing at the tip of a chance flaw. The success of the WLF reduction principle for fracture energy G in tearing thus implies that it will also hold for tensile rupture properties. Indeed, σ_b and e_b may be calculated from the appropriate value of G at each rate and temperature, using relations analogous to Eq. (11-6) and Eq. (11-7).

The rate of extension at the crack tip will, however, be much greater than the rate of extension of the whole test piece, and this discrepancy in rates must be taken into account.

In addition, it is clear from the derivation of Eq. (11-5) that U represents the energy obtainable from the deformed material rather than the energy put into deforming it. For a material with energy-dissipating properties, the energy available for fracture is only a fraction of that supplied. Such a material will therefore appear doubly strong in a tensile test or in any other fracture process in which the tear energy is supplied indirectly by the relief of deformations elsewhere.

11.6.2 The Failure Envelope

An alternative representation of tensile rupture data over wide ranges of temperature and rate of elongation is obtained by plotting the breaking stress σ_b against the corresponding breaking extension e_b. Tensile results on which Fig. 11.20～Fig. 11.22 are based are replotted in this way in Fig. 11.23. They yield a single curve, termed the failure envelope, which has a characteristic parabolic shape. Following around the curve in an anticlockwise sense corresponds to increasing the rate of extension or to decreasing the temperature, although these two variables do not appear explicitly. Thus, at the lower extreme, the breaking stress and elongation are both small. These conditions are found at low rates of strain and at high temperatures. Conversely, the upper extreme corresponds to a high breaking stress and low extensibility. These conditions obtain at high rates of strain and low temperatures, when the material responds in a glasslike way.

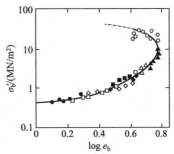

Fig. 11.23 Failure envelope for an SBR vulcanizate.

The principal advantages of the "failure envelope" representation of data seem to be twofold. First, it clearly indicates the maximum possible breaking elongation $e_{b,max}$ for the material. This is found to be well correlated with the degree of crosslinking, specifically with the molecular mass between crosslinks M_c, as predicted by elasticity theory:

$$\lambda_{max} = 1 + e_{b,max} \propto M_c^{1/2} \qquad (11\text{-}24)$$

Second, the failure envelope can be generalized to deal with different degrees of crosslinking, as discussed later. It has therefore been employed to distinguish between changes in crosslinking and other changes that affect the response to rate of elongation and temperature but do not necessarily affect M_c. Examples of this are plasticization and addition of reinforcing fillers.

11.6.3 Effect of Degree of Crosslinking

The breaking stress is usually found to pass through a sharp maximum as the degree of crosslinking is increased from zero. An example is shown in Fig. 11.24. This maximum is due primarily to changes in viscoelastic properties with crosslinking and not to changes in intrinsic strength. For example, it is much less pronounced at lower rates of extension, and it is not shown at all by swollen specimens. Bueche and Dudek and Smith and Chu therefore conclude that it would not exist under conditions of elastic equilibrium.

Two different reduction schemes have been employed to construct failure envelopes for materials having different degrees of crosslinking. The first, shown in Fig. 11.25, consists of scaling the breaking elongation e_b in terms of its maximum value, which is, of course, dependent on the degree of crosslinking [Eq. (11-24)]. Also, the breaking stress σ_b is converted into a true stress at break, rather than the nominal stress employed until now. (The nominal tensile stress is given by the tensile force divided by the unstrained cross-sectional area of the specimen. It has been commonly used in the literature dealing with deformation and fracture of elastomers.) This reduction scheme is clearly quite successful in dealing with a wide range of crosslinking (see Fig. 11.25).

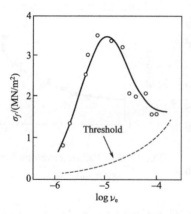
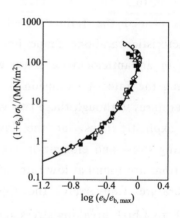

Fig. 11. 24　Tensile strength of SBR vulcanizates versus degree of crosslinking, represented by ν_e. Broken curve: author's estimate of threshold strength under non-dissipative conditions.

Fig. 11. 25　Failure envelope for Viton A-HV materials crosslinked to various extents ($e_{b,max}$ ranging from 2.5 to 18).

The second method consists of scaling the stress axis by dividing the nominal stress at break by a measure of the density of crosslinking. This method also appears to bring data from differently crosslinked materials into a common relationship.

11. 6. 4　Strain-Crystallizing Elastomers

Whereas amorphous elastomers show a steady fall in tensile strength as the temperature is raised (see Fig. 11. 20), strain-crystallizing elastomers show a rather sudden drop at a critical temperature T_c (Fig. 11. 26). This temperature depends strongly on the degree of crosslinking, as shown in Fig. 11. 26.

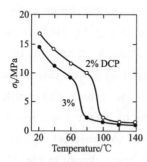
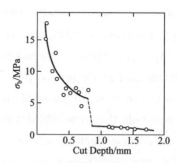

Fig. 11. 26　Tensile strength of natural rubber crosslinked with dicumyl peroxide (DCP) versus temperature.

Fig. 11. 27　Tensile strength of natural rubber crosslinked with 2% dicumyl peroxide versus depth of initial edge cut.

It is clearly associated with failure to crystallize at high temperatures; however, although the bulk of the specimen is amorphous above T_c, the highly strained material at the flaw tip probably continues to crystallize. Thomas and Whittle draw a parallel between the drop in strength at the critical temperature T_c and the similar sharp drop at a critical depth l_c of an edge cut (Fig. 11.27), for strength measurements made at room temperature.

Two other aspects of the critical temperature are noteworthy. First, it is substantially the same for compounds reinforced with fillers. Second, it depends strongly on the type of crosslinking, being highest for long, polysulfidic crosslinks and lowest for carbon-carbon crosslinks. Apparently, labile crosslinks are an important factor in promoting crystallization.

11.6.5 Energy Dissipation and Strength

A general correlation between tensile strength and the temperature interval $(T - T_g)$ between the test temperature T and the glass transition temperature T_g has been recognized for many years, as discussed in Section 6.2 The Failure Envelope. An example is shown in Fig. 11.28, where the strengths of polyurethane elastomers with T_g values ranging from $-67°C$ to $-17°C$ are plotted against $T - T_g$. All the results fall on a single curve in this representation, indicating once more that segmental viscosity governs the observed strength.

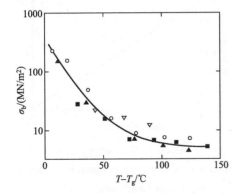

Fig. 11.28 Tensile strength of polyurethane elastomers versus $T - T_g$ (T_g ranging from $-67°C$ to $-17°C$).

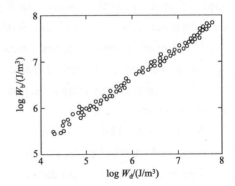

Fig. 11.29 Work-to-break (W_b) versus energy dissipated (W_d) on stretching almost to the breaking elongation.

A more striking demonstration of the close connection between energy dissipation and strength has been given by Grosch et al. They showed that a direct relationship exists between the energy density required to break elastomers U_b and the energy density dissipated on stretching them almost to the breaking elongation U_d. This relationship held irrespective of the mechanism of energy loss, i.e., for filled and unfilled, strain-crystallizing and amorphous elastomers (Fig. 11.29). Their empirical relation is

$$U_b = 410 U_d^{2/3} \tag{11-25}$$

W_b and W_d are expressed in joules per cubic meter. Those materials that require the most energy to bring about rupture, i.e., the strongest elastomers, are precisely those in which the major part of the energy is dissipated before rupture.

11.7 Repeated Stressing: Mechanical Fatigue

Under repeated tensile deformations, cracks appear, generally in the edges of the specimen, and grow across it in an accelerating way. This process is known as fatigue failure. It has been treated quantitatively in terms of step-wise tearing from an initial nick or flaw, as follows: Every time a deformation is imposed, energy G becomes available in the form of strain energy to cause growth by tearing of a small nick in the edge of the specimen.

The value of G for tensile test pieces is given by Eq. (11-5). The corresponding growth step Dl is assumed to obey Eq. (11-22), i.e., to be proportional to G_α, so that the crack growth law becomes

$$\Delta l / l^\alpha = (2kU)^\alpha B' \Delta n \tag{11-26}$$

where n is the number of times the deformation is imposed and k is a numerical constant, about 2. The depth of the crack after N strain cycles is then obtained by integration,

$$l_0^{(1-\alpha)} - l^{(1-\alpha)} = (2kU)^\alpha B' N \tag{11-27}$$

where l_0 is the initial depth of the nick. An example of crack growth is shown in Fig. 11.30; it conforms closely to Eq. (11-27) with $\alpha = 2.0$.

If the crack grows to many times its original depth, so that $l \gg l_0$ before fracture ensues, then the corresponding fatigue life may be obtained by setting

$l = \infty$ in Eq. (11-27) yielding

$$1/N = (2kU)^\alpha B' l_0^{(\alpha-1)} \tag{11-28}$$

This is a quantitative prediction for the fatigue life N in terms of the strain energy U and

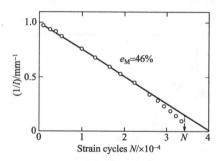

Fig. 11.30 Growth of an edge crack in a test piece of a natural rubber vulcanizate stretched repeatedly to 46% extension.

two material properties, the crack growth exponent α and the characteristic dimension B', which can be determined in a separate experiment as described earlier. Measured fatigue lives for specimens with initial cuts of different length (see Fig. 11.31) and for imposed deformations of different magnitude have been found to be in good agreement with the predictions of Eq. (11-28).

Examples of the dependence of fatigue life on initial cut size are shown in Fig. 11.31 and Fig. 11.32. Lives for test pieces which contain no deliberately introduced cuts, represented by horizontal broken lines in Fig. 11.32, may be interpreted as stepwise tearing from a hypothetical nick or flaw, about 20mm deep, as discussed previously. It is particularly noteworthy that closely similar sizes are deduced for natural flaws for both strain-crystallizing and noncrystallizing elastomers by such extrapolations, because for a noncrystallizing elastomer (SBR), the crack growth law is quite different over the main tearing region.

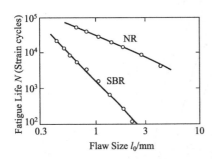
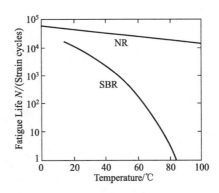

Fig. 11.31 Fatigue life versus depth of initial cut for test pieces of natural rubber and SBR stretched repeatedly to 50% extension.

Fig. 11.32 Fatigue life versus temperature for test pieces of natural rubber and SBR stretched repeatedly to 175% extension.

The exponent a in the fatigue life relation, Eq. (11-28), becomes 4 in place of 2. Measured fatigue lives for an unfilled SBR compound have been found to be in good accord with this relation for a wide range of initial cut depths l_0 and deformation amplitudes.

The different crack growth laws for strain-crystallizing and noncrystallizing elastomers thus lead to quite different fatigue life relations. For a non-crystallizing elastomer, the fatigue life is much more dependent on the size of the initial flaw (see Fig. 11.31) and the magnitude of the imposed deformation, so that such elastomers are generally longer-lived at small deformations and with no accidental cuts, but much shorter-lived under more severe conditions.

The fatigue life is also drastically lowered at high temperatures as a result of the sharp increase in cut growth rate as the internal viscosity is decreased (Fig. 11.32). In contrast, the hysteresis associated with strain-induced crystallization is retained, provided that the temperature does not become so high (about 100°C for natural rubber) that crystallization no longer occurs. The fatigue life for natural rubber is therefore not greatly affected by a moderate rise in temperature.

A more striking difference is found between strain-crystallizing and non-crystallizing elastomers when the stress is not relaxed to zero during each cycle. As shown in Fig. 11.33, the fatigue life of a natural rubber vulcanizate is greatly increased when the minimum strain is raised from zero to, say, 100% because the crystalline barrier to tearing at the tip of a crack does not then disappear in the minimum-strain state. As a result, the growth of flaws is virtually stopped unless the total applied strain is very large, about 400%. No comparable strengthening effect on raising the minimum strain level is found for noncrystallizing elastomers.

Corresponding to the threshold value G_0 of tearing energy, below which no crack growth occurs by mechanical rupture, there is a minimum tensile strain e_0 below which normal-sized flaws do not grow under fatigue conditions. For typical elastomers, this me-

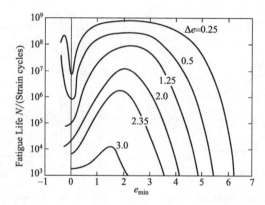

Fig. 11.33　Fatigue life for test pieces of natural rubber versus minimum extension e_{min}.
Δe denotes the additional strain imposed repeatedly.

chanical fatigue limit is found to be about 50% to 100% extension, by calculation from Eq. (11-7) and by direct observation.

At extensions below this level, the fatigue life is infinite in the absence of chemical attack, for example, by ozone in the surrounding atmosphere. Reinforcing fillers greatly enhance the tear strength and tensile strength of elastomers but do not cause an equivalent improvement in the crack growth and fatigue properties. At a given strain energy input, the measured lives are appreciably longer, but if compared at equal available energy levels, they are not much increased. The initial flaw size and threshold tear energy G_0 are therefore deduced to be similar to those for unfilled materials. The growth steps are apparently too small for pronounced deviation of the tear, and hence "reinforcement" against fatigue failure by this mechanism is not so pronounced.

11.8　Surface Cracking by Ozone

In an atmosphere containing ozone, stretched samples of unsaturated elastomers develop surface cracks which grow in length and depth until they eventually sever the test piece. Even when they are quite small, they can cause a serious reduction in strength and fatigue life. The applied tensile stress necessary for an ozone crack to appear may be calculated approximately from Eq. (11-6). The fracture energy G is only about $0.1 J/m^2$, representing the small amount of energy needed for "fracture" of a liquid medium, i.e., about twice the surface energy for a hydrocarbon liquid. Molecular scission apparently occurs readily by reaction with ozone, and does not require mechanical energy to be induced. Taking a representative value for E for a soft rubber of 2MPa and a value of 40mm for the effective depth l of a chance surface flaw, Eq. (11-6) yields a critical tensile stress for ozone cracking of about 50kPa and a critical tensile strain of about 5%. These predictions are in reasonably good agreement with experimentally observed minimum values for ozone attack.

As the stress level is raised above the minimum value, numerous weaker stress raisers become effective and more cracks form. Actually, a large number of small, mutually interfering cracks are less harmful than a few widely separated cracks which develop into deep cuts, so that the most harmful condition is just above the critical stress.

The rate at which a crack grows when the critical energy condition is satisfied depends on two factors: the rate of incidence of ozone at the crack tip and the rate of segmental motion in the tip region. When either of these processes is sufficiently slow, it becomes rate controlling. The overall rate R of crack growth is thus given approximately by

$$R^{-1}(\sec/m) = 8 \times 10^{13} \phi_T^{-1} + 1.2 \times 10^5 C^{-1} \tag{11-29}$$

where ϕ_T ($\sec^{-1}$) is the natural frequency of Brownian motion of molecular segments at the temperature T, given by the WLF relation [Eq. (11-18) and Eq. (11-19)], in terms of (where $\phi_{T_g} = 0.1\sec^{-1}$), and C(mg/L) is the concentration of ozone in the surrounding air.

For a typical outdoor atmosphere, C is of the order of 10^{-4} mg/L, and the second term in Eq. (11-30) is then dominant for values of ϕ_T greater than about 10^4 sec^{-1}, that is, at temperatures more than 25℃ above T_g.

11.9 Abrasive Wear

11.9.1 Mechanics of Wear

Abrasive wear consists of the rupture of small particles of elastomer under the action of frictional forces, when sliding takes place between the elastomer surface and a substrate. A suitable measure of the rate of wear is provided by the ratio A/μ, where A is the volume of rubber abraded away per unit normal load and per unit sliding distance, and μ is the coefficient of friction. This ratio, termed abradability, represents the abraded volume per unit of energy dissipated in sliding. Master curves for the dependence of abradability on the speed of sliding, reduced to a convenient reference temperature by means of the WLF relation [Eq. (11-18) and Eq. (11-19)], are shown in Fig. 11.34. The abradability is seen to decrease

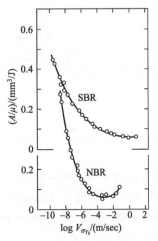

Fig. 11.34 Abradability A/μ versus speed of sliding V, reduced to 20℃, for SBR and NBR vulcanizates.

with increasing speed, pass through a minimum, and then rise again at high speeds as the material becomes glasslike in response. This behavior resembles the variation of the reciprocal of the breaking energy U_b with rate of deformation (a reciprocal relationship because high abradability corresponds to low strength). Indeed, Grosch and Schallamach found a general parallel between A/μ and l/U_b. For this comparison values of the breaking energy U_b were determined at high rates of extension, about 10000% per second, to bring them into agreement with measurements of abradability carried out at a sliding speed of 10mm/sec. This scaling relation indicates that the size of the rubber elements involved in deformation and wear was of the order of 0.1mm, comparable to the size of the abrasive asperities on the particular track employed in the experiments. Moreover, the coefficient of proportionality C between abradability and breaking energy was found to be similar, about 10^{-3}, for all the unfilled elastomers examined. The magnitude of C represents the volume A of rubber abraded away by unit energy applied frictionally to a material for which unit energy U_b per unit volume is necessary to cause tensile rupture. It may be regarded as a measure of the inefficiency of rupture by tangential surface tractions; large volumes are deformed but only small volumes are removed. Apparently the ratio is similar throughout the rubber-to-glass transition and for a variety of elastomers.

The abradabilities A/μ were found to be generally about twice as large for carbon black-filled elastomers as for corresponding unfilled materials. This surprising observation that "reinforced" materials wear away faster can be partially accounted for in terms of the tear strength measurements referred to in a previous section. Under conditions of relatively smooth tearing, it was concluded that the intrinsic strength of reinforced materials is not particularly high; instead, it was found to be comparable to that of unfilled elastomers. The measurements of abradability considered here suggest that it is actually somewhat lower under abrasion conditions.

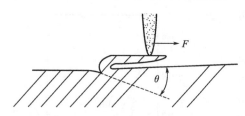

Fig. 11.35 Sketch of a single surface ridge subjected to a frictional (tearing) force F.

Southern and Thomas have related the rate of wear A to the crack growth resistance of the rubber by a simple theoretical treatment. A pattern of lateral ridges is generated in steady-state wear, known as the Schallamach abrasion pattern. A single ridge is shown in Fig. 11.35. The frictional force F pulls laterally on the ridge crest and tends to tear the rubber in the direction indicated by a broken line, at an angle θ to the surface. Fracture energy G is made available for tearing in this direction, given by $F(1 + \cos\theta)$. The crack will therefore advance by a distance Δl, given by Eq. (11-23). This leads to a loss in thickness of rubber of $\Delta l \, (\sin\theta)$. Thus,

$$A = B'/G_0^\alpha F^\alpha (1 + \cos\theta)^\alpha \sin\theta \qquad (11\text{-}30)$$

where α is either 2 or 4, depending on elastomer type. The angle θ may be estimated by direct inspection of the way in which abrasion patterns move over the surface during wear. All other terms in Eq. (11-30) can be determined from tear growth measurements. Thus, the theory does not involve any fitting constants. In these circumstances, it is remarkably successful in accounting for the rate of wear of several unfilled elastomers under severe conditions of pattern abrasion (Fig. 11.36).

Two difficulties must be mentioned, however. The agreement is unsatisfactory for unfilled natural rubber, which wears away much more rapidly than crack growth measurements would predict (see Fig. 11. 36). It has been suggested that this material may not undergo strain-induced crystallization under abrasive conditions, i. e. , under rapidly applied compressive and shearing stresses, and therefore does not show the high resistance to crack growth associated with crystallization. Also, the wear of reinforced rubber is much slower than would be predicted on the basis of crack growth measurements. Further work is needed to clarify this point, which is of great practical importance.

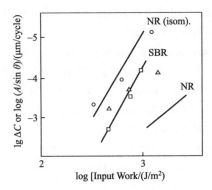

Fig. 11. 36 Rate of wear A versus frictional work input for NR, △; SBR, ▫; and isomerized (non-crystallizing) NR, ○. Solid lines represent crack growth properties Δl ($= \Delta C$) of the same materials under repeated stressing.

11. 9. 2 Chemical Effects

Under mild abrasion conditions, chemical changes within the elastomer become important in wear. The scale of wearing remains small, but the particles of debris are often sticky and agglutinate to form larger particles, several millimeters in size. Indeed, cis-polyisoprene and poly(ethylene-co-propylene), for which molecular rupture under shearing conditions is particularly pronounced, both develop a tarry liquid surface during abrasion. In contrast, cis-polybutadiene shows no signs of structural deterioration. The debris appears to be unchanged chemically and nonadhering. Evidently, different chemical changes are undergone by different elastomers and are responsible for the different types of wear.

Two reactions can occur during wear: oxidative degradation as a result of frictional heating in the contact zone and mechanochemical degradation initiated by shear-induced rupture of chemical bonds. Present evidence favors the latter process. For example, in the absence of oxygen, the wear of cis-polyisoprene changes to resemble that of cis-polybutadiene, whereas the wear of poly(ethylene-co-propylene) is unaltered. These results are in accord with the response of these materials to free radical reactions.

Surprisingly, the products of chemical changes within the elastomer are capable of causing rapid wear of metals used as abraders. For example, when a knife blade is used as a scraper to abrade rubber, the blade itself wears away and becomes blunted, and the volume of metal removed is substantially greater when the broken elastomer molecules form relatively stable free radicals. Thus, wear of the metal is attributed to direct chemical attack by reactive polymeric species, probably free radicals, during frictional contact. Similar wear has been observed with polymers in other physical states; it appears to be a quite general phenomenon when molecular rupture takes place during sliding.

References

[1] C. E. Inglis, Trans. Inst. Naval Architects (London) 55, 219 (1913).

[2] R. J. Eldred, J. Polym. Sci. B 10, 391 (1972).

[3] J. P. Berry, in "Fracture: An Advanced Treatise," Vol. 7: "Fracture of Non-metals and Composites," H. Liebowitz (Ed.), Academic Press, New York, 1972, Chap. 2.

[4] M. Braden and A. N. Gent, Kautsch. Gummi 14, WT157 (1961); E. H. Andrews, D. Barnard, M. Braden, and A. N. Gent, in "The Chemistry and Physics of Rubberlike Substances," L. Bateman (Ed.), Wiley, New York, 1963, Chap. 12.

[5] A. G. Thomas, J. Polym. Sci. 31, 467 (1958).

[6] G. R. Irwin, "Fracturing of Metals," Am. Soc. Metals, Cleveland, 1948.

[7] G. R. Irwin, J. Appl. Mech. 24, 361 (1957).

[8] A. A. Griffith, Philos. Trans. R. Soc. (London), Ser. A 221, 163 (1921).

[9] A. A. Griffith, "Proceedings, 1st International Congress on Applied Mechanics, Delft, 1924," pp. 55-63.

[10] E. Orowan, Rep. Progr. Phys. 12, 185 (1949).

[11] R. S. Rivlin and A. G. Thomas, J. Polym. Sci. 10, 291 (1953).

[12] A. G. Thomas, J. Appl. Polym. Sci. 3, 168 (1960).

[13] H. W. Greensmith, J. Appl. Polym. Sci. 7, 993 (1963); G. J. Lake, in "Proceedings, International Conference on Yield Deformation Fracture Polymers, Cambridge," Institute of Physics, London, 1970, p. 53.

[14] A. G. Thomas, J. Polym. Sci. 18, 177 (1955).

[15] W. G. Knauss, Int. J. Fracture Mech. 6, 183 (1970).

[16] A. Ahagon, A. N. Gent, H.-W. Kim, and Y. Kumagai, Rubber Chem. Technol. 48, 896 (1975).

[17] H. W. Greensmith and A. G. Thomas, J. Polym. Sci. 18, 189 (1955).

[18] H. W. Greensmith, J. Polym. Sci. 21, 175 (1956).

[19] A. G. Veith, Rubber Chem. Technol. 38, 700 (1965).

[20] A. Ahagon and A. N. Gent, J. Polym. Sci., Polym. Phys. Ed. 13, 1903 (1975).

[21] A. N. Gent and R. H. Tobias, J. Polym. Sci., Polym. Phys. Ed. 20, 2051 (1982).

[22] A. K. Bhowmick, A. N. Gent, and C. T. R. Pulford, Rubber Chem. Technol. 56, 226 (1983).

[23] H. Tarkow, J. Polym. Sci. 28, 35 (1958).

[24] G. J. Lake and P. B. Lindley, J. Appl. Polym. Sci. 9, 1233 (1965).

[25] H. K. Mueller and W. G. Knauss, Trans. Soc. Rheol. 15, 217 (1971).

[26] H. W. Greensmith, L. Mullins, and A. G. Thomas, in "The Chemistry and Physics of Rubber-like Substances," L. Bateman (Ed.), Wiley, New York, 1963, Chap. 10.

[27] G. J. Lake and A. G. Thomas, Proc. R. Soc. (London) A 300, 108 (1967).

[28] J. E. Mark and M. Y. Tang, J. Polym. Sci., Polym. Phys. Ed. 22, 1849 (1984).

[29] G. J. Lake and O. H. Yeoh, Int. J. Fracture Mech. 14, 509 (1978); Rubber Chem. Technol. 53, 210

(1980).

[30] A. Stevenson, Int. J. Fracture Mech. 23, 47 (1983).
[31] P. B. Lindley and S. C. Teo, Plast. Rubber Mater. Appl. 4, 29 (1979).
[32] R. A. Dickie and T. L. Smith, J. Polym. Sci. A-2 7, 687 (1969).
[33] C. W. Extrand and A. N. Gent, Int. J. Fracture Mech. 48, 281 (1991).
[34] A. E. Green and W. Zerna, "Theoretical Elasticity," Sect. 3.10. Oxford Univ. Press, London, 1954.
[35] A. N. Gent and P. B. Lindley, Proc. R. Soc. (London) A 249, 195 (1958).
[36] G. H. Lindsay, J. Appl. Phys. 38, 4843 (1967).
[37] M. L. Williams and R. A. Schapery, Int. J. Fracture Mech. 1, 64 (1965).
[38] A. N. Gent and D. A. Tompkins, J. Polym. Sci. A-2 7, 1483 (1969).
[39] A. N. Gent and C. Wang, J. Mater. Sci. 26, 3392 (1991).
[40] A. E. Oberth and R. S. Bruenner, Trans. Soc. Rheol. 9, 165 (1965).
[41] A. N. Gent and B. Park, J. Mater. Sci. 19, 1947 (1984).
[42] B. J. Briscoe and S. Zakaria, Polymer 31, 440 (1990).
[43] L. Mullins, Trans. Inst. Rubber Ind. 35, 213 (1959).
[44] A. N. Gent and S. M. Lai, J. Polym. Sci.: Part B: Polym. Phys. 32, 1543 (1994).
[45] J. D. Ferry, "Viscoelastic Properties of Polymers," 2nd ed., Wiley, New York, 1970.
[46] R. P. Kambour, J. Polym. Sci., Macromol. Rev. 7, 1 (1973).
[47] W. G. Knauss, in "Deformation and Fracture of Polymers," H. H. Kausch, J. A. Hassell, and R. I. Jaffee (Eds.), Plenum Press, New York, 1974, pp. 501-540; J. M. Bowen and W. G. Knauss, J. Adhesion 39, 43 (1992).
[48] G. J. Lake and O. H. Yeoh, J. Polym. Sci. 25, 1157 (1987).
[49] A. N. Gent, S. M. Lai, C. Nah, and C. Wang, Rubber Chem. Technol. 67, 610 (1994).
[50] E. H. Andrews, J. Appl. Phys. 32, 542 (1961).
[51] D. De and A. N. Gent, Rubber Chem. Technol. 69, 834 (1996).
[52] A. N. Gent and A. W. Henry, in "Proceedings, International Rubber Conference, London, 1967," Maclaren, London, 1968, pp. 193-204.
[53] R. Houwink and H. H. J. Janssen, Rubber Chem. Technol. 29, 4 (1956).
[54] A. N. Gent, M. Razzaghi-Kashani, and G. R. Hamed, Rubber Chem. Technol. 76, 122 (2003).
[55] G. J. Lake and P. B. Lindley, Rubber J. Int. Plast. 146, 24; 146, 30 (1964).
[56] G. J. Lake and P. B. Lindley, J. Appl. Polym. Sci. 10, 343 (1966).
[57] E. H. Andrews, J. Mech. Phys. Solids 11, 231 (1963).
[58] E. H. Andrews and Y. Fukahori, J. Mater. Sci. 12, 1307 (1977).
[59] A. N. Gent and M. Hindi, Rubber Chem. Technol. 63, 123 (1990).
[60] A. N. Gent, G. L. Liu, and T. Sueyasu, Rubber Chem. Technol. 64, 96 (1991).
[61] T. L. Smith, J. Polym. Sci. 32, 99 (1958).
[62] F. Bueche, J. Appl. Phys. 26, 1133 (1955).
[63] F. Bueche and J. C. Halpin, J. Appl. Phys. 35, 36 (1964); J. C. Halpin, Rubber Chem. Technol. 38, 1007 (1965).
[64] T. L. Smith, J. Polym. Sci. A 1, 3597 (1963).
[65] T. L. Smith, "Rheology," F. R. Eirich (Ed.), Academic Press, New York, Vol. 5, 1969, Chap. 4.
[66] F. Bueche and T. J. Dudek, Rubber Chem. Technol. 36, 1 (1963).
[67] L. M. Epstein and R. P. Smith, Trans. Soc. Rheol. 2, 219 (1958).
[68] T. L. Smith and W. H. Chu, J. Polym. Sci. A-2 10, 133 (1972).

[69] R. F. Landel and R. F. Fedors, in "Proceedings 1st International Conference on Fracture, Sendai," Vol. 2, p. 1247, Japan. Soc. Strength and Fracture of Materials, Tokyo, 1966.

[70] B. B. S. T. Boonstra, India Rubber World 121, 299 (1949).

[71] J. A. C. Harwood, A. R. Payne, and R. E. Whittaker, J. Appl. Polym. Sci. 14, 2183 (1970).

[72] A. G. Thomas and J. M. Whittle, Rubber Chem. Technol. 43, 222 (1970).

[73] A. N. Gent and L. Q. Zhang, Rubber Chem. Technol. 75, 923 (2002).

[74] A. M. Borders and R. D. Juve, Ind. Eng. Chem. 38, 1066 (1946).

[75] K. A. Grosch, J. A. C. Harwood, and A. R. Payne, Nature 212, 497 (1966).

[76] P. B. Lindley and A. G. Thomas, in "Proceedings, Fourth International Rubber Conference, London, 1962," pp. 428-442.

[77] A. N. Gent, P. B. Lindley, and A. G. Thomas, J. Appl. Polym. Sci. 8, 455 (1964).

[78] S. M. Cadwell, R. A. Merrill, C. M. Sloman, and F. L. Yost, Ind. Eng. Chem., Anal. Ed. 12, 19 (1940).

[79] A. N. Gent and J. E. McGrath, J. Polym. Sci. A 3, 1473 (1965).

[80] K. A. Grosch and A. Schallamach, Trans. Inst. Rubber Ind. 41, 80 (1965).

[81] E. Southern and A. G. Thomas, Plast. Rubber Mater. Appl. 3, 133 (1978).

[82] G. I. Brodskii, N. L. Sakhnovskii, M. M. Reznikovskii, and V. F. Evstratov, Sov. Rubber Technol. 18, 22 (1960).

[83] A. Schallamach, J. Appl. Polym. Sci. 12, 281 (1968).

[84] A. N. Gent and C. T. R. Pulford, J. Appl. Polym. Sci. 28, 943 (1983).

[85] G. A. Gorokhovskii, P. A. Chernenko, and V. A. Smirnov, Sov. Mater. Sci. 8, 557 (1972).

[86] Y. A. Evdokimov, S. S. Sanches, and N. A. Sukhorukov, Polym. Mech. 9, 460 (1973).

[87] A. N. Gent and C. T. R. Pulford, J. Mater Sci. 14, 1301 (1979).

[88] G. V. Vinogradov, V. A. Mustafaev, and Y. Y. Podolsky, Wear 8, 358 (1965).

[89] E. H. Andrews, "Fracture in Polymers," American Elsevier, New York, 1968.

[90] A. N. Gent, in "Fracture: An Advanced Treatise," Vol. 7: "Fracture of Non-metals and Composites," H. Liebowitz (Ed.), Academic Press, New York, 1972, Chap. 6.

[91] F. R. Eirich and T. L. Smith, in "Fracture: An Advanced Treatise," Vol. 7: "Fracture of Non-metals and Composites," H. Liebowitz (Ed.), Academic Press, New York, 1972, Chap. 7.

[92] A. N. Gent and C. T. R. Pulford, in "Developments in Polymer Fracture—1," E. H. Andrews (Ed.), Applied Science Publishers, London, 1979, Chap. 5.

[93] G. J. Lake and A. G. Thomas, Strength, in "Engineering with Rubber," 2nd ed., A. N. Gent (Ed.), Hanser Publishers, Munich, 2001, Chap. 5.

Chapter 12
Tire Engineering

A tire or tyre is a ring-shaped component that surrounds a wheel's rim to transfer a vehicle's load from the axle through the wheel to the ground and to provide traction on the surface traveled over. Most tires, such as those for automobiles and bicycles, are pneumatically inflated structures, which also provide a flexible cushion that absorbs shock as the tire rolls over rough features on the surface. Tires provide a footprint that is designed to match the weight of the vehicle with the bearing strength of the surface that it rolls over by providing a bearing pressure that will not deform the surface excessively.

The materials of modern pneumatic tires are synthetic rubber, natural rubber, fabric and wire, along with carbon black and other chemical compounds. They consist of a tread and a body. The tread provides traction while the body provides containment for a quantity of compressed air. Before rubber was developed, the first versions of tires were simply bands of metal fitted around wooden wheels to prevent wear and tear. Early rubber tires were solid (not pneumatic). Pneumatic tires are used on many types of vehicles, including cars, bicycles, motorcycles, buses, trucks, heavy equipment, and aircraft. Metal tires are still used on locomotives and rail cars, and solid rubber (or other polymer) tires are still used in various non-automotive applications, such as some casters, carts, lawnmowers, and wheelbarrows.

Pneumatic tires are manufactured in about 450 tire factories around the world. Tire production starts with bulk raw materials such as rubber (60% ~70% synthetic), carbon black, and chemicals and produces numerous specialized components that are assembled and cured. Many kinds of rubber are used, the most common being styrene-butadiene copolymer. The article Tire manufacturing describes the components assembled to make a tire, the various materials used, the manufacturing processes and machinery, and the overall business model.

Styrene-butadiene copolymer is the most popular material used in the production of rubber tires. In 2004, $80 billion of tires were sold worldwide, in 2010 it was $140 billion (approximately 34% growth adjusting for inflation), and is expected to grow to $258 billion per year by 2019. In 2015, the US manufactured almost 170 million tires. Over 2.5 billion tires are manufactured annually, making the tire industry a major consumer of natural rubber. It is estimated that by 2019, 3 billion tires will be to be sold globally every year.

As of 2011, the top three tire manufacturing companies by revenue were Bridgestone (manufacturing 190 million tires), Michelin (184 million), Goodyear (181 million); they were followed by Continental, and Pirelli. The Lego group produced over 318 million toy tires in 2011 and was recognized by Guinness World Records as having the highest annual production of tires by any manufacturer.

12.1 Introduction

In human history, the wheel is considered one of the most important inventions, because it found use in a wide range of applications such as transportation vehicles, construction equipment, and internal parts of machinery.

Like most inventions, the wheel was a development of earlier devices such as rollers, dating to the Bronze age over 5000 years ago, used to move heavy objects. Wheeled vehicles were recorded in Sumeria in 3500 BC, Assyria in 3000 BC, and central Europe toward 1000 BC. Four-wheeled wagons using a swiveling front axle for steering were recorded in 1500 BC.

With the introduction of swifter horses from the Asiatic Steppes into Mesopotamia, the cart was adapted for military applications. The spoked wheel was introduced and then given its first "tire," consisting of first a leather and then copper and iron binding to prevent damage to the wooden wheel frame.

The next most important event was probably in 1846 when Thompson was granted a patent for an elastomeric air tube, to be fixed onto a wheel to reduce the power to haul a carriage, make motion easier, and reduce noise. This concept was much refined in the 1880s when the first pneumatic tire was developed for use on tricycles.

The discovery of vulcanization by Charles Goodyear in 1839 and the industrialization of Europe and North America enabled the tire to evolve from a rubberized canvas covering a rubber tube to a complex fabric, steel, and elastomeric composite.

There are 2 aspects to how pneumatic tires support the rim of the wheel on which they are mounted. First, tension in the cords pull on the bead uniformly around the wheel, except where it is reduced above the contact patch. Second, the bead transfers that net force to the rim.

Air pressure, via the ply cords, exerts tensile force on the entire bead surrounding the wheel rim on which the tire is mounted, pulling outward in a 360 degree pattern. Thus the bead must have high tensile strength. With no force applied to the outer tread, the bead is pulled equally in all directions, thus no additional net force is applied to the tire bead and wheel rim. However, when the tread is pushed inward on one side, this releases some tension in the corresponding sidewall ply pulling on the bead. Yet the sidewall ply on the other side continues to pull the bead in the opposite direction. Thus the still fully tensioned sidewall ply pulls the tire bead and wheel rim in the same direction as the tread displacement with equal force as that applied to push the tread inward.

This sidewall ply to bead tension support was a big reason for older cross-ply cord tire construction using the materials available in the early 19th century. The cross-ply cord arrangement orients the cords to more directly support the bead & wheel rim (like a sling: bead to cord and around below the tread back to the opposite bead, both ways, thus crossing plies of cords). However, with improved combinations of cord, bead, rim materials, and manufacturing techniques, combined with ongoing focus and research on tire efficiency and durability, it became both feasible and desirable to manufacture radial-ply cord tires, which, for many applications (despite higher costs), outperform and more than outlast (reliable-usual-service-life/cost ratio) similar cross-ply cord tire designs by (a) facilitating a flatter contact pattern with more evenly distributed pressure on the momentarily stationary area of contact between tread & ground and (b) lower operating costs over time: due to reduced tire temperature, decreased rolling resistance, lower puncture rates, greater longevity, etc.

12.2 Tire Types and Performance

In terms of both volume production and consumer awareness, pneumatic tires fall into essentially nine categories which are based on vehicle application. There are tires for racing vehicles, passenger vehicles, and light trucks where gross vehicle weights typically do not exceed 7250kg. In such tires, significant quantities of fabric are used as a reinforcement. Larger tires such as those for heavy trucks, farm and agricultural vehicles, earthmoving equipment, and large aircraft tend to comprise both steel wire and fabric reinforcements.

Finally there are a range of specialty tires which include those used on fork lift trucks, light aircraft, light construction equipment, and golf cars. Regardless of the design or application of the tire, all pneumatic tires must fulfill a fundamental set of functions:

① Provide load-carrying capacity;
② Provide cushioning and dampening;
③ Transmit driving and braking torque;
④ Provide cornering force;
⑤ Provide dimensional stability;
⑥ Resist abrasion;
⑦ Generate steering response;
⑧ Have low rolling resistance;
⑨ Provide minimum noise and minimum vibration;
⑩ Be durable throughout the expected life span.

Dampening characteristics, elastic properties of rubber, and unique deformability and recovery combine to make tires the only product which satisfies all of these functions. Essentially three performance parameters govern a tire's functions. These are (1) vehicle mission profile; (2) mechanical properties and performance such as wear resistance and casing durability; and (3) esthetics, comfort, and behavioral characteristics such as vehicle steering precision.

The mechanical properties of a tire describe the tire's characteristics in response to the application of load, torque, and steering input, resulting in the generation of external forces and deflection. Such mechanical properties are interrelated, and thus a design decision affecting one factor will influence the other factors, either positively or negatively. The result is a complex set of forces acting on a rolling tire on a vehicle (Fig. 12.1).

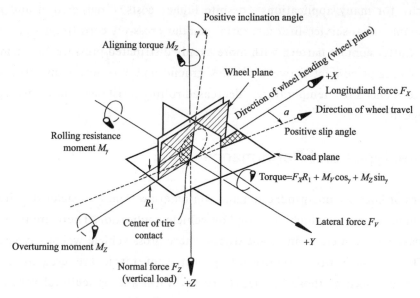

Fig 12.1 Tire forces and moments acting at the center of tire contact.

The tire axis system is the center of the tire-road surface contact shown in Fig. 12.1. The X axis is the intersection of the wheel plane with the road plane and with the positive direction forward. The Z axis is perpendicular to the road plane with a positive direction downward. Thus, the normal force exerted by the tire is positive downward, and the vertical load reaction, which is the road pushing on the tire, is considered negative. The Y axis is in the road plane and is directed so as to make the system right handed and orthogonal. This representation of the tire axis system corresponds closely to the vehicle axis system when the tire is in the right front position of an automobile. In this case, the vehicle would be undergoing a left turn with positive self-aligning torque (M_z) and negative lateral force vector. The forces acting on a tire can thus be broken down into three fundamental vectors: the vertical forces control vehicle esthetics and comfort, the lateral forces impact vehicle control, and the longitudinal or forward forces control performance such as rolling resistance (Fig. 12.2).

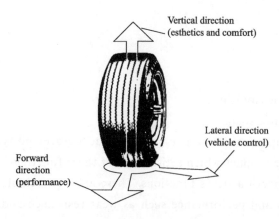

Fig. 12.2 Tire functions.

12.3 Basic Tire Design

A tire is essentially a cord-rubber composite. Tires have plies of reinforcing cords extending transversely from bead to bead, on top of which is a belt located below the tread. The belt cords have low extensibility and are made of steel and fabric depending on the tire application. The belt cords are at a relatively low angle, between 12° and 25°, and serve as restrictions to the 90° casing plies.

12.3.1 Tire Construction

A range of specialized components are found in a tire, which serve to ensure that the product meets its intended design and performance requirements. For example, high-performance passenger tires can have a nylon overlay, also called a cap ply, located over the belt package. Nylon overlays restrict and control tire growth resulting from centrifugal forces created at high speeds. The ply turn up, which describes the manner in which the body ply wraps around the bead wire and turns up the sidewall, anchors the body ply to the bead bundle and further reinforces the lower sidewall region. A toe guard, which is a nylon-reinforced rubber component, protects the bead toe from tearing during mounting and dismounting. Above the toe guard, the chafer further protects the bead area from abrading against the rim flange during use (Fig. 12.3).

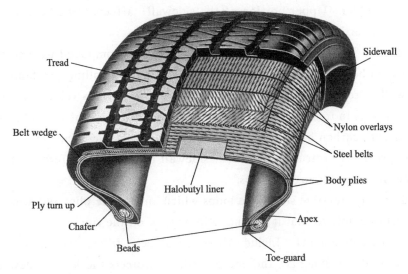

Fig. 12.3 Cross-section of a high-performance passenger tire

12.3.2 Tire Components

A tire is an assembly of a series of parts or subassemblies, each of which has a specific function in the service and performance of the product. Fig. 12.4 illustrates the key components of a truck tire. The following are the parts of a tire:

① Tread The wear resistance component of the tire in contact with the road. It must

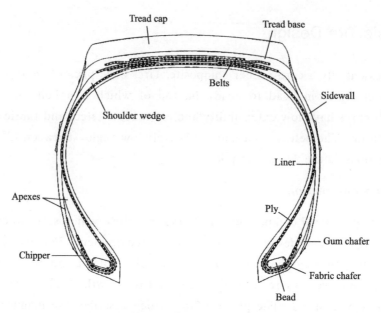

Fig. 12.4 Components of a radial truck tire.

also provide traction, wet skid, and good cornering characteristics with minimum noise generation and low heat buildup. Tread components can consist of blends of natural rubber, polybutadiene (BR), and styrene-butadiene rubber (SBR), compounded with carbon black, silica, oils, and vulcanizing chemicals.

② Tread shoulder　Upper portion of the sidewall; affects tread heat dissipation and tire cornering properties.

③ Tread base　Also termed the cushion; the rubber compound used to ensure good adhesion between belts and tread, heat dissipation, and low rolling resistance.

④ Sidewall　Protects the casing from side scuffing, controls vehicle-tire ride characteristics, and assists in tread support. Sidewall compounds consist of natural rubber, SBR, and BR along with carbon black and a series of oils and organic chemicals.

⑤ Curb guard　A protrusion of rubber sidewall running circumferentially around the tire to protect it from scuffing on curbs.

⑥ Beads　Nonextensible steel wire loops which anchor the plies and also lock the tire onto the wheel assembly so that it will not slip or rock on the rim.

⑦ Bead area components　Include the apex or bead filler; the chafer, which protects the wire bead components; the chipper which protects the lower sidewall; and the flipper, which helps hold the bead in place.

⑧ Plies　Textile or steel cords extending from bead to bead and thereby serving as the primary reinforcing material in the tire casing.

⑨ Belts　Layers of textile or steel wire lying under the tread and serving to stiffen the casing, thereby allowing improved wear performance and handling response, better damage resistance, and protection of the ply cords from road hazards.

⑩ Shoulder belt wedge　High-adhesive rubber compound in the shoulder region between the belts and casing; improves tread wear and durability.

⑪ Liner Butyl rubber or halogenated derivatives of such polymers which retains the compressed air inside the tire.

12.4 Tire Engineering

In essence, a tire is a composite of complex elastomer formulations, fibers, textiles, and steel cord. The term tire structure defines the number, location, and dimensions of the various components used in its composition. The primary components that govern the performance of a tire are the casing plies, bead construction, belts, sidewall, inner liner, and tread. Chafers, flippers, and overlays, which are strips of rubberized fabric located in the bead and crown area, of the tire are termed secondary components because they protect the primary components by minimizing stress concentrations.

12.4.1 Tire Nomenclature and Dimensions

Terminology used to describe tire and rim dimensions is explained in Fig. 12.5. These dimensions are commonly used throughout the tire industry to describe size, growth, and wheel well clearance factors in addition to computation of variables such as load capacity and revolutions per unit distance traveled. In addition, the following definitions should be noted:

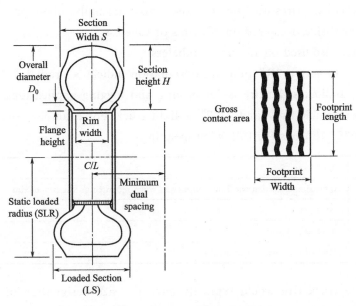

Fig. 12.5 Tire and rim dimensions.

① Aspect ratio A numerical term which expresses the relationship between tire section height and cross-section width. An aspect ratio of 70 indicates the tire section is approximately 70% as high as it is wide.

② Cord angle Angle of the cord path to the centerline the tire; the predominant factor affecting the tire shape or contour. In radial tires the cord angle is 90°.

③ Overall diameter   Unloaded diameter of a new tire-rim combination.

④ Section width   Width of a new tire section excluding side ribs, lettering, and decorations.

⑤ Static loaded radius   Distance from the road surface to the horizontal center line of the rim.

⑥ Minimum dual spacing   Minimum dimension recommended from rim centerline to rim center-line for optimum performance of a dual-wheel installation.

⑦ Footprint length   Length of a loaded footprint.

⑧ Footprint width   Width of the loaded footprint.

⑨ Gross contact area   Total area under the loaded footprint.

⑩ Net contact area   Area of the tread, excluding voids, under the loaded footprint. Also abbreviated as the percent net-to-gross.

⑪ Asymmetrical Tire   design in which the tread pattern on one side of the centerline differs from that on the other side.

⑫ Load rating   Maximum load a tire is rated to carry for a given usage at a specified inflation. The "Load Range" is also used to define the load-carrying capability of a tire. Load ranges are specified in Tire & Rim Association tables. Table 12.1 illustrates load values for a single mounted 11R24.5 heavy-duty truck tire.

Three basic tire size designations are used:

a. Conventional-size tires used on flat base rims, normally tube-type tires.

b. Conventional-size tires used on 15° rims or tubeless.

c. Metric sizes also used on 15° rims; tubeless.

In addition, the letter "R" in a size designation indicates "radial," "D" or (−) is bias, and "ML" indicates the tire is for mining and logging applications. The letter "P" denotes passenger vehicle tire, and "LT" is light truck. For example, the tire size designation "LT 235/75R15" has the following meaning:

LT	Light truck
235	Approximate section width in millimeters when mounted on the proper rim
75	Aspect ratio
R	Radial construction
15	Nominal rim diameter in inches

A heavy-duty truck tire would typically have the size designation of 11R24.5 for a conventional tubeless tire; a low-profile metric tire typically could be sized 295/75R22.5. For the conventional tire, 11 Nominal section diameter in inches.

R	Radial construction
24.5	Rim diameter in inches

Such conventional-size tires have a standard aspect ratio of 80. The initial step in designing a tire is determination of the required size. Size determination is governed by rim dimensions, wheel well envelope, service load, service speed, and inflation. For a metric heavy-duty truck tire these factors are related by the equation

$$L = (6.075 \times 10^{-5}) K \times P^{0.7} \times S_d^{1.1} (D_r + S_d) \tag{12-1}$$

where L = load at 100 kph (kg); P = pressure (kPa); S_d = dimensional factor, section width adjusted for aspect ratio; D_r = rim diameter; and K = constant dependent on vehicle speed. Using equations such as (1), developed by the Tire & Rim Association, the tire engineer can determine the optimum tire size for a specific application. Normally, load requirements are known so Eq. (12-1) can thus be used to calculate required service pressure. This process is then used in size and load range selection. Tire industry standards are used extensively in the design process. Load specification are tabulated in Table 12.1 and Table 12.2.

Table 12.1 Tire and Rim Association Load Values for a Single Mounted Radial Medium Truck Tire

Load range requirement	Ply rating	Example of tire size	Loaded tire limit/kg	Inflation pressure
F	12	295/75R22.5	2500	90
G	14	295/75R22.5	2350	100
H	16	295/75R22.5	3300	120
H	16	285/75R24.5	3370	120
J	18	385/65R22.5	4570	120
L	20	425/65R22.5	5600	120

Table 12.2 European Tire and Rim Technical Organization ETRTO Load Index Numerical Code for Tire Load Carrying Capability

Load index	Load carrying capability/kg	Load index	Load carrying capability/kg
144	2800	152	3550
145	2900	155	3875
146	3000	160	4500
149	3250		

12.4.2 B. Tire Mold Design

Tire mold design initially begins with determination of the inflated dimensions of the required tire size. By use of inflated tire and growth characteristics of the tire, preliminary plyline and mold dimensions are computed (Fig. 12.6).

Once the mold boundary dimensions, location of the plyline, and tread width and depth are known, the contours of the tread, shoulder, sidewall, and bead components can be established. These dimensions and contours are developed using computer-aided engineering techniques (Fig. 12.7).

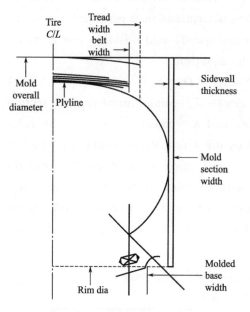
Fig. 12.6 Plyline boundaries.

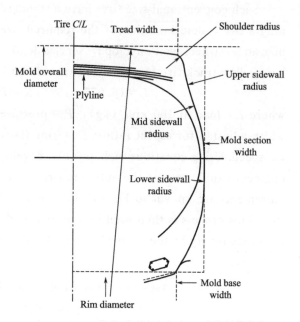
Fig. 12.7 CADAM-developed mold cavity.

The primary interest in designing a tire lies in the belt area, bead area, and belt and ply cord tension. As radial tires contain multiple belts, these layups must be viewed as a package. The stiffness of the radial tire belt package is a function of belt wire angles, wire gauge, belt gauge, and compound stiffness. Fig. 12.8 illustrates a four-belt configuration of a radial truck tire. As an empirical guide, an increase in the stiffness of the belt package, while maintaining belt width, will improve tread wear performance.

The stiffness of a tire belt package can be quantified by determination of the "Gough stiffness" (S), which is a measure of the in-plan bending stiffness of the rigid belt wire, cord, and rubber compound laminate. High radial tire belt rigidity and crown stiffness can significantly improve tire tread wear through provision of a solid foundation for the tread compound. Elaborating, Gough determined that a simple beam model of tire construction features could be used to predict relative tread wear performance. Using both shearing and bending moments in the computation, an empirical equation was derived defining the stiffness parameter S for a simple laminate

$$S = P/d \tag{12-2}$$

where load, P, is the force applied to the crown area layup or laminate to give a deflection d.

In a tire, Gough stiffness is a function of the circumferential modulus and shear modulus of the composite.

$$S = (E \times G)(C_1 \times E) + (C_2 \times G) \tag{12-3}$$

where S = Gough stiffness, E = circumferential modulus of the cord-rubber laminate, G = shear modulus of the cord-rubber laminate, C_1 = constant dependent on the tire size, and

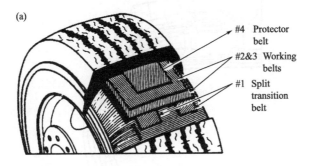

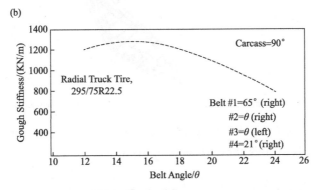

Fig. 12.8 (a) Typical four-belt layup for a truck tire. (b) Impact of Gough stiffness.

C_2 = constant dependent on the tire size. In essence, Eq. (12-3) can be simplified to a model consisting of a simple supported beam of length L with elastic constants E and G deflected distance d by force P as

$$S = PL^3/48EI + 2PL/8AG \qquad (12\text{-}4)$$

where A is the in-plane cross-sectional area to which the force P is applied, and I is the moment of inertia of the beam.

Computer analysis of the effect of the belt angle on Gough stiffness of the four-belt layup is illustrated in Fig. 12.8(b).

For a right/right/left/right belt configuration, the belt angle (θ) of the second and third belts was varied from 10° to 26°. Computation of an optimum belt layup will permit achievement of the required Gough stiffness.

Structural mechanical calculations such as finite-element analysis (FEA) are used to analyze both the inflated and loaded deflected shapes of a tire cross-section and the resulting stress-strain relationships in the belt area. Such studies permit both quantitative analysis and qualitative comparisons of the range of belt configuration options. Fig. 12.9 shows a heavy-duty truck tire in the loaded and unloaded states. The density of grids is designed so as to preserve the essential features of the tire cross-section geometry while maintaining the total number of grid points.

In an analysis of the belt package, three conditions can be evaluated which allow computation of the range of strain energy densities, i.e., inflated tire condition, loaded

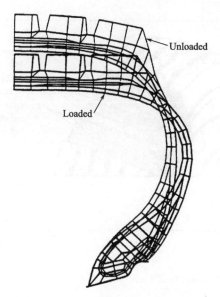

Fig. 12.9 Finite-element structure of a heavy-duty truck tire.

tire condition 180° away from the footprint, and loaded tire condition at the center of the footprint. Fig. 12.10 illustrates these three conditions.

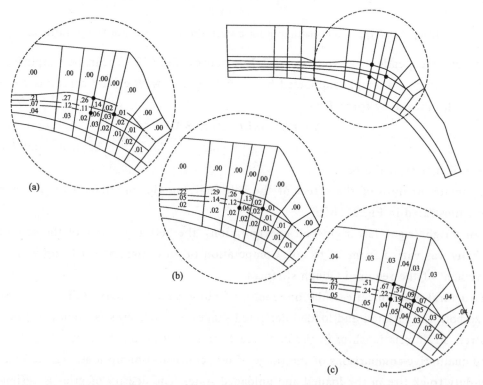

Fig. 12.10 Finite-element analysis of belt area showing strain energy density (in MPa).
(a) Inflation. (b) Loaded 180° away from footprint. (c) Loaded at center of footprint.

It shows that the strain energy density in megapascals in an inflated tire is similar to that in a load tire 180°away from the footprint. At the center of the footprint in a loaded condition, however, the strain energy density at the belt edge has increased from the range 0.01～0.27MPa to 0.05～0.51MPa. These high-strain areas correspond to typical failure-sensitive regions in actual tires.

Similar to the belt area of the tire, the bead region also lends itself to finite-element analysis. Switching grid details to the bead enables analysis of the ply end strains on inflation as well as in a loaded state as the tire makes a complete revolution. By viewing Fig. 12.11, it is possible to evaluate strains caused by tire inflation and cyclic deformation via FEA quantitatively, whereas without such tools, tire building and testing are required.

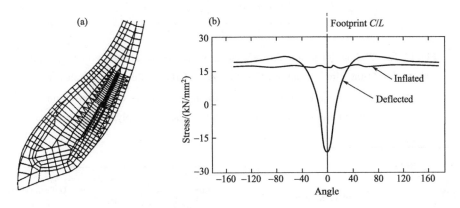

Fig. 12.11 (a) Finite-element analysis of a bead region. (b) Ply end stress versus rotation from the center of footprint.

12.4.3 Cord Tension

The selection of cord materials for belt and plies in a tire and the associated physical locations lend themselves to further FEA analysis. If one takes an inflated nonbelted tire, the ply cords, whether they are in a single-ply casing or multiple-bias-ply construction, will assume a configuration which minimizes strain within the composite. The resulting cord path is termed the neutral contour. Belted tires introduce a restriction to the inflated diameter of the tire, and the neutral contour or plyline of such systems is consequently altered.

The plyline is determined as shown in Fig. 12.12. The principles of plyline determination were developed by Purdy, a pioneer mathematician who derived the basic mathematical equations for cord path and tire properties.

Typical inflated cord tension plots for a truck tire are shown in Fig. 12.13. In an unloaded state the cord tension for the belts tends to be at the tire centerline, and the ply tension is greatest at the point corresponding to the side-wall location; however, on application of a load to the inflated tire and consequent deflection, the cord tensions increase at the belt edges away from the centerline and also in the bead zone. As reviewed earlier, these two regions tend to be the failure zones in a tire construction (Fig. 12.14).

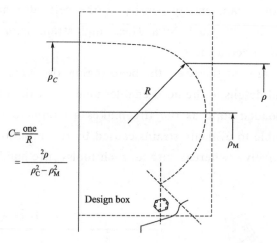

Fig. 12.12 Plyline definition. Purdy's equation defines ideal inflated or "natural" shape for "thin-film" structure. C = curvature of the plyline, ρ_C = radius from the center of the axle to the center of the plyline or tire centerline, and ρ_M = radius from the center of the axle to the center plyline width.

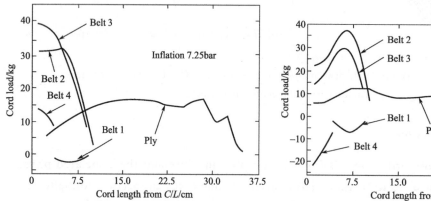

Fig. 12.13 Inflated tire unloaded.

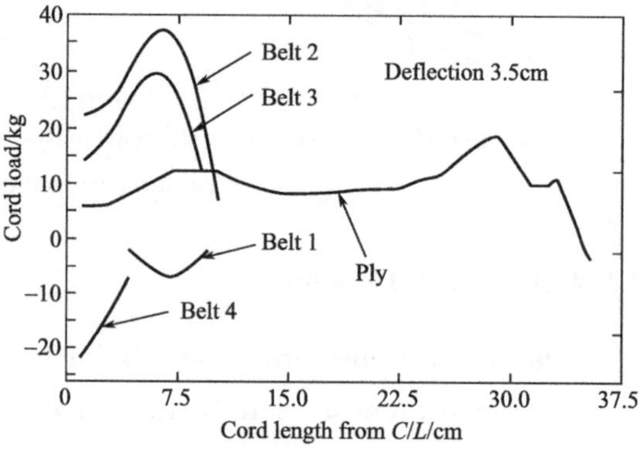

Fig. 12.14 Inflated tire loaded and deflected.

12.4.4 Tread Design Patterns

The tread is the part of the tire that comes in contact with the road surface. The portion that is in contact with the road at a given instant in time is the contact patch. The tread is a thick rubber, or rubber/composite compound formulated to provide an appropriate level of traction that does not wear away too quickly. The tread pattern is characterized by the geometrical shape of the grooves, lugs, voids and sipes. Grooves run circumferentially around the tire, and are needed to channel away water. Lugs are that portion of the tread design that contacts the road surface. Voids are spaces between lugs that allow the lugs to flex and evacuate water. Tread patterns feature non-symmetrical (or non-uniform) lug sizes circumferentially to minimize noise levels at discrete frequencies. Sipes are slits cut across the tire, usually perpendicular to the grooves, which allow the water from the grooves to escape to the sides in an effort to prevent hydroplaning.

Treads are often designed to meet specific product marketing positions. High-performance tires have small void ratios to provide more rubber in contact with the road for higher traction, but may be compounded with softer rubber that provides better traction, but wears quickly. Mud and snow (M&S) tires are designed with higher void ratios to channel away rain and mud, while providing better gripping performance. For a truck tire, tread patterns can be classified into five basic types (Fig. 12.15):

① Highway rib.

② Highway rib/lug combination, where the rib can be either inboard or outboard in the shoulder area (the lugs will be correspondingly opposite to the ribs, outboard or inboard).

③ Highway lug design.

④ On/off highway (special service).

⑤ Off highway.

The tread pattern influences the ability of the tire to transmit driving forces, braking, and lateral forces while operating on a broad range of highway and off-road surfaces. The design of a tread pattern is essentially the separation or division of a smooth tread into smaller elements or blocks. These elements are arranged in a repetitive pattern of voids, ribs, lugs, slots, and grooves.

The pattern is first described in terms of length, width, percentage of void of the various elements, and angles. The tread elements are then arranged to give the tread traction characteristics, optimized ratio of net-to-gross contact area, and minimum noise creation.

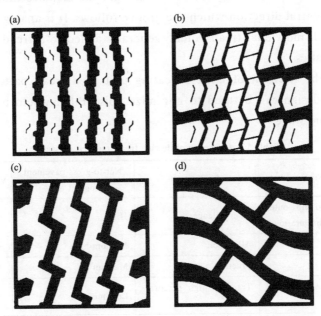

Fig. 12.15 Basic tread patterns: (a) highway rib; (b) highway rib/lug design; (c) on/off highway (mixed service); (d) off highway.

Rib designs with design elements principally in the circumferential direction are the most common type of tread pattern and show overall good service for all-wheel-position summer service. On heavy trucks, they are used nearly exclusively on steer and trailer axles because of their lateral traction and uniform wear characteristics. Rib/lug combinations tend to find use on all-season tires, which require a balance of good tread wear, traction, and wet skid.

On heavy-duty truck drive axles, where forward traction is a prime requirement and where fast tread wear occurs as a result of torque-induced slip, the highway lug design is required. For off-highway service conditions the tread pattern assumes a staggered joint lateral circumferential direction for both lateral and forward traction. Grooves tend to be larger and deeper, with the rib walls angled to prevent stone retention.

Table 12.3 Net-to-gross contact area with nature of service

Type of service	Net to gross/%	Relative tread depth	Traction handling	Wear
Highway steer tires	70~75	+ +	+	+ + +
Highway drive axle tire	70~80	+ + +	+ +	+ + + +
Highway trailer tires	75~85	+	+	+ + + +
On/off road (mixed service)	60~70	+ + +	+ + +	+ +
Off highway	55~65	+ + + + +	+ + +	+ +

The ratio of net contact to gross tread surface area decreases as wet traction becomes more important. Table 12.3 and Table 12.4 illustrate the net-to-gross percentage for various tire tread patterns. A number of additional terms are used to describe a tread pattern:

① Groove amplitude In staggered groove designs the distance the groove pattern oscillates about the central direction which the groove follows. It is analogous to a sine wave.

② Sipes Small individual tread voids, generally added to a tread design to improve traction characteristics.

③ Pitch length Length of each repeating unit in a tread pattern. Variable pitch lengths in a tread design can be used to minimize noise. Pitch would be analogous to the wavelength of a sine wave.

Table 12.4 Automobile tire tread pattern classes

Category	Design	Application	Net-to-gross pavement contact area	Vehicle handling comfort
1	Central solid rib, outer rib block configuration	High mileage	High	+ + + + +
2	All block	All season	Medium-high	+ + + +
3	Block-rib	Traction	Medium	+ + +
4	Central groove outboard blocks	Traction, high performance	Low-medium	+ + +
5	Direction —Asymmetric —Symmetrical	High performance	Low-medium	+ +

④ Blade A protrusion in a curing mold that forms part of the tread design. The protrusion forms a corresponding depression in the finished tire.

⑤ Stone ejection rib Portion of the tread rib designed to throw off stones with the aid of normal tire flexing. Located up to 75% down in the tread grooves to prevent small stones from locking down at the base of the grooves where they cannot be ejected.

12.5 Tire Materials

Tire engineering is the study of the stresses created within a tire and includes such factors as straining of components while the tire is being built, the tire stresses while mounted on a wheel of a moving loaded vehicle, quantification of such stresses, and minimization of such stresses through effective distribution of load and proper selection of materials. Consequently, tire technology groups tend to be multidisciplinary teams consisting of mechanical engineers, computer scientists, chemical engineers, chemists, and mathematicians. To understand tire engineering, it is necessary to have knowledge of the function of each of the types of materials used in a tire structure. This could be considered essentially in two aspects: the tire reinforcing system and rubber compounding.

12.5.1 Tire Reinforcement

A tire is a textile-steel-rubber composite; the steel and textile cords reinforce the rubber and are the primary load-carrying structures within the tire. Because of the performance demands of fatigue resistance, tensile strength, durability, and resilience, seven principal materials have been found suitable for tire application: cotton, rayon, nylon, polyester, steel, fiberglass, and aramid; the latter three materials find primary usage in the tire crown or belt region.

The science of tire reinforcement employs a specialized terminology which the tire engineer must understand:

① Brass weight Typically 3.65 gms/kg of cable; brass coat thickness is of the order of 0.3 microns.

② Breaking Tensile strength.

③ Cord Structure Consisting of two or more strands when used as plied yarn or an end product.

④ Denier The weight of cord expressed in grams per 9000 meters.

⑤ EPI Ends of cord per inch width of fabric.

⑥ Fibers Linear macromolecules orientated along the length of the fiber axis.

⑦ Filaments Smallest continuous element of textile, or steel, in a strand.

⑧ Filling Light threads that run right angles to the warp (also referred to as the "pick") that serves to hold the fabric together.

⑨ LASE Load at a specified elongation or strain.

⑩ Length of lay Axial distance an element or strand requires to make a 360° revolution in a cord.

⑪ Ply twisting Twisting of the tire yarn onto itself the required number of turns per inch; two or more spools of twisted yarn are then twisted again into a cord: for example, if two 840-denier nylon cords are twisted together, an 840/2 nylon cord construction is formed; if three 1300-denier polyester cords are twisted together, they give a 1300/3 cord construction.

⑫ Rivit Distance between cords in a fabric; high rivit typically describes a fabric with a low EPI.

⑬ Tenacity Cord strength, frequently expressed in grams per denier.

⑭ Tex Cord weight expressed in grams per 1000 meters.

⑮ Twist Number of turns per unit length in a cord or yarn; direction of twist can be either clockwise ("S" twist) or counterclockwise ("Z" twist); twist imparts durability and fatigue resistance to the cord, though tensile strength can be reduced.

⑯ Warp Cords in a tire fabric that run lengthwise.

⑰ Weft Cords in a fabric running crosswise.

⑱ Yarn Assembly of filaments.

Fibers and steel cord are the primary reinforcement and load-carrying material in the tire. It is thus appropriate to review the properties of such materials for application in tires.

12.5.2 Steel Cord

Steel wire used in tires are of various configurations, but all are brass-coated wire strands wrapped together to give cords of different characteristics, depending on the application. Steel tire cord is manufactured from high-carbon-steel rod which is first drawn down to a diameter of approximately 1.2mm. A brass plating is then added to the wire before a final drawing to 0.15~0.40mm. These filaments are next stranded to form a cord construction which is designed and optimized for a specific service requirement.

Table 12.5 Composition of steel tire cord

Element	Composition/%	Function
Carbon	0.65	Strength
Chromium	0.05	Strength
Copper	0.02	Strength
Manganese	0.60	Deoxidation
Silicon	0.25	Deoxidation
Sulfur	0.03	Machinability

Steel tire cord is manufactured from high-quality steel which is necessary because of the performance demands to which tires are subjected. The composition of a typical steel cord is illustrated in Table 12.5. The key mechanical properties governing a steel cord or wire are its tensile strength, elongation, and bending stiffness. A tire cord construction is normally defined by the structure, the length of lay, and the direction of lay. The full description of a steel cord is given by

$$(N \times F) \times D + (N \times F) \times D + (F \times D) \tag{12-5}$$

where N = number of strands, F = number of filaments, and D = nominal diameter of filaments (in millimeters).

An example of a steel cord specification would therefore take the form

$$(1 \times 4) \times 0.175 + (6 \times 4) \times 0.175 + (1 \times 0.15) \tag{12-6}$$

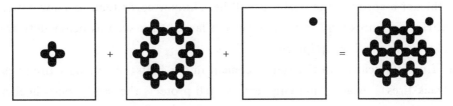

When N or F equals 1, the nomenclature system allows their exclusion. Thus, (12-6) is reduced to

$$4 \times 0.175 + (6 \times 4) \times 0.175 + 0.15 \tag{12-7}$$

A number of additional conventions are used in defining a steel tire cord:

1) If the diameter D is the same for two or more parts in a sequence, then the diameter is specified only at the end of the sequence.

2) The diameter of the spiral wrap is specified separately.

3) When the innermost strand or wire is identical to the adjacent strand or wire, the definition of the wire can be further simplified by specifying only the sum of the identical components. Then (12-7) becomes

$$7 \times 4 \times 0.175 + 0.15 \tag{12-8}$$

4) The sequence or order in a wire designation follows the sequence of manufacture. The length of lay and direction of lay for the cord in (12.8) can be described as illustrated in Fig. 12.16.

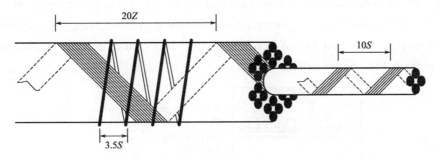

Fig. 12.16　Length of lay and direction of lay.

A number of empirical guidelines govern the construction of a wire for use in tires. For example, if the wire is used in a ply rather than in belts, it will undergo a greater amount of flexing. Hence fatigue performance will be important. If application is in belts, then stiffness becomes a primary design parameter. Thus, key design properties can be specified and are highlighted in Table 12.6.

Table 12.6　Design of a tire cord

Design parameter	Design factor
Stiffness	Number of strands in cord, gauge
Strength	Gauge, number of strands in cord, steel rod composition
Fatigue	Filament gauge
Elongation	Twist, lay

The thicker a cord is, the stiffer it will be. Thinner cords tend to show better fatigue resistance. Heavier cords tend to find use in the larger tires such as heavy-duty truck tires and tires for earth moving equipment.

Wire finds application in tire reinforcement in tire belts, heavy-duty tire plies (e.g., in large truck tires), beads, and chippers (which protects the bead from wheel rim damage).

12.5.3　Mechanism of Rubber: Brass Wire Adhesion

The thin coating of brass on the steel cord is the primary adhesive used in steel-to-rubber bonding. The quality of this bonding system built up during vulcanization of, for example, a radial tire will influence the performance of the steel ply or steel belt in the tire and, ultimately, the durability of the product. Though the mechanism of bond formation in rubber-steel cord adhesion is very complex, a brief review of the current understanding of wire to rubber adhesion is presented.

Natural rubber typically used in wire coat compounds forms a strong bond with brass as a result of the formation of an interfacial copper sulfide (CuS) film during vulcanization. Copper sulfide domains are created on the surface of the brass film during the vulcanization reaction. Such domains have a high specific surface area and grow within the wire coat compound before the viscous polymer phase is crosslinked into a elastomeric network. Thus the polymer molecules become locked into the crystalline copper sulfide lattice (Fig. 12.17). Important factors governing this bonding are formation of copper sulfide, cohesive strength, adhesion to the brass substrate, and rate of secondary corrosion reactions underneath the copper sulfide film.

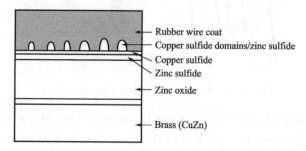

Fig. 12.17　Interfacial copper sulfide film in rubber-brass bonding.

Zinc sulfide and iron sulfide do not bond because they do not grow rapidly enough during vulcanization, do not form porous domains, and thus cannot interlock with the polymer. As the primary requirement is the formation of a copper sulfide domain before the initiation of crosslinking, reduction of compound scorch time, consequently, can adversely affect bond formation.

Mechanical stability of the copper sulfide domains is essential to retain long-term durability of the rubber-to-wire adhesion. However, corrosion of the wire-rubber adhesive bond is catalyzed by Zn^{2+} ions that diffuse through the interfacial CuS layer. This will eventually result in an excess of either ZnS or $ZnO/Zn(OH)_2$. Under dry conditions, this process is slow. Nevertheless, Zn^{2+} will migrate to the surface with a consequent drop in mechanical interlocking of the CuS domains and rubber followed by adhesion loss.

Migration of Zn^{2+} ions is a function of the electrical conductivity of the brass coating. Addition of Co^{2+} or Ni^{2+} ions will reduce this conductivity. Cobalt salts in the wire coat compound act to accelerate the vulcanization rate of high-sulfur compounds which wire coat compounds can be. The increase in crosslink density increases the pullout force of the wire in the rubber. More important, cobalt salts form Co^{2+} ions at the interface of the brass surface during vulcanization, and this will affect copper sulfide formation.

Differences in efficiencies between cobalt adhesion promoters is due to the ease with which Co^{2+} ions can be formed. For example, zinc or brass reacts more easily with cobalt boron decanoate complexes than with cobalt naphthenate or stearate. The Co^{2+} and Co^{3+} ions are incorporated into the ZnO film before the sulfide film has been built up. Both di- and trivalent cobalt ions reduce the electrical conductivity of the ZnO lattice, thereby reducing the diffusion of Zn^{2+} ions through the semiconducting film.

Diffusion of metallic copper domains to the surface following oxidation by $R-S_x$ is not affected, as Cu^{2+} ions migrate along grain boundaries of the ZnO layer. Thus if a cobalt salt is used, formation of copper sulfide at the cord surface will be accelerated, whereas ZnS generation will be hindered (Fig. 12.18). This review is necessarily brief, and the reader is encouraged to consult additional references for further detail on the chemistry of rubber-brass adhesion.

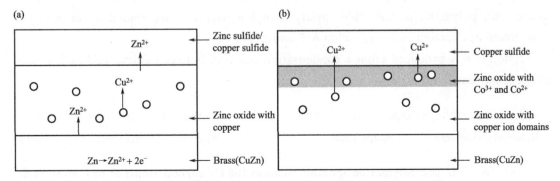

Fig. 12.18 Copper sulfide formation at low cobalt concentration.
(a) Absence of cobalt allows formation of zinc sulfide. (b) Cobalt ions in zinc oxide film hinder zinc ion migration and zinc sulfide formation.

12.5.4 Rayon

The first synthetic fiber for tires was rayon. Cellulose is initially treated with sodium hydroxide to form an alkali cellulose. It is then shredded and allowed to age in air where it is oxidized and undergoes molecular mass reduction to enable subsequent spinning operations. Treatment with carbon disulfide produces cellulose xanthate which is then dissolved in sodium hydroxide to form viscose. The material undergoes further hydrolysis and is then fed into spinnerets to produce the fiber. This fiber is passed through a bath of sulfuric acid and sodium sulfate where the viscose fibers are further coagulated. Washing, bleaching, and twisting into cords follow. The rayon fibers can be drawn or stretched up to 100% of their original length to enable crystalline orientation to produce a high-tenacity rayon suitable for tires.

$$\text{Cellulose} \xrightarrow{NaOH} \text{Alkali Cellulose} \xrightarrow{CS_2} \text{Cellulose Xanthate} \xrightarrow{H^+} \text{Regenerated Cellulose} \tag{12-9}$$

12.5.5 Nylon

Nylon is a polyamide polymer characterized by the presence of amide groups —(CO—NH)— in the main polymer chain. A wide variety of nylon polymers are available but only two have found application in tires: nylon 6,6 and nylon 6.

Nylon 6,6 is produced from a condensation reaction between adipic acid and hexamethylenediamine,

$$H_2N-(CH_2)_6-NH_2 + HOOC-(CH_2)_4-COOH \longrightarrow +NH-(CH_2)_6-NH-CO-(CH_2)_4-CO+_n \tag{12-10}$$

Hexamethylenediamine   Adipic acid   Nylon 6,6 polymer chain

The "6,6" in the polymer designation denotes the six carbon atoms of hexamethylenediamine and the six carbons of adipic acid. Nylon 6 is produced from caprolactam by a ring-opening polymerization. As caprolactam contains six carbon atoms and only one monomer are used, the polymer is thus designated nylon 6.

$$\text{Caprolactam} \longrightarrow [-NH-(CH_2)_5-CO-NH-(CH_2)_5-CO-]_n \qquad (12\text{-}11)$$

Nylon 6 polymer chain

After the polymerization stage, the material is passed through a spinneret to form filaments, cooled, and then twisted to form a yarn. This is then drawn by up to 500% to orient the polymer chains, create polymer crystallite zones, and increase tensile strength (Fig. 12.19).

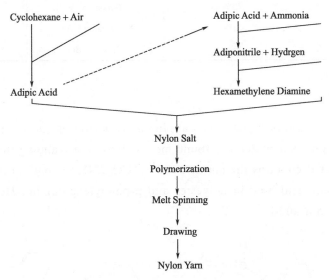

Fig. 12.19 Nylon 6,6 manufacture.

12.5.6 Polyester

Like nylon, a range of polyesters are available commercially for use in tires. Polyethylene terephthalate (PET) is the most important. Also, like nylon, polyester is formed by a condensation polymerization but with the monomers ethylene glycol and dimethyl terephthalate.

$$HO-CO_2-CH_2-OH + H_3COOC-\!\!\!\bigcirc\!\!\!-COOCH_3$$

Ethylene glycol   Dimethyl terephalate

$$\longrightarrow [O-(CH_2)_2-OOC-\!\!\!\bigcirc\!\!\!-CO]_n \qquad (12\text{-}12)$$

Polyethylene terepthalate

The polymerized material is then extruded through a spinneret to form filaments about 0.025mm in diameter. These filaments are cooled, spun into a yarn, and drawn to give the required orientation and crystallinity.

12.5.7 Fiberglass

Like steel wire, fiberglass is an inorganic fiber and is essentially a lime-alumina-borosilicate glass (Table 12.7). It is manufactured by blending sand, clay, limestone, and borax, melting the blend at 1700℃, and drawing the filaments of glass through a platinum-rhodium filter. The filaments are then treated with an adhesive precoat and compiled into a yarn with a low twist.

Table 12.7 Fiberglass Cord Composition

Composition	Percentage/%	Composition	Percentage/%
Silicon dioxide	53%	Boron oxide	9%
Calcium oxide	21%	Magnesium oxide	0.30%
Aluminum oxide	15%	Other oxides	1.7%

12.5.8 Aramid

Another class of fibers which finds application in tires is the aramids. Kevlar is the trade name of the polymer which has found most extensive use among the aramids. Aramid is like nylon in that it contains the amide bond —(CO—NH)— but is produced by copolymerizing terephthalic acid used in polyester and p-phenylenediamine. Hence this aromatic polyamide is termed aramid.

$$\text{(structure of aramid)} \tag{12-13}$$

Table 12.8 illustrates the evolution of the range of reinforcement materials which have been used in tires.

Table 12.8 Trends in Reinforcements

Time	Cord	Time	Cord
1900—1956	Cotton	1970 to date	Steel cord
1939—1975	Rayon	1975—1985	Fiberglass
1950 to date	Nylon	1980 to date	Aramid
1965 to date	Polyester	1990 to date	Hybrid cords

12.5.9 Tire Cord Construction

To use the range of fibers for tire applications, the yarns must be twisted and processed into cords. First, yarn is twisted on itself to give a defined number of turns per inch, i.e., ply twisting. Two or more spools of twisted yarn are then twisted into a cord. Generally the direction of twist is opposite that of the yarn; this is termed a balanced twist. There are a number of reasons for twist in a tire cord:

1) Twist imparts durability and fatigue resistance to the cord, though tensile strength can be reduced.

2) Without twist, the compressive forces would cause the cord outer filaments to buckle.

3) Increasing twist in a cord further reduces filament buckling by increasing the extensibility of the filament bundle.

Durability reaches a maximum and then begins to decrease with increasing twist. This can be explained by considering the effect of stresses on the cord as the twist increases. As twist increases, the helix angle (the angle between the filament axis and cord axis) increases (Fig. 12.20). Thus tension stresses normal to the cord axis result in greater-force parting filaments. The reason for the reduction in strength is also evident in this figure. As cord twist increases, the force in the direction of the yarn axis increases, causing a lower overall breaking strength.

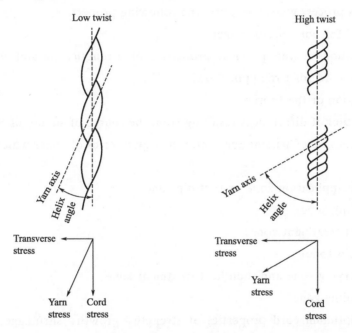

Fig. 12.20 Cord geometry.

In addition to twist, the cord size may be varied to allow for different applications (Table 12.9). Generally three-ply cords have the best durability.

After cable twisting, the cords are woven into a fabric using small fill threads. These threads are also referred to as picks. This weaving process introduces an additional construction variable, i.e., the number of cords per inch or EPI ends per inch that are woven into the fabric. High-end-count fabric gives greater plunger strength or penetration resistance. Low-end-count fabrics have more rivet (distance between cords) and give better separation resistance because of the greater rubber penetration around the cords.

Table 12.9 Tire cord applications

Fiber type	Tenacity(strength) /g·Den	Ultimate elongation/%	Modulus/g·Den	Relative durability	Tire applications
Rayon	5.0	16	50	300	Passenger
Nylon	9.0	19	32	1200	Truck tires, off-road
Polyester	6.5	18	65	400	Passenger, truck, farm
Fiberglass	9.0	4.8	260	5	Passenger
Wire	3.8	2.5	200	3	Truck, off-road
Aramid	20.1	4	350	400	Passenger

12.5.10 Fabric Processing

The most critical stage in preparing a cord or fabric for use in tires is fabric treatment, which consists of applying an adhesive under controlled conditions of time, temperature, and tension. This process gives the fabric the following properties:

1) Adhesion for bonding to rubber;

2) Optimization of the physical properties of strength, durability, growth, and elongation of the cord for tire application;

3) Stabilization of the fabric;

4) Equalization of differences resulting from the source of supply of the fiber.

Processing consists of passing the fabric through a series of zones which can be viewed as follows:

1) Adhesive application zone or first dip zone;

2) First drying zone;

3) First heat treatment zone;

4) Second dip zone;

5) Second drying zone and then heat treatment zone;

6) Final cooling zone.

To obtain optimum cord properties of strength, growth, shrinkage, and modulus, specific temperatures and tensions are set at various exposure times within the fabric processing unit. The temperature and tensions in part determine the ratio of crystalline and amorphous areas within the fiber and the orientation of the crystallites, which in turn determine the physical properties of the cord. For example, polyester, when heated, tends to revert to its unorientated form, and the cord shrinks. Stretching the cord in the first heating zone and then allowing the cord to relax in a controlled manner in the second heat treatment zone, i.e., stretch relaxation, will control shrinkage.

Another variable is increase in processing temperature which can decrease cord tensile strength and modulus but will improve fatigue life. The general relationship between cord properties and processing temperatures and tensions is illustrated in Table 12.10.

Table 12.10 Relation of cord properties to processing tensions and temperature

Cord property	Change effected by first increase in		Change effected by second increase in	
	Tension time	Temperature	Tension time	Temperature
Tensile strength	Slightly Decrease	Decrease	Slightly Decrease	Decrease
Loaded at specified elongation(5%)	Decrease	Decrease	Decrease	Decrease
Ultimate elongation	Increase	Increase	Increase	Increase
Shrinkage	Decrease	Decrease	Decrease	Decrease
Rubber coverage	Increase	Increase	Increase	Increase
Fatigue	Decrease	Decrease	Decrease	Decrease
SCEF	Decrease	Decrease	Decrease	Decrease
Voids	Decrease	Decrease	Decrease	Decrease

Note that not all cord properties behave similarly with varying processing conditions. It is thus necessary to determine the processing conditions that optimize cord properties for the desired tire end use. When two or more diametrically opposite properties have to be optimized, more complex mathematical methods must be used.

Wire and fiberglass, being high-modulus inorganic belt cords, are not processed like textile cords. Steel cord is brass plated at the foundry and, thus, can be used directly at the calendars. Glass yarn is treated with adhesive dip and then used directly in the weaving operation.

12.5.11 Function of Adhesive

There are three aspects to adhesion of tire cord to the elastomer treatment: molecular, chemical, and mechanical. Molecular bonding is due to absorption of adhesive chemicals from the adhesive dip or elastomer coating onto the fiber surface by diffusion and could be described by hydrogen bonding and van der Waals forces. Chemical bonding is achieved through chemical reactions between the adhesive and the fabric and rubber, i.e., crosslinking and resin network formation. Mechanical adhesion is a function of the quality of coverage of the cord by the rubber coat compound. The greater the coverage, the better the adhesion.

The fiber properties of primary importance to adhesion are reactivity, surface characteristics, and finish. Rayon has many reactive hydroxyl groups. Nylon is less reactive but contains highly polar amide linkages. Polyester is relatively inert. Thus an adhesive system must be designed for each type of fiber. Regardless of the fiber, each adhesive system must conform to a rigid set of requirements:

  1) Rapid rate of adhesion formation;
  2) Compatibility with many types of compounds;
  3) No adverse effect on cord properties;

4) Heat resistance;

5) Aging resistance;

6) Good tack;

7) Mechanical stability.

The adhesive bond between the rubber and cord is achieved during the tire vulcanization cycle. The rate of adhesive formation should give maximum adhesion at the point of pressure release in the cure cycle.

12.5.12 Rubber Compounding

The principles of compounding were reviewed earlier in this text and cover the fundamental characteristics of polymers, filler systems, and the basics of vulcanization in the context of compound development for tire applications.

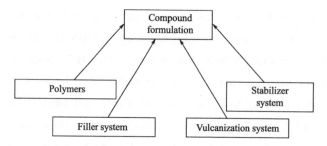

Fig. 12.21 Basic compound components.

A compound formulation consists of four basic components: the polymer network, the filler or particulate reinforcing system, the stabilizer system, and the vulcanization system (Fig. 12.21). In addition a series of secondary materials such as resins, processing oils, and short fiber reinforcements may be included in a formula.

Elastomers used in radial tires are basically of four types:

1) Natural rubber;

2) Styrene-butadiene copolymer;

3) Polybutadiene;

4) Butyl rubber (polyisobutylene with approximately 2% isoprene) and halogenated butyl rubber.

Carbon blacks, clays, and silicas constitute the filler or compound reinforcement system. Optimization of these materials in a formulation depends on the application for which the component is intended, e.g., tread, sidewall. The stabilizer system or antioxidant system protects the compounds from aging and oxidation and improves the long-term durability of the tire.

Design of the vulcanization system is probably the most challenging aspect in designing a compound formulation for application in a tire. Knowledge of accelerator activity, reaction kinetics, and nature of the resulting crosslink network is important in constructing such systems.

Optimizing rolling resistance in the elastomer material is a key challenge for reducing fuel consumption in the transportation sector. It is estimated that passenger vehicles consume approximately 5%~15% of its fuel to overcome rolling resistance, while the estimate is understood to be higher for heavy trucks. However, there is a trade-off between rolling resistance and wet traction and grip: while low rolling resistance can be achieved by reducing the viscoelastic properties of the rubber compound (low tangent δ), it comes at the cost of wet traction and grip, which requires hysteresis and energy dissipation (high tangent δ). A low tangent δ value at 60℃ is used as an indicator of low rolling resistance, while a high tangent δ value at 0℃ is used as an indicator of high wet traction. Designing an elastomer material that can achieve both high wet traction and low rolling resistance is key in achieving safety and fuel efficiency in the transportation sector.

The most common elastomer material used today is a styrene-butadiene copolymer. It combines the properties of polybutadiene, which is a highly rubbery polymer ($T_g = -100℃$) having high hysteresis and thus offering good wet grip properties, with the properties of polystyrene, which is a glassy polymer ($T_g = 100℃$) having low hysteresis and thus offering low rolling resistance in addition to wear resistance. Therefore, the ratio the two monomers in the styrene-butadiene copolymer is considered key in determining the glass transition temperature of the material, which is correlated to its grip and resistance properties.

Materials science research efforts are underway to improve such properties of elastomers. For instance, this involves modifying the microstructure of the copolymer (for instance, using solution styrene butadiene rubber (S-SBR) to control the addition of vinyl butadiene units) as well as the macrostructure of the polymer [such as the width of molecular mass distribution (MWD)]. Current investigation also involves looking at the functionalization of the elastomer through the addition of filler materials such as silica and carbon black, as well as testing other nano-fillers such as nanocellulose crystals, carbon nanotubes, and graphene. Trialkoxymercaptoalkyl-silanes are a class of silane bonding agents that offer advantages in reduced rolling resistance and emission of volatile substances.

12.5.13 Tire building

Building is the process of assembling all the components onto a tire building drum. Tire-building machines (TBM) can be manually operated or fully automatic. Typical TBM operations include the first-stage operation, where inner liner, body plies, and sidewalls are wrapped around the drum, the beads are placed, and the assembly turned up over the bead. In the second stage operation, the carcass of the tire is inflated, then the belt package and tread are applied.

All components require splicing. Inner liner and body plies are spliced with a square-ended overlap. Tread and sidewall are joined with a skived splice, where the joining ends are bevel-cut. Belts are spliced end to end with no overlap. Splices that are too heavy or non-symmetrical will generate defects in force variation, balance, or bulge parameters. Splices that are too light or open can lead to visual defects and in some cases tire failure. The final product of the TBM process is called a green tire, where green refers to the uncured state.

Pirelli Tire developed a special process called MIRS that uses robots to position and rotate the building drums under stations that apply the various components, usually via extrusion and strip winding methods. This permits the equipment to build different tire sizes in consecutive operations without the need to change tooling or setups. This process is well suited to small volume production with frequent size changes.

The largest tire makers have internally developed automated tire-assembly machines in an effort to create competitive advantages in tire construction precision, high production yield, and reduced labor. Nevertheless, there is a large base of machine builders who produce tire-building machines.

12.5.14 Tire curing

Curing is the process of applying pressure to the green tire in a mold in order to give it its final shape, and applying heat energy to stimulate the chemical reaction between the rubber and other materials. In this process the green tire is automatically transferred onto the lower mold bead seat, a rubber bladder is inserted into the green tire, and the mold closes while the bladder inflates. As the mold closes and is locked the bladder pressure increases so as to make the green tire flow into the mold, taking on the tread pattern and sidewall lettering engraved into the mold. The bladder is filled with a recirculating heat transfer medium, such as steam, hot water, or inert gas. Temperatures are in the area of 180℃ with pressures around 2.5 MPa. Passenger tires cure in approximately 16 minutes. At the end of cure the pressure is bled down, the mold opened, and the tire stripped out of the mold. The tire may be placed on a PCI, or post-cure inflator, that will hold the tire fully inflated while it cools. There are two generic curing press types, mechanical and hydraulic. Mechanical presses hold the mold closed via toggle linkages, while hydraulic presses use hydraulic oil as the prime mover for machine motion, and lock the mold with a breech-lock mechanism. Hydraulic presses have emerged as the most cost-effective because the press structure does not have to withstand the mold-opening pressure and can therefore be relatively lightweight. There are two generic mold types, two-piece molds and segmental molds. Large off-road tires are often cured in ovens with cure times approaching 24 hours.

12.6 Tire Testing

Having developed a tire construction for a defined service requirement, the tire engineer must now subject the design to a series of test. This testing falls into two performance categories:

(1) Mechanical characteristics of the tire

• Load deflection of a vertically loaded mounted tire, load-carrying capacity, and load rating;

• Steering properties such as aligning torque, cornering characteristics, and tire lateral and tangential stiffness;

• Traction and wet skid performance;

• Rolling resistance, which affects vehicle fuel economy;

• Noise.

(2) Durability

• Tread wear, which encompasses slow wear rates, fast wear rate, and uniformity of wear;

• Casing fatigue resistance;

• Tire heat buildup under loaded dynamic conditions;

• Chipping, cutting, and tearing resistance of the tread and sidewall.

This testing is broken down into three phases: tire laboratory testing, proving grounds, and commercial evaluation.

12.6.1 Laboratory Testing

Laboratory testing of tires is preceded by testing of raw materials. For compounds such as the tread, sidewall, wire coat compound, and liner, such tests would include determination of the kinetics of vulcanization, tensile strength, tear strength, resilience, and dynamic properties (e.g., storage modulus and loss modulus). Reinforcements such as the ply cord are similarly tested for tensile strength, but also creep behavior, stability (shrink behavior), and fatigue resistance. Much of the physical testing of compounds in a modern laboratory is done by robotics. Compounds are vulcanized under defined conditions, and samples are cut out depending on the type of testing to be done. Robots then load the samples onto test equipment, and the data generated are collected by computer for the test engineer to access.

Materials that meet the appropriate physical property targets are then used in tire building. Testing of these tires depends on the application for which the tire was designed. It is thus appropriate to introduce another series of definitions:

① Balance   Weight distribution around the circumference of a tire; poor balance is due to components having irregular dimensions. Balance can also be affected by irregular component splice widths and poor application of component parts.

② Conicity   Tendency of a tire to pull a vehicle to one side or another; it is caused by off-centered or misplaced components during the tire building process.

③ Cornering coefficient   Lateral force divided by the vertical load at a defined slip angle. Stiffer tread compounds would tend to improved cornering coefficient.

④ Force variation   Periodic variation in normal vertical force of a loaded free-rolling tire, which repeats with every revolution.

⑤ Harmonic   Periodic or rhythmic force variations occurring in a sinusoidal manner around tire. One phase is described as the 1st harmonic. When two phases are noted, it is described as a 2nd harmonic. Lateral force variation 1st harmonic is typically due to a tread splice. Radial harmonic may be due to irregular placement of the belt layup.

⑥ Lateral force   Side force that is exerted by a tire as it rotates under a load.

⑦ Lateral force coefficient   The lateral force divided by the vertical load.

⑧ Lateral force variation   Change in force from one side of the tire to the other as it rotates under a load. It may cause the tire to wobble and is due to irregular tire component dimensions. Lateral force variation is a summation of the lateral 1st, 2nd, 3rd, etc., harmonic.

⑨ Lateral runout   Difference between the maximum and minimum measurements parallel to the spin axis at the widest point of each tire sidewall when the tire is mounted on a wheel.

⑩ Radial runout   Difference between the maximum and minimum measurements on the tread surface and in a plane perpendicular to the spin axis while the tire is mounted on a wheel. It is a measure of the out-of-roundness of the tire. It can also be termed "Centerline Runout."

⑪ Radial force   Force acting on a tire perpendicular to the centerline of rotation or direction of axle. It is caused by heavy tire component splices and will increase radial force. Soft spots in the tire, such as those due to stretched ply cords, cause a decrease in radial force.

⑫ Radial force variation   Summation of the radial 1st, 2nd, 3rd, etc., harmonic. It is the change in radial force as the tire is rotated. Radial force variation will cause the vehicle to have a rough ride (as if on a poor surfaced road).

⑬ Rolling resistance   Resistance of a tire to rolling. It has a direct impact on vehicle fuel economy and is influenced most by compound hysteretic properties.

⑭ Runout   Differential between the maximum and minimum lateral or radial forces.

⑮ Self-aligning torque   Stabilizing reaction to slip angle that helps the tire and vehicle to return to neutral conditions at the completion of a maneuver.

⑯ Slip angle   Angle between the vehicle's direction of travel and the direction in which the front wheels are pointing.

⑰ Speed rating   Alphabetic ratings that define the design speed capability of the tire. The letter is incorporated into the size description of the tire, for example a 195/75SR14 has a speed rating of "S". Tables of speed ratings and corresponding alphabetic designations are published by the Tire & Rim Association (Table 12.11).

⑱ Uniformity   Measure of the tire's ability to run smoothly and vibration free; sometimes measured as tire balance, radial force variation, or lateral force variation.

Table 12.11 Tire Speed Ratings

Speed symbol	Maximum speed/kph	Typical application
D	50	Farm tractors
L	120	Commercial truck
M	130	Commercial truck
S	180	Passenger cars & light trucks
T	190	Passenger
H	210	Luxury passenger cars
V	240	High-performance cars
Z	270	High-performance sports cars
W	Above 270	Super-high-performance cars

The laboratory equipment designed to test the aforementioned tire properties are based mostly on a steel flywheel with the appropriate monitoring devices such as transducers and infrared temperature monitors. Data are collected directly into computers for real-time analysis and downloading to the tire engineer's work station. Many of these flywheels are also computer controlled so as to simulate service conditions. For example, aircraft tires can undergo a complete cycle of taxi, takeoff, and landing.

12.6.2 Proving Grounds

The most definitive method of determining the behavior of a tire is to examine its performance when subjected to road testing. Proving ground testing allows all types of tires such as passenger car, truck, earthmover, and farm to be tested under closely monitored conditions. An industry proving ground will generally have the following test tracks and road courses available:

(1) High-speed tracks, either circular or oval;
(2) Interstate highway simulation;
(3) Gravel and unimproved roads;
(4) Cobblestone;
(5) Cutting, chipping, and tearing courses;
(6) Wet and dry skid pads, serpentine and slalom courses for esthetics, and handling tests;
(7) Tethered tracks for farm tire durability;
(8) Glass roads for footprint monitoring.

For example the 30km^2 Goodyear Proving Grounds in Texas contain a 9km high-speed circle, 13km of simulated interstate highway, gravel and rock courses for a range of tire type testing, skid pads with spray equipment, and a glass road facility for tire footprint observations and evaluation of water dispersion.

12.6.3 Commercial Evaluation

When a new tire design meets the performance targets identified in the laboratory and proving grounds, commercial evaluation is the next stage in the product development cycle. For truck tires for highway service, a quantity of tires will be placed with a commercial fleet, and their performance will be monitored continuously. Data collected from such tests include tread wear, uniformity of wear, casing durability, and driver assessment. Tires are placed with a range of commercial fleets in both short-haul and long-distance service. The collected data are then entered into computers for detailed analysis, calculation of regression equations, and performance evaluation. Customer input is one of the key parameters in such studies.

12.7　Tire Manufacturing

The manufacture of tires consists of six basic processes:

1) Mixing elastomers, carbon blacks, and chemicals to form the rubber compound;

2) Processing the fabrics and steel cord and coating them with rubber at the calendaring operation;

3) Extruding treads, sidewalls, and other rubber components;

4) Assembling components on the tire building machine;

5) Curing the tire under heat and pressure;

6) Finishing, making the final inspection, and shipping.

The processes involved in the tire manufacturing operation are illustrated in Fig. 12.22.

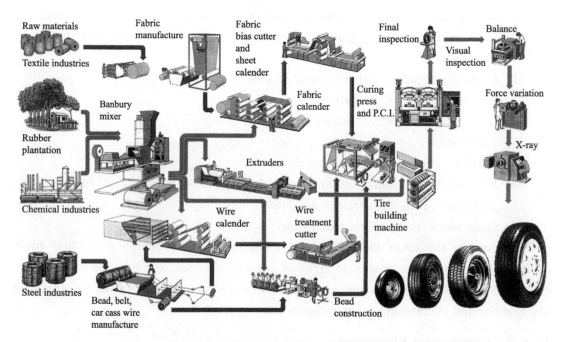

Fig. 12.22　The tire manufacturing process.

12.7.1 Compound Processing

Polymers are first broken down in an internal mixer where, in addition to the polymer, a peptizer may also be added. This stage is essentially a polymer molecular mass reduction phase. After initial breakdown of the polymer, carbon black, rubber chemicals, and oils can be added to the polymer at intervals to complete the compound formulation. Polymer breakdown and mixing generally occur at a high temperature, up to 180℃. Compounds may also be mixed on open mills, but this takes considerably more time, batch weights are lower, and it is thus less efficient than use of internal mixers.

Degree of breakdown with both types of equipment is dependent on the "friction ratio" or the difference between the operating speeds of the front and back rolls (or rotors, for internal mixers). In addition, clearance, conditions of the rotor surfaces, pressure, and speed influence breakdown.

The mixing operation is designed to obtain uniform dispersion of all the compounding materials in a formulation. For every batch there is a defined mix period, temperature, mill or mixing speed, and sequence of material addition. Though the general guidelines on compound preparation for both mill mixing and internal mixers are similar, mill mixing has been replaced by internal mixers because of efficiency, automation, quality, and uniformity. For example, internal mixers can be computer controlled, allowing monitoring of power consumption, mix times, and batch drop temperatures and control of temperature gradients through a mixing compound at any point in the mix cycle.

The sequence of addition in an internal mixer is typically

1) polymers and peptizer;
2) plasticizers, most carbon black or silica, and oils;
3) balance of fillers and antioxidants;
4) vulcanization system components.

Construction of a mix cycle is governed by a set of empirical rules:

1) Keep high-tack resins separate from dry powders;
2) Hold batch drop temperatures above the softening point of hard resins;
3) Contain liquids to prevent leakage;
4) Make use of the shear properties of rubber to accelerating mixing;
5) Avoid scorch and subsequent formation of cured particles and crumb.

These five guidelines are adhered to by reducing internal friction through the use of plasticizers, adequate breakdown of the rubber, holding of the curing system ingredients to the final stage of the mix cycle, and use of master batches where possible. After mixing, compounds can either be sheeted and water cooled or passed into pelletizer and then air cooled.

12.7.2 Calendaring

Calendaring is the forming operation in which the rubber compound is sheeted or spread evenly onto fabric. The calendar is a heavy-duty machine equipped with three or more chrome-plated steel rolls that revolve in opposite directions (Fig. 12. 23). The rolls are heated with steam or circulated water; the gearing allows the rollers to operate at variable speeds like the mill rolls.

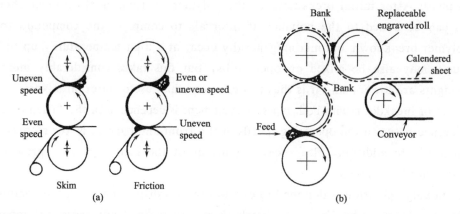

Fig. 12. 23 (a) Applying rubber to fabrics. (b) Profiling by means of a four-roll engraving calendar.

Fabric or wire is passed through the calendar rolls, and compound is applied above and below to fully cover the material. The amount of compound deposited onto a fabric or steel cord is determined by the distance between rollers and is monitored by Beta Gauges. Each cord is insulated on all sides with rubber. The finished treatment or calendared steel cord sheet is cut to length and angle for tire building. Key compound requirements in calendaring operations are minimum shrinkage, optimum tack for component-to-component adhesion, and sufficient scorch resistance to enable long dwell times at high processing temperatures.

12. 7. 3　Extrusion

Many of the components of a tire, such as tread, sidewall, and apex, are formed by extrusion of the uncured or "green" rubber. Extruders in a tire manufacturing operation are conventional screw-types systems that fall into two categories:

1) Hot feed systems Strips of green compound are fed from warmup mills into a extruder feed box and then into the barrel.

2) Cold feed systems Mixed compound is fed directly into the extruder system.

Extruders consist of a barrel into which the material is fed. The mixed compound is then pushed forward by a specially designed screw to a filter system, where any foreign material is removed, and then onto a die, where the required compound profile is produced.

The extruder screw consists of three primary zones: feed zone, compression zone, and metering zone. In the feed zone located under the extruder feed box, the screw flights are well spaced to optimize compound flow. The rubber is passed into a compression zone, where the compound is heated through shear; then into a metering zone, where the compound is further heated to reduce viscosity; and finally into the die for profile formation.

Hot feed extruder systems (Fig. 12.24) normally have a short barrel and screw with short compression and metering zones because the compound is already hot and has a low viscosity as it enters the feed hopper. Barrel length-to-diameter ratios from 4 : 1 to 6 : 1 are typical.

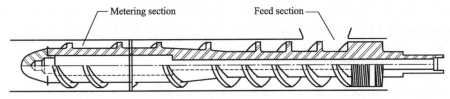

Fig. 12.24 Hot feed extruder.

Cold feed extruders (Fig. 12.25) have much larger length-to-diameter ratios because of the requirements of reducing green compound plasticity, heat buildup in the compound, and pressure buildup required to produce the extruded profile. The length-to-diameter ratio is typically 24 : 1. The modern cold feed extruder is also computer controlled, which enables adjustment of the compound temperature profile through the barrel, pressure control, flow rate, and feed rate. This provides accurate control over die swell, extrudate surface quality, and buildability of the extruded product. Periodic monitoring of screw flight-to-barrel distances are important to ensure minimum back pressure and compound swirl or turbulence in the flow channel.

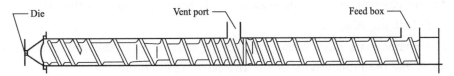

Fig. 12.25 Cold feed extruder.

12.7.4 Tire Building

All tire components, i.e., extruded parts, calendared plies and belts, and beads, are assembled at the tire-building machine. Traditionally, radial tires are built on a flat drum. The initial stage entails application of the inner liner on the cylindrical drum, followed by the ply and any other additional barrier components that the tire engineer has designed into the manufacturing specification. The beads are then positioned and the ply is turned up over them. The structure is transferred to a second-stage machine, where the belts and tread are applied. The building drum is collapsed, and the "green tire" illustrated in Fig. 12.26 is transferred to the curing or vulcaniza-

Fig. 12.26 Radial green tire.

tion presses.

During vulcanization, or curing, the green tire is molded into a high-quality engineered product. In the curing press, the compound flows into the mold shape to give a design to the tread and the desired thickness to the side-wall. Good flow depends on the plasticity of the uncured stock. To flow, the compound must resist scorching. Flow must be complete before cure begins, or distortion may result. Proper flow is achieved by effective use of compounding fillers, adjustment of acceleration levels, and use of plasticizers.

12.7.5 Final Tire Inspection

After the tire has been cured, there are several additional operations. Tire uniformity measurement is a test where the tire is automatically mounted on wheel halves, inflated, run against a simulated road surface, and measured for force variation. Tire balance measurement is a test where the tire is automatically placed on wheel halves, rotated at a high speed and measured for imbalance.

Large commercial truck/bus tires, as well as some passenger and light truck tires, are inspected by X-ray or magnetic induction based inspection machines, that can penetrate the rubber to analyze the steel cord structure.

In the final step, tires are inspected by human eyes for numerous visual defects such as incomplete mold fill, exposed cords, blisters, blemishes, and others. The manufactured tire must meet the following criteria:

1) All tire components have been successfully processed through the production facility to the tire building machine.

2) The components have come together satisfactorily at the tire building machine without hindering the productivity of the tire building operators.

3) The tire meets the quality and performance goals set by the consumer.

Quality assurance is the last stage in the manufacturing cycle. Here the product is checked to ensure customer satisfaction. Quality checks on a tire include:

1) Buffing and trimming off mold flash from the tire.

2) Visual inspection of each tire for defects.

3) X-ray checks on tires for ply cord spacing and belt layup.

4) Statistical sampling of tires for durability testing, uniformity, and dynamic balancing. This testing includes many of the development tests reviewed earlier such as conicity, radial runout, and lateral force variation.

After the tire passes the quality checks, it is ready for shipment to the ware house for distribution.

12.8 Summary

In many ways, a tire is an engineering marvel. Geometrically, a tire is a torus. Mechanically, a tire is a flexible, high-pressure container. Structurally, a tire is a high-performance composite built using elastomers, fibers, steel, and a range of organic and inorganic chemicals.

Tire technology is a complex combination of science and engineering that brings together a variety of disciplines. In the development of a tire, knowledge in the areas of tire geometry, dynamic tire behavior, chemistry of component materials, and technology of composite structures is essential. The result is a broad range of products which satisfy vehicle manufacturers as well as end-consumer needs for optimum performance under a variety of service conditions.

Tires that are fully worn can be re-manufactured to replace the worn tread. This is known as retreading or recapping, a process of buffing away the worn tread and applying a new tread. Retreading is economical for truck tires because the cost of replacing the tread is less than the price of a new tire. Retreading passenger tires is less economical because the cost of retreading is high compared to the price of new cheap tires, but favorable compared to high-end brands.

Worn tires can be retreaded by two methods, the mold or hot cure method and the pre-cure or cold one. The mold cure method involves the application of raw rubber on the previously buffed and prepared casing, which is later cured in matrices. During the curing period, vulcanization takes place and the raw rubber bonds to the casing, taking the tread shape of the matrix. On the other hand, the pre-cure method involves the application of a ready-made tread band on the buffed and prepared casing, which later is cured in an autoclave so that vulcanization can occur. During the retreading process, retread technicians must ensure the casing is in the best condition possible to minimize the possibility of a casing failure. Casings with problems such as capped tread, tread separation, irreparable cuts, corroded belts or sidewall damage, or any run-flat or skidded tires, will be rejected.

In most situations, retread tires can be driven under the same conditions and at the same speeds as new tires with no loss in safety or comfort. The percentage of retread failures should be about the same as for new tire failures, but many drivers, including truckers, are guilty of not maintaining proper air pressure on a regular basis, and, if a tire is abused (overloaded, underinflated, or mismatched to the other tire on a set of duals), then that tire (new or recapped) will fail. Many commercial trucking companies put retreads only on trailers, using only new tires on their steering and drive wheels. This procedure increases the driver's chance of maintaining control in case of problems with a retreaded tire.

Once tires are discarded, they are considered scrap tires. Scrap tires are often re-used for things from bumper car barriers to weights to hold down tarps. Some facilities are permitted to recycle scrap tires through chipping, and processing into new products, or selling the material to licensed power plants for fuel. Some tires may also be retreaded for reuse. One group did "a study to evaluate the possibility of using scrap tires as a crash cushion system. The objective of this study was to evaluate the material properties of used tires and recycled tire-derived materials for use in low-cost, reusable crash cushions".

Americans generate about 285 million scrap tires per year. Many states have regulations as to the number of scrap tires that can be held on site, due to concerns with dumping, fire hazards, and mosquitoes. In the past, millions of tires have been discarded into open fields. This creates a breeding ground for mosquitoes, since the tires often hold water inside and remain warm enough for mosquito breeding. Mosquitoes create a nuisance and may increase the likelihood of spreading disease. It also creates a fire danger, since such a large tire pile is a lot of fuel. Some tire fires have burned for months, since water does not adequately penetrate or cool the burning tires. Tires have been known to liquefy, releasing hydrocarbons and other contaminants to the ground and even ground water, under extreme heat and temperatures from a fire. The black smoke from a tire fire causes air pollution and is a hazard to down wind properties.

The use of scrap tire chips for landscaping has become controversial, due to the leaching of metals and other contaminants from the tire pieces. Zinc is concentrated (up to 2% by weight) to levels high enough to be highly toxic to aquatic life and plants. Of particular concern is evidence that some of the compounds that leach from tires into water, contain hormone disruptors and cause liver lesions.

References

[1] T. L. Ford, F. S. Charles, "Heavy Duty Truck Tire Engineering," SAE's 34th L. Ray Buckingdale Lecture, SP729, 1988.

[2] J. A. Davison, "Design and Application of Commercial Type Tires," SAE's 15th L. Ray Buckingdale Lecture, SP344, 1969.

[3] The Tire & Rim Association, Inc. Year Book, 175 Montrose West Ave., Suite 150, Copley, OH 44321.

[4] European Tyre and Rim Technical Organization, Standards Manual, 32/2 Avenue Brugmann-B1060 Brussels, Belgium.

[5] Engineering Data Book, "Over-The-Road Truck Tires," The Goodyear Tire & Rubber Company, 2001.

[6] F. J. Kovac, "Tire Technology," 5th ed., The Goodyear Tire & Rubber Company, 1978.

[7] V. E. Gough, Rubber Chem. Technol. 44, 988 (1968).

[8] S. K. Clark, "Mechanics of Pneumatic Tires," U. S. Department of Transportation, Washington, D. C., 1982.

[9] J. E. Purdy, "Mathematics Underlying the Design of Pneumatic Tires," 2nd ed., Hiney Printing, Akron, OH, 1970.

[10] J. W. Fitch, "Motor Truck Engineering Handbook," 4th ed., Society of Automotive Engineers, Warrendale, PA, 1994.

[11] W. H. Waddell, M. B. Rodgers, and D. S. Tracey, "Tire Applications of Elastomers, Part 1. Treads," Rubber Division Meeting, Grand Rapids, MI, Paper H, 2004.

[12] Bekaert Corporation, Steel Cord Catalogue, Akron, OH, 1991.

[13] W. J. van Ooij, Rubber Chem. Technol. 57, 421 (1984).

[14] J. Duddey, in "Rubber Compounding, Chemistry and Applications," Marcel Dekker Inc., New York, 2004.

[15] B. Rodgers and W. H. Waddell, Rubber Compounding, in "Kirk-Othmer Encyclopedia of Chemical Technology," 5th ed., John Wiley & Sons, New York, 2004.

[16] H. Long, "Basic Compounding and Processing of Rubber," Rubber Division, American Chemical Society, Lancaster Press, Lancaster, PA, 1985.

[17] K. C. Baranwal and H. S. Stephens, "Basic Elastomer Technology," Rubber Division, American Chemical Society, Akron, OH, 2001.

Chapter 13
Elastomer Recycling

Tire recycling, or rubber recycling, is the process of recycling waste tires that are no longer suitable for use on vehicles due to wear or irreparable damage. These tires are a challenging source of waste, due to the large volume produced, the durability of the tires, and the components in the tire that are ecologically problematic. Because tires are highly durable and non-biodegradable, they can consume valued space in landfills. In 1990, it was estimated that over 1 billion scrap tires were in stockpiles in the United States. As of 2015, only 67 million tires remain in stockpiles. From 1994 to 2010, the European Union increased the amount of tires recycled from 25% of annual discards to nearly 95%, with roughly half of the end-of-life tires used for energy, mostly in cement manufacturing. Newer technology, such as pyrolysis and devulcanization, has made tires suitable targets for recycling despite their bulk and resilience. Aside from use as fuel, the main end use for tires remains ground rubber.

Due to their heavy metal and other pollutant content, tires pose a risk for the leaching of toxins into the groundwater when placed in wet soils. Research has shown that very little leaching occurs when shredded tires are used as light fill material; however, limitations have been put on use of this material; each site should be individually assessed determining if this product is appropriate for given conditions.

For both above and below water table applications, the preponderance of evidence shows that TDA (tire derived aggregate, or shredded tires) will not cause primary drinking water standards to be exceeded for metals. Moreover, TDA is unlikely to increase levels of metals with primary drinking water standards above naturally occurring background levels.

13.1 Introduction

Manufacturing of tire and other rubber products involves vulcanization process, an irreversible reaction between the elastomer, sulfur, and other chemicals producing crosslinks between the elastomer molecular chains and leading to the formation of a three-dimensional chemical network. The crosslinked elastomers are solid, insoluble, and infusible thermoset materials. This makes the direct reprocessing and recycling of used tires and waste rubbers impossible. Therefore, the environmental problems caused by used tires and other waste rubber products have become serious in recent years. In fact, Charles Goodyear, who invented the sulfur vulcanization process more than 150 years ago, was also the first to initiate efforts to recycle cured rubber wastes through a grinding method. Even after so many years of efforts in recycling, the development of a suitable technology to utilize waste rubbers is an important issue facing the rubber industry.

According to a recent survey of the Scrap Tire Management Council of the Rubber Manufacturers Association, approximately 281 million scrap tires were generated in the United States in 2001 alone. The market for scrap tires is consuming about 77.6% of that total amount while the rest is added to an existing stockpile of an estimated 300 million scrap tires located around the United States. These stockpiled tires create serious fire dangers and provide breeding grounds for rodents, snakes, mosquitoes, and other pests, causing health hazards and environmental problems. Moreover, the major use of scrap tires in the United States is to generate the so-called tire-derived energy by burning used tires. However, burning tires may create a danger of air pollution. About 53% of the consumed scrap tires were burned in 2001, and only 19% of the total consumed amount was turned into ground tire rubber (GRT), which is the initial material for the tire rubber recycling processes. Also, the management of other waste rubbers has become a growing problem in the rubber industry since over 150000 tons or more of rubber are scrapped from the production of non-tire goods in the form of runners, trim, and pads.

Waste tires and rubbers, being made of high quality rubbers, represent a large potential source of raw material for the rubber industry. The main reasons for the low-scale current application of tire and rubber recycling include the following: more stringent requirements for quality of rubber articles, and hence for that of reclaimed rubber; the substitution of other materials for raw rubber (e.g., plastics in some cases); rising costs of recovery from tires and rubber waste due to the more stringent regulations for environmental protection; a comparatively high labor input into reclaim production; and, as a result of all this, the high cost of reclaimed rubber. However, the increasing legislation restricting land fills is demanding the search for economical and environmentally sound methods of recycling discarded tires and waste rubbers. Recent aggressive policies of the automotive industry are aimed at increasing the use of recycled plastic and rubber materials. This may serve as an example of the growing industrial demand for such technologies.

The main objective of this chapter is to provide the up-to-date account on recycling of used tires and waste rubbers including existing methods and emerging technologies of grinding, reclaiming, and devulcanization, and also the possibility for recycled rubber utilization into products. Rubber devulcanization is a process in which the scrap rubber or vulcanized waste product is converted, using mechanical, thermal, or chemical energy, into the state in which it can be mixed, processed, and vulcanized again. Strictly speaking, devulcanization in sulfur-cured rubber can be defined as the process of cleaving, totally or partially, poly-, di-, and monosulfidic crosslinks formed during the initial vulcanization. Devulcanization of peroxide-and resin-cured rubber can be defined as the process of cleaving carbon-carbon or other, stronger crosslinks. However, in the present concept, devulcanization is defined as a process that causes the breakup of the chemical network along with the breakup of the macromolecular chains.

A number of methods have been applied in an attempt to solve the problem and to find more effective ways of tire rubber recycling and waste rubber utilization. These methods include retreading, reclaiming, grinding, pulverization, microwave and ultrasonic processes, pyrolysis, and incineration.

Processes for utilization of recycled rubber are also being developed, including the use of reclaimed rubber to manufacture rubber products and thermoplastic rubber blends, and the use of GRT to modify asphalt and cement.

13.2　Retreading of Tire

Tires that are fully worn can be re-manufactured to replace the worn tread. This is known as retreading or recapping, a process of buffing away the worn tread and applying a new tread. Retreading is economical for truck tires because the cost of replacing the tread is less than the price of a new tire. Retreading passenger tires is less economical because the cost of retreading is high compared to the price of new cheap tires, but favorable compared to high-end brands.

Worn tires can be retreaded by two methods, the mold or hot cure method and the pre-cure or cold one. The mold cure method involves the application of raw rubber on the previously buffed and prepared casing, which is later cured in matrices. During the curing period, vulcanization takes place and the raw rubber bonds to the casing, taking the tread shape of the matrix. On the other hand, the pre-cure method involves the application of a ready-made tread band on the buffed and prepared casing, which later is cured in an autoclave so that vulcanization can occur.

During the retreading process, retread technicians must ensure the casing is in the best condition possible to minimize the possibility of a casing failure. Casings with problems such as capped tread, tread separation, irreparable cuts, corroded belts or sidewall damage, or any run-flat or skidded tires, will be rejected.

In most situations, retread tires can be driven under the same conditions and at the same speeds as new tires with no loss in safety or comfort. The percentage of retread failures should be about the same as for new tire failures, but many drivers, including truckers, are guilty of not maintaining proper air pressure on a regular basis, and, if a tire is abused (overloaded, under-inflated, or mismatched to the other tire on a set of duals), then that tire (new or recapped) will fail. Many commercial trucking companies put retreads only on trailers, using only new tires on their steering and drive wheels. This procedure increases the driver's chance of maintaining control in case of problems with a retreaded tire.

Retreading is one way to recycle. Also, it saves energy. It takes about 83 liters of oil to manufacture one new truck tire whereas a retread tire requires only about 26 liters. The cost of a retread tire can be from 30% to 50% less than new tire. Approximately 24.2 million retreaded tires were sold in North America in 2001, with sales totaling more than $2 billion. Mostly medium-and heavy-duty truck tires, off-the-road vehicles, and aircraft tires were retreaded. However, high labor costs and the potential for tougher safety regulations may hurt the retreading business.

Although tires are usually burnt, not recycled, efforts are continuing to find value. Tires can be reclaimed into, among other things, the hot melt asphalt, typically as crumb rubber modifier-recycled asphalt pavement (CRM-RAP), and as an aggregate in portland cement concrete. Efforts have been made to use recycled tires as raw material for new tires, but such tires may integrate recycled materials no more than 5% by weight, and tires that contain recycled material are inferior to new tires, suffering from reduced tread life and lower traction. Tires have also been cut up and used in garden beds as bark mulch to hold in the water and to prevent weeds from growing. Some "green" buildings, both private and public, have been made from old tires.

13.3 Recycling of Rubber Vulcanizates

13.3.1 Reclaiming Technology

Reclaiming is a procedure in which the scrap tire rubber or vulcanized rubber waste is converted, using mechanical and thermal energy and chemicals, into a state in which it can be mixed, processed, and vulcanized again. The principle of the process is devulcanization. In devulcanization, it is assumed that the cleavage of intermolecular bonds of the chemical network, such as carbon-sulfur and/or sulfur-sulfur bonds, takes place, with further shortening of the chains occuring.

Many different reclaiming processes have been applied through the years in an attempt to solve the problem of rubber recycling. Generally, ground rubber scrap is, in most cases, the feedstock for the devulcanization step. It has presented reviews of the existing literature that is relevant to various methods of devulcanization. Reclaiming is the most important process in rubber recycling. Many different reclaiming processes have been used through the years depending on scrap characteristics and economics. Generally, ground rubber scrap is, in most cases, the feedstock for the reclaiming. The pan process, digester process (either wet or dry), and mechanical or reclaimator processes are currently the common processes used for reclaiming.

The digester process uses a steam vessel equipped with a paddle agitator for continuous stirring of the crumb rubber while steam is being applied. The wet process may use caustic and water mixed with the rubber crumb, while the dry process uses steam only. If necessary, various reclaiming oils may be added to the mixer in the vessel. The dry digester has the advantage of less pollution being generated and was adopted after the Clean Air and

Water Act was enacted.

A mechanical or reclaimator process has been used for the continuous reclaiming of whole tire scrap. Fine rubber crumb (typically 30 mesh) mixed with various reclaiming oils is subjected to high temperature with intense mechanical working in a modified extruder for reclaiming the rubber scrap.

Scrap rubber containing natural and synthetic rubbers can be reclaimed by digester process with the use of reclaiming oil having molecular weight between 200 and 1000 and consisting of benzene, alkyl benzene, and alkylate indanes. The composition of this reclaiming oil and the improved digester process using such reclaiming oil has been patented.

Recently, a new technology for the devulcanization of sulfur-cured scrap elastomers was reported using a material termed "Delink". Such technique of devulcanization was designated as Delink process. In this process, 100 parts of 40-mesh or finer crumb is mixed with 2 to 6 parts of Delink reactant in an open two-roll mixing mill. Delink reactant is a proprietary material, and its nature and composition was not disclosed.

A simple process for reclaiming of rubber with a vegetable product that is a renewable resource material (RRM) was developed. The major constituent of RRM is diallyl disulfide. Other constituents of RRM are different disulfides, monosulfides, polysulfides, and thiol compounds.

It is known that sulfur-vulcanized Natural Rubber (NR) can be completely recycled at 200℃ to 225℃ by using diphenyldisulphide. Recently, the efficacy of various disulphides as recycling agents for NR and EPDM vulcanizates were reported. While complete devulcanization was observed on sulfur-cured NR at 200℃, a decrease in crosslink density of 90% was found when EPDM sulfur vulcanizates with diphenyldisulphide were heated to 275℃ in a closed mold for 2 hours. At the same time, EPDM cured by peroxide showed a decrease in crosslink density of about 40% under the same conditions.

Another chemical method was recently proposed. It is based on the use of 2-butanol solvent as a devulcanizing agent for sulfur-cured rubber under high temperature and pressure. It is claimed that the molecular weight of the rubber is retained, and its microstructure is not significantly altered during the devulcanization process. However, the process is extremely slow and requires separation of the devulcanized rubber from the solvent.

In addition to the use of organic chemicals, rubbers can be devulcanized by means of inorganic compounds. Discarded tires and tire factory waste were devulcanized by desulfurization of suspended rubber vulcanizate crumb (10 to 30 mesh) in a solvent such as toluene, naphtha, benzene, cyclohexane, etc., in presence of sodium. The alkali metal cleaves mono-, di-, and polysulfidic crosslinks of the swollen and suspended vulcanized rubber crumb at around 300℃ in absence of oxygen. However, this process may not be economical because the process involves swelling of the vulcanized rubber crumb in an organic solvent where the metallic sodium in molten condition should reach the sulfidic crosslink sites in the rubber crumb. Also, the solvent may cause pollution and be hazardous. A technology was also proposed to reclaim powder rubbers using an iron oxide phenyl hydrazine based catalyst and a copper (I) chloride-tributyl amine catalyst.

Depending on the specification of the finished products, fillers may be added to the devulcanized product before further processing. The devulcanized rubber from each process is then strained and refined as dictated by the specification of the finished product before being powdered, baled, sheeted, or extruded into the finished form.

Chemical reclaiming process is a possible method for devulcanizing the vulcanized network through the use of chemical agents that attack the C—S or S—S bonds. However, this process of devulcanization is very slow and creates further problems with the removal of the solvents, and additional waste is generated in the form of sludges. Also, some processes require elaborate chemical process techniques, therefore handling and safety become a concern.

13.3.2 Surface Treatment

Surface treatment technology uses a solvent to treat (devulcanize) the surface of rubber crumb particles of sizes within about 20 to 325 meshes. It is a variation of earlier disclosed technology. The process is carried out at a temperature range of between 150℃ and 300℃ at a pressure of at least 3.4MPa in the presence of a solvent selected from the group consisting of alcohols and ketones. Among various solvents, the 2-butanol exhibited the best ability to devulcanize sulfur-cured styrene-butadiene rubber (SBR). Duration of the process is above 20 minutes. Reported data on surface devulcanization experiments were obtained by treating small amounts of rubber crumb in the gas chromatography column. The solvent suitable for this process should have a critical temperature in the range of about 200℃ to 350℃. The process produces slurry of the surface devulcanized rubber crumb that has to be separated from the solvent. It is claimed that in this process a preferential breakage of S—S and C—S bonds takes place with little breakage of the main chains. The obtained surface modified rubber crumb was subjected to vulcanization as obtained and also in blends with virgin rubber. The vucanizates exhibited a good retention of mechanical properties in blends with virgin rubber. However, this process has been tested on a small laboratory scale.

13.3.3 Grinding and Pulverization Technology

Use of waste rubber in a vulcanized state most often requires reduction of particle size and/or surface area. One of the most widely used methods for doing this with scrap tires and rubber wastes is a grinding process. This method was invented and put forward by Goodyear about 150 years ago. Presently, there are three methods of grinding waste rubber: ambient grinding, cryogenic grinding, and wet-ambient grinding. There are a number of ways to reduce tires. The primary reduction of whole tires down to a manageable form is done using the guillotine, the cracker mill, the high-impact hammer mill, and the rotary shear shredder. Vulcanized scrap rubber is first reduced to a 5-by-5cm^2 or 2.5-by-2.5cm^2 chip. Then a magnetic separator and a fiber separator (cyclone) remove all the steel and polyester fragments. This can then be further reduced using an ambient ground mill or ground into fine particles while frozen using cryogenic grinding.

A method for obtaining fine-mesh rubber is cooling scrap tires in liquid nitrogen below their glass transition temperature and then pulverizing the brittle material in a hammer mill. Cryogenically ground rubber has a much finer particle size, varying from 30 to 100 mesh. But for inexpensive rubbers such as tire rubbers, the process is not economical because of the amount of liquid nitrogen or other cryogen liquids needed to freeze the rubber.

However, the process may be economical for expensive rubbers such as fluorocarbon rubbers. Little or no heat is generated in the process; this results in less degradation of the rubber. In addition, the most significant feature of the process is that almost all fiber or steel is liberated from the rubber, resulting in a yield of usable product and little loss of rubber.

Because of the high cost of cryogenic size reduction at liquid nitrogen temperature, mechanical size reduction by chopping and grinding is used often. The ambient process often uses a conventional high-powered rubber mill set at close nip. The vulcanized rubber is sheared and ground into small particles.

Using this relatively inexpensive method, it is common to produce 10-to 30-mesh material and relatively large crumb. In addition, multiple grinds can be used to further reduce the particle size. Ambient grinding produces an irregular-shaped particle with many small hairlike appendages that attach to the virgin rubber matrix, producing an intimate bonded mixture. The lower particle limit for the process is the production of 40-mesh material. The process, however, generates a significant amount of heat. Excess heat can degrade the rubber, and if not cooled properly, combustion can occur upon storage.

Other suggested recycling processes include mechanical and thermomechanical methods, which only comminute the vulcanizates in rubber and do not devulcanize them. A process using a wet grinding method to achieve a crumb fineness of approximately 200 mesh has been reported. When this product, which had a much higher surface-to-mass ratio, was devulcanized, no chemicals and only minimal heating and mechanical processing were required.

Wet or solution process grinding may yield the smallest particle size, ranging from 400 to 500 mesh. The advantage of fine particle wet ground rubber is that it allows good processing, producing relatively smooth extrudates and calendered sheets.

The pulverization techniques for rubbers are also being developed based on the concept of polymer pulverization originally proposed for plastics. The process manufactures polymer powder using a twin-screw extruder imposing compressive shear on the polymer at specific temperatures that depend on the polymer. Based on this method, the solid-state shear extrusion pulverization method of rubber waste was also proposed. The obtained rubber particles were fluffy and exhibited unique elongated shape.

Recently, this process was further developed to carry out pulverization of rubbers in a single screw extruder to obtain particles varied in size from 40mm to 1700mm. A schematic diagram of the pulverization technique based on a single screw extruder is shown in Fig. 13.1(a) and (b). As indicated in Fig. 13.1(a), the extruder consists of three zones: feeding (Zone 1), compression (Zone 2), and pulverization (Zone 3). The screw is a square pitched with the compression zone having a uniform taper to create a compression ratio of 5.

The water-cooling channel is located in the barrel in order to remove the heat generated by the pulverization of rubber. Experimental studies showed that during the pulverization of vulcanized scrap rubber in the extruder, due to friction, a significant amount of heat is generated, leading to partial degradation of the rubber. The rubber granulates are fed into the hopper of the extruder and conveyed into the compression zone, where they are subjected to high compressive shear. Under simultaneous action of this compressive shear and torsion due to the screw rotation, granulates are pulverized and emerge from the pulverization zone as rubber powder having a smaller particle size. Surface oxidation of the rubber particles and initiation of agglomeration of a fraction of the produced particles may take place. The produced particles exhibit irregular shapes with rough surfaces and have a porous structure. The crosslink density and gel fraction of the particles are reduced in comparison with those of the initial rubber granulates. This would indicate the occurrence of partial devulcanization. Due to this effect, the particles obtained in this process can be molded into products after an exposure to high heat and high pressure for a period of at least 1 hour.

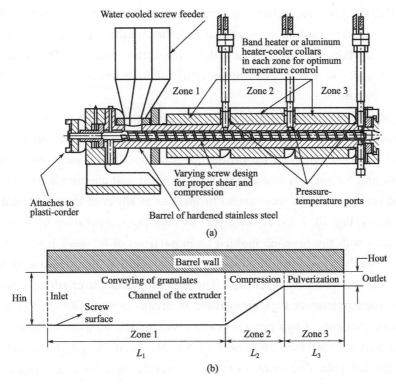

Fig. 13.1 Schematic diagram of (a) the single-screw extruder for pulverization of rubbers and (b) geometry of the screw channel with variable depth.

Table 13.1 shows the dependence of the elongation at break, ε_b, tensile strength, σ_b, and crosslink density, v, of compression-molded slabs of the original rubber vulcanizate and the vulcanizates prepared from particles of size in the range 250mm to 425mm obtained by solid-state shear extrusion pulverization from discarded by-product of natural rubber (SMR-20) vulcanizates. Approximate composition of the rubber compound was about 54wt% of SMR-20, 27wt% carbon black (SRF), 11wt% aromatic oil, and 8wt% vulcanizing ingredients. Molding temperature and pressure were 157℃ and 5.11MPa, respectively. Slab F1, produced without adding sulfur curatives, exhibited the best failure properties among all slabs produced from the rubber powder.

Table 13.1 Properties of the Slabs of Pulverized Rubber Waste

Slab code	Revulcanizing system		ε_b/%	σ_b/MPa	v/(mol/m^3)
	Sulfur(phr)	TBBS(phr)			
F1	—	—	360	10.3	50.4
F2	1.0	0.5	350	7.0	73.9
F3	1.0	—	320	8.2	69.5
Original	—	—	470	16.5	66.9

In Sample F1 oil, vulcanizing residue and the sol fraction of the rubber were removed by toluene extraction. This, according to the authors, enhanced particle bonding, leading to improvement of the failure properties. On the other hand, the slabs F2 and F3 produced by adding sulfur curatives to particles showed inferior failure properties than those of slab F1 due to less particle bonding at increased crosslink density during the revulcanization. Furthermore, the slabs F1～F3 showed failure properties inferior to the original slab indicating the inadequacy of compression molding of rubber particles to achieve the properties of the original vulcanizate.

The particles obtained by other grinding processes can also be compression molded into slabs by means of high-pressure, high-temperature sintering. In particular, rubber particles of several rubbers, obtained by various grinding methods, were compression molded into slabs with and without an addition of various acids and chemicals. The effect of time, pressure, and temperature on mechanical properties of sintered slabs was studied.

In particular, Fig. 13.2 shows the effect of molding temperature on mechanical properties of NR-SBR slab compression molded from particles of 80 mesh for 1 hour at pressure of 8.5MPa. It clearly shows the importance of the molding temperature. Below approximately 80℃, this process does not work. The highest tensile strength of about 4MPa was achieved with the sufficiently high elongation at break (about 800%). The mechanism of consolidation of particles in this process is the result of the breakup of bonds into radicals that cross the particle interface and react with other radicals and thus create a chemical bond across the interface. The authors explained that the inferior properties of the sintered NR particle slabs in comparison with the original one was due to the energetics between void propagation and strain-induced crystallization. Less energy is required to generate voids in the sintered slabs than in the original slab, and this does not allow one to achieve a strain-induced crystallization in the sintered slabs.

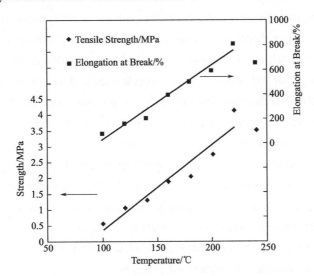

Fig. 13.2 Effect of the molding temperature on the mechanical properties of NR-SBR slab compression molded from particles of 80 meshes for 1 hour at pressure of 8.5MPa.

13.3.4 Devulcanization Technology

(1) Microwave Method

Microwave technology has also been proposed to devulcanize waste rubber. This process applies the heat very quickly and uniformly on the waste rubber. The method employs the application of a controlled amount of microwave energy to devulcanize a sulfur-vulcanized elastomer, containing polar groups or components, to a state in which it could be compounded and revulcanized to useful products, such as hoses, requiring significant physical properties. On the basis of the relative bond energies of carbon-carbon, carbon-sulfur, and sulfur-sulfur bonds, it was presumed that the scission of the sulfur-sulfur and sulfur-carbon crosslinks actually occurred. However, the material to be used in the microwave process must be polar enough to accept energy at a rate sufficient to generate the heat necessary for devulcanization. This method is a batch process and requires expensive equipment.

Recently, thermogravimetry was employed to study the changes occurring in rubber vulcanizates during devulcanization carried out by microwave treatment. The degree of degradation of the polymer chains in response to microwave treatment was obtained, allowing the establishment of conditions of devulcanization in order to obtain the best properties of rubber devulcanizates for reuse in rubber processing.

(2) Ultrasonic Method

Numerous publications in recent literature are devoted to the study of the effect of ultrasound on polymer solutions and on polymer melts during extrusion. Significant efforts have also been made to understand the mechanism of the effect of ultrasound on fluids and degradation of polymer in solution.

The application of ultrasonic waves to the process of devulcanizing rubber is an attractive field of study. Most references indicate that rubber is vulcanized by ultrasound rather than devulcanized. Rubber devulcanization by using ultrasonic energy was first discussed. It was a batch process in which a vulcanized rubber was devulcanized at 50kHz ultrasonic waves after treatment for 20 minutes. The process claimed to break down carbon-sulfur bonds and sulfur-sulfur bonds, but not carbon-carbon bonds. The properties of the revulcanized rubber were found to be very similar to those of the original vulcanizates.

A continuous process was developed for devulcanization of rubbers as a suitable way to recycle used tires and waste rubbers. This technology is based on the use of high-power ultrasounds. The ultrasonic waves of certain levels, in the presence of pressure and heat, can quickly break up the three-dimensional network in crosslinked rubber. The process of ultrasonic devulcanization is very fast, simple, efficient, and solvent and chemical free. Devulcanization occurs at the order of a second and may lead to the preferential breakage of sulfidic crosslinks in vulcanized rubbers.

The process is also suitable for decrosslinking of the peroxide-cured rubbers and plastics. A schematic diagram of the various devulcanization reactors suitable to carry out this process is shown in Fig. 13.3. Initially, the so-called coaxial devulcanization reactor [see Fig. 13.3(a)] was developed in our laboratory. The reactor consists of a single screw rubber extruder and an ultrasonic die attachment. A cone-shaped die and the ultrasonic horn have sealed inner cavities for running water for cooling. The shredded rubber is fed into the extruder by a feeder with adjustable output. Thus, the rubber flow rate in the process is controlled by the feed rate. An ultrasonic power supply, an acoustic converter, booster, and a cone-tipped horn are used. The horn vibrates longitudinally with a frequency of 20kHz and various amplitudes. The ultrasonic unit is mounted onto the extruder flange. The convex tip of the horn matches the concave surface of the die so that the clearance between the horn and the die is uniform. The clearance is controlled. The rubber flows through the clearance and under the action of ultrasonic waves, propagating perpendicular to the flow direction, and it is devulcanized. The die plate and the horn are cooled with tap water.

Later, the barrel [see Fig. 13.3(b)] and the grooved barrel [see Fig. 13.3(c)] ultrasonic reactors were developed. In the barrel reactor, two ultrasonic water-cooled horns of rectangular cross-sections were inserted into the barrel through two ports. Two restrictors made of bronze were placed in the barrel.

These restrictors blocked the flow of rubber and forced the rubber to flow through the gap created between the rotating screw and the tip of the horn. In the devulcanization section, the larger diameter provided the converging flow of the rubber to the devulcanization zone. The latter may enhance the devulcanization process. In the grooved barrel ultrasonic reactor, two helical channels were made on the barrel surface (grooved barrel). Rubber flows into the helical channel and passes through the gap created between the rota-

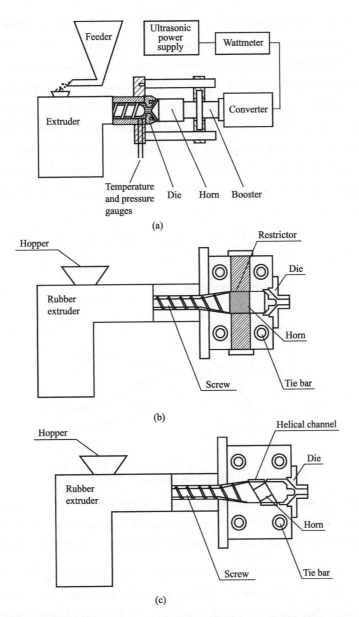

Fig. 13.3 Schematic diagram of (a) coaxial reactor, (b) barrel reactor, and (c) grooved barrel reactor built for devulcanization of rubbers.

ting shaft and the tip of the horns, where devulcanization takes place.

Under the license from the University of Akron for the ultrasonic devulcanization technology, NFM Co. of Massillon, Ohio, has built a prototype of the machine for ultrasonic devulcanization of tire and rubber products. It was reported that retreaded truck tires containing 15wt% and 30wt% of ultrasonically devulcanized carbon-black-filled SBR had passed the preliminary dynamic endurance test.

Extensive studies on the ultrasonic devulcanization of rubbers and some preliminary studies on ultrasonic decrosslinking of crosslinked plastics were carried out. It was shown that this continuous process allows one to recycle various types of rubbers and thermosets. As a most desirable consequence, ultrasonically devulcanized rubber becomes soft, therefore making it possible for this material to be reprocessed, shaped, and revulcanized in very much the same way as the virgin rubber. This new technology has been used successfully in the laboratory to devulcanize a ground tire rubber (GRT), unfilled and filled NR, guayule rubber, unfilled and filled SBR, unfilled and filled peroxide-cured silicone rubber, unfilled and filled EPDM and EPDM roofing membrane, unfilled polyurethane, unfilled resin-cured butyl rubber, used tire curing bladder, fluoroelastomer, ethylene vinyl acetate foam, and crosslinked polyethylene. After revulcanization, rubber samples exhibit good mechanical properties, which in some cases are comparable to or exceeding those of virgin vulcanizates.

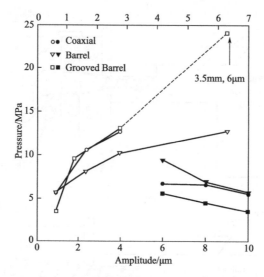

Fig. 13.4 The entrance pressure of devulcanization zone of different reactors vs. amplitude of ultrasound at a flow rate of 0.63g/s, and the entrance pressure of devulcanization zone vs. flow rate at the amplitude of 10mm and clearance of 2mm during devulcanization of GRT.

Ultrasonic devulcanization studies were concerned with finding the effect of processing parameters such as the pressure, power consumption, die gap, temperature, flow rate, and ultrasonic amplitude on devulcanization; structural changes occurring in various rubbers; rheological properties and curing kinetics of devulcanized rubbers; mechanical properties of revulcanized rubbers; and the effect of design of the devulcanization reactor. Fig. 13.4 shows the entrance pressure of devulcanization zone vs. amplitude of ultrasound at a flow rate of 0.63g/s, and the entrance pressure of devulcanization zone vs. flow rate at the amplitude of 10mm and clearance of 2mm during devulcanization of GRT.

The entrance pressure of the devulcanization zone was substantially reduced as the amplitude of ultrasound was increased. Ultrasound facilitated the flow of rubber through the gap not only because of reduction of the friction in the presence of ultrasonic waves but also because of the devulcanization taking place as GRT particles entered the devulcanization zone. The barrel reactor showed a higher pressure in the devulcanization zone than the coaxial reactor, and the grooved barrel reactor showed the lowest pressure at low amplitude of ultrasound and a flow rate of 0.63g/s. The barrel reactor had a converging zone before the devulcanization zone. The GRT flow was essentially blocked by the restrictor of the devulcanization zone at low amplitude of ultrasound. However, at the ultrasound amplitude of 10mm, the entrance pressure of the devulcanization zone for the coaxial and barrel reactors was almost the same due to a reduction of restrictor effect at high amplitude. The highest flow rate achieved in the barrel and grooved barrel reactors was 6.3g/s. The devulcanized sample of flow rate of 6.3g/s for the coaxial reactor could not be obtained due to an overload of the ultrasonic generator. Furthermore, in the grooved barrel reactor, at a flow rate of 6.3g/s, the gap size needed to be increased to 3.5mm, and ultrasonic amplitude needed to be decreased to 6mm due to an overload of the ultrasound unit. It was natural that the entrance pressure of the devulcanization zone rises with increasing flow rate as indicated in Fig. 13.4 for all three reactors. Nevertheless, at a high flow rate the barrel reactor had lower entrance pressure at the devulcanization zone than that of the other reactors at the ultrasound amplitude of 10mm. The difference in die characteristics (pressure vs. flow rate) among the three reactors having the devulcanization zone thickness of 2mm was possibly related to the difference in power consumption and the difference in shearing conditions. In the barrel and grooved barrel reactor, the GRT in the devulcanization zone was subjected to a pressure and drag flow while in the coaxial reactor to a pressure flow alone.

The comparison of stress-strain behavior of the vulcanizates prepared from devulcanized GRT produced by the three reactors at the maximum flow rate is shown in Fig. 13.5. The revulcanized sample obtained from the barrel reactor, having a flow rate of 6.3g/s, shows a tensile strength of 8.7MPa, elongation at break of 217%, and modulus at 100% elongation of 2.6MPa. In addition, the revulcanized sample obtained from the grooved barrel reactor having, a flow rate of 6.3g/s, shows the tensile strength of 8.3MPa, the elongation at break of 184%, and modulus at 100% elongation of 3.3MPa. The output of the barrel and grooved barrel reactors was higher than that of the coaxial reactor. In addition, the mechanical properties of the sample obtained using the barrel reactor at the higher flow rate, which could not be achieved in the coaxial reactor, were higher. These properties met the higher level of specification made for tire reclaim. The samples showing inferior performance were considered overtreated. The overtreatment meant a higher degree of devulcanization along with a signiflcant degradation of the backbone molecular chains. The overtreated samples were usually softer and stickier.

It is believed that the process of ultrasonic devulcanization is based on a phenomenon called cavitation. In this case, acoustic cavitation occurs in a solid body. This is in contrast to the cavitation typically known to occur in liquids in the regions subjected to rapidly alternating pressures of high amplitude generated by high-power ultrasonics. During the negative half of the pressure cycle, the liquid is subjected to a tensile stress, and during the positive half cycle, it experiences a compression. Any bubble present in the liquid will thus expand and contract alternately. The bubble can also collapse suddenly during the compres-

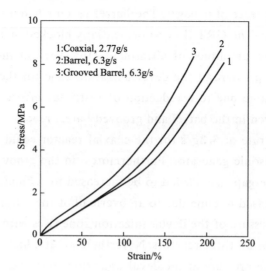

Fig. 13.5 The stress-strain curves of the vulcanizates prepared from devulcanized GRT produced by the three reactors at the maximum flow rate as shown.

sion. This sudden collapse is known as cavitation and can result in almost instantaneous release of a comparatively large amount of energy. The magnitude of the energy released in this way depends on the value of the acoustic pressure amplitude and, hence, the acoustic intensity.

Although the presence of bubbles facilitates the onset of cavitation, it can also occur in gas-free liquids when the acoustic pressure amplitude exceeds the hydrostatic pressure in the liquid. For a part of the negative half of the pressure cycle, the liquid is in a state of tension. Where this occurs, the forces of cohesion between neighboring molecules are opposed, and voids are formed at weak points in the structure of the liquid. These voids grow in size and then collapse in the same way gas-filled bubbles do. Cavitation may be induced in a gas-free liquid by introducing defects, such as impurities, in its lattice structure. In the case of polymer solutions, it is well known that the irradiation of a solution by ultrasound waves produces cavitation of bubbles. The formation and collapse of the bubble plays an important role in the degradation of polymers in solution. Most of the physical and chemical effects caused by ultrasound are usually attributed to cavitation: the growth and very rapid, explosive collapse of microbubbles as the ultrasound wave propagates through the solution. The intense shock wave radiating from a cavitating bubble at the final stage of the collapse is undoubtedly the cause of the most severe reactions. This shock wave is capable of causing the scission of macromolecules that lie in its path. The degradation arises as a result of the effect of the ultrasound on the solvent. In any medium, cavities, voids, and density fluctuations exist. It is believed that these induce cavitation, leading to molecular rupture.

In solid polymers, the microvoids, present intrinsically, are responsible for cavitation when they are subjected to a hydrostatic pressure in the manner of an impulse. One of the main causes of microvoid generation in polymer materials is the interatomic bond rupture when they are subjected to mechanical and thermal stresses. Extensive studies showing microvoid formation in stressed polymers have been carried out. When applied to rubbers, the cavitation usually corresponds to the effect of formation and unrestricted growth of voids in gas-saturated rubber samples after a sudden depressurization. In general, this has a broader sense and may be understood as the phenomena related to the formation and dynamics of cavities in continuous media. In materials science, for example, it means a fracture mode characterized by formation of internal cavities. In acoustics, the cavitation denotes the phenomena related to the dynamics of bubbles in sonically irradiated liquids.

Structural studies of ultrasonically treated rubber show that the breakup of chemical crosslinks is accompanied by the partial degradation of rubber chains. The mechanism of rubber devulcanization under ultrasonic treatment is presently not well understood, unlike the mechanism of the degradation of long-chain polymer in solutions irradiated with ultrasound. Specially, the mechanisms governing the conversion of mechanical ultrasonic energy to chemical energy are not clear. However, it has been shown that devulcanization of rubber under ultrasonic treatment requires local energy concentration, since uniformly distributed ultrasonic energy among all chemical bonds is not capable of rubber devulcanization.

It is well known that some amounts of cavities or small bubbles are present in rubber during any type of rubber processing. The formation of bubbles can be nucleated by precursor cavities of appropriate size. The proposed models were based on a mechanism of rubber network break-down caused by cavitation, which is created by high-intensity ultrasonic waves in the presence of pressure and heat. Driven by ultrasound, the cavities pulsate with amplitude depending mostly on the ratio between ambient and ultrasonic pressures (acoustic cavitation).

It is known that, in contrast to plastics, rubber chains break down only when they are fully stretched. An ultrasonic field creates high-frequency extension-contraction stresses in crosslinked media. Therefore, the effects of rubber viscoelasticity have been incorporated into the description of dynamics of cavitation. The devulcanization of the rubber network can occur primarily around pulsating cavities due to the highest level of strain produced by the powerful ultrasound.

Generally, cleavage in polymer chains results in the production of macroradicals, the existence of which have been confirmed spectroscopically by the use of radical scavengers such as diphenyl picrylhydracyl (DPPH). Obviously, in the absence of scavengers, the macroradicals are free to combine by either disproportionation or combination termination, the former leading to smaller-sized macromolecules while the latter will give a distribution dependent on the size of the combining fragments.

It was reported that under some devulcanization conditions the tensile strength of unfilled revulcanized SBR was found to be much higher than that of the original vulcanizate, with elongation at break being practically intact. In particular, Fig. 13.6 shows the stress-strain curves of unfilled virgin vulcanizates and revulcanized SBR obtained from devulcanized rubbers at various values of ultrasonic amplitudes, A. The devulcanized rubbers were obtained by using the coaxial ultrasonic reactor depicted in Fig. 13. 3 (a) at the barrel temperature of 120℃, screw speed of 20 r/min, and flow rate of 0. 63g/s.

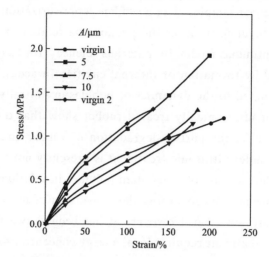

Fig. 13. 6 The stress-strain curves of unfilled virgin vulcanizates and revulcanized SBR obtained from rubbers devulcanized in coaxial reactor at various values of ultrasonic amplitude.

The ultrasonic horn diameter was 76. 2mm. In contrast to usual findings that the mechanical properties of reclaimed rubber obtained by using different techniques are inferior to those of virgin vulcanizates, the present data are rather unexpected. It was proposed that the improvement in the mechanical properties of revulcanized SBR was primarily due to the extent of nonaffine deformation of the bimodal network that appears in the process of revulcanization of ultrasonically devulcanized rubber. The superior properties of revulcanized rubbers were also observed in the case of unfilled EPDM and silicone rubbers. Unfilled revulcanized NR rubber also shows good properties, with the elongation at break remaining similar to that of the original NR vulcanizates but with the ultimate strength being about 70% of the original NR.

Interestingly, the strain-induced crystallization typical for the original NR vulcanizate remained intact in revulcanized NR, as indicated in Fig. 13. 7, where upturn of the stress-strain curves is observed in both the original and revulcanized rubbers. These samples were devulcanized in a coaxial reactor at different flow rates and ultrasonic amplitudes and revulcanized with a recipe consisting of 2. 5 phr ZnO, 0. 5 phr stearic acid, and 2 phr sulfur.

Fillers play an interesting role in the devulcanization process. Fig. 13. 8 shows the stress-strain curves for virgin and devulcanized 35 phr carbon-black-filled NR vulcanizates. Virgin vulcanizates were cured using 5 phr ZnO, 1 phr stearic acid, 1 phr CBS, and 2 phr sulfur. The revulcanization recipe contained 2. 5 phr ZnO, 0. 5 phr stearic acid, 0. 5 CBS, and 2 phr sulfur. The experiments have shown that upon filling rubbers with carbon

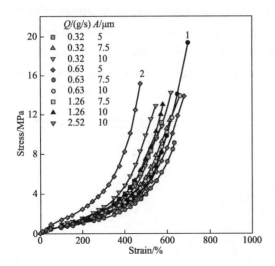

Fig. 13. 7 The stress-strain curves for unfilled NR vulcanizates prepared from ultrasonically devulcanized NR in coaxial reactor at various flow rates and amplitudes at a die gap of 2. 54mm and a barrel temperature of 120℃.

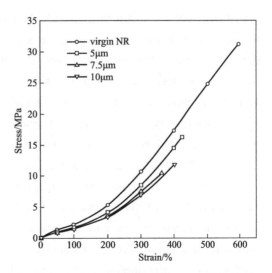

Fig. 13. 8 The stress-strain curves for 35 phr carbon-black-filled virgin NR and revulcanized NR devulcanized in coaxial reactor at a barrel temperature of 120℃, a gap of 2. 54mm, a flow rate of 0. 63g/s, and various ultrasonic amplitudes.

black, after devulcanization the mechanical properties of revulcanized rubbers typically deteriorate, with the level of deterioration depending on the devulcanization conditions. This is clearly evident from Fig. 13. 8. It was suggested that ultrasonic devulcanization causes a partial deactivation of filler due to the breakup of the macromolecular chains attached to the surface of carbon black. In many cases, this effect leads to inferior properties of revulcanized carbon-black-filled rubbers. Thus, ultrasonically devulcanized rubber was blended with virgin rubber. The blend vulcanizates indicated significantly improved properties. Also, attempts were made to add a certain amount of a fresh carbon black into the devulcanized rubber. It was shown that the vulcanizates containing a fresh carbon black exhibited better properties than the revulcanized rubber that did not contain an addition of fresh carbon black. However, in some cases, even carbon-black-filled devulcanized rubber shows mechanical properties similar to or better than the original rubber. In particular, this was shown for an EPDM roofing membrane containing carbon black and a significant amount of oil. Apparently, oil plays an important role in the devulcanization process. Possibly, the presence of oil prevents a deactivation of the filler that was observed in vulcanizates not containing oil. But in order to prove this hypothesis, further experiments are required.

Ultrasonic devulcanization also alters revulcanization kinetics of rubbers. It was shown that the revulcanization process of devulcanized SBR was essentially different from that of the virgin SBR. The induction period is shorter or absent for revulcanization of the devulcanized SBR. This is also true for other unfilled and carbon-black-filled rubbers such as GRT, SBR, NR, EPDM, and BR cured by sulfur-containing curative systems, but not for silicone rubber cured by peroxide. It was suggested that a decrease or disappearance of the induction period in the case of the sulfur-cured rubbers is due to an interaction between the rubber molecules chemically modified in the course of devulcanization and unmodified rubber molecules, resulting in crosslinking. It was shown that approximately 85% of the accelerator remained in the ultrasonically devulcanized SBR rubber.

Ultrasonically devulcanized rubbers consist of sol and gel. The gel portion is typically soft and has significantly lower crosslink density than that of the original vulcanizate. Due to the presence of sol, the devulcanized rubber can flow and is subjected to shaping. Crosslink density and gel fraction of ultrasonically devulcanized rubbers was found to correlate by a universal master curve. This curve is unique for every elastomer due to its unique chemical structure. Fig. 13.9 presents the normalized gel fraction as a function of normalized crosslink density of devulcanized GRT obtained from three different reactors. The gel fraction and crosslink density of GRT was 0.82 and 9.9×10^{-2} kmol/m^3, respectively.

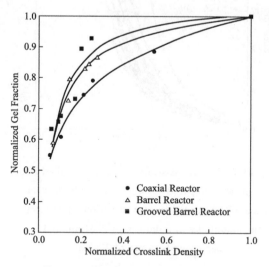

Fig. 13.9 Normalized gel fraction vs. normalized crosslink density of devulcanized GRT obtained at various devulcanization conditions using the coaxial, barrel, and grooved barrel reactors.

For each reactor, the dependence of gel fraction on crosslink density was described by a unique master curve that was independent of processing conditions such as flow rate (residence time) and amplitude. The unique correlation between gel fraction and crosslink density obtained in the barrel and grooved barrel reactors was shifted toward lower crosslink density than those obtained in the coaxial reactor, indicating a better efficiency of devulcanization. It is considered possible that the additional shearing effect caused by the screw rotation in the barrel and grooved barrel reactors had a positive influence on improving the efficiency of devulcanization.

13.4 Use of Recycled Rubber

13.4.1 General Remarks

There are certain technical limitations in the devulcanization of rubbers, and vulcanization is, in fact, not truly reversible. The partial devulcanization of scrap rubber will result in a degradation of physical properties. In many cases, this may limit the amount of substitution levels in high-tech applications such as passenger tires. But it can provide the compounder of less stringent products with an excellent low-cost rubber that can be used as the prime rubber or at very high substitution levels. Reclaim cannot be used for tread compounds in tires because every addition may decrease their resistance to wear. However, this statement has not been checked in the case of rubber devulcanized without an addition of chemicals.

Considerable amounts of reclaim are consumed for carcasses of bias ply tires for cars if the compounds are of NR; for carcasses of radial tires no reclaim is added. On the other hand, reclaim is added to compounds for bead wires, and it may also be added to sidewalls. Within the framework of direct recycling options, a number of applications for GRT outside the rubber industry have been proposed. Such applications include the use as a filler in asphalt for the surface treatment of roads and as a rubberized surface for sport facilities. The ground scrap rubber can be used as fillers in raw material and plastic compounds. However, the problem of compatibility with the matrix and size of the filler, as well as the discontinuity at the interface between the two phases, should be considered. Rubber products containing ground rubber have low tensile properties due to insufficient bonding between the ground rubber and the virgin matrix. However, this bonding can be improved in the case of the addition of devulcanized rubber.

13.4.2 Use in New Tires

The tire is a complicated composite product consisting of tread, under-tread, carcass, inner liner, bead, and sidewall. Many different types of rubber and carbon-black reinforcement are used in manufacturing tires. Therefore, GRT is a blend of various rubbers and carbon blacks. Accordingly, in using GRT powder and devulcanized GRT in new tire manufacturing, many factors should be considered. Evidently, scrap tire powder can be used as a filler for virgin rubbers, and devulcanized GRT can be used in blends with virgin rubbers. This market consumed approximately 50 million lb of scrap tire rubber in 2001.

Until recently, it was generally understood that only a few percent of ground rubber can be used in new tires. The Scrap Tire Management Council reports that 5% of recycled tire rubber is used in an original equipment tire for the Ford Windstar. Although no other information on the amount of devulcanized rubber used in new tires is available in open literature, a possibility exists for the use of up to 10wt% of recycled tire rubber in new tire compounds. Recently, it was reported that actual road tests of a truck tire containing 10wt% of the devulcanized rubber in the tread exhibited tread wear behavior almost equal to that for the standard type with the new rubber compound. The increase in the amount of recycled rubber in tires is growing, but it is likely that results will not be available for a number of years.

13.4.3 Rubber/Recycled Rubber Blends

The rubber particles from scrap tires can be incorporated into a virgin rubber as a filler. However, in this case, the compatibility with the matrix is a significant issue. Rubber products containing ground rubber typically have lower tensile properties due to insufficient bonding between the ground rubber and the virgin matrix. The effect of GRT particles of different sizes incorporated in a NR compound on its mechanical properties was reported. Table 13.2 shows the tensile strength, elongation at break, and tear strength before and after aging for a virgin vulcanizate and vulcanizates containing GRT particles of various sizes.

Table 13.2　Properties of GRT-filled NR vulcanizates

Properties	Compound									
	Control		A		B		C		D	
	Before aging	After aging	Before aging	After aging	Before aging	After aging	Before aging	After aging	Before aging	After aging
Tensile strength/MPa*	14.0	8.8 (63)	2.2	1.6 (71)	4.2	2.5 (60)	7.3	3.3 (45)	8.0	2.5 (31)
Elongation at break/%	1175	770 (66)	430	230 (53)	620	360 (58)	780	410 (53)	860	400 (47)
Tear strength/(kN/m)*	28.2	20.3 (72)	12.4	9.7 (78)	18.5	10.6 (57)	23.8	11.5 (48)	21.2	9.7 (46)
Sol%	2.1	2.2	5.8	6.4	5.2	6.2	4.9	6.0	4.3	6.6
%increase in sol after aging	—	4.8	—	10.3	—	19.2	—	22.4	—	53.5

* Values in parentheses show the % retention of properties after aging at 100℃ for 36 h.

The vulcanizates A, B, C, and D contained 30phr of GRT particles of sizes in the range of 650~450mm, 300~215mm, 205~160mm, and 150~100mm, respectively. The control sample was the virgin NR vulcanizate. The curing recipe contained 6 phr ZnO, 0.5 phr stearic acid, 3.5 phr sulfur, and 0.5 phr MBT. Incorporation of GRT in a NR compound decreased the physical properties of the vulcanizate, with the effect being larger in the case of large particles. However, the NR vulcanizate containing GRT exhibited better properties retention upon aging. It was shown that smaller particles contain less the amount of rubber but higher the amount of fillers and metals.

Recycled tire rubber in the form of large crumb particles is also used for making prepackaged pour-in-place surfacing product. GRT is combined with premixed polyurethane to produce a soft, pliable energy-absorbing rubber surface for playground and other recreational surfaces and is intended for placing over compacted gravel, concrete, or asphalt.

In recycling of rubbers, it is customary to add various proportions of ground rubbers to the virgin material. Therefore, blends of both filled and unfilled ground and ultrasonically devulcanized rubbers with virgin rubber have been prepared. The mechanical properties of blends of unfilled Polyurethane (PUR) virgin and devulcanized rubber have been measured. Curatives were added in the blends of devulcanized rubber based on the total rubber, while for blends of ground rubber curatives were added based on virgin rubber content. Fig. 13.10 shows the tensile properties of the two types of blends of PUR. From a comparison of these two figures, it is quite apparent that the blends of the devulcanized samples [see Fig. 13.10(a)] have better tensile properties compared to the blends of ground samples [see Fig. 13. 10 (b)]. For the latter, only at a ground concentration of 25%, the tensile properties are similar to that of the original, whereas for the former, the properties are superior to the original at this concentration and comparable at concentrations of 50%. It may be thought that ultrasonic devulcanization causes a better bonding of the devulcanized rubber to the virgin rubber in the blends than in the case of ground-virgin blends.

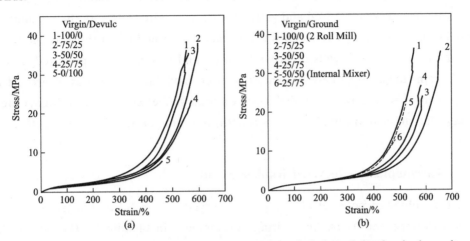

Fig. 13. 10 Stress vs. strain curves of (a) vulcanized blends of virgin and devulcanized sample, and (b) ground and virgin sample. The conditions of devulcanization are a flow rate of 1. 26g/s, a gap size of 3mm, an amplitude of 7. 5mm, and a barrel temperature of 120℃ using the coaxial reactor.

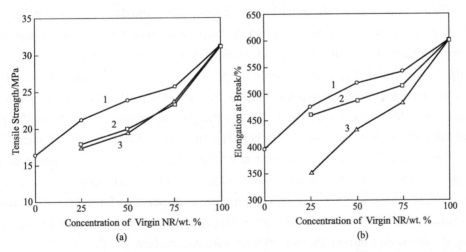

Fig. 13. 11 (a) Tensile strength and (b) elongation at break of 35 phr CB-filled virgin and devulcanized NR blends: (1) curatives were added to total rubber content; (2) blends of ground NR vulcanizates and virgin NR; curatives were added to virgin rubber; and (3) blends of ground NR vulcanizates and virgin NR; curatives were added to the total rubber content. NR was devulcanized at a flow rate of 0. 63g/s, a gap size of 2. 54mm, an amplitude of 5mm, and a barrel temperature of 120℃, using the coaxial reactor.

Ultrasonically devulcanized and ground-carbon black (CB) -filled NR was blended with virgin CB-filled material. Fig. 13. 11 gives the tensile strength of these NR-NR blends containing 35 phr CB. The blends of devulcanized NR and virgin NR have much better tensile properties than blends with fully cured ground rubber. As the proportion of the virgin NR in the blends of devulcanized NR was increased, the mechanical properties progressively increased at or above the rule of mixture. Therefore, the mechanical properties of these kinds of blends can be improved significantly by ultrasonic devulcanization of ground vulcanizates. The modulus at 100% strain of NR-NR blends containing 35 phr CB was measured. Blends of devulcanized and virgin NR showed a little drop in the modulus, while the modulus of ground and virgin NR was reduced. It is thought that this was due to the migration of curatives; the blends of ground rubber and virgin rubber prepared using curatives for virgin rubber are likely to exhibit reduced modulus. The blends of ground rubber and virgin rubber with curatives added based on total rubber content show higher modulus. However, they indicated lower tensile strength and elongation of break due to excess of curatives.

13.4.4 Thermoplastic-Recycled Rubber Blend

The technology of polymer blending has emerged as a useful tool in tailoring polymers to the needs of the end users. An exciting development in blending is the introduction of thermoplastic elastomers (TPEs) based on plastic-rubber blends. These TPEs are becoming increasingly important because of their elastomeric properties and easy processability of the blends and their lower cost.

The blending of waste rubber with thermoplastics is important from the point of view of both disposal of waste and the reduction in the product cost. More attention has been focused on compounding ground tire rubber (GRT) with thermoplastics, which can be subsequently remelted and shaped into a wide range of molded and extruded products. The mechanical properties of such compounds depend on the concentration of GRT, polymer matrix type, and adhesion between the GRT and the polymer matrix, as well as the particle size and their dispersion and interaction between the GRT and the matrix. Generally, adhesion between the GRT and the polymer matrix and the size of the GRT particles are the two major factors controlling the mechanical properties of such composites. Also, dynamic vulcanization techniques can be used to improve the properties of the blends of ultrasonically devulcanized GRT and thermoplastics. Dynamic vulcanization is the process of vulcanizing the elastomer during its melt mixing with the molten plastic.

The use of GRT, instead of virgin elastomers, however, results in significant deterioration in the mechanical properties of these composites. It was reported that the GRT has a detrimental effect on most of the physical properties of cured rubber, the extent of deterioration increasing with the amount and size of the GRT. There have been several investigations with the aim of improving the adhesion between the GRT and the polymer matrix.

The effects of various compatibilizers to promote the adhesion of polyethylene (PE)-GRT blends were studied. The authors reported that it was possible to achieve a partial recovery of the same properties through a melt blending process where each component was first conditioned with compatibilizers of similar structure before the actual blending was carried out. Among the various compatibilizers, epoxydized natural rubber, ethylene-*co*-acrylic acid copolymer, and ethylene-*co*-glycidyl methacrylate polymer were found to be effective in improving the impact properties of PE-GRT composites. It was also reported that smaller GRT particle size results in a small increase in the impact property of the composite and has a greater influence on the melt processability of the composites. The percent improvement in the impact energy for linear low-density polyethylene (LLDPE) and the composites prepared from them was greater than that for the corresponding high-density polyethylene (HDPE) composites. It was also suggested that the low polarity and/or low crystallinity of the matrix polymer appeared to favor the compatibility with GRT.

Modification of GRT particle surface has been studied to improve the compatibility of GRT and polymer. The polymeric surface modification can be carried out by chemical treatments like chromic acid etching and thermal oxidation, or by mechanical means. Use of maleic anhydride-grafted and chlorinated GRT, respectively, improved physical properties of GRT-EPDM-acrylated high-density polyethylene and GRT-polyvinyl chloride blends. It was found that surface treatment of ground rubber with a matrix of unsaturated curable polymer and a curing agent could also improve the performance of the blends. The effect of cryogenically ground rubber (CGR) (approximately 250 microns) from old tires on some mechanical properties of an unsaturated polyester resin was investigated. Composites made from silane-treated ground rubber showed better mechanical properties than composite made from untreated CGR. However, the particle size of the ground rubber was apparently too large to produce a toughening effect on the filled materials.

High-energy treatments including plasma, corona discharge, and electron-beam radiation were used to modify the surface of GRT. The oxidation on the surface of GRT generated by chemical and physical treatment, such as occurs in plasma and autoclave in oxygen atmosphere, was shown to improve adhesion between GRT and polyamide in the blend.

An epoxy resin compounded with tire rubber particles modified by plasma surface treatment was also studied. An improvement in mechanical properties of the resulting material over those containing the untreated rubber was observed. The effects of corona discharge treatment of GRT on the impact property of the thermoplastic composite containing the GRT were investigated. X-ray photoelectron spectroscopy analysis showed that the corona discharge treatment of GRT increased the oxygen-containing groups on the ground rubber surface. In some composites, it has been found that treated GRT marginally improved the impact property of the composites. However, prolonged times of treatment and higher power inputs for corona discharge of GRT reduced the impact strength of the composites.

The phenolic resin cure system and maleic anhydride grafted PP compatibilizer significantly improved mechanical properties of PP-ultrasonically devulcanized GRT blends prepared by dynamic vulcanization. Also, a new ultrasonic reactor was built, and an improvement in the efficiency of the process and better properties of PP-GRT blends were achieved while carrying ultrasonic treatment during extrusion in this reactor. In the reactor, two horns were placed in a slit die attached to a plastic extruder. Mechanical properties of PP-GRT, PP-devulcanized GRT (DGRT), and PP-revulcanized GRT (RGRT) that mixed in proportion of 40 : 60 are shown in Table 13.3.

Table 13.3 The Mechanical Properties of PP-GRT, PP-DGRT, and PP-RGRT Blends

Blend		Tensile strength/MPa	Young's modulus/MPa	Elongation at break/%
PP-GRT		6.7	116	16.6
PP-DGRT	2 horns,5μm	6.6	98	21.5
	2 horns,7.5μm	6.5	100	20.6
	2 horns,10μm	5.15	102	7.6
	1 horn,5μm	6.9	110	19.3
	1 horn,7.5μm	6.6	104	17.4
	1 horn,10μm	7.0	116	21.0
	earlier work	5.2	108	20.7
PP-RGRT	2 horns,5μm	6.7	109	13.2
	2 horns,7.5μm	6.7	110	13.3
	2 horns,10μm	5.9	111	7.0
	1 horn,5μm	6.4	110	11.8
	1 horn,7.5μm	6.6	110	13.3
	1 horn,10μm	7.2	122	14.8
	earlier work	6.7	110	18.0

Comparison indicates that the tensile strength, Young's modulus, and elongation at break of the blend prepared by this reactor under certain conditions are higher than those obtained earlier. Also, properties of PP-RGRT at 10mm are higher than those of PP-GRT at 10mm. Evidently, the ultrasonic treatment of PP-GRT blends led to a certain level of compatibilization at the interface between the plastic and rubber phases. This was due to mechanochemical reactions induced by ultrasound at the interface. In addition, GRT was blended with ultrasonically devulcanized GRT with HDPE using a Brabender internal mixer and a twin-screw extruder. These blends were dynamically vulcanized in these mixers. Also, HDPE and GRT blends mixed earlier by using a twin-screw extruder were passed through the ultrasonic devulcanization extruder and subsequently dynamically vulcanized by means of the internal mixer and the twin-screw extruder. The blends mixed by using the twin-screw extruder prior to devulcanization were found to have better tensile properties and impact strength than any other blends. Rheological properties of these blends were also studied.

Novel blends of GRT and recycled HDPE from used milk containers were studied and patented. The effects of GRT particle size and concentration on mechanical and rheological properties were determined. The blend systems were optimized by a soft rubber-plastic binder produced from a mixture of HDPE and EPDM, wherein EPDM is dynamically vulcanized during its mixing with the HDPE. It was concluded that the softening of the HDPE binder provides compositions of improved ultimate mechanical properties.

13.4.5 Concrete Modified by Recycled Rubber

Production of rubber-filled concrete compositions is an area in which further expansion of GRT usage is possible. The advantages of using GRT in the cement-concrete structure are an increased crack, freeze-thaw, and impact resistance; shock wave absorption; reduced heat conductivity; and increased resistance to acid rain. However, an addition of rubber particles to concrete has shown to reduce the compressive and flexural strength.

Concerning the effect of the size of rubber particles on the compressive strength, the results are contradictory. The compressive strength of concrete was lowered upon an addition of coarse-graded rubber particles while increased with an addition of the fine-graded particles. In contrast, tests have shown opposite results.

This contradiction can possibly be explained by the difference in rubber source, the geometry of the particles, and the way in which the particles were prepared.

The influence of the shape of rubber particles on mechanical properties, workability, and chemical stability of rubber-filled cement was studied.

The composite containing rubber shred was able to bridge the crack and to prevent catastrophic failure of the specimen, while the composite containing granular rubber particles was unable to bridge the crack. The pull-out test indicated poor interfacial bonding between the granular rubber particles and the matrix. In fact, many studies in this area indicated that the interface between the rubber and cement is weak. Attempts to improve the interface were made by washing the rubber particles. Some improvement has been achieved by washing the particles with water, with water and carbon tetrachloride mixture, and with water and latex mixture, leading to an enhanced adhesion in rubber-filled cement. Several surface modifications have also been proposed, including the treatment of rubber with sulfuric acid and nitric acid to chemically oxidize rubber and introduce polar groups. Contrary to expectation, the treatment with nitric acid led to a decrease of the strength of the composite. On the other hand, the treatment with sulfuric acid improved the adhesion of rubber to concrete. Using a combination of chemical and surface probing techniques, it was shown that the hydrophilicity of the rubber surface is greatly improved by acid or base treatment.

The rubber surface is typically hydrophobic. This is due to the fact that rubber typically contains zinc stearate that diffuses to the surface and causes hydrophobicity. By acid treatment, zinc stearate can be hydrolyzed to stearic acid. Treating the rubber with base, the zinc ions are converted into sodium ions of NaOH. These conversions create the soluble sodium stearate. It was also found that addition of rubber particles to the mortar led to a decrease in their compressive and flexural strengths due to pull out of particles. However, the treatment of rubber particles before mixing with a bifunctional silane-coupling agent, such as gamma mercapto trimethoxy, improved the interface and led to increased ductility. It is suggested that this research be expanded to find the effects of the type of coupling agent on the adhesion and the fracture behavior of the rubber-filled cement paste, mortar, and concrete. These materials can be utilized in highway pavement overlays, sidewalks, medians, sound barriers, and other transportation nonstructural uses.

13.4.6 Asphalt Modified by Recycled Rubber

Asphalt can be blended with tire rubber to modify the properties of the asphalt. This remains the largest single market for ground rubber. In 2001, an estimated 0.11 million tons, or approximately 12 million tires, were used in the United States. Mostly, these were consumed in California, Arizona, and Florida. However, other states are beginning to recognize the benefits of the modified asphalt. It was reported that the asphalt industry can absorb up to 40% of scrap tires.

Utilization of scrap tire rubber in asphalt has advantages in the performance of roads and their longevity. These include enhanced ductility, crack resistance, skid resistance, and noise reduction. Disadvantages of the rubber-modified asphalt are its cost and a possibility of toxic emissions into the air.

Tests indicated that rubber-modified asphalt increases the cost of road construction by about 50% in comparison with conventional asphalt. The requirement for an additional step of hot mixing during processing of the rubber-asphalt mix may possibly cause toxic emissions into the air. Two processes are used in preparing the rubberized asphalt: dry and wet process. In both processes, the GRT particle size ranges from 6.35mm to 40mesh. In the wet process, asphalt is blended with GRT particles and then added into the hot mix. In the dry process, GRT is mixed with aggregate, and the resulting mix is blended with asphalt.

Asphalt blended with GRT has been used for quite some time. Depending on the type of tire, the composition of GRT may include different rubbers. These crosslinked rubbers are mostly immiscible in bitumen. The blends show an improvement in basic asphalt properties as well as in rubber-like characteristics. The blend is thought of as a dispersion of undissolved swollen rubber particles acting as an elastic aggregate within asphalt, modified by the portion of the rubber particles that have dissolved. The blending of GRT with asphalt was begun in the 1960s, developed and patented a patching material consisted of 25wt% scrap and asphalt blended at 190℃ for 20 minutes. MacDonald continued his work by expanding it to actual road pavement test sections as a seal coat, and in 1968, Sahmaro Petroleum and Asphalt conducted application of a blend of GRT and asphalt as a binder for hot premix. The hot premix is a mixture of stone aggregate, sand, and the tire-asphalt binder all premixed in a batch-or drum-type mixer. This material is then applied as a carpet on top of the road by a paving machine, followed by a steel roller used to compact the material. Thereafter, many variations of this basic process of hot premix with the blend of GRT and asphalt as the binder have been proposed; most of them involve replacing stone aggregate with GRT. However, this method does not truly modify the asphalt binder. Thus, the process of blending GRT and asphalt before preparing a mixture is the most efficient in improving properties. Typical use levels range from 15wt% to 30wt%. A limited amount of work has been done on the characterization of the blends of GRT and asphalt. Blends are typically mixed at temperatures of 300°C to 400°C for a period of 0.5 to 2 hours. The mix increases in viscosity and has a consistency of slurry with discernable rubber particles spread throughout.

At room temperatures, the resulting composition is a tough rubbery elastic like material. This mixing period is often referred to as reaction or digestion time. It was suggested that the elastic quality of the blend is caused by the mechanical action of the undissolved rubber particles performing as a completely elastic aggregate within asphalt, which is modified by a portion of the rubber particles that have dissolved. The mixing time of 2 hours, as opposed to times of 0.5 and 1 hour, significantly improves elastic recovery, while increasing the mixing time reduces the amount of solid rubber in the mixture and increases both high and low molecular weight fractions of dissolved rubber in the asphalt. This brings up a very interesting point for further investigation: whether the increase in asphalt-soluble-rubber fractions that occurs with longer mixing times can be attributed to elastic properties. The addition of GRT at 10wt%, 20wt%, and 30wt% levels has significantly increased the softening points and strain recovery over the base asphalt, with a viscosity increase of similar magnitude for all three blends.

Rheological properties as affected by asphalt composition, rubber dissolution, and temperature were studied. Rubber content, rubber particle size, and base asphalt composition were found to be the main factors affecting the rheology of the asphalt-rubber binder. By controlling these variables, binders with improved cracking resistance and rutting resistance can be produced. Finally, scrap tires, used as a crumb rubber modifier for asphalt, improve paving performance and safety by being an excellent and cost-effective modifier for the highway pavement industry.

13.4.7 Use of Crumb Rubber in Soil

The patented soil amendment method of using tire crumb can decrease the negative impacts associated with compaction. The resiliency of the turf is not a direct factor of the elastic nature of rubber but rather the result of increased aeration. Surface hardness characteristics were evaluated. Crumb rubber significantly reduced soil hardness, soil shear strength, and water content.

13.4.8 Products Made from Recycled Rubber

Compounds containing devulcanized or ground tire rubbers can be used by various manufacturers to make a variety of rubber products. In particular, these compounds can be used or already are used for the production of shoe heels and soles, tubes, conveyor belts, technical rubber moldings, automobile floor mats, and flaps, livestock stall mattresses, playground and track surfacing, railroad track crossing, lower layers of floor coverings, various molded and extruded profiles, sealing plates, battery boxes, and other hard rubber goods. Since tire rubber is typically black, it therefore cannot be used for light and colored compounds unless additional measures have been taken to change the color. Obviously, for every such use, the recycled tire rubber must undergo extensive testing.

13.5 Pyrolysis and Incineration of Rubber

13.5.1 Recovery of Hydrocarbon Liquid and Carbon Black

One method suitable for use in recycling used tires is pyrolysis. Pyrolysis is the thermal decomposition of rubbers in the absence of air and oxygen to produce oils and gases for reuse by petrochemical industries. Carbon black and other solid content remaining after pyrolysis can be used as fillers. Pyrolysis is typically carried out in boilers, autoclaves, rotary kilns, screw conveyors, and fluidized beds. Also, hydrogenation has been performed using a tubing bomb reactor. Research activities in tire rubber pyrolysis to recovery hydrocarbon liquid and carbon black were quite extensive in the 1960s and 1970s and led to plant construction for the pyrolysis of scrap tires in the 1970s. Since then, significant studies have been carried out on tire pyrolysis concerning the evolution of volatile and the utilization of oil and carbon black. However, these attempts proved to be economically unsuccessful due to the low price of crude oil. Also, pyrolysis plants are believed to produce toxic waste as a by-product of operation. During the last 20 years, significant research has been carried out and various pyrolysis processes have been developed. However, despite this progress, pyrolysis of scrap tires is done on a limited scale. This is mainly due to the absence of a wide market for the oil and the carbon black derived by means of the pyrolysis process.

13.5.2 Tire-Derived Fuel

Tire rubber can be transformed into energy via the incineration method. This process is advocated by a number of the major tire and rubber companies and also by the major utility companies. Tire-derived fuel can be in the form of rubber chips containing or not containing the inherent wire. The nominal size of the chips is usually about 5 cm to 10 cm. The larger the size, the greater the content of the wire and the less likely it is to be able to be handled and metered. Scrap tires are used as a fuel supplement for coal or gas in kilns for manufacturing Portland cement, lime, and steel. This reduces by 25% the amount of coal consumption by cement industries. Scrap tires are also burned to generate electricity. Tire-derived fuel may reduce sulfur emissions of power plants and may improve the combustion efficiency by adjusting proper stoichiometry in combination with various coals, wood wastes, and household garbage. The consumption of tire-derived fuel was 115 million scrap tires in the United States in 2001. However, in burning rubber for fuel, valuable rubber materials are lost. In fact, 120MJ of energy is consumed to make 1 kg of synthetic tire rubber. In contrast, caloric value recovery by burning is 26MJ to 32MJ per kg of rubber, which is not much higher than that of burning much cheaper coal. Moreover, the burning of tires for energy may lead to air pollution.

13.6 Concluding Remarks

Waste tires and rubber present a problem of international significance. The present work describes some routes available to solve this problem. Numerous technologies are being developed. Among them, in addition to the well known grinding techniques, are continuous pulverization methods based on a single-or twin-screw extruder that may serve as a possible route to supply rubber powder as a feedstock for various present and future devulcanization and recycling technologies. These include reclaiming, surface treatment, ultrasonic devulcanization, and utilization of rubber particles for making composites with other materials. Some success has been achieved in development of ultrasonic technology, which is considered one of the promising methods, making devulcanized rubbers suitable for making rubber products from 100% recycled rubber as well as for adding to virgin rubber, virgin and recycled plastics, asphalt, concrete, and cement. Equipment has to be developed that is capable of achieving high enough output to make the process economically feasible. The major challenge in industrial implementation of this process is the development of a high-power ultrasonic generator capable of continuously operating under high pressure and high temperature. Clearly, there is also a lack of scientific understanding of the various processes governing the recycling of rubbers. Development of science-based technologies and processes for rubber recycling, and the use of recycled rubbers in varied end-products, would significantly reduce worldwide energy consumption, provide renewable rubbers from scrap tires and rubber waste, and lead to less pollution of the environment.

References

[1] C. Goodyear, inventor, U.S. Patent 3, 633 (1844).
[2] C. Goodyear, inventor, U.K. Patent 2, 933 (1853).
[3] U.S. Scrap Tire Markets 2001, Scrap Tire Management Council.
[4] Scrap Tire Stockpile Abatement, May 2002, Rubber Manufacturers Association.
[5] R. H. Snyder, "Scrap Tyres: Disposal and Reuse," Society of Automotive Engineers, Inc., Warrendale, PA, 1998.
[6] P. Hous, H. Bartelds, and E. Smit, Rubber World 212, 36 (1995).
[7] W. Klingensmith and K. Baranwal, Rubber World 218, 41 (1998).
[8] V. M. Makarov and V. F. Drozdovski, "Reprocessing of Tyres and Rubber Wastes," Ellis Horwood, New York, 1991.
[9] Scrap Tire News, 13, 2, 6 (1999).
[10] Scrap Tire News, 13, 6, 7 (1999).
[11] J. Pryweller, European Rubber Journal 181, 17 (1999).
[12] I. Franta (Ed.), "Elastomers and Rubber Compounding Materials," Elsevier, New York, 1989.

[13] W. C. Warner, Rubber Chem. Technol 67, 559 (1994).

[14] C. P. Rader (Ed.), "Plastic, Rubber and Paper Recycling," American Chemical Society, Washington, D. C., 1995.

[15] S. R. Fix, Elastomerics 112, 38 (1980).

[16] P. P. Nicholas, Rubber Chem. Technol. 55, 1499 (1982).

[17] M. Myhre and D. A. MacKillop, Rubber Chem. Technol. 75, 429 (2002).

[18] Fact Sheet, Tire Retread Information Bureau (2002).

[19] A. A. Phadke, A. K. Bhattacharya, S. K. Chakraborty, and S. K. De, Rubber Chem. Technol. 56, 726 (1983).

[20] B. Siuru, Scrap Tire News 11, 14 (1997).

[21] A. Accetta and J. M. Vergnaud, Rubber Chem. Technol. 55, 961 (1982).

[22] D. De, S. Maiti, and B. Adhikari, J. Appl. Polym. Sci. 73, 2951 (1999).

[23] B. Adhikari, D. De, and S. Maiti, Prog. Polym. Sci. 25, 909 (2000).

[24] A. I. Isayev, in "Rubber Technologist's Handbook," J. R. White and S. K. De (Eds.), RAPRA, Shawbury, U. K., 2001, Chap. 15, pp. 511-547.

[25] B. D. LaGrone, Conservation and Recycling 9, 359 (1986).

[26] B. Bowers, D. Barber, and R. Allinger, Oct., 1986, Paper #82 presented at the meeting of the ACS Rubber Division, Atlanta, GA.

[27] K. Knoerr, Oct., 1995, Paper #5 presented at the ACS Rubber Division Meeting, Cleveland, OH.

[28] P. M. Lewis, NR Technology 17, 57 (1986).

[29] R. Schaefer, Oct., 1986, Paper #79 presented at the ACS Rubber Division Meeting, Atlanta, GA.

[30] R. Schaefer and R. Berneking, Oct., 1986, Paper #80 presented at the ACS Rubber Division Meeting, Atlanta, GA.

[31] E. M. Solov'ev, V. B. Pavlov, and N. S. Enikolopov, Intern. Polym. Sci. Technol. 14, 10 (1987).

[32] J. A. Szilard, "Reclaiming Rubber and Other Polymers," Noyes Data Corporation, London, 1973.

[33] W. Kongensmith, Rubber World 203, 16 (1991).

[34] J. J. Leyden, Rubber World 203, 28 (1991).

[35] J. G. Bryson, U. S. Patent 4, 148, 763 (1979).

[36] R. Kohler and J. O'Neill, Rubber World 216, 32, 34 (1997).

[37] B. C. Sekhar and V. A. Kormer, European Patent Application, EP 0 690 091 A1 (1995).

[38] D. De, B. Adhikari, and S. Maiti, J. Polym Material 14, 333 (1997).

[39] D. De, S. Maiti, and B. Adhikari, Kautchuk Gummi Kunststoffe 53, 346 (2000).

[40] D. De, A. K. Ghosh, S. Maiti, and B. Adhikari, Polym. Recycling 4, 15 (1999).

[41] K. Knorr, Kautchuk Gummi Kunststoffe 47 1, 54 (1994).

[42] M. A. L. Verbruggen, L. van der Does, J. W. M. Noordermeer, M. van Duin, and H. J. Manuel, Rubber Chem. Technol. 72, 731 (1999).

[43] L. K. Hunt and R. R. Kovalak, inventors, Assignee Goodyear Tire and Rubber Company, U. S. Patent 5, 891, 926 (1999).

[44] R. D. Myers, P. Nicholson, J. B. MacLeod, and M. E. Moir, U. S. Patent, 5, 602, 186 (1997).

[45] N. Kawabata, B. Okuyama, and S. Yamashita, J. Appl Polym Sci. 26, 1417 (1981).

[46] N. Kawabata, T. Murakami, and S. Yamashita, Nippon Gomu Kyokaishi 52, 768 (1979).

[47] D. A. Benko and R. N. Beers, inventors, Assignee Goodyear Tire and Rubber Company, U. S. Patent 5, 380, 269 (2002).

[48] D. A. Benko and R. N. Beers, inventors, Assignee Goodyear Tire and Rubber Company, U. S. Patent 6,

387, 965 (2002).

[49] D. A. Benko and R. N. Beers, inventors, Assignee Goodyear Tire and Rubber Comapny, U. S. Patent 6, 462, 099 (2002).

[50] A. A. Hershaft, Environ. Sci. Technol. 6, 412 (1972).

[51] A. A. Hershaft, Elastomerics 109, 39 (1977).

[52] A. Ratcliffe, Chem. Eng. 79, 62 (1972).

[53] J. Lynch and B. LaGrone, "Ultrafine Crumb Rubber," Paper #37 presented at a meeting of the Rubber Division, ACS, Atlanta, GA, October 1986.

[54] N. S. Enikolopian, Pure Appl. Chem. 57, 1707 (1985).

[55] K. Khait and J. M. Torkelson, Polym. Plast. Technol. Eng. 38, 445 (1999).

[56] K. Khait, Paper #24 presented at the ACS Rubber Division Meeting, Chicago, IL, 1994.

[57] E. Bilgili, H. Arastoopour, and B. Bernstein, Rubber Chem. Technol. 73, 340 (2000).

[58] E. Bilgili, B. Berstein, H. Arastoopour, AIChE Symp. Ser. 95, 83 (1999).

[59] E. Bilgili, H. Arastoopour, and B. Bernstein, Powder Technol. 115, 265 (2001).

[60] E. Bilgili, H. Arastoopour, and B. Bernstein, Powder Technol. 115, 277 (2001).

[61] H. Arastoopour, D. A. Schocke, B. Bernstein, and E. Bilgili, U. S. Patent 5, 904, 885 (1999).

[62] E. Bigili, A. Dybek, H. Arastoopour, and B. Bernstein, J. Elast. Plastics 35, 235 (2003).

[63] J. E. Morin, D. E. Williams, and R. J. Farris, Rubber Chem. Technol. 75, 955 (2002).

[64] A. R. Tripathy, J. E. Morin, D. E. Williams, S. J. Eyles, and R. J. Farris, Macromol. 35, 4616 (2002).

[65] D. S. Novotny, R. L. Marsh, F. C. Masters, and D. N. Tally, U. S. Patent 4, 104, 205 (1978).

[66] T. Kleps, M. Piaskiewicz, and W. Parasiewicz, J. Thermal Analysis Calorimetry 60, 271 (2000).

[67] A. M. Basedow and K. Ebert, Adv. Polym. Sci. 22, 83 (1987).

[68] G. J. Price, in "Advances in Sonochemistry," T. J. Mason (Ed.), JAI Press Ltd., Greenwich, CT, Vol. 1, 1990, p. 231.

[69] G. J. Price, D. J. Norris, and P. J. West, Macromolecules 25, 6447 (1992).

[70] G. Schmid and O. Rommel, Z. Elektrochem 45, 659 (1939).

[71] G. Schmid, Physik. Z. 41, 326 (1940).

[72] H. H. G. Jellinek and G. White, J. Polymer Sci. 7, 21 (1951).

[73] A. I. Isayev, C. Wong, and X. Zeng, SPE ANTEC Tech. Papers 33, 207 (1987).

[74] A. I. Isayev, C. Wong, and X. Zeng, Adv. Polym. Technol. 10, 31 (1990).

[75] A. I. Isayev, Proceeding of the 23rd Israel Conference on Mechanical Engineering, Technion, Haifa, Paper #5.2.3, 1990.

[76] A. I. Isayev and S. Mandelbaum, Polym. Eng. Sci. 31, 1051 (1991).

[77] S. L. Peshkovsky, M. L. Friedman, A. I. Tukachinsky, G. V. Vinogradov, and N. S. Enikolopian, Polym. Compos 4, 126 (1983).

[78] R. Garcia Ramirez and A. I. Isayev, SPE ANTEC Tech. Papers 37, 1084 (1991).

[79] K. S. Suslick, Scientific American 260, 80 (1989).

[80] K. S. Suslick, S. J. Doctycz, and E. B. Flint, Ultrasonics 28, 280 (1990).

[81] G. Gooberman, J. Polymer Sci. 42, 25 (1960).

[82] M. Okuda and Y. Hatano, Japanese patent application 62, 121, 741 (1987).

[83] A. I. Isayev, inventor, Assignee the University of Akron, U. S. Patent 5, 258, 413 (1993).

[84] A. I. Isayev and J. Chen, inventors, Assignee the University of Akron, U. S. Patent 5, 284, 625 (1994).

[85] A. I. Isayev, J. Chen, and A. Tukachinsky, Rubber Chem. Tech. 68, 267 (1995).

[86] A. Tukachinsky, D. Schworm, and A. I. Isayev, Rubber Chem. Tech. 69, 92 (1996).

[87] V. Yu Levin, S. H. Kim, A. I. Isayev, J. Massey, and E. von Meerwall, Rubber Chem. Technol. 69, 104 (1996).

[88] A. I. Isayev, S. P. Yushanov, and J. Chen, J. Appl. Polym. Sci. 59, 803 (1996).

[89] A. I. Isayev, S. P. Yushanov, and J. Chen, J. Appl. Polym. Sci. 59, 815 (1996).

[90] A. I. Isayev, S. P. Yushanov, D. Schworm, and A. Tukachinsky, Plast. Rubber and Compos. Process. and Appl. 25, 1 (1996).

[91] S. P. Yushanov, A. I. Isayev, and V. Y. Levin, J. Polym. Sci.: Part B: Polymer Physics 34, 2409 (1996).

[92] A. I. Isayev, S. P. Yushanov, S. H. Kim, and V. Yu Levin, Rheol. Acta 35, 616 (1996).

[93] V. Yu Levin, S. H. Kim, and A. I. Isayev, Rubber Chem. Technol. 70, 120 (1997).

[94] S. T. Johnston, J. Massey, E. von Meerwall, S. H. Kim, V. Yu Levin, and A. I. Isayev, Rubber Chem. Tech. 70, 183 (1997).

[95] A. I. Isayev, S. H. Kim, and V. Yu. Levin, Rubber Chem. Tech. 70, 194 (1997).

[96] V. Yu Levin, S. H. Kim, and A. I. Isayev, Rubber Chem. Tech. 70, 641 (1997).

[97] S. P. Yushanov, A. I. Isayev, and S. H. Kim, Rubber Chem. Technol. 71, 168 (1998).

[98] B. Diao, A. I. Isayev, V. Yu Levin, and S. H. Kim, J. Appl. Polym. Sci. 69, 2691 (1998).

[99] M. Tapale and A. I. Isayev, J. Appl. Polym. Sci. 70, 2007 (1998).

[100] B. Diao, A. I. Isayev, and V. Yu Levin, Rubber Chem. Technol. 72, 152 (1999).

[101] V. V. Yashin and A. I. Isayev, Rubber Chem. Technol. 72, 741 (1999).

[102] V. V. Yashin and A. I. Isayev, Rubber Chem. Technol. 73, 325 (2000).

[103] C. K. Hong and A. I. Isayev, J. Appl. Polym. Sci. 79, 2340 (2001).

[104] S. E. Shim and A. I. Isayev, Rubber Chem. Tech. 74, 303 (2001).

[105] J. Yun, J. S. Oh, and A. I. Isayev, Rubber Chem. Technol. 74, 317 (2001).

[106] A. I. Isayev, in "Rubber Technologist's Handbook," J. R. White and S. K. De (Eds.), RAPRA Technology Ltd., U. K., 2001.

[107] A. I. Isayev, in "Encyclopedia of Materials: Science and Technology," K. H. J. Buschow (Ed.), Elsevier, Amsterdam, Vol. 3, 2001.

[108] C. K. Hong and A. I. Isayev, Rubber Chem. Technol. 75, 617 (2002).

[109] C. K. Hong and A. I. Isayev, Rubber Chem. Technol. 75, 133 (2002).

[110] J. Yun and A. I. Isayev, Rubber Chem. Technol. 76, 253 (2003).

[111] S. E. Shim, S. Ghose, and A. I. Isayev, Polymer 43, 5535 (2002).

[112] S. E. Shim, J. C. Parr, E. von Meerwall, and A. I. Isayev, J. Phys. Chem. B 106, 12072 (2002).

[113] S. E. Shim, A. I. Isayev, and E. von Meerwall, J. Polym. Sci.: Part B: Polym. Phys. 41, 454 (2003).

[114] J. Yun, A. I. Isayev, S. H. Kim, and M. Tapale, J. Appl. Polym. Sci. 88, 434 (2003).

[115] S. E. Shim, and A. I. Isayev, J. Appl. Polym. Sci. 88, 2630 (2003).

[116] S. Ghose and A. I. Isayev, J. Appl. Polym. Sci. 88, 980 (2003).

[117] J. Yun and A. I. Isayev, Polym. Eng. Sci. 43, 809 (2003).

[118] W. Feng and A. I. Isayev, J. Appl. Polym. Sci. (in press).

[119] W. Feng and A. I. Isayev, Paper presented at the ACS Rubber Division Meeting, Columbus, OH, October 2004.

[120] T. Boron, P. Roberson, and W. Klingensmith, Tire Technology International 96, 82 (1996).

[121] T. Boron, W. Klingensmith, C. Forest, and S. Shringarpurey, 156th Meeting of the ACS Rubber Division, Orlando, FL, 1999, Paper #136.

[122] E. A. Gonzalez de Los Santas, F. Sorieno-Corral, Ma J. Lozano Gonzalez, and R. Cedillo Garcia,

Rubber Chem. Technol. 72, 854 (1999).

[123] J. Blitz, "Fundamentals of Ultrasonics," 2nd ed., Butterworth, London, 1967, Chap. 8.

[124] K. S. Suslick, "Ultrasound: Its Chemical, Physical and Biological Effects," VCH, New York, 1988.

[125] S. N. Zhurkov, V. A. Zakrevskii, V. E. Korsukov, and V. S. Kuksenko, 1972, Soviet Physics Solid State 13, 1680 (1972).

[126] A. N. Gent and D. A. Tompkins, J. Appl. Phys. 40, 2520 (1969).

[127] A. N. Gent, Rubber Chem. Technol. 63, G49 (1990).

[128] M. B. Bever (Ed.), "Encyclopedia of Materials Science and Engineering," Pergamon Press, Oxford, Vol. 4, 1986, p. 2934.

[129] F. R. Young, "Cavitation," McGraw-Hill Co., London, 1989, Chap. 2.

[130] A. I. Isayev, J. Chen, and S. P. Yushanov, in "Simulation of Materials Processing: Theory, Methods and Application," S. F. Shen and P. Dawson (Eds.), Balkema, Rotterdam, 1995, pp. 77-85.

[131] A. I. Kasner and E. A. Meinecke, Rubber Chem. Technol. 69, 424 (1996).

[132] A. J. Kinloch and R. J. Young, "Fracture Behavior of Polymers," Applied Science Publishers, London, 1983.

[133] H. H. Kausch, "Polymer Fracture," Springer-Verlag, Berlin, 1987.

[134] M. Tabata and J. Sohma, Chem. Phys. Lett. 73, 178 (1980).

[135] M. Tabata and J. Sohma, Eur. Polym. J. 16, 589 (1980).

[136] J. P. Lorimer, in "Chemistry with Ultrasound," T. J. Mason (Ed.), Elsevier, New York, 1990, Chap. 4.

[137] C. K. Hong and A. I. Isayev, J. Mater. Sci. 37, 385 (2002).

[138] A. P. Yushanov, A. I. Isayev, and S. H. Kim, Rubber Chem. Technol. 71, 168 (1998).

[139] P. Rajalingam and W. E. Baker, Rubber Chem. Technol. 65, 908 (1992).

[140] N. P. Chopey, Chem. Eng. 80, 54 (1973).

[141] D. Gibala and G. R. Hamed, Rubber Chem. Technol. 67, 636 (1994).

[142] D. Gibala, K. Laohapisitpanich, D. Thomas, and G. R. Hamed, Rubber Chem. Technol. 69, 115 (1996).

[143] R. H. Wolk, Rubber Age 104, 103 (1973).

[144] A. N. Theodore, R. A. Pett, and D. Jackson, Rubber World 218, 23 (1998).

[145] H. J. Radusch, T. Luepke, S. Poltersdorf, and E. Laemmer, Kautschuk Gummi 43, 767 (1990).

[146] K. Fukumori, M. Matsushita, H. Okamoto, N. Sato, Y. Suzuki and K. Takeuchi, JSAE Review 23, 259 (2002).

[147] A. K. Naskar, P. K. Pramanik, R. Mukhopadhyay, S. K. De, and A. K. Bhowmick, Rubber Chem. Technol. 73, 902 (2000).

[148] Scrap Tire News 15, No. 10, Oct. 2001, pp. 1-3.

[149] S. Ghose and A. I. Isayev, Polym. Eng. Sci. 44, 794 (2003).

[150] J. R. M. Duhaime and W. E. Baker, Plast. Rubb. Comp. Proc. Appl. 15, 87 (1991).

[151] P. Rajalingam, J. Sharpe, and W. E. Baker, Rubber Chem. Technol. 66, 664 (1993).

[152] R. D. Deanin and S. M. Hashemiolya, Polym. Mater. Sci. Eng. 57, 212 (1987).

[153] T. Luo and A. I. Isayev, J. Elastomers and Plastics 30, 133 (1998).

[154] J. S. Oh and A. I. Isayev, Rubber Chem. Tech. 75, 617 (2002).

[155] C. K. Hong and A. I. Isayev, J. Elast. and Plast. 33, 47 (2001).

[156] H. Michael, H. Scholz, and G. Mennig, Kautchuk Gummi Kunststoffe 52 510 (1999).

[157] A. Y. Coran and R. P. Patel, in "Thermoplastic Elastomers," 2nd ed., G. Holden, N. R. Legge,

R. Quirk, and H. E. Schroeder (Eds.), Hanser Publishers, New York, 1996.

[158] N. C. Hilyward, S. G. Tong, and K. Harrison, Plast. Rubb. Proc. Appl. 3, 315 (1983).

[159] A. A. Phadke, S. K. Chakraborty, and S. K. De, Rubber Chem. Technol. 57, 19 (1984).

[160] A. A. Phadke and S. K. De, Polym. Eng. Sci. 26, 1079 (1986).

[161] K. Oliphant and W. E. Baker, Polym. Eng. Sci. 33, 166 (1993).

[162] P. K. Pramanik and W. E. Baker, J. Elast. Plast. 27, 253 (1995).

[163] P. K. Pramanik and W. E. Baker, Plast. Rubb. Comp. Proc. Appl. 24, 229 (1995).

[164] D. Briggs, Surf. Interface Anal. 2, 107 (1980).

[165] D. Briggs, Euro. Polym. J. 14, 1 (1978).

[166] A. K. Naskar, S. K. De, and A. K. Bhowmick, J. Appl. Polym. Sci. 84, 370 (2002).

[167] A. K. Naskar, A. K. Bhowmick, and S. K. De, J. Appl. Polym. Sci. 84, 622 (2002).

[168] F. J. Stark Jr., A. Leigton, and D. Wagner, Rubber World 188, 36 (1983).

[169] E. L. Rodriguez, Polym. Eng. Sci. 28, 1455 (1988).

[170] A. Chidambaram and K. Min, SPE ANTEC Tech. Papers 40, 2927 (1994).

[171] R. H. Campbell and R. W. Wise, Rubber Chem. Technol. 37, 635 (1964).

[172] Z. Xu, N. S. Losure, and S. D. Gardner, J. Avd. Mater. 30, 11 (1998).

[173] A. Y. Coran and F. Howard, U. S. Patent 5, 889, 119 (1999).

[174] F. Howard and A. Y. Coran, Paper presented at the ITEC, September, Akron, OH, 2000.

[175] D. Raghavan, H. Huynh, and C. F. Ferraris, J. Mater. Sci. 33, 1745 (1998).

[176] D. Raghavan, K. Tratt, and R. P. Wool, Mater. Research Soc. Sym. 344, 177 (1994).

[177] H. Goldstein, Civ. Eng. 65, 60 (1995).

[178] N. N. Eldin and A. B. Senouci, ASCE J. Const. Eng. Mgmt. 188, 561 (1992).

[179] N. N. Eldin and A. B. Senouci, Cem. Concr. Agg. 15, 74 (1993).

[180] N. N. Eldin and A. B. Senouci, ASCE J. Mater. Civ. Eng. 5, 478 (1993).

[181] B. I. Lee, L. Burnett, T. Miller, B. Postage, and J. Cuneo, J. Mater. Sci. Lett. 12, 967 (1993).

[182] F. Shutov, G. Ivanov, H. Arastoopour, and S. Volfson, Polym. Mater. Sci. Eng. 67, 404 (1992).

[183] I. B. Topcu, Cement Concrete Research 25, 304 (1995).

[184] N. N. Eldin and A. B. Senouci, Cement Concrete Compos. 16, 287 (1994).

[185] N. A. Ali, A. D. Amos, and M. Roberts, in Proceed. Intern. Conf. Concrete 2000, University of Dundee, U. K., 2000, pp. 379-390.

[186] H. Rostami, J. Lepore, T. Silverstram, and I. Zandi, Proceed. Intern. Conf. Concrete 2000, Dundee, U. K., 1993, pp. 391-399.

[187] J. A. Lepore and M. W. Tantala, In Proceed. Concr. Inst. Australia, Concrete 97, 1997, pp. 623-627.

[188] N. Segre and I. Joekes, Cement Concrete Research 30, 1421 (2000).

[189] N. Segre, P. J. M. Monteiro, and G. Sposito, J. Coll. Interface Sci. 248, 521 (2002).

[190] D. Raghavan, J. Appl. Polym. Sci. 77, 934 (2000).

[191] Anonymous, Biocycle 34, 9 (1993).

[192] J. L. McQuillen Jr., H. B. Takallou, R. G. Hicks, and D. Esch, ASCE J. Transport. Eng. 114, 259 (1988).

[193] F. L. Roberts and R. L. Lytton, Transportation Research Record 115, 216 (1987).

[194] R. H. Renshaw, "Rubber in Roads," Plastics and Rubber Institute, South African Section, S. Africa, 1985, p. 1.

[195] H. B. Takallou and M. B. Takallou, Elastomerics 123, 19 (1991).

[196] I. S. Schuller, Ph. D. Dissertation, Texas A&M University, 1991.

[197] H. Al-Abdual-Wahhab and G. Al-Amri, J. Mater. Civil Eng. 3, 189 (1991).
[198] T. C. Billiter, R. R. Davison, C. J. Glover, and J. A. Bullin, Petrol. Sci. Technol. 15, 205 (1997).
[199] M. Rouse, Rubber World, p. 23, May (1995).
[200] R. C. Malmgren, N. Parviz, P. N. Soltanpour, and J. E. Cipra, U. S. Patent 5, 014, 562 (1991).
[201] G. Logsdon, Biocycle 31, 44, 84 (1990).
[202] J. N. Rogers III and D. V. Waddington, Agronomy J. 84, 203 (1992).
[203] P. H. Groenevelt and P. E. Grunthal, Soil and Tillage Res. 47, 169 (1998).
[204] M. Moore, Crain's Tire Business 9, 15 (1991).
[205] W. Kaminsky and H. Sinn, in "Book Recycling and Recovery of Plastics," J. Brandrup, M. Bittner, W. Michaeli, and G. Menges (Eds.), Hanser Publishers, Munich, 1996.
[206] A. M. Mastral, R. Murillo, M. S. Callen, and T. Garcia, Resources Conserv. Recycling 29, 263 (2000).
[207] S. Kawakami, K. Inoue, H. Tanaka, and T. Sakai, in "Thermal Conversion of Solid Wastes and Biomass," J. L. Jones and S. B. Radding (Eds.), Symposium Series 130, ACS Publishers, Washington, D. C., 1980.
[208] J. A. Conesa, A. Fullana, and R. Font, Energy Fuel 14, 409 (2000).
[209] C. Roy, A. Chaala, and H. Darmstadt, J. Anal. Appl. Pyrol. 51, 201 (1999).
[210] M. R. Beck and W. Klingensmith, ACS Sympos. Series 609, 254 (1995).
[211] J. D. Osborn, Rubber World p. 34, May (1995).
[212] W. Kaminsky and C. Mennerich, J. Anal. Appl. Pyrolysis 58, 803 (2001).
[213] P. T. William, Chem. Review 12, 17 (2002).
[214] P. T. William and A. J. Brindle, J. Anal. Appl. Pyrolysis 67, 143 (2003).
[215] P. T. William and F. Ferrer, Resources Conserv. Recycling 19, 221 (1997).
[216] A. M. Mastral, M. S. Callen, R. Murillo, and T. Garcia, Environ. Sci. Technol. 33, 4155 (1999).

Appendix I:
Demonstration

Experiment: Characterization of uncrosslinked natural rubber from rubber tree latex and of crosslinked natural rubber.

Aim: To study the stress-strain properties (and effects of time and temperature) of uncrosslinked natural rubber synthesized in the lab from latex and commercial cross-linked rubber.

Materials and Apparatus: Latex uncross-linked rubber; Commercial vulcanized rubber pads; Ice; Ring stand; Clamps; Thermometer; Dial micrometer (0.001″); Force gauge; Stop watch.

Brief Background: Natural rubber: It has been established especially by nuclear magnetic resonance spectroscopy that "natural rubber" from the tree Hevea Braziliensis is more than 98% poly (1,4-*cis*-isoprene). This configuration is shown below:

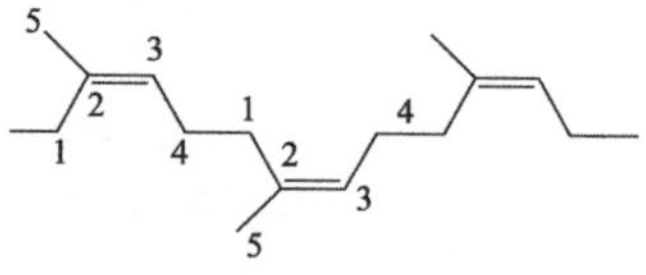

$$\tag{1}$$

Rubber is drawn from the rubber tree in the form of an emulsion, called "latex". Latex from the tree Havea Braziliensis consists of ~60% poly (1,4-*cis*-isoprene), 40% water, soap and NH_3. Upon evaporation of water, the sub-micron globules of rubber coalesce, yielding a material that is strong mechanically, yet tacky (sticky). GPC has established that the molecular weight distribution is very broad: some fractions run into few million g/mol in molecular weight (this fraction is the origin of strength) and others are only a few thousand g/mol (this fraction is the origin of "tackiness").

Other configurations of polyisoprene: In the case of polydienes, several possibilities of addition of the monomer onto the growing chain are possible. In the case of isoprene, for instance, the polymerization can involve only one double bond (1,2-addition, or 3,4-addition), or it can involve both together (1,4-addition). In the former cases, the stereo placement of the substituent in 1,2-or 3,4-units can be controlled in some instances. In the latter case, the double bonds in the chain also involve stereoisomerism: They can be 1,4-*cis* or 1,4-*trans*, the properties of the corresponding polymers being widely different, ranging from a soft, amorphous gum to a hard, crystalline, non-elastomeric solid.

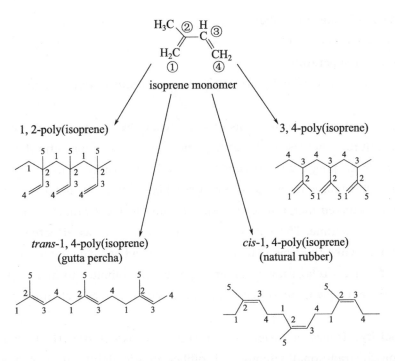

Condition for exhibiting rubber-like properties: Any material with rubber-like properties is characterized by chains which exhibit significant flexibility and mobility at the temperature under consideration. Hence, for polybutadiene and polyisoprene to exhibit rubber-like properties, the molecules must be preferably *cis*-1,4 or at least a random copolymer of cis and trans units. The all 1,4-*trans* arrangement results in a very regularly packed organization of the molecules into crystals with high melting points (well above room temperature). The second requirement is the presence of crosslinks (permanent covalent crosslinks and chain entanglements), which hold the material together at specific points during deformation.

Mechanical properties vs. time and temperature for rubbers: A rubber sample (of initial length, L_o and cross sectional area, A_o) under tensile stress (f/A_o) stretches to a length, L and exhibits a resistance against further deformation. One unique aspect of this resistance in rubbers is that it originates from the tendency of the chains to increase their entropy again following the reduction of entropy of the chains on stretching. This is very different from what happens in other materials where the tendency to reduce total energy (not entropy) is the source of resistance. The following relationship is found to hold for small elongation ratios ($\alpha \approx 1$) between stress and elongation ratio ($\alpha = L/L_o$) in rubber:

$$\tau = \frac{f}{A_o} = E_o \left(\alpha - \frac{1}{\alpha^2} \right) \tag{2}$$

where,

$$E_o = RT \left(\frac{v_e}{V_o} \right) \tag{3}$$

where,

E_o = Tensile modulus of rubber.

R = Universal gas constant.

T = Absolute temperature.

(v_e/V_o) = moles of elastically effective chains per unit initial volume = network chain density

Using this equation, what happens to stress when the elongation ratio is kept constant and the temperature is increased? What happens to the elongation ratio if the temperature is increased while keeping the stress constant? (Rubber shrinks on increasing the temperature). Contrast this with what happens when you heat a piece of metal.

The non-crosslinked material is viscoelastic—not an ideal rubber—and hence "flows" (or "creeps") with time. The chain entanglements can act as effective crosslinks over short time frames (like the time it takes to "snap" a rubber band) but not over the long time frames of a stress relaxation or creep experiment (minutes to hours). Under the applied stress, the entanglements slowly "unwind" and/or polymer chains slip and flow past each other.

Viscoelasticity: It may be easier to understand the viscoelastic effect—which is akin to a combination of traditional viscous and rubber elastic deformation behavior—using a spring and dashpot model (called the Maxwell model). The spring is elastic; stress is directly proportional to strain. The dashpot is plastic; it behaves like a liquid and flows in response to a stress.

Stress Relaxation experiment and creep experiment: In a relaxation experiment, the strain is kept constant and the variation of stress with time is monitored (i.e., the stress relaxation is observed). In a creep experiment, the stress is kept constant and the variation of strain with time is monitored (i.e., the creep of the material is observed).

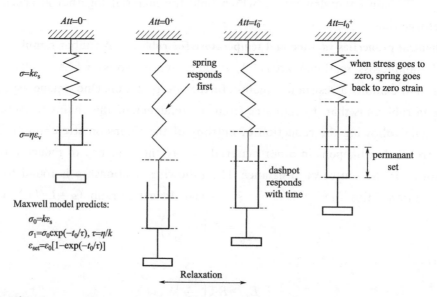

Precautions:

No special precautions.

Procedure:

The session before this lab

You will need at least one rubber band for this experiment. Rubber band preparation will be done by carefully syringing the unvulcanized (uncrosslinked) latex into O-shaped molds. Syringing of the polymer must be done slowly to maintain a continuous stream out of the needle, and you must avoid introducing holes or bubbles that may serve as mechanical defects. The mold should be filled as close to the top as possible. The bands will be allowed to "dry" under the hood. You will also be working with cross-linked rubber bands that are available commercially.

These bands will be provided in lab (they are the traditional rubber bands used in offices).

Note: Calculate stresses and elongation ratios as you move along in these experiments. Also note that experiments A. 2-3 and B. 2-3 may be done simultaneously in the same ice bath.

A. Experiments with non-crosslinked rubber:

1. You have already prepared non-crosslinked rubber bands (prepared in the lab from latex by evaporation of water) which look like O-rings, are somewhat tacky, and are cast from D cm diameter circles. The circumference is πD in cm. When this ring is stretched, it effectively behaves as two equal length specimens, each of initial length $L_o = \pi D/2 =$ (half of the circumference). If the ring had an initial cross sectional area of A_o, then the cross sectional area bearing the stress is $2 A_o$.

2. Strain at low temperature: Set up the ice-water mix and measure the temperature. Holding the "O-ring" under the ice-water mix, pull it as far as you can. Measure the length to which it gets stretched. Release the rubber band, and measure the length (if different) upon release.

3. Strain at high temperature: Now remove the ice cubes, while keeping a thermometer immersed along with the rubber. Slowly add tepid water with stirring. Observe:

a) the temperature at which the rubber band starts to shrink from its stretched state and

b) the temperature at which it has regained its initial circular state (D cm diameter) or the temperature at which no further shrinking occurs.

4. Stress relaxation experiment: Using the force gauge, a ring-stand, and suitable clamps so as to pull the band to an elongation ratio of L/L_o of about 3 at room temperature (you may need to use lower strain ratios depending on the thickness of the band you made).

Whatever the elongation ratio is, record it, and keep it fixed. Record ambient temperature. Record force as a function of time at 5 minute intervals for 30 minutes. After 30 minutes, remove the load, let the band relax, and measure its new diameter D'. The original diameter being D in cm, $(D' - D)/D$ is the fractional permanent "set" or fractional creep.

B. Experiments with commercial vulcanized rubber bands:

1. You are given commercial rubber bands that are thinner in cross section than the non-crosslinked rubber bands, are non-tacky and usually race track shaped (two straight runs connected to semicircles).

2. Strain at low temperature: Set up an ice-water bath and measure the temperature. Holding the commercial rubber band under the ice-water mix, pull it as far as you can. Measure the length to which it gets stretched. Release the rubber band, and measure the length (if different) upon release.

3. Strain at high temperature: Now remove the ice cubes, while keeping a thermometer immersed along with the rubber. Slowly add tepid water with stirring. Observe the temperature at which the rubber band starts to shrink from its stretched state and the temperature at which no further shrinking occurs.

4. Using a hand held force gauge, determine the tensile stress-strain at room temperature. Record the temperature. Recall that two strands of rubber support the load equally, so that the cross-sectional area $= 2 A_o$. A_o is the cross-sectional area of the rubber band.

5. Thermoelastic Property Experiment: Arrange a set-up like that discussed in section A. 4 of the procedure, so that you can impose an elongation ratio L/L_o of 3. Record the tensile force and the temperature.

6. Keeping the elongation ratio at 3.0, pour ice water (near 0℃) over the band until you get a constant force reading. Record it.

7. Remove the ice water and substituting hot water of known temperature (say 60～70℃). Again try to obtain a constant force reading. Record the temperature.

8. Stress relaxation experiment: This experiment is the same as in section A. 4.

Using the force gauge, a ring-stand, and suitable clamps, pull the band to an elongation ratio of L/L_o of about 3 at room temperature (you may need to use lower strain ratios depending on the thickness of the band you made). Whatever the elongation ratio is, record it, and keep it fixed. Record ambient temperature. Record force as a function of time at 5 minute intervals for 30 minutes. After 30 minutes, remove the load, let the band relax, and measure its new length L'. The original diameter being L inches, $(L' - L)/L$ is the fractional permanent "set" or fractional creep.

Your report:

Observations and Calculations:

1. Record required observations in the procedure/brief background in the lab notebook. Be sure to record all relevant dimensions of the samples.

2. Report all relevant stresses and elongation ratios.

Discussion:

1. Non-crosslinked rubber:

a) Plot the circumference of the band as a function of temperature temperature for the ice bath experiment.

b) After reporting exact elongation ratio and other relevant conditions, plot the tensile stress vs. time for the relaxation experiment. Choose appropriate units. Compare your results with what the Maxwell model predicts.

c) Report fractional creep. Explain why you observe creep in these samples.

2. Crosslinked rubber:

a) Plot tensile stress vs. elongation ratio, α.

b) Plot the reduced modulus, t^*, vs. $1/\alpha$ where,

$$\tau^* = \frac{\tau}{(\alpha - \alpha^{-2})} \tag{4}$$

c) Use plot in part b) to calculate the tensile modulus using equation 5. (The tensile modulus is the reduced modulus as elongation ratio approaches 1.) Use the tensile modulus to calculate the network chain density.

d) Plot tensile stress vs. temperature (in Kelvin) and discuss the results with respect to the background given in this handout and the lecture.

3. Discuss experimental difficulties or inaccuracies, sources of error etc. (For example, were there air bubbles or did the rings cast from latex have uniform thickness? What will be the effect of such errors on your result? Or was temperature control a problem?)

Appendix II:
Acronyms for Common Elastomers

ABR	acrylate-butadiene rubber
ACM	copolymer of ethylacrylate and a comonomer (acrylic rubber)
ANM	ethylacrylate-acrylonitrile copolymer (acrylate rubber)
BIMS	brominated isobutylene paramethyl styrene rubber
BR	butadiene rubber (polybutadiene)
BIIR	bromobutyl rubber
CIIR	chlorobutyl rubber
CFM	polychlorotrituoroethylene (fluoro rubber)
CM	chloropolyethylene (previous designation CPE)
CO	epichlorohydrin homopolymer rubber (polychloromethyloxiran)
CR	chloroprene rubber (polychloroprene)
CSM	chlorosulfonylpolyethylene
EAM	ethylene-ethyl acrylate copolymer (e. g., Vamac)
ECO	copolymer of ethylene oxide (oxiran) and chloromethyloxiran
ENM or H-NBR	proposed code for hydrogenated NBR
ENR	epoxidized NR
EPDM	ethylene-propylene-diene terpolymer
EPM	ethylene-propylene copolymer
EVM	ethylene-vinylacetate copolymer (previous code: EVA or EVAC)
FMQ	methyl silicone rubber with fluoro groups (previous designation FSI)
FPM / FPM	rubber having fluoro and fluoroalkyl or fluoroalkoxy substituent group
IIR	isobutylene-isoprene rubber (butyl rubber)
IR	isoprene rubber (synthetic)
MQ (PVMQ)	methyl silicone rubber (with vinyl and phenyl end groups)
NBR	acrylonitrile-butadiene rubber (nitrile rubber)
NR	isoprene rubber (natural rubber)
PUR	generic code for urethane rubbers
Q	generic code of silicone rubbers
SBR	styrene-butadiene rubber
TM	polysulfide rubbers
TOR	*trans*-polyoctenamer
VMQ	methyl silicone rubber with vinyl groups